THE SARCOLEMMA

Proceedings of the Seventh Annual Meeting of the
International Study Group for Research
In Cardiac Metabolism
18–21 June, 1974
Québec, CANADA

Jointly held by the
European and American Research Groups

Organized by the
Department of Pathology
and Medicine
of Laval University
and by the Institute of Cardiology
of Laval Hospital

Chariman of the Congress
ROY, Paul-E., Québec

Members of the Organization Committee:
Roy, Paul-E., Québec
Leblanc, Jacques, Québec
Dagenais, Gilles R., Québec
Rona, George, Montréal

Management:
Pageau, Laurette, Québec

Recent Advances in Studies on Cardiac Structure and Metabolism

Series Editor: G. Rona (*formerly with the late E. Bajusz*)

Volume 1: **MYOCARDIOLOGY** (edited by E. Bajusz and G. Rona)
Volume 2: **CARDIOMYOPATHIES** (edited by E. Bajusz and G. Rona, with A. J. Brink and A. Lochner)
Volume 3: **MYOCARDIAL METABOLISM** (edited by N. S. Dhalla)
Volume 4: **MYOCARDIAL BIOLOGY** (edited by N. S. Dhalla)
Volume 5: **MYOCARDIAL CELL DAMAGE** (edited by A. Fleckenstein and N. S. Dhalla)
Volume 6: **PATHOPHYSIOLOGY AND MORPHOLOGY OF MYOCARDIAL CELL ALTERATIONS** (edited by A. Fleckenstein and G. Rona)
Volume 7: **BIOCHEMISTRY AND PHARMACOLOGY OF MYOCARDIAL HYPERTROPHY, HYPOXIA, AND INFARCTION** (edited by P. Harris, R. J. Bing, and A. Fleckenstein)
Volume 8: **THE CARDIAC SARCOPLASM** (edited by P.-E. Roy and P. Harris)
Volume 9: **THE SARCOLEMMA** (edited by P.-E. Roy and N. S. Dhalla)
Volume 10: **THE METABOLISM OF CONTRACTION** (edited by P.-E. Roy and G. Rona)

Recent Advances in Studies on
Cardiac Structure and Metabolism, Volume 9

THE SARCOLEMMA

Edited by

Paul-Emile Roy, M.D.
Institut de Cardiologie de Québec
Hopital Laval, Chemin Ste-Foy
Québec, Canada

and

Naranjan S. Dhalla, Ph.D.
Department of Physiology
Faculty of Medicine, University of Manitoba
Winnipeg, Canada

UNIVERSITY PARK PRESS
Baltimore · London · Tokyo

UNIVERSITY PARK PRESS
International Publishers in Science and Medicine
Chamber of Commerce Building
Baltimore, Maryland 21202

Typeset by The Composing Room of Michigan, Inc.

Manufactured in the United States of America by Universal Lithographers, Inc., and the Maple Press.

Library of Congress Cataloging in Publication Data
Main entry under title:

The Sarcolemma.

(Recent advances in studies on cardiac structure and metabolism ; v. 9)
"Proceedings of the Seventh Annual Meeting of the International Study Group for Research in Cardiac Metabolism, 18-21 June 1974, Quebec, Canada."
Includes index.
1. Heart–Muscle–Congresses. 2. Sarcolemma–Congresses. I. Roy, Paul-Emile, 1929– II. Dhalla, Naranjan S. III. International Study-Group for Research in Cardiac Metabolism. IV. Series. [DNLM: 1. Sarcolemma–Congresses. 2. Myocardium Congresses. W3 RE354 v. 9 1974 / WG200 I615s 1974]
RC667.R4 vol. 9 [QP113.2] 616.1'2'008s [612'.17]
ISBN 0-8391-0669-6 76-1956

Contents

Sodium-Potassium Pump and Cardiac Glycosides

Pathophysiological Aspects

Contributing Authors

Akera, T., *East Lansing, Mich.* (USA)
Alousi, A. A., *Rensselaer, N.Y.* (USA)
Anand, M. B., *Winnipeg, Man.* (Canada)
Bär, H. P., *Edmonton, Alta.* (Canada)
Beardsley, A. C., *Charlottesville, Va.* (USA)
Beller, G. A., *Boston, Mass.* (USA)
Bhakthan, N. M. G., *Burnaby, B.C.* (Canada)
Brody, T. M., *East Lansing, Mich.* (USA)
Buja, L. M., *Bethesda, Md.* (USA)
Busuttil, R. W., *New Orleans, La.* (USA)
Chabi, E., *Houston, Tex.* (USA)
de Champlain, J., *Montreal, P.Q.* (Canada)
De Robertis, E., *Buenos Aires* (Argentina)
Dhalla, N. S., *Winnipeg, Man.* (Canada)
Drabikowski, W., *Warsaw* (Poland)
Drummond, G. I., *Vancouver, B.C.* (Canada)
Drummond, M., *Vancouver, B.C.* (Canada)
Dunnett, J., *London* (England)
Džurba, A., *Bratislava* (Czechoslovakia)
Erdmann, E., *Gottingen* (Germany)
Fabiato, A., *Boston, Mass.* (USA)
Fabiato, F., *Boston, Mass.* (USA)
Fedelešova, M., *Bratislava* (Czechoslovakia)
Ferrans, V. J., *Bethesda, Md.* (USA)
Fiszer de Plazas, S., *Buenos Aires* (Argentina)
George, W. J., *New Orleans, Lu.* (USA)
Harrow, J. A. C., *Winnipeg, Man.* (Canada)
Hoeschen, R. J., *Winnipeg, Man.* (Canada)
Hood, W. B., Jr., *Boston, Mass* (USA)
Imbach, A., *Montréal, P.Q.* (Canada)
Katz, A. M., *New York, N.Y.* (USA)
Kennedy, J. H., *Houston, Tex.* (USA)
Kidwai, A. M., *Edmonton, Alta.* (Canada)
Kirchberger, M. A., *New York, N.Y.* (USA)
Ku, D., *East Lansing, Mich.* (USA)
Kutchai, H. C., *Charlottesville, Va.* (USA)
Langer, G. A., *Los Angeles, Calif.* (USA)
Lee, S. L., *Winnipeg, Man.* (Canada)
Lefer, A. M., *Charlottesville, Va.* (USA)
Limet, R., *Houston, Tex.* (USA)
Lüllmann, H., *Kiel* (Germany)
MacWilliam, L. D., *Burnaby, B.C.* (Canada)
Maron, B. J., *Bethesda, Md.* (USA)
McNamara, D. B., *Winnipeg, Man.* (Canada)
Moffet, F. J., *Edmonton, Alta.* (Canada)
Musso, E., *Brooklyn, N.Y.* (USA)
Nadeau, R., *Montréal, P.Q.* (Canada)
Nayler, W. G., *London* (England)
Paddock, R. J., *New Orleans, La.* (USA)
Patzelt, R., *Gottingen* (Germany)
Peters, T., *Kiel* (Germany)
Reader, T. A., *Quebec, P.Q.* (Canada)
Rich, T. L., *Philadelphia, Pa.* (USA)
Salas, P. J. I., *Buenos Aires* (Argentina)
Santa Barbara, B. L., *Rensselaer, N.Y.* (USA)
Schirpke, B., *Berlin-Buch* (German Democratic Republic)
Schneider, J. A., *Charlottesville, Va.* (USA)
Schoner, W., *Gottingen* (Germany)
Schulze, W., *Berlin-Buch* (German Democratic Republic)
Shigenobu, K., *Charlottsville, Va.* (USA)
Shine, K. I., *Los Angeles, Calif.* (USA)
Singh, J. N., *Winnipeg, Man.* (Canada)
Smith, T. W., *Boston, Mass.* (USA)
Sperelakis, N., *Charlottesville, Va.* (USA)
Sulakhe, P. V., *Winnipeg, Man.* (Canada)
Sullivan, A., *London* (England)
Tada, M., *Osaka* (Japan)
Tomlinson, C. W., *Winnipeg, Man.* (Canada)
Vallières, J., *Vancouver, B.C.* (Canada)
Vassalle, M., *Brooklyn, N.Y.* (USA)
Welch, K. M. A., *Houston, Tex.* (USA)
Will, H., *Berlin-Buch* (German Democratic Republic)
Will-Shahab, L., *Berlin-Buch* (German Democratic Republic)
Williamson, J. R., *Phildaelphia, Pa.* (USA)
Wollenberger, A., *Berlin-Buch* (German Democratic Republic)
Yamaguchi, N., *Montrèal, P.Q.* (Canada)
Yates, J. C., *Winnipeg, Man.* (Canada)
Ziegelhöffer, A., *Bratislava* (Czechoslovakia)
Zubrzycka, E., *Warsaw* (Poland)

Editorial Foreword

The ability of sarcolemma to maintain ionic gradients between extracellular and intracellular compartments of the myocardial cell has fascinated very many investigators. This function of sarcolemma has been studied by employing different electrophysiological techniques; the influx of sodium as well as of calcium and the efflux of potassium have been shown to occur during depolarization of the myocardium. On the other hand, efflux of sodium, influx of potassium, and efflux of calcium for recovery are considered to be attributable to the presence of "sodium-potassium pump" (possibly Na^+-K^+ATPase) and "calcium pump" (possibly Ca^{++} stimulated ATPase) in the heart sarcolemma. Cytochemical and biochemical techniques have provided evidence concerning the localization of Na^+-K^+ ATPase in sarcolemma in addition to another enzyme system, adenylate cyclase, which catalyzes the formation of cyclic AMP from ATP. Cyclic AMP is considered to be a key substance that not only controls the movements of ions across sarcolemma but also regulates myocardial function and metabolism by phosphorylating different enzyme systems, membrane systems, and contractile as well as regulatory proteins. Thus, sarcolemma can be conceived to play a crucial role in events leading to contraction and relaxation of myocardium by raising and lowering the cytoplasmic concentrations of different cations and cyclic AMP.

Various drugs and hormones are considered to produce changes in myocardial function and metabolism by interacting with specific receptors on the sarcolemma. For example, digitalis glycosides have been claimed to produce positive inotropic action by inhibiting Na^+-K^+ATPase in the heart sarcolemma and thus raising the intracellular concentration of free calcium either directly or indirectly. On the other hand, catecholamines combine with adrenergic receptors and increase the permeability of cell to calcium as well as increase the formation of cyclic AMP because of the activation of adenylate cyclase. Thus, changes in myocardium because of catecholamines are the consequences of integrated functions of calcium and cyclic AMP that originate at the level of sarcolemma. It should also be recognized that the function and metabolism of myocardium are controlled by the central nervous system primarily because of the release of catecholamines from the adrenergic nerve terminals present in the vicinity of heart sarcolemma. Other different hormones circulating in the body as well as neurotransmitters released from their storage sites also produce changes in the sarcolemmal functions and thereby influence the myocardial contractile behavior and metabolism. In addition, sarcolemma is believed to control the permeability of different substrates selectively and prevents the leakage of important enzymes from the cytoplasm. All these properties of sarcolemma are attributed to the presence of phospholipids, proteins, and cholesterol in a special proportion in this membrane system. Thus any changes in these sarcolemmal constituents can be considered to result in pathophysiology of myocardium.

The present book is an attempt to gather information on sarcolemma and associated features obtained through multidisciplinary approaches. It contains a part of the proceedings of the Seventh Meeting of the International Study Group for Research in Cardiac Metabolism held in Quebec City, June 18 to 21, 1974. Selected papers have been grouped under different sections such as Physiological and Biochemical aspects, Adenylate Cyclase and Catecholamines, Na^+-K^+ Pump and Cardiac Glycosides, and Pathophysiological Aspects for the purpose of clarity and convenience to our readers. We believe that both the contents and concepts contained in Volume 9 of the *Recent Advances in Studies on Cardiac Structure and Metabolism* will stimulate further research in the field of myocardial biology.

We wish to thank the Canadian Heart Foundation, Medical Research Council of Canada, Government of Canada, Government of Quebec, University of Laval, University Park Press, and Canadian Pharmaceutical Agencies for their financial support. The advice and cooperation of the members of the International Executive and Coordination Committee for planning of the symposium are gratefully acknowledged. We also express our sincere gratitude to the editorial staff of the University Park Press for their assistance in preparing this book.

THE SARCOLEMMA

Physiological and Biochemical Aspects

Recent Advances in Studies on Cardiac Structure and Metabolism
Volume 9
The Sarcolemma
Edited by Paul-Emile Roy and Naranjan S. Dhalla

HEART SARCOLEMMA AS A DYNAMIC EXCITABLE MEMBRANE[1]

N. S. DHALLA, D. B. McNAMARA, and M. B. ANAND

Experimental Cardiology Laboratory, Department of Physiology,
Faculty of Medicine, University of Manitoba,
Winnipeg, Manitoba, Canada

SUMMARY

Sarcolemma consists of plasma and basement membranes and constitutes the real permeability barrier to the heart cell. It is considered to provide high electrical resistance and capacitance to heart cell and its properties are essentially similar to those of the other excitable membranes. Methods are now available for isolating heart sarcolemma with high specific activities of adenylate cyclase, (Na^+-K^+) ATPase, Ca^{++} ATPase, and Mg^{++} APTase. These enzymes are considered to play an important role in heart function by regulating ion movements across sarcolemma as well as by providing signals for various metabolic processes. Sarcolemma possesses different hormone and drug receptors and any alteration in its composition could result in abnormal responses of the myocardium. We believe that heart failure is associated with sarcolemmal defects which can be detected by monitoring the activities of different membrane-bound enzymes and other related processes.

GENERAL ASPECT

Cardiac muscle sarcolemma shares many common features with cell membrane in general and excitable membranes in particular. Although the cell membrane was believed to be merely protective, there is now much evidence that it contains highly specific and sensitive receptors for drug or hormonal signals (19). Various excellent reviews concerning the membrane molecular organization, ultrastructure, bioelectric properties, and transport have appeared in the literature (35, 40, 43, 77). The plasma membrane constitutes the real permeability barrier of the cell and gives the membrane its high electrical resistance and capacitance. It is now apparent that different membranes may vary substantially in molecular composition, enzymatic activity, transport functions, thickness, and in the type of image yielded by high resolution electron microscopic techniques, such as negative contrast and freeze-etching (33, 35, 75, 91).

[1] The work reported in this paper was supported by Grant MT-3394 from the Medical Research Council of Canada.

Furthermore, the plasma membrane of any given cell type is not necessarily uniform over its whole surface and may be differentiated locally at desmosomes, tight junctions, and synapses, and may possess a microscopic nonuniformity as in a two-dimensional molecular mosaic.

Each type of membrane contains a characteristic set of complex lipids in a specific molar ratio, which appears to be genetically determined. Plasma membranes, because of their relatively high cholesterol content, are more rigidly structured and "tighter" than other types of membranes (16). Furthermore, different molecular species of structural proteins in a given membrane are undoubtedly genetically coded and they may in turn specify the content and ratio of the various lipids of the membrane. It has been suggested (49) that each species of membrane protein may be able to bind selectively a single type of membrane phospholipid.

The most accepted molecular model of the structure of plasma membranes is that first proposed by Danielli and Davson (18) and later refined by Robertson (74) as the unit membrane hypothesis. Accordingly, the lipids of the membrane, which comprise 40–50 percent of its mass, are arranged in a bilayer, with the hydrocarbon chains of the two lipid layers apposed to form a continuous, nonpolar hydrocarbon phase. On either side are monolayers of protein, which comprise 50–60 percent of the membrane mass. The unit membrane hypothesis was originally postulated to account for the basic structure of all types of membranes; however, membranes may change locally in structure as a function of activity. There is one feature of the unit membrane structure, namely the lipid bilayer, which still best accounts for the characteristic permeability and electrical properties of plasma membranes. Nuclear magnetic resonance measurements have revealed that the proteins are relatively fixed in the membrane structure but that the lipids have considerable freedom of movement (14, 15). One of the strongest pieces of evidence in support of the lipid bilayer model is the fact that, in the complete absence of protein, phospholipids in aqueous systems form bilayers spontaneously in the form of flat micelles as large closed vesicles. Most pertinent are the studies by Mueller *et al.* (66) and by Maddy and coworkers (55), who have shown that phospholipid bilayers may be formed in apertures separating two aqueous phases. Such bilayers have low permeability to polar solutes, high permeability to water, and extremely high electrical resistance and capacitance comparable with that of the plasma membrane. Such lipid bilayers can also be made electrically excitable in the presence of an ion gradient (65).

For many years it has been thought that movements of Na^+ and K^+ are specific elements in the excitation of the cell and the generation of the action potential. The classical work of Hodgkin (42) showed that the sum of the electrical currents carried by the entry of Na^+ and the exit of K^+ can be quantitatively related to the transmembrane potential and the transmembrane conductance at all points in the action potential. It has been assumed that the first result of stimulation is the entry of Na^+; however, more recent experiments by Tasaki (110) are opening a new chapter in our understanding of the role of cations in the excitation process. The most surprising finding is that neither Na^+ or K^+ is required on the inside or outside of the membrane to support excitability. Phosphate salts of the univalent cations, Li^+, Rb^+, or Cs^+, or substituted ammonium ions, can maintain excitability of cells and show an action potential if they are bathed externally with divalent cations such as Ca^{++}, Ba^{++}, or Sr^{++}, despite the completely "unphysiological" nature of the internal and external ions. Na^+ or K^+ are thus not specific requirements for exciting the membrane. Tasaki and his colleagues (110) have concluded that their data are not in accord with any hypothesis for the excitation process that requires as the first event a specific increase in Na^+ permeability and influx of Na^+ into the cell. It was therefore proposed that in the resting state the outer region of the cell membrane contains bound divalent cations, probably Ca^{++} derived from the external medium. On stimulation, some of the bound Ca^{++} is replaced by K^+ derived from the internal medium. It is suggested that this ion exchange process triggers a change in the conformation of membrane subunits to a second state in which membrane conductance is increased, permitting Na^+ to diffuse in and K^+ to diffuse out. At the end of the action potential the membrane-bound K^+ is displaced by Ca^{++} again, returning it to its original resting state of conformation. It is possible that the Ca^{++} binding sites, which are few in number, may be provided by specific sialic acid residues of the membrane gangliosides. Actually, only a very small fraction of the membrane area needs to undergo loss of Ca^{++} to cause its all-or-none excitation (110).

Whatever events may determine the process of excitation, two properties of the excitable cell membrane suggest that it undergoes cooperative transitions. One is the finding that interaction of only a few molecular sites in the membrane is required to trigger a change of the whole membrane. The other is the all-or-none nature of the response of the excitable cell to increasing stimulus intensity. These properties are suggestive of the behavior of allosteric or regulatory proteins and enzymes (63, 74). At least some investigators (17, 67) have postulated that electrical or chemical perturbation of a few specific membrane subunits causes a change in their conformation and that this change in conformation is then physically transmitted to neighboring subunits in a cooperative manner. In the excitable membrane the cooperativity is of a much higher order, so that it shows all-or-none and amplification characteristics.

Depolarization is usually associated with an increase in membrane permeability and there seems to be general agreement that the cellular permeability is regulated by certain enzymes such as (Na^+-K^+)-ATPase, Ca^{++}ATPase, Mg^{++}ATPase, and adenylate cyclase which are bound to the cell membrane. For example, (Na^+-K^+)-ATPase has been implicated in the active transport of Na^+

Table I Various parameters for monitoring sarcolemmal status in heart

1. Electrical properties of myocardium
2. Electron microscopic and cytochemical examination of myocardial cell
3. Transport of different substrates such as glucose and free fatty acids
4. Cationic fluxes in myocardium
5. Coronary sinus plasma enzymes and electrolytes
6. Electrolyte composition of myocardium
7. Composition of sarcolemmal lipids and proteins
8. Activities of sarcolemmal adenylate cyclase, $(Na^+\text{-}K^+)$-ATPase, Ca^{++} ATPase, Mg^{++} ATPase, and other enzymes
9. Sarcolemmal binding of hormones, drug, and calcium
10. Conformational changes by sarcolemmal interaction with different fluorescent and other probes

and K^+ across the cell membrane (86, 93). The presence of Ca^{++}ATPase and Mg^{++}ATPase has been demonstrated in membranes isolated from various tissues (61, 78, 101) and these enzymes have been presumed to be involved in the transport of Ca^{++} and Mg^{++} respectively. Adenylate cyclase, an enzyme which catalyzes the formation of adenosine 3′,5′-monophosphate (cyclic AMP), has been speculated to regulate the cell permeability in addition to its well established role in cellular metabolic events such as glycolysis and lipolysis (29, 70, 76, 108). Since membrane integrity and particularly the lipoprotein composition of the membrane are essential for the full expression of the activities of membrane-bound enzymes, we feel that the status of membrane function can be assessed by monitoring the activities of adenylate cyclase, Mg^{++}ATPase, Ca^{++}ATPase, and $(Na^+\text{-}K^+)$-ATPase. In addition, various other parameters can be used for gaining information on heart cell membrane or sarcolemma (table I). The significance of sarcolemma in maintaining heart function is quite apparent from the fact that any damage to its integrity leads to an alteration in the intracellular environment. Some of the possible sites and consequence of sarcolemmal defects associated with heart failure are given in table II.

MORPHOLOGICAL AND BIOCHEMICAL ASPECT

Although considerable information is available on the electrical behavior of the heart cell membrane, very little is yet known of its molecular composition and structure because of the difficulties in obtaining sufficient material uncontaminated by the intracellular membranes such as mitochondria and sarcoplasmic reticulum for biochemical studies. The sarcolemma has two components: a thin, electron dense plasma membrane and a much thicker amorphous coating, basement membrane which surrounds the entire cell and is its outermost layer. The basement membrane is a mucopolysaccharide and has a multiplicity of negative

Table II Possible sites and consequences of sarcolemmal defects in failing hearts

Changes in parameters	Consequences
1. Membrane integrity	Leakage of enzymes and other cytoplasmic constituents and myocardial cell damage
2. Phospholipid to cholesterol ratio	Alterations in electrical properties, membrane bound activities, substrate transport, and electrolyte composition
3. (Na^+-K^+)-ATPase activity	Elevation in Na^+ and depression in K^+ levels
4. Ca^{++}- or Mg^{++}ATPase activity	Changes in Ca^{++} and Mg^{++} contents
5. Calcium binding	Intracellular Ca^{++} overload or deficiency
6. Adenylate cyclase activity	Changes in cyclic AMP levels and metabolic derangements
7. Hormone binding	Metabolic disturbances and functional abnormalities

charges by which it can bind cations. The sarcolemma sends penetrating finger-like invaginations into the myocardial cell, thus creating an intricate pattern of tubules which are known as the transverse tubular system (T system). Both the basement membrane and the plasma membrane participate in the formation of the T system. The tubules come into intimate contact with the lateral sacs of the sarcoplasmic reticulum, the main site of calcium storage. It is generally believed that the excitatory impulse along the sarcolemma travels down the T system and promptly releases calcium ion from the lateral sac of the sarcoplasmic reticulum in the vicinity of the myofibrils and thus initiates contraction. Since myocardial cells such as the atrial and Purkinje cells (97), as well as the ventricular cells of embryonic or neonatal chickens have no T system (41), it appears that the T system is not essential for the rapid propagation of excitation throughout the cell or for the coupling of excitation to contraction. It should be noted that the time between excitation and the beginning of contraction is about 20 msec in cardiac muscle. Thus it is not necessary that the exciting impulse release calcium from the sarcoplasmic reticulum almost simultaneously at the level of each sarcomere, as thought previously. On the other hand, current evidence points to the sarcolemmal basement membrane as the source of excitation-contraction coupling in the myofiber (89). Accordingly, the calcium which is presented to the area of the contractile filaments is released from the superficial cell membrane and travels inward to the area of the sarcomeres by a simple process of diffusion. Calcium ion diffuses in myoplasm at a rate of 1 μ/msec and would take 5 msec to reach the contractile unit of the cardiac cell with an average diameter of 10μ. Whether or not sarcolemmal calcium is sufficient for the purpose of contraction is not clear at present. The role of extracellular calcium in this regard cannot be ignored.

The myocardial cell is capable of excitation and can depolarize spontaneously in a periodic fashion. The process of excitation and depolarization depends upon the properties of the sarcolemma, the T system, and the intercalated disc. The excitation is coupled to contraction so that the cell shortens in reponse to stimulation. One of the most important areas currently under investigation in cardiac cell physiology is the mechanism by which excitation is coupled to contraction in the myofiber. There is growing evidence that Ca^{++} enters the myocardium during the plateau of the cardiac action potential (10, 39) and augmentation of this Ca^{++} influx can increase myocardial contractility (73, 115). Some cations such as Ni^{++} and Co^{++} have been found to reduce or even completely suppress the myocardial contractility, although the height and the shape of the action potential remain nearly unchanged (47). On the other hand, Mn^{++} ions are known to have similar effects on cardiac contractility without any pronounced decrease in the plateau of the action potential (13, 112). Procaine, an agent which blocks the movement of Ca^{++} through membranes, prevented the rise in exchangeable Ca^{++} by depolarization while leaving the depolarization unchanged (32). These observations lend support to the general concept that Ca^{++} ion is not only essential for the contractile events but also plays a crucial role in membrane events associated with excitation and depolarization of the cardiac cell.

Several investigators have attempted to isolate heart sarcolemma for biochemical, pharmacological, and pathological studies. Stam *et al.* (99) were the first to report the isolation of myocardial sarcolemma, afer extraction with NaI or KI. Morphologically, the preparation progressed from fibers to fiber-sized, membrane-enclosed, striation-free sacs. This preparation contained both Mg^{++}ATPase and (Na^+-K^+)-ATPase activities and was found to be associated with phospholipases A and A_2 as well as lysophospholipase (113). In addition, this preparation has been found to contain Mg^{++}-dependent, Ca^{++}-stimulated, ouabain-sensitive ATPase (46). Other workers, however, have demonstrated the presence of Mg^{++}-independent, Ca^{++}-stimulated, ouabain-insensitive ATPase in heart sarcolemma containing Mg^{++}-activated ATPase, Na^+-K^+-stimulated, ouabain-sensitive ATPase (101). The presence of this Mg^{++}-independent, Ca^{++}-stimulated ATPase in heart sarcolemma has been confirmed (20). The application of sucrose density gradient technique has also been reported to yield plasma membrane containing Mg^{++}ATPase and (Na^+-K^+)-ATPase (45). On the other hand, plasma membranes prepared by exposing the myocardial cell segments to osmotic shock followed by extraction in KCl have been found to contain adenylate cyclase in addition to (Na^+-K^+)-ATPase activity (109). Recently we isolated heart sarcolemma by hypotonic shock and LiBr treatment and this preparation was found to contain different membrane-bound enzymes exhibiting good specific activities (59, 60). The activities of different heart sarcolemmal preparations are shown in table III for the purpose of comparison. Various

Table III Sarcolemmal enzymes activities reported by various investigators in hearts from different animals

Animals used by investigators	Adenylate cyclase (pmoles cyclic AMP/ mg protein/min)	ATP hydrolysis (μmoles Pi/mg protein/hr)		
		$(Na^+$-$K^+)$-ATPase	Mg^{++}ATPase	Ca^{++}ATPase
Dog				
Stam *et al.*, 1970 (99)		5.3	3.6	3.0
Sulakhe and Dhalla, 1971 (101)		7.4	4.1	3.4
McNamara *et al.*, 1974 (59, 60)	297	10.0	15.9	15.0
Rat				
Dietze and Hepp, 1972 (20)		4.0	1.9	2.0
Kidwai *et al.*, 1971 (46)		13.0	13.0	
Guinea pig		5.4		6.9
Tada *et al.*, 1972 (109)	246			

reports concerning the isolation and properties of skeletal muscle sarcolemma have also appeared in the literature (34, 46, 61, 88, 105–107, 111).

The heart sarcolemmal activities of adenylate cyclase, Ca^{++}ATPase, Mg^{++}ATPase, and $(Na^{+}\text{-}K^{+})$ATPase appear to be species dependent (table IV). Directly acting sympathomimatics such as norepinephrine and isoproterenol were found to increase heart sarcolemmal adenylate cyclase activity, whereas ouabain, another well known positive inotropic agent, decreased $(Na^{+}\text{-}K^{+})$-ATPase activity without affecting Ca^{++}ATPase or Mg^{++}ATPase activities (table V). On the other hand, indirectly acting sympathomimatic agents such as tyramine, metaraminol, ephedrine, and nicotine as well as cyclic AMP had no effect on the dog heart sarcolemma (table V). These results suggest the presence of receptors for adrenergic amines and cardiac glycosides in the heart sarcolemma.

Recently some investigators (105, 106) have reported the presence of "calcium pump" mechanisms in the skeletal muscle sarcolemma like that present in the sarcoplasmic reticulum. However, our efforts to demonstrate "calcium pump" in heart sarcolemma obtained by the hypotomic shock and LiBr treatment have not been successful thus far. The results shown in table VI indicate that calcium binding by mitochondria and sarcoplasmic reticulum is increased by MgATP, whereas it is decreased in the case of heart sarcolemma. Furthermore, unlike sarcoplasmic reticulum, both sarcolemmal and mitochondrial calcium binding is not influenced by the presence of potassium oxalate. We have also been unable to show Ca^{++}-stimulated, Mg^{++}-dependent ATPase activity in heart sarcolemma unlike that in the heart sarcoplasmic reticulum. Our inability to detect "calcium pump" in heart sarcolemma may be attributable to rupture of this mechanism during the isolation procedure. It is also possible that such a system is not present in heart sarcolemma.

STUDIES ON ADENYLATE CYCLASE

Several hormones and neurotransmitters have been shown to alter adenylate cyclase activity in the myocardium (29, 70, 76, 108). The cytochemical studies have revealed that the adenylate cyclase activity in the cardiac muscle is not only localized at the sarcolemma but is also present in the sarcotubular system within the cell (81). Association of adenylate cyclase with the sarcoplasmic reticulum has also been demonstrated biochemically (22, 24, 28, 104). The properties of heart adenylate cyclase in washed cell particles and in sarcotubular vesicles have been studied extensively by different investigators (24, 26, 27, 104). Both catecholamines and NaF have been shown to activate adenylate cyclase by increasing V_{max} without any changes in K_m. The true substrate for this enzyme appears to be MgATP. The integrity of the lipoprotein complex seems to be essential for the full expression of the enzyme activity (104) and phospholipids

Table IV Activities of different sarcolemmal enzymes in hearts from various animals*

Animals	Adenylate cyclase in the presence of 2 mM NaF (pmoles cyclic AMP/ mg protein/min	ATP hydrolysis (μmoles Pi released/mg protein/hr)		
		Ca^{--}ATPase	Mg^{++}ATPase	$(Na^{+}$-$K^{+})$-ATPase
Mongrel dogs	360	15.4	13.7	10.5
Albino rabbits	950	26.5	22.7	9.8
Albino rats	973	38.2	30.4	14.3
Syrian hamsters	580	24.2	24.9	8.7
Chinese hamsters		23.1	17.1	10.0

*Sarcolemmal preparations were isolated by hypotonic shock and LiBr treatment according to the method described earlier (59, 60). The membranes were extracted with 0.6 M KCl for 15 min, washed, and assayed for different enzyme activities as described elsewhere (59, 60). Each value is an average of at least four experiments.

Table V Effect of various sympathomimetic agents, cyclic AMP, and ouabain on dog heart sarcolemmal enzyme activities*

Agents	Adenylate cyclase (pmoles cyclic AMP/ mg protein/min)	ATP hydrolysis (μmoles Pi released/mg protein/hr)		
		Ca^{++}ATPase	Mg^{++}ATPase	$(Na^{+}-K^{+})$-ATPase
None	287 ± 18	11.6 ± 1.2	13.0 ± 1.1	9.7 ± 0.5
Norepinephrine	368 ± 25	11.2 ± 1.3	12.9 ± 1.6	10.2 ± 0.7
Isoproterenol	400 ± 27	11.8 ± 1.3	12.6 ± 1.5	9.9 ± 0.7
Tyramine	278 ± 21	10.8 ± 1.3	12.6 ± 1.4	9.7 ± 0.8
Metaraminol	301 ± 23	10.3 ± 1.1	12.2 ± 1.8	9.0 ± 0.6
Ephedrine	298 ± 17	11.6 ± 1.7	13.2 ± 1.3	9.8 ± 0.8
Nicotine	291 ± 17	12.7 ± 1.8	13.4 ± 1.6	10.3 ± 0.5
Cyclic AMP	280 ± 15	11.6 ± 1.7	11.9 ± 1.7	9.8 ± 0.8
Ouabain	277 ± 23	11.3 ± 1.1	12.7 ± 1.4	4.5 ± 0.4

*The methods for preparing dog heart sarcolemma and assaying enzyme activities were same as those described by McNamara *et al.* (59, 60). All the agents used in this study were in the range of 50–150 μM and were added to the medium before starting the incubation. Each value is a mean ± S.E. of eight experiments.

Table VI Calcium accumulation by different rat heart cell membranes in absence of presence of Mg^{++}, ATP, and oxalate*

Additions	Calcium accumulation (nmoles Ca^{2+}/mg protein/5 min)		
	Sarcolemma	Sarcoplasmic reticulum	Mitochondria
None	46	5	7
5 mM Mg^{2+}	21	6	6
5 mM ATP	14	8	9
5 mM Mg^{2+} + 2 mM ATP	13	39	47
5 mM Mg^{2+} + 2 mM ATP + 2 mM potassium oxalate	14	280	45

*The membrane proteins (0.2–0.5 mg/ml) were incubated in a medium containing 100 mM KCl, 20 mM Tris-HCl, pH 6.8, and 0.1 mM $^{45}CaCl_2$ at 25°C in the presence or absence of the above mentioned agents. The reaction was started by the addition of $^{45}CaCl_2$ and terminated by Millipore filtration. Each value is an average from three experiments. The heart membranes were isolated by the methods described earlier (59, 60).

are necessary components of this enzyme system (50). The presence of adenylate cyclase in the heart plasma membrane has been demonstrated (59, 60, 109) and its properties are similar to those mentioned above.

Conflicting reports concerning changes in adenylate cyclase activity in heart failure have appeared in the literature. Sobel *et al.* (96) have reported a reduction in adenylate cyclase activity in failing guinea pig hearts, whereas other investigators have failed to observe such a change (36). Although Gold *et al.* (38) did not observe any alteration in adenylate cyclase activity in the absence or presence of norepinephrine or NaF in chronic heart failure induced in cats by occluding the pulmonary artery, glucagon was found to be unable to activate this enzyme. The isolated right ventricle failure led not only to a decreased capacity of right ventricular adenylate cyclase to respond to glucagon stimulation, but also to a similar impairment of adenylate cyclase derived from the unstressed left ventricle (51). A marked reduction in the adenylate cyclase activity in the presence of epinephrine and NaF was observed in hearts which failed to generate contractility because of substrate lack; however, no alteration in the enzyme activity was noted in hearts which failed to generate contractile force by about 50 percent of the control value (23). Likewise, no appreciable changes in adenylate cyclase were seen in myopathic hamsters (BIO 14.6) with a moderate degree of heart failure, whereas its responses to norepinephrine and NaF were markedly altered in late stages of failure (102). The adenylate cyclase activity in the isolated perfused rat heart was increased upon inducing hypoxia which markedly reduced the ability of heart to generate contractile force (58).

The discrepancy in results from various laboratories may be due to either the difference in the type of heart failure or the degree of heart failure.

In spite of a great deal of effort the precise role of adenylate cyclase in myocardial function remains unclear. Although denervation supersensitivity in myocardium has been recognized, the adenylate cyclase activity in the absence and presence of epinephrine in the chronically denervated cat heart was found to be normal (94). Likewise, normal myocardial adenylate cyclase activity has been reported in hyperthyroid cats (95). On the other hand, Levey *et al.* (52) have observed a decrease in adenylate cyclase activity in hypothyroidism. An increase in adenylate cyclase activity in washed cell particles obtained from overloaded left ventricles of the perfused guinea pig heart has been observed to be associated with an increased protein synthesis (80). On the other hand, depressed cardiac adenylate cyclase activity in both hypertrophied and failing hearts due to aortic constriction in rabbits has been demonstrated (116). Such conflicting results point out the need for further studies on adenylate cyclase with respect to changes in myocardial function.

STUDIES ON (Na^+-K^+)-ATPase

As in the case of adenylate cyclase, cytochemical evidence (82, 114) has also accumulated to support the localization of (Na^+-K^+)-ATPase in the sarcolemma after the demonstration of this enzyme in cardiac tissue (6, 7, 11, 48, 84, 85, 92). The properties of (Na^+-K^+)-ATPase from various tissues have been extensively studied (37, 56, 71, 79, 86, 93, 98). This enzyme requires both Na^+ and K^+ for full activation and uses MgATP as substrate. Both ouabain and Ca^{++} are well known inhibitors of this enzyme. The kinetic as well as cardiac glycoside binding data suggest this is an allosteric enzyme possessing multiple ligand binding sites (3, 5, 86). There is also a sufficient body of evidence to indicate a specific inhibitory effect by digitalis glycosides on the (Na^+-K^+)-ATPase activity of hearts from various species (2, 4, 53, 57, 68, 72, 83, 87). The well known differences in the sensitivity of various species to cardiac glycosides have been explained on the basis of relative affinity of (Na^+-K^+)-ATPase to the drug. It should be pointed out that most of the above work with (Na^+-K^+)-ATPase has been carried out with a solubilized enzyme preparation; however, the properties of the enzyme in its native state in the cardiac sarcolemma (109) suggest many similarities.

In addition to the suggested role of (Na^+-K^+)-ATPase in the transport of Na^+ and K^+ across the cell membrane, this enzyme has been considered to have an important function in the transport of calcium, sugars, amino acids, and biogenic amines (86). The positive inotropic effect of ouabain has been shown to occur concomitantly with an inhibitory effect on (Na^+-K^+)-ATPase (1, 9). Studies by Prasad (69) have demonstrated a relation between the inhibition of (Na^+-K^+)-ATPase and shortening of action potential duration. Recently, increased

Ca^{++} entry during the plateau of the action potential has been proposed to occur as a possible mechanism of cardiac glycoside action on (Na^+-K^+)-ATPase (44). All of these studies reflect the importance of (Na^+-K^+)-ATPase in a wide variety of physiological and pharmacological events in the myocardium but do not clearly elucidate the exact mechanisms of its involvement.

Some reports indicating alterations in (Na^+-K^+)-ATPase activity in different models of failing hearts have appeared in the literature. For example, an increase in (Na^+-K^+)-ATPase activity has been observed in experimental cardiac insufficiency induced by Co^{++} or vitamin E deficiency (25, 31). A marked elevation of the enzyme activity was also observed in failing hearts of genetically myopathic hamsters (BIO 14.6) (103). Similarly, cardiac (Na^+-K^+)-ATPase activity has been found to increase in K^+ deficiency (30). On the other hand, a decrease in (Na^+-K^+)-ATPase activity has been reported in failing hearts because of aortic constriction, hypoxia, or substrate lack (8, 21, 117). Reduction in cardiac (Na^+-K^+)-ATPase activity has also been observed in animals with hypoadrenalism (12). Furthermore, low (Na^+-K^+)-ATPase activity was reported in failing human hearts and these preparations were also less sensitive to inhibition by ouabain (54). It should be noted that Mead *et al.* (62) did not find any change in (Na^+-K^+)-ATPase in failing dog hearts because of pulmonary artery stenosis. These apparently contradictory results on the (Na^+-K^+)-ATPase activity in failing hearts could be due to differences in the degree and the type of heart failure.

Recent experiments from our laboratory have revealed that not only was (Na^+-K^+)-ATPase activity of the heart sarcolemma from myopathic hamsters (UM - X7.1) at late stages of heart failure decreased, but also the activities of other membrane-bound enzymes, Ca^{++}ATPase and Mg^{++}ATPase, were depressed (90). Although basal adenylate cyclase activity of the failing heart sarcolemma in these animals was not different from the control, the activity of this enzyme in the presence of catecholamines and NaF was lower. From these results and the foregoing discussion we conclude that heart dysfunction is associated with defects in sarcolemma.

REFERENCES

1. AKERA, T., LARSEN, F. S., and BRODY, T. M. 1970. Correlation of cardiac sodium-and-potassium-activated adenosine triphosphatase activty with ouabain-induced inotropic stimulation. J. Pharmacol. Exp. Therap. 173: 145–151.
2. ALLEN, J. C., BESCH, H. R., Jr., GLICK, G., and SCHWARTZ, A. 1970. The binding of tritiated ouabain to sodium and potassium activated adenosine triphosphatase and cardiac relaxing system of perfused dog heart. Molec. Pharmacol. 6: 441–443.
3. ALLEN, J. C., LINDENMAYER, G. E., and SCHWARTZ, A. 1970. An allosteric explanation for ouabain-induced time-dependent inhibition of sodium, potassium-adenosine triphosphatase. Arch. Biochem. Biophys. 141: 322–328.
4. ALLEN, J. C., and SCHWARTZ, A. 1969. A possible biochemical explanation for the insensitivity of the rat to cardiac glycosides. J. Pharmacol. Exp. Therap. 168: 42–46.

5. ALLEN, J. C., and SCHWARTZ, A. 1970. Effects of potassium, temperature and time on ouabain interaction with cardiac Na^+, K^+ ATPase: Further evidence supporting an allosteric site. J. Molec. Cell. Cardiol. 1: 39–45.
6. AUDITORE, J. V. 1962. Sodium-potassium activated G-strophanthin sensitive ATPase in cardiac muscle. Proc. Soc. Exp. Biol. Med. 110: 595–597.
7. AUDITORE, J. V., and MURRAY, L. 1962. Cardiac (microsomal) Na + K adenosinetriphosphatase and its possible relationship to the active Na + K transport system. Arch. Biochem. Biophys. 99: 372–382.
8. BALASUBRAMANIAN, V., McNAMARA, D. B., SINGH, J. N., and DHALLA, N. S. 1973. Biochemical basis of heart function. X. Reduction in the Na^+ - K^+ stimulated ATPase activity in failing rat heart due to hypoxia. Can. J. Physiol. Pharmacol. 51: 504–510.
9. BESCH, H. R., Jr., ALLEN, J. C., GLICK, G., and SCHWARTZ, A. 1970. Correlation between the inotropic action of ouabain and its effects on subcellular enzyme systems from canine myocardium J. Pharmacol. Exp. Therap. 171: 1–12.
10. BESLER, G. W., Jr., and RUETER, H. 1970. Membrane calcium current in ventricular myocardial fibers. J. Physiol. 207: 191–209.
11. BONTING, S. L., SIMON, K. A., and HAWKINS, N. M. 1961. Studies on sodium-potassium-activated adenosine triphosphatase. 1. Quantitative distribution in several tissues of the cat. Arch. Biochem. Biophys. 95: 416–423.
12. BORSCH-GALETKE, E., DRANSFELD, H., and GREEFF, K. 1972. Specific activity and sensitivity to strophanthin of the Na^+ + K^+-activated ATPase in rats and guinea pigs with hypoadrenalism. Naunyn-Schmeideberg's Arch. Pharmacol. 274: 74–80.
13. CARMELIET, E., and VEREECKE, E. 1969. Adrenaline and the plateau phase of the cardiac action potential. Pfluger's Arch. ges. Physiol. 313: 300–315.
14. CHAPMAN, D., KAMAT, V. B., de GIER, J., and PENKETT, S. A. 1967. Nuclear magnetic resonance spectroscopic studies of erythrocyte membranes. Nature 213: 74–75.
15. CHAPMAN, D., KAMAT, V. B., and LEVERNE, R. J. 1968. Infra red spectra and the chain organization of erythrocyte membranes. Science 160: 314–316.
16. CHAPMAN, D., and PENKETT, S. A. 1966. Nuclear magnetic resonance spectroscopic studies of the interaction of phospholipids with cholesterol. Nature 211: 1304–1305.
17. CHANGEUX, J. P., THIERY, J., TUNG, Y., and KITTEL, C. 1967. On the cooperativity of biological membranes. Proc. Nat. Acad. Sci. U.S.A. 57: 335–341.
18. DANIELLI, J. F., and DAVSON, H. 1935. A contribution to the theory of permeability of thin films. J. Cell. Comp. Physiol. 5: 495–514.
19. DAVIS, B. D., and WARREN, L. (eds.) 1967. The Specificity of Cell Surfaces. Prentice-Hall, Englewood Cliffs, N.J.
20. DIETZE, G., and HEPP, K. D. 1972. Effect of 3′, 5′-AMP on calcium-activated ATPase in rat heart sarcolemma. Biochem. Biophys. Res. Commun. 46: 269–277.
21. DHALLA, N. S., SINGH, J. N., FEDELESOVA, M., BALASUBRAMANIAN, V., and McNAMARA, D. B. 1974. Biochemical basis of heart function. XII. Sodium-potassium stimulated adenosine triphosphatase activity in the perfused rat heart made to fail by substrate-lack. Cardiovasc. Res. 8: 227–237.
22. DHALLA, N. S., SULAKHE, P. V., KHANDELWAL, R. L., and HAMILTON, I. R. 1970. Excitation contraction coupling in heart. II. Studies on the role of adenyl cyclase in the calcium transport by dog heart sarcoplasmic reticulum. Life Sci. 9: 625–632.
23. DHALLA, N. S., SULAKHE, P. V., KHANDELWAL, R. L., and OLSON, R. E. 1962. Adenyl cyclase activity in the perfused rat heart made to fail by substrate-lack. Cardiovasc. Res. 6: 344–352.
24. DHALLA, N. S., SULAKHE, P. V., and McNAMARA, D. B. 1973. Studies on the relationship between adenylate cyclase activity and calcium transport by cardiac sarcotubular membranes. Biochim. Biophys. Acta 323: 276–284.

25. DRANSFELD, H., LIPINSKI, J., and BORSCH-GALETKE, E. 1971. Die Na^+ - K^+ activierte transport-ATPase bei experimenteller Herzinsuffizienzy duich Kobaltchloride. Naunyn - Schmeideberg's Arch. Pharmakol. 270: 335–342.
26. DRUMMOND, G. I., and DUNCAN, L. 1970. Adenyl cyclase in cardiac tissue. J. Biol. Chem. 245: 976–983.
27. DRUMMOND, G. I., SEVERSON, D. L., and DUNCAN, L. 1971. Adenyl cyclase. Kinetic properties and nature of fluoride and hormone stimulation. J. Biol. Chem. 246: 4166–4173.
28. ENTMAN, M. L., LEVEY, G. S., and EPSTEIN, S. E. 1969. Demonstration of adenyl cyclase activity in canine cardiac sarcoplasmic reticulum. Biochem. Biophys. Res. Commun. 35: 728–733.
29. EPSTEIN, S. E., LEVEY, G. S., and SKELTON, C. L. 1971. Adenyl cyclase and cyclic AMP. Biochemical links in the regulation of myocardial contractility. Circulation 43: 437–450.
30. ERDMAN, E., BOLTE, H. D., and LUDERITZ, B. 1971. The Na^+ - K^+ ATPase activity of guinea pig heart muscle in potassium deficiency. Arch. Biochem. Biophys. 145: 121–125.
31. FEDELESOVA, M., SULAKHE, P. V., YATES, J. C., and DHALLA, N. S. 1971. Biochemical basis of heart function. IV. Energy metabolism and calcium transport in hearts of vitamin E deficient rats. Can. J. Physiol. Pharmacol. 49: 909–918.
32. FEINSTEIN, M. B. 1963. Inhibition of caffeine rigor and radio-calcium movements by local anesthetics in frog sartorius muscle. J. Gen. Physiol. 47: 151–172.
33. FERANDEZ-MORAN, H. 1967. Membrane ultrastructure in nerve cells, pp. 281–304. *In* G. D. Quarton, T. Melnechuk, and F. O. Schmitt (eds.), The Neurosciences Study. Rockefeller University Press, New York.
34. FERDMAN, D. L., GUMMEL'REIKH, N. G., and DYADYUSHA, G. P. 1969. Enzyme activity of the sarcolemma membrane of rabbit skeletal muscles. Biochemistry 34: 402–405.
35. FINEAN, J. B. 1966. The molecular organization of cell membranes, pp. 143–170. *In* J. A. V. Butlerand and H. E. Huxley (eds.), Progress in Biophysics and Molecular Biology, Vol. 16. Pergamon Press, Oxford.
36. GERTLER, M. M., SALUSTE, E., and SPENCER, F. 1970. Biochemical analysis of human papillary muscles and guinea pig heart ventricles in failure. Proc. Soc. Exp. Biol. Med. 135: 817–824.
37. GIBSON, K., and HARRIS, P. 1968. Some properties of human myocardial microsomal (Na^+ - K^+) - ATPase. Cardiovasc. Res. 2: 367–370.
38. GOLD, H. K., PRINDLE, K. H., LEVEY, G. S., and EPSTEIN, S. E. 1970. Effects of experimental heart failure on the capacity of glucagon to augment myocardial contractility and activate adenyl cyclase. J. Clin. Invest. 49: 999–1006.
39. HAGIWARA, S., and NAKAJIMA, S. 1966. Difference in Na^+ and Ca^{++} spikes as examined by application of tetradotoxin, procaine and manganese ions. J. Gen. Physiol. 49: 793–806.
40. HENDLER, R. W. 1971. Biological membrane ultrastructure. Physiol. Rev. 51: 66–97.
41. HIRAKOW, R. 1970. Ultrastructural characteristics of the mammalian and sauropsidan heart. Amer. J. Cardiol. 25: 195–203.
42. HODGKIN, A. L. 1964. The Conduction of the Nerve Impulse. Charles C Thomas, Springfield, Ill.
43. HOFMANN, W. W. 1968. Membrane potential and muscle function. Ann. Rev. Med. 20: 101–129.
44. KATZ, A. M. 1972. Increased Ca^{++} entry during the plateau of the action potential: A possible mechanism of cardiac glycoside action. J. Molec. Cell. Cardiol. 4: 87–89.
45. KIDWAI, A. M., RADCLIFFE, M. A., DUCHON, G., and DANIEL, E. E. 1971. Isolation of plasma membrane from cardiac muscle. Biochem. Biophys. Res. Commun. 45: 901–910.
46. KIDWAI, A. M., RADCLIFFE, M. A., LEE, E. Y., and DANIEL, E. E. 1973. Isolation

and properties of skeletal muscle plasma membrane. Biochim. Biophys. Acta 298: 593–607.

47. KAUFMANN, R., and FLECKENSTEIN, A. 1970. Ca^{++}-kompetitive elektromechanische Entkoppelung durch Ni^{++}-und Co^{++}-Ionen am Warmbluter-Myokard. Pfluger's Arch. ges. Physiol. 282: 290–297.

48. LEE, K. S., and YU, D. H. 1963. A study of the sodium-and potassium-activated adenosinetriphosphatase activity of heart microsomal fraction. Biochem. Pharmacol. 12: 1253–1264.

49. LEHNINGER, A. L. 1966. Supramolecular organization of enzyme and membrane systems. Naturwissenchaften 53: 57–63.

50. LEVEY, G. S. 1972. Phospholipids, adenylate cyclase and the heart. J. Molec. Cell. Cardiol. 4: 283–285.

51. LEVEY, G. S., PRINDLE, K. H., Jr., and EPSTEIN, S. E. 1970. Effects of glucagon on adenyl cyclase activity in the left and right ventricles and liver in experimentally-produced isolated right ventricular failure. J. Molec. Cell. Cardiol. 1: 403–419.

52. LEVEY, G. S., SKELTON, C. L., and EPSTEIN, S. E. 1969. Decreased myocardial adenyl cyclase activity in hypothyroidism. J. Clin. Invest. 48: 2244–2250.

53. LINDENMAYER, G. E., and SCHWARTZ, A. 1970. Conformational changes induced in Na^+, K^+-ATPase by ouabain through a K^+ sensitive reaction: Kinetic and spectroscopic studies. Arch. Biochem. Biophys. 140: 371–378.

54. LINDENMAYER, G. E., SORDAHL, L. S., HARIGAYA, S., ALLEN, J. C., BESCH, H. R., and SCHWARTZ, A. 1971. Some biochemical studies on subcellular systems isolated from fresh recipient human cardiac tissue obtained during transplantation. Amer. J. Cardiol. 27: 277–283.

55. MADDY, A. H., HUANG, C., and THOMPSON, T. E. 1966. Studies on lipid bilayer membranes: A model for the plasma membranes. Fed. Proc. 25: 933–936.

56. MATSUI, H., and SCHWARTZ, A. 1966. Purification and properties of a highly active ouabain sensitive Na^+, K^+ - dependent adenosinetriphosphatase from cardiac tissue. Biochim. Biophys. Acta 128: 380–390.

57. MATSUI, H., and SCHWARTZ, A. 1968. Mechanism of cardiac glycoside inhibition of the (Na^+-K^+)-dependent ATPase from cardiac tissue. Biochim. Biophys. Acta 151: 655–663.

58. McNAMARA, D. B., SULAKHE, P. V., SINGH, J. N., and DHALLA, N. S. 1973. Adenyl cyclase activity in the hypoxic heart in myocardial metabolism, pp. 303–309. *In* N. S. Dhalla (ed.), Recent Advances in Studies on Cardiac Structure and Metabolism, Vol. 3. University Park Press, Baltimore.

59. McNAMARA, D. B., SULAKHE, P. V., SINGH, J. N., and DHALLA, N. S. 1974. Properties of heart sarcolemmal Na^+ - K^+ ATPase. J. Biochem. 75: 795–803.

60. McNAMARA, D. B., SINGH, J. N., and DHALLA, N. S. 1974. Properties of some heart sarcolemmal-bound enzymes. J. Biochem. 76: 603–609.

61. McNAMARA, D. B., SULAKHE, P. V., and DHALLA, N. S. 1971. Properties of the sarcolemmal calcium ion-stimulated adenosine triphosphatase of hamster skeletal muscle. Biochem. J. 125: 525–530.

62. MEAD, R. J., PETERSON, M. B., and WELTY, J. D. 1971. Sarcolemmal and sarcoplasmic reticular ATPase activities in the failing canine heart. Circ. Res. 24: 14–19.

63. MONOD, J., CHANGEUX, J. P., and JACOB, F. 1963. Allosteric proteins and cellular control systems. J. Molec. Biol. 6: 306–329.

64. MONOD, J., WYMAN, J., and CHANGEUX, J. P. 1965. On the nature of allosteric transitions: A plausible model. J. Mol. Biol. 12: 88–118.

65. MUELLER, P., and RUDIN, D. O. 1968. Action potentials induced in biomolecular lipid membranes. Nature 217: 713–719.

66. MUELLER, P., RUDIN, D. O., TIEN, H. T., and WESTCOTT, W. C. 1963. Methods for the formation of single biomolecular lipid membranes in aqueous solutions. J. Phys. Chem. 67: 534–535.

67. NACHMANSOHN, D. 1966. Chemical control of the permeability cycle in excitable membranes during electrical activity. Ann. N. Y. Acad. Sci. 137: 877–900.
68. NAGAI, K., LINDENMAYER, G., and SCHWARTZ, A. 1970. Direct evidence for the conformational nature of the Na^+, K^+-ATPase system: Fluorescence and circular dichroism studies. Arch. Biochem. Biophys. 139: 252–254.
69. PRASAD, K. 1972. Transmembrane potential, contraction and ATPase activity of human heart in relation to ouabain. Jap. Heart J. 13: 59–72.
70. RASMUSSEN, H., GOODMAN, D. B. P., and TENENHOUSE, A. 1972. The role of cyclic AMP and calcium in cell activation. Crit. Rev. Biochem. 1: 95–148.
71. REPKE, K. 1965. Effect of digitalis on membrane adenosine triphosphatase of cardiac muscle, pp. 65–87. *In* C. A. M. Hogben (ed.), Drugs and Membranes, Vol. IV. Pergamon Press, New York.
72. REPKE, K., EST, M., and PORTIUS, H. J. 1965. Uber die ursache der speciesunterschiede in der digitalisempfindlich kleit Biochem. Pharmacol. 14: 1785–1802.
73. REUTER, H. 1967. Dependence of slow inward current in Purkinje fibers on the extracellular calcium concentration. J. Physiol. 207: 479–492.
74. ROBERTSON, J. D. 1959. The ultrastructure of cell membranes and their derivatives. Biochim. Soc. Symp. 16: 3–43.
75. ROBERTSON, J. D. 1966. Granulo-fibrillar and globular substructure in unit membranes. Ann. N. Y. Acad. Sci. 137: 421–440.
76. ROBISON, G. A., BUTCHER, R. W., and SUTHERLAND, E. W. 1968. Cyclic AMP. Ann. Rev. Biochem. 37: 149–174.
77. ROTHSTEIN, A. 1968. Membrane phenomena. Ann. Rev. Physiol. 30: 15–72.
78. SCHATZMANN, H. J., and VINCENZI, F. F. 1969. Calcium movements across the membrane of human red cells. J. Physiol. 201: 369–395.
79. SCHON, R., MENKE, K. H., and REPKE, K. R. H. 1971. Bedeutung der Praparations methode fur die Eigenschaften der (Mg^{++})- und $(Na^+ - K^+)$-ATPase der Herzmuskel-Zellmembran vom Meerschwein cheir. Acta Biol. Med. Germ. 26: 51–65.
80. SCHREIBER, S. S., KLEIN, I. L., ORATZ, M., and ROTHSCHILD, M. A. 1971. Adenyl cyclase activity and cyclic AMP in acute cardiac overload: A method for measuring cyclic AMP production based on ATP specific activity. J. Molec. Cell. Cardiol. 2: 55–65.
81. SCHULZE, W., KRAUSE, E. G., and WOLLENBERGER, A. 1972. Cytochemical demonstration and localization of adenyl cyclase in skeletal and cardiac muscle, pp. 249–260. *In* P. Greengard, R. Padetti, and G. A. Robison (eds.), Advances in Cyclic Nucleotide Research, Vol. 1. Raven Press, New York.
82. SCHULZE, W., and WOLLENBERGER, A. 1969. Zytochemische Lokalisation und Differenzierung von $Na^+ - K^+$ und anderer membranstandiger ATPase - Aktivitat im Herzmuskel. Hitochemistry 19: 302–318.
83. SCHWARTZ, A., ALLEN, J. C., and HARIGAYA, S. 1969. Possible involvement of cardiac Na^+, K^+ - adenosine triphosphatase in the mechanism of action of cardiac glycosides. J. Pharmacol. Exp. Therap. 168: 31–41.
84. SCHWARTZ, A., and LASETER, A. H. 1964. A sodium-and potassium-stimulated adenosine-triphosphatase from cardiac tissues. II. The effects of ouabain and other agents that modify enzyme activity. Biochem. Pharmacol. 13: 337–348.
85. SCHWARTZ, A., and LASETER, A. H. 1963. Observations on the mechanism of action of ouabain on an active transport enzyme system. Life Sci. 2: 363–367.
86. SCHWARTZ, A., LINDENMAYER, G. E., and ALLEN, J. C. 1972. The Na^+,K^+-ATPase membrane transport system: Importance in cellular function, pp. 1–83. *In* F. Bonner and A. Kleinzeller, (eds.), Current Topics in Membranes and Transport, Vol. 3. Academic Press, New York.
87. SCHWARTZ, A., MATSUI, H., and LAUGHTER, A. H. 1968. Tritiated digoxin binding to $(Na^+ + K^+)$-activated adenosine triphosphatase: Possible allosteric site. Science 160: 323–325.
88. SEVERSON, D. L., DRUMMOND, G. I., and SULAKHE, P. V. 1972. Adenylate

cyclase in skeletal muscle: Kinetic properties and hormonal stimulation. J. Biol. Chem. 247: 2949–2958.

89. SHINE, K. I., SERENA, S. D., and LANGER, G. A. 1971. Kinetic localization of contractile calcium in rabbit myocardium. Amer. J. Physiol. 221: 1408–1417.

90. SINGH, J. N., DHALLA, N. S., McNAMARA, D. B., BAJUSZ, E., and JASMIN, G. 1975. Membrane alteration in failing hearts of cardiomyopathic hamsters, pp. 259–268. *In* A. Fleckenstein and I. Rona, (eds.), Recent Advances in Studies on Cardiac Structure and Metabolism, Vol. 6. University Park Press, Baltimore.

91. SJOSTRAND, F. S. 1968. Molecular structure and function of cellular membranes, pp. 15–36. *In* H. Peeters (ed.), Protides of the Biological Fluids, Vol. 15. Elsevier, Amsterdam.

92. SKOU, J. C. 1962. Preparation from mammalian brain and kidney of the enzyme system involved in active transport of Na^+ and K^+. Biochim. Biophys. Acta 58: 314–325.

93. SKOU, J. C. 1965. Enzymatic basis for active transport of Na^+ and K^+ across cell membranes. Physiol. Rev. 45: 596–617.

94. SOBEL, B. E., DEMPSEY, P. J., and COOPER, T. 1968. Adenyl cyclase activity in the chronically denervated cat heart. Biochem. Biophys. Res. Commun. 33: 758–762.

95. SOBEL, B. E., DEMPSEY, P. J., and COOPER, T. 1969. Normal myocardial adenyl cyclase activity in hyperthyroid cats. Proc. Soc. Exp. Biol. Med. 132: 6–9.

96. SOBEL, B. E., HENRY, P. D., ROBISON, A., BLOOR, C., and ROSS, J., Jr. 1969. Depressed adenyl cyclase activity in the failing guinea pig heart. Circ. Res. 24: 507–512.

97. SOMMER, J. R., and JOHNSON, E. A. 1968. Cardiac muscle. A comparative study of Perkinje fibers and ventricle fibers. J. Cell Biol. 36: 497–526.

98. SPERELAKIS, N., and LEE, E. C. 1971. Characterization of (Na^+, K^+)-ATPase isolated from embryonic chick hearts and cultured chick heart cells. Biochim. Biophys. Acta 233: 562–579.

99. STAM, A. C., Jr., WEGLICKI, W. B., Jr., FELDMAN, D., SHELBURNE, J. C., and SONNENBLICK, E. H. 1970. Canine myocardial sarcolemma–Its preparation and enzymatic activity. J. Molec. Cell. Cardiol. 1: 117–130.

100. STAM, A. C., Jr., WEGLICKI, W. B. GERTZ, E. W., and SONNENBLICK, E. H. 1973. A calcium-stimulated, ouabain-inhibited ATPase in a myocardial fraction enriched with sarcolemma. Biochim. Biophys. Acta 298: 927–931.

101. SULAKHE, P. V., and DHALLA, N. S. 1971. Excitation contraction coupling in heart VI. Demonstration of calcium activated ATPase in the dog heart sarcolemma. Life Sci. 10: 185–191.

102. SULAKHE, P. V., and DHALLA, N. S. 1972. Adenyl cyclase activity in failing hearts of genetically myopathic hamsters. Biochem. Med. 6: 471–482.

103. SULAKHE, P. V., and DHALLA, N. S. 1972. Alterations in the activity of cardiac Na^+-K^+-stimulated ATPase in congestive heart failure. Exp. Molec. Path. 18: 100–111.

104. SULAKHE, P. V., and DHALLA, N. S. 1973. Adenylate cyclase of heart sarcotubular membranes. Biochim. Biophys. Acta 293: 379–396.

105. SULAKHE, P. V., DRUMMOND, G. I., and NG, D. C. 1973. Calcium binding by skeletal muscle sarcolemma. J. Biol. Chem. 248: 4150–4157.

106. SULAKHE, P. V., DRUMMOND, G. I., and NG, D. C. 1973. Adenosine triphosphatase activities of muscle sarcolemma. J. Biol. Chem. 248: 4157–4162.

107. SULAKHE, P. V., FEDELESOVA, M., McNAMARA, D. B., and DHALLA, N. S. 1971. Isolation of skeletal muscle membrane fragments containing active Na^+ - K^+ stimulated ATPase: Comparison of normal and dystrophic muscle sarcolemma. Biochem. Biophys. Res. Commun. 42: 793–799.

108. SUTHERLAND, E. W., ROBISON, G. A., and BUTCHER, R. W. 1968. Some aspects of the biological role of adenosine 3′,5′-monophosphate (cyclic AMP). Circulation 37: 279–306.

109. TADA, M., FINNEY, J. O., Jr., SWARTZ, M. H., and KATZ, A. M. 1972. Preparation

and properties of plasma membranes from guinea pig hearts. J. Molec. Cell. Cardiol. 4: 417–426.
110. TASAKI, I. 1968. Nerve Excitation, Charles C Thomas, Springfield, Ill.
111. THORPE, W. R., and SEEMAN, P. 1971. Effect of denervating muscle on calcium binding by isolated sarcolemma. Exp. Nuerol. 30: 277–290.
112. VITEK, M., and TRAUTWEIN, W. 1971. Slow inward current in Purkinje fibers. Pfluger's Arch. ges. Physiol. 323: 204–218.
113. WEGLICKI, W. B., WAITE, B. M., and STAM, A. C. 1972. Association of phospholipase A with a myocardial membrane preparation containing the (Na^+ + K^+)-Mg^{++}ATPase. J. Molec. Cell. Cardiol. 4: 195–201.
114. WOLLENBERGER, A., and SCHULZE, W. 1966. Darstellung der Na^+- und K^+ abhangigen glykosidempfindlichen Adenosinetriphosphatase des Herzmuskelsandu Plasmamembran des Sarkolemms. Naturwissenchaften 53: 134.
115. WOOD, E. H., HEPPNER, R. L., and WEIDMANN, S. 1969. Inotropic effects of electric currents. I. Positive and negative effects of constant electric currents or current pulses applied during cardiac action potential. II. Hypothesis: Calcium movements, excitation-contraction coupling and inotropic effects. Circ. Res. 24: 409–445.
116. YAMAZAKI, M., and OGAWA, K. 1971. Cardiac catecholamine metabolism in heart failure. Jap. Circ. J. 35: 965–971.
117. YAZAKI, Y., and FUJII, J. 1972. Depressed Na^+-K^+-ATPase activity in the failing rabbit heart. Jap. Heart J. 13: 73–83.

Mailing address: Naranjan S. Dhalla, Ph.D., F.I.C.,
Professor of Physiology, Faculty of Medicine, University of Manitoba,
Winnipeg, Manitoba R3E 0W3 (Canada).

Recent Advances in Studies on Cardiac Structure and Metabolism
Volume 9
The Sarcolemma
Edited by Paul-Emile Roy and Naranjan S. Dhalla

CONTROL OF ION MOVEMENT BY CARDIAC SARCOLEMMA[1]

K. I. SHINE and G. A. LANGER

The Departments of Physiology and Medicine and
the Los Angeles County Heart Association Cardiovascular Research Laboratory,
University of California at Los Angeles,
Center for the Health Sciences, Los Angeles, California, USA

SUMMARY

The cation affinity sequence based on ^{45}Ca displacement from neonatal rat heart tissue culture cells was La > Cd > Zn > Mn > Mg. This sequence correlated with the rate at which these ions diminished contractile tension in neonatal rat papillary muscles. In adult rat interventricular septa elevated extracellular Mg diminished tension development concomitant with a decrease in ^{45}Ca influx. The results support the hypothesis that these cations interfere with contraction by occupying sarcolemmal sites ordinarily reserved for calcium or by interfering with transmembrane Ca movements, both of which are crucial for tension development. The data of Bockman *et al.* (2) provides further support for this hypothesis.

Elevated extracellular Mg reduced potassium exchange by 30% in adult rat interventricular septa but had no effect on ^{42}K exchange in control rabbit interventricular septa. Elevated extracellular Mg did inhibit the net ^{42}K loss induced by digitalis glycoside without affecting tension development or glycoside inhibition of ^{42}K influx in the rabbit heart. High Mg also attenuated the increase in ^{42}K exchange produced by elevated extracellular potassium. These data demonstrate that glycoside inotropy is not dependent on potassium loss from myocardium but are consistent with the hypothesis that increased intracellular sodium mediates glycoside inotropy. A model for the effects of Mg in rat and rabbit heart is presented.

ION MOVEMENT AND CARDIAC SARCOLEMMA

The physiological burdens placed upon cardiac muscle require this tissue to make almost instantaneous changes in the vigor and rate of its contractile force while assuring an adequate opportunity for diastolic filling. In contrast to skeletal muscle, which can recruit additional muscle bundles or employ a tetany to maximize and sustain force, the heart must employ all of its fibers with each

[1] This study was supported by Grant HL-11351-07 from the United States Public Health Service and Grant 481-C1 from the Los Angeles County Heart Association and by a grant from the Anna Borun and Harry Borun Foundation.

beat and relax adequately between beats. Since the early experiments of Ringer (10) it has been clear that cardiac muscle can make such extremely rapid adaptations to changes in the extracellular ionic milieu. This responsiveness is far greater than that exhibited by skeletal muscle (9). The sarcolemmal membrane occupies a pivotal role in cardiac muscle not only for propagation of electrical events but also as the avenue by which ionic events in the extracellular environment are communicated rapidly to the contractile apparatus. In the present discussion we shall dwell primarily upon ionic movements which influence force of contraction directly rather than those related primarily to the propagation of electrical activity.

Calcium plays the central role in the induction of the contractile event at the troponin and tropomyosin sites (5). It is a thesis of this discussion that, under most circumstances, calcium ion bound to the sarcolemmal membrane and passing through that membrane is responsible for the contractile event (14). A number of ions interfere with contraction by virtue of their ability to occupy sarcolemmal sites ordinarily reserved for calcium or by interfering with transmembrane calcium movements.

The sarcolemmal membrane of cardiac muscle is extensive, and in comparison to skeletal muscle, develops into large capacious T tubules well described anatomically by Fawcett and McNutt (4). The T tubules are lined with a mucopolysaccharide-like substance which is negatively charged. Because the cell diameter of mammalian myocardium averages 10–15 μm, in contrast to approximately 100 μm for skeletal muscle, and because the transverse tubules are 4–5 times larger in cross-sectional diameter, the relative proportion of sarcolemmal membrane to sarcoplasmic reticulum and to cell volume is substantially greater in cardiac tissue than in skeletal muscle. Sanborn and Langer (11) showed that the trivalent cation lanthanum, in concentrations as little as 40 μM, reduced tension development in cardiac muscle to 2 percent of control value with only minimal alterations in the action potential. Since observations in cardiac tissue indicate that lanthanum does not penetrate intracellularly, as long as the sarcolemmal membrane is intact, the inference drawn from these experiments predicted a critical role of sarcolemmal anionic binding sites in the control of that calcium responsible for the contractile event.

Langer and Frank (7) characterized the role of such binding sites employing a technique for following isotopic activity of rat heart cells grown in tissue culture. A rapid displacement of ^{45}Ca from the heart cells, which eliminated further ^{45}Ca exchange, was produced by 1.0 mM La (fig. 1). Localization of the La was determined in electron microscopical studies of the same tissue (7). A thick basement membrane averaging 0.06 – 0.1 μm in thickness, was seen external to the sarcolemmal membrane of the tissue culture cells. In figure 2 particles of electron dense lanthanum are shown adherent to this basement

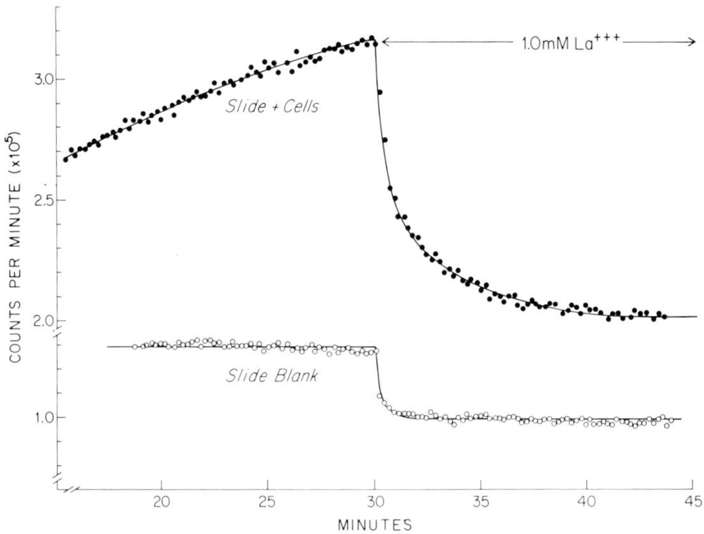

Figure 1 Pattern of ^{45}Ca displacement from slide + cells (•) and slide blank (○) by 1.0 mM La. The scintillator slide with a monolayer of heart cells on both surfaces was labeled for 30 min with ^{45}Ca solution containing a total Ca concentration of 2.5 μM. La, 1.0 nM, was added to the perfusate at 30 min. Note the abrupt displacement of ^{45}Ca with flattening of the slide + cell curve after 40 min. The slide blank indicates the response of the scintillator slide after the heart cells were removed to exactly the same experimental sequence. Note that the ^{45}Ca displacement by 1.0 mM La from the blank slide surface is completed by 2 min. The ^{45}Ca displacement at 2.2 min from the cells is obtained by: (slide + cell displacement) – (blank displacement) = 58,000 c.p.m. (Reproduced from Langer *et al.* (8) by permission of *Journal of Molecular and Cellular Cardiology.*

membrane. In intact cells no lanthanum was identified inside the sarcolemmal membrane. In further studies by Langer *et al.* (7) the effect of cadmium, zinc, manganese, and magnesium were examined. Of particular interest was the response to 1.0 mM Mg (fig. 3), in which there was an initial decrease in ^{45}Ca uptake followed by a gradual increase in ^{45}Ca. When the concentration of Mg was increased to 20 mM, however, the same type of persistent displacement as described for La was produced. Cd, Zn, and Mn occupied intermediate positions between Mg and La in the degree of displacement of ^{45}Ca from these preparations. The final affinity sequence based on ^{45}Ca displacement was La > Cd > Zn > Mn > Mg. This sequence correlated with the rate with which these ions diminished contractile tension in the neonatal rat papillary muscles. La and Cd were the most potent, Zn and Mn presented an intermediate position, and Mg was the least potent, in the ratio of 17:7:1. Langer *et al.* (8) suggested that the ability of an ion to displace Ca from ionized carboxyl groups localized to the

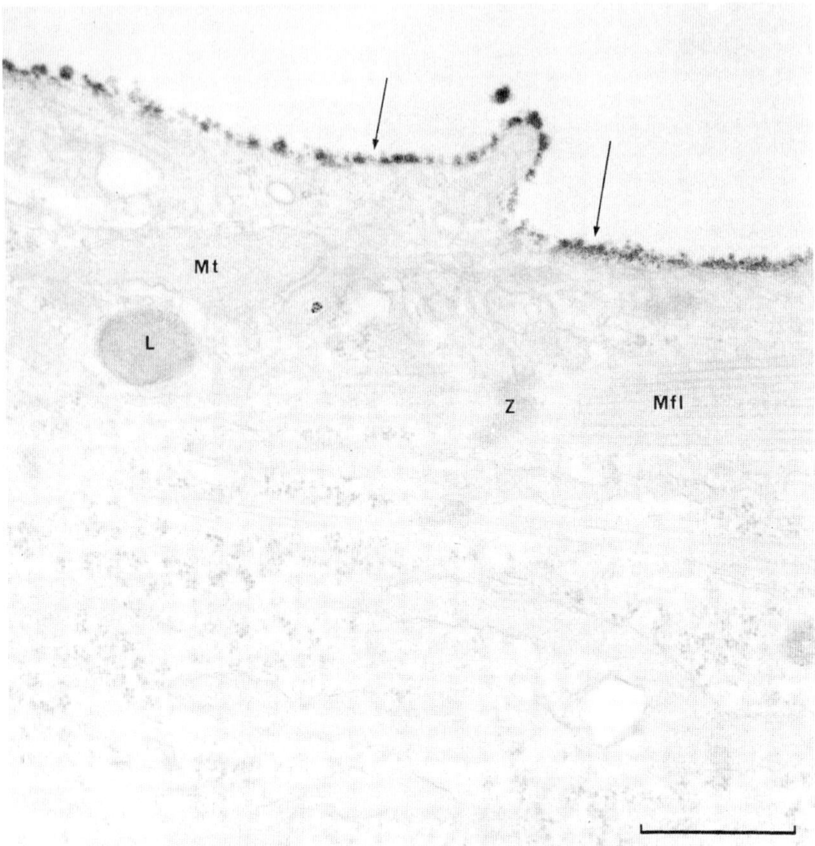

Figure 2 A typical myoblast cell from a 2-day tissue culture preparation that was exposed to 1.0 mM $LaCl_3$ for 5 min. This section was unstained. Note the accumulation of lanthanum exclusively on the surface of the cell (*arrows*). Mitochondria (*Mt*), myofilaments (*Mfl*), Z line (*Z*), and lipid droplets (*L*) are evident. Scale, 0.5 μm; × 50,000. (Reproduced from Langer and Frank (7) by permission of *Journal of Cellular Biology.*)

basement membrane, and the subsequent inhibition of cellular uptake of ^{45}Ca, was causally related to the diminished contractile force produced by these substances.

Additional support for the importance of extracellular and sarcolemmal calcium in control of contractile events has come from the experiments of Bockman *et al.* (2). These authors showed that 30 μM La prevented isoproterenol-stimulated contraction in rat and guinea pig ventricle but did not completely abolish isoproterenol-induced activation of phosphorylase in the same prepara-

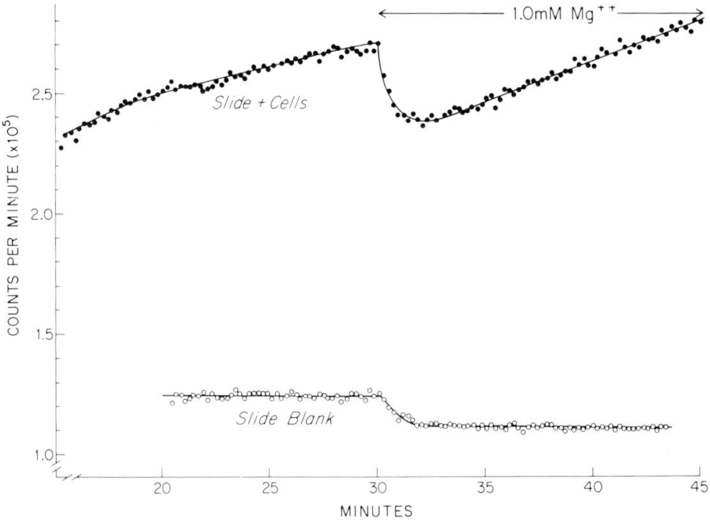

Figure 3 ^{45}Ca displacement by 1.0 mM. Note the relatively small initial net displacement which persists only for as long as required for the slide surface to come to steady state. Net ^{45}Ca uptake then resumes. This indicates a weak displacement from rapidly exchangeable sites followed by inability of Mg to inhibit subsequent uptake or displace ^{45}Ca from more slowly exchangeable sites. (Reproduced from Langer *et al.* (8) by permission of *Journal of Molecular and Cellular Cardiology.*)

tions. Since the K_m for calcium of the kinase involved in conversion of phosphorylase *b* to *a* is about 10-fold less than the K_m for calcium of the contractile proteins, the authors argued that if the amount of intracellular calcium released were large enough to activate the contractile proteins phosphorylase activation should have been nearly complete. Although intracellular Ca levels were adequate to allow partial conversion of phosphorylase *b*, an interruption of Ca influx by lanthanum prevented a response by the contractile proteins and complete phosphorylase activation.

We (12, 13) have subsequently examined further the effects of Mg on the movement of Ca and K ions across the sarcolemmal membrane to further define the possible role of K in control of contractile force in isolated heart. These studies, in isolated blood perfused interventricular septa also allowed comparison of Mg effects in neonatal tissue culture and adult heart. Qualitative and quantitative differences were found in the two species studied, rat and rabbit. At least two and perhaps three sites of interaction between Mg and other ions were identified.

In rat septa, Mg ion produced an abrupt and rapid decline in tension development with a rate constant of 1.27 ± 0.16 min^{-1}. This effect was rapidly reversible and was antagonized by increases in extracellular Ca. As demonstrated

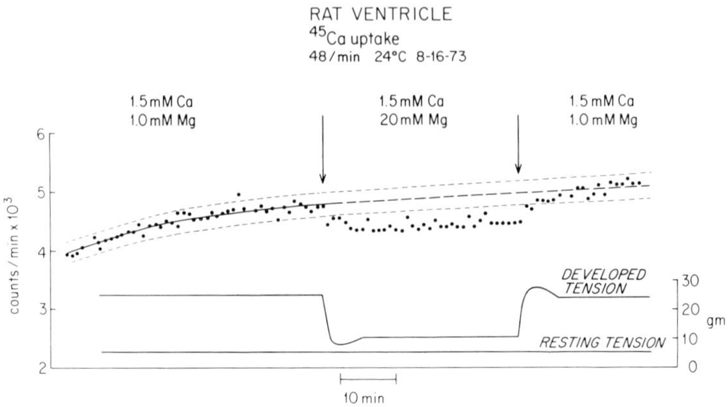

Figure 4 ^{45}Ca uptake. The probe counts observed during ^{45}Ca labeling of a rabbit interventricular septum in 1.5 mM Ca, 1.0 mM Mg, and 60 mM sucrose are shown. After 70 min of labeling the sucrose was replaced with 20 mM Mg. An abrupt decline in tension, dP/dt, and ^{45}Ca counts occurred, all of which persisted until return to the 1.8 mM Mc, 1.0 mM Mg, 60 mM sucrose perfusate. (Reproduced from Shine and Douglas (13) by permission of *American Journal of Physiology*).

in figure 4, a series of muscles was labled with ^{45}Ca during which 10 mM Mg was introduced into the extracellular perfusate. Simultaneous with the decline in tension and dP/dt there was an abrupt decline in the uptake of ^{45}Ca. When 20 mM Mg was introduced during ^{45}Ca efflux experiments no effect of Mg upon ^{45}Ca efflux was observed. These data emphasize that in rat ventricle, elevated extracellular Mg impedes tension development concomitantly with a decrease in ^{45}Ca influx.

Simultaneous with the decline of tension, elevated extracellular Mg also produced an abrupt decline in the ^{42}K efflux from rat ventricle (13). A concentration of 20 mM Mg produced a 30 percent decrease in the ^{42}K efflux with an accompanying flattening of the tissue probe indicative of slowed exchange. That this was due to an Mg effect upon ^{42}K exchange was demonstrated in experiments in which the rat septa were labeled with ^{42}K and washouts conducted first in 1 mM Mg and then, after relabeling, in 20 mM Mg. A 27 percent decrease in the exchange rate in 20 mM Mg was demonstrated. The effects of Mg upon K and Ca movement could be separated as demonstrated in figure 5. Here ^{42}K washout was conducted in 1.5 mM Ca, 1 mM Mg. When the extracellular Mg concentration was increased to 20 mM, Ca was simultaneously increased to 5 mM. Note that tension and dP/dt rose slightly as a consequence of the increased extracellular Ca concentration, although a decline in ^{42}K efflux persisted. These data emphasize the presence in rat ventricles of two distinguishable effects of Mg, one upon tension development and ^{45}Ca uptake, and a

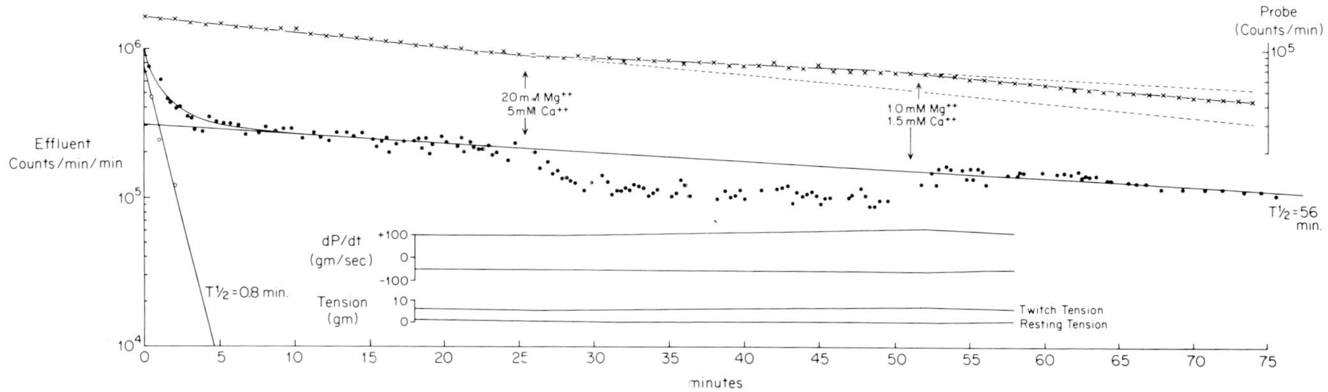

Figure 5 High calcium and high magnesium perfusion during ^{42}K washout. A rat interventricular septum was labeled for 50 min with ^{42}K at 24°C and 48 beats/min and washout with unlabeled perfusate containing 1.5 mM Ca and 1.0 mM Mg was initiated. After 25 min the perfusate was changed to 20 mM Mg and 5 mM Ca for the next 25 min. During high magnesium, high calcium perfusion dP/dt and tension rose to 125 percent of the control values. During this time ^{42}K efflux declined abruptly. The slope of the effluent curve declined from 0.018 min^{-1} to 0.012 min whereas the tissue probe flattened from 0.20 min^{-1} to 0.005 min^{-1}. The marked depression of ^{42}K efflux in the presence of well maintained mechanical function is striking. (Reproduced from Shine and Douglas (13) by permission of *American Journal of Physiology.*)

second upon ^{42}K exchange. Studies of the action potential in the rat septa revealed only a slight decrease in conduction velocity and overshoot of the action potential when Mg was increased from 1 to 20 mM. This could represent a third site of action of Mg on a sodium channel. However, this interpretation remains speculative.

A more interesting set of results were obtained when the effects of Mg were studied in the rabbit ventricle (12). A concentration of 20 mM Mg had no effect on ^{42}K exchange in rabbit ventricle. Although tension development and indeed ^{45}Ca uptake could be inhibited by 20 mM Mg in the rabbit, the degree of tension depression was substantially less than that observed in the rat. However, an effect of Mg upon ^{42}K exchange was easily demonstrable if the rabbit septa were first treated with digitalis glycoside. Acetylstrophanthidin produced a prompt increase in tension development and net ^{42}K loss when perfused in the presence of 1 mM Mg. Introduction of 10 mM Mg, however, decreased the net K loss to the control level but did not attenuate the contractile response to glycoside. Each of eight septa were exposed to acetylstrophanthidin, 2.5×10^{-6} M, on two occasions, once in 1 mM Mg and on another occasion in the presence of 20 mM Mg. In four septa glycoside exposure took place in low Mg first and then in high Mg and in the remaining four septa the order of exposure was reversed. A concentration of 20 mM Mg reduced the net K loss by two-thirds from an average of 155 μM/kg tissue water to 44 μM/kg tissue water but produced no significant effect on the isotopic response to glycoside. These data emphasized that Mg could diminish the net K loss produced by digitalis without affecting force development.

Two lines of data suggested that Mg was not reversing digitalis inhibition of Na,K-ATPase. When rabbit septa were labeled with ^{42}K while the tissue counts of ^{42}K were monitored with a closely apposed geiger probe, introduction of digitalis produced a prompt flattening of the tissue uptake. The introduction of 10 or 20 mM Mg did not effect the uptake and ^{42}K uptake increased only when digitalis was removed; *i.e.*, Mg then did not reverse the inhibition of ^{42}K influx produced by digitalis.

After it was shown that the effects of ouabain upon ^{42}K exchange were also reversed by 20 mM Mg, septa were labeled with tritiated ouabain and washed out. No displacement of ouabain was demonstrated during 20 mM Mg perfusion.

The data from rat and rabbit indicated that Mg decreased ^{42}K exchange in the untreated rat septum, but only when the digitalis glycoside had been applied did the rabbit become sensitive to Mg. There was a qualitative difference between the mechanical response to Mg in the two species, the rabbit being considerably less susceptible than the rat. It seemed desirable to see whether Mg had an effect upon ^{42}K exchange when a perturbation was produced by an intervention other than glycoside. This was accomplished by increasing perfusate potassium from 4 mM to 16 mM. As predicted from previous work of Blesa *et al.* (1) and Carmeliet (3), increase in extracellular K produced a 2-fold increase in exchange rate as well as a prompt increase in the number of counts appearing in the effluent. When each septum was relabeled, the same maneuver was carried out a second time with the addition of 20 mM Mg. There was an abrupt decline in the counts appearing in the effluent. The effect was reminiscent of that seen after digitalis glycoside. In six septa a control washout was carried with exposure to 16 mM K and 1 mM Mg followed by a second washout and exposure to 16 mM K and 20 mM Mg. In every septum the increase in effluent count produced by 16 mM K was diminished in 20 mM Mg.

These data led to the model shown in figure 6 to describe the relationship between Mg, K, and glycoside. In the figure each rectangle represents a myocardial cell enclosed by sarcolemmal membrane. As shown in *A,* two separate sites for Mg action at the sarcolemmal membrane were suggested for rat ventricle, one of which produced Ca-Mg antagonism and a second of which involved Mg inhibition of K exchange at a site which was not identical with that inhibited by glycoside. It should be noted that in rat septa no effect of Mg upon ^{24}Na exchange could be demonstrated. In *B* the steady state situation in rabbit septum is shown. The Ca-Mg interaction was of less significance and the Mg-sensitive K site was also little affected. However, when digitalis glycoside inhibited the Na,K-ATPase as shown in *C,* or when elevated extracellular K resulted in an increase in K exchange, the Mg-sensitive site emerged. In this last regard it should be pointed out that, in the absence of reversal of Na,K-ATPase inhibition by Mg, one would still expect an accumulation of intracellular Na to occur. If Na-Ca exchange, as suggested by Langer (6), is indeed the mechanism of glycoside inotropy, such inotropy would be expected to continue unabated despite the effect on K exchange. These data demonstrate that inotropy induced by glycoside is not necessarily correlated with net K losses from rabbit septa.

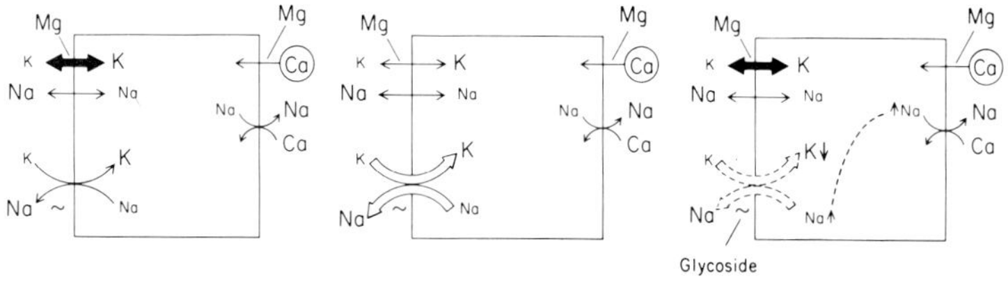

Figure 6 Model for magnesium interactions in rat and rabbit. Proposed sites of Mg-K and Mg-Ca interactions are indicated. *Large solid arrows* indicate relative importance of magnesium sensitive potassium channels. Glycoside required enhanced ^{42}K efflux through the Mg-sensitive channel. Increased extracellular potassium presumably also activates the same channel to account for increased ^{42}K exchange in 16 mM K.

Moreover, they emphasize the role which a physiologically present ion, Mg, can play at the cell membrane in attenuating the movement of important ions which control tension development and electrical activity.

At the present time our data are most consistent with the hypothesis that calcium ions bound to the sarcolemmal membrane, and in rapid equilibrium with extracellular space, are responsible for contraction in mammalian myocardium under normal circumstances. The basement membrane of cardiac sarcolemma includes important anionic binding sites which can be occupied by other cations at the expense of calcium ions and contractile force. Magnesium ion competes at these sites to impede calcium binding and calcium influx, but magnesium also acts upon a separate potassium channel. The inotropy accompanying glycoside action is not related primarily to potassium loss from myocardium but more likely is a consequence of intracellular Na accumulation with subsequent enhanced calcium influx.

REFERENCES

1. BLESA, E. S., LANGER, G. A., BRADY, A. J., and SERENA, S. D. 1970. Potassium exchange in rat ventricular myocardium: Its relation to rate of stimulation. Amer. J. Physiol. 219: 747–754.
2. BOCKMAN, E. L., RUBIO, R., and BERNE, R. M. 1973. Effect of lanthanum on isoproterenol-induced activation of myocardial phosphorylase. Amer. J. Physiol. 225: 438–443.
3. CARMELIET, E. E. 1961. The potassium chloride permeability in Purkinje fibers. *In* A. Paes de Carvalho, W. C. de Mello, and B. F. Hoffman (eds.), The Specialized Tissues of the Heart. Elsevier, New York.
4. FAWCETT, D. W., and McNUTT, N. S. 1969. The ultrastructure of the cat myocardium. I. Ventricular papillary muscle. J. Cell Biol. 42: 1–45.
5. LANGER, G. A. 1968. Ion fluxes in cardiac excitation and contraction and their relation to myocardial contractility. Physiol. Rev. 48: 708–757.
6. LANGER, G. A. 1971. Physiology in medicine: The intrinsic control of myocardial contraction. New Engl. J. Med. 285: 1065–1071.
7. LANGER, G. A., and FRANK, J. S. 1972. Lanthanum in heart cell culture: Effect on calcium exchange correlated with its localization. J. Cell Biol. 54: 441–455.
8. LANGER, G. A., SERENA, S. D., and NUDD, L. N. 1974. Cation exchange in heart cell culture: Correlation with effects on contractile force. J. Molec. Cell Cardiol. 6: 2, 149–161.
9. LUTTGAU, H. C. 1964. Electrophysiology of the heart, pp. 87–95. Proceedings of the Meeting at the Instituto Cardiologie Sperimentale Serv. Sci. Semes, Milan, Italy, October 10–13, 1963.
10. RINGER, S. A. 1883. A further contribution regarding the influence of different constituents of the blood on the contraction of the heart. J. Physiol. London 4: 29–42.
11. SANBORN, W. G., and LANGER, G. A. 1970. Specific uncoupling of excitation and contraction in mammalian myocardium by La^3: Kinetic studies. J. Gen. Physiol. 56: 191–217.
12. SHINE, K. I. 1975. Magnesium effects upon mechanical function, ionic exchange and arrhythmias in rabbit ventricle. Amer. J. Physiol. 228: 1545–1554.
13. SHINE, K. I., and DOUGLAS, A. M. 1974. Magnesium effects on ionic exchange and mechanical function in rat ventricle. Amer. J. Physiol. 227: 317–324.

14. SHINE, K. I., SERENA, S. D., and LANGER, G. A. 1971. Kinetic localization of contractile calcium in rabbit myocardium. Amer. J. Physiol. 221: 1408–1417.

Mailing address: Dr. Kenneth I. Shine,
UCLA Medical Center, Room A3-381, BRI,
Los Angeles, California 90024 (USA).

Recent Advances in Studies on Cardiac Structure and Metabolism
Volume 9
The Sarcolemma
Edited by Paul-Emile Roy and Naranjan S. Dhalla

VALINOMYCIN INHIBITION OF THE INWARD SLOW CURRENT OF CARDIAC MUSCLE[1]

J. A. SCHNEIDER,[2] K. SHIGENOBU,[3] and N. SPERELAKIS

Departments of Physiology and Surgery
University of Virginia School of Medicine,
Charlottesville, Virginia, USA

SUMMARY

Valinomycin (10^{-6} M), a K^+ ionophore and uncoupler of oxidative phosphorylation, produced pronounced shortening of the action potential plateau in young (3 day) and old (15–19 day) embryonic chick hearts and isolated perfused guinea pig hearts. The rate of rise of the action potential was usually not affected. Dimethylsulfoxide (DMSO) (2 percent), the vehicle for valinomycin, did not cause plateau shortening. In chick hearts, diminution of the plateau by valinomycin at 10^{-6} M usually occurred by 1 hr; larger doses produced prominent effects within 15 min. Young embryonic hearts were more resistant to valinomycin, and 20×10^{-6} M for 1–2 hr was necessary to produce a significant diminution in plateau. Several agents were able to partially relengthen the plateau shortened by valinomycin. In both young and old hearts, the resting potential was not increased by valinomycin, even though young hearts have a low resting potential because of a low K^+ conductance (g_K). In guinea pig, 10^{-6} M valinomycin markedly shortened the plateau within 5–10 min. Contractions were abolished, *i.e.*, the hearts became electromechanically uncoupled. When hearts were perfused with 27 mM K^+ to depolarize the cells to about −40 mV to inactivate fast Na^+ channels, catecholamines or methylxanthines rapidly induced slowly rising (about 10 V/sec), overshooting, plateau-like responses with accompanying contractions. Valinomycin rapidly (within 2–5 min) abolished these slow responses and contractions. Hearts perfused with 10^{-6} M valinomycin or 10^{-5} dinitrophenol became nearly entirely depleted of ATP within 2–10 min. These results suggest that valinomycin blocks the slow inward current indirectly through ATP depletion, the slow cation channels requiring metabolic energy. The possibility that resting g_K or kinetics of changes in g_K during the plateau is also affected by valinomycin cannot be excluded.

The macrocyclic antibiotic, valinomycin, facilitates K^+ transport in mitochondria and acts as an uncoupler of oxidative phosphorylation (6, 11). Valinomycin also

---

[1] The work reviewed in this article and the new experiments were supported by Grant HL-11155 from the United States Public Health Service.

[2] Dr. Schneider is a Surgery Research Fellow under the National Institute of Health Academic Surgery Training Program.

[3] Dr. Shigenobu's present address is: Research Institute for Chemical Hazards, Faculty of Pharmaceutical Sciences, University of Tokyo, Tokyo, Japan.

markedly reduces the very high electrical resistance of lipid bilayer membranes by increasing K^+ conductance (g_K) selectively (6, 10). The valinomycin molecule is shaped as a donut, and the central hole (diameter of about 6 Å) can accommodate a hydrated K^+ ion, *i.e.*, valinomycin acts as a cage for K^+ (fig. 1). There are six coordination bonds between the K^+ ion and carbonyl groups projecting from valine residues in the ring into the central hole. The hole surface is polar, whereas the outer rim of the ring-shaped molecule is nonpolar and hence lipid soluble. Thus, valinomycin acts as a carrier or ionophore for K^+ in the lipid bilayer membrane.

Several years ago we added valinomycin to cultured chick heart cell monolayers, but we could not detect any changes in the membrane potentials during observation periods of up to 20 min (unpublished observations). Since young embryonic chick hearts have a low g_K (compared with old hearts), causing a low resting potential (22), this preparation would be better for determining whether valinomycin can become incorporated into the cell membrane of a living cell and increase the resting or steady state g_K. That is, if valinomycin were to increase

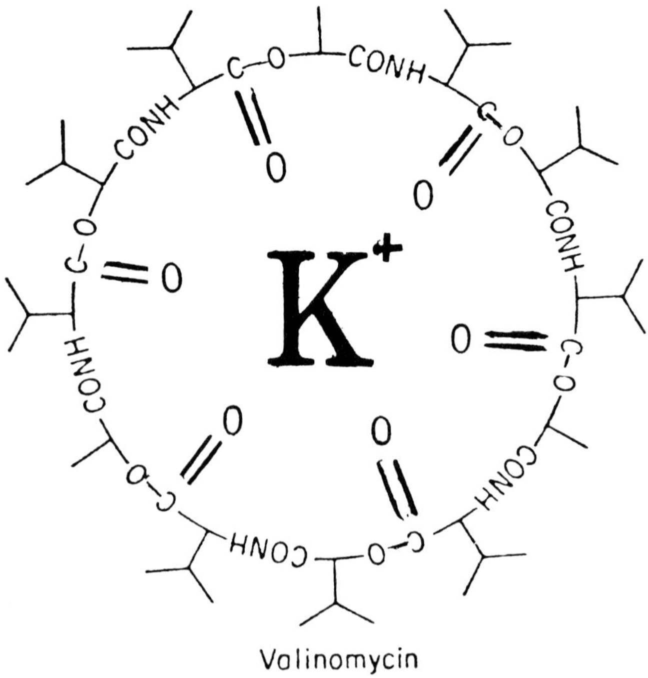

Figure 1 Chemical structure of the macrocyclic antibiotic, valinomycin. This compound acts as a K^+ ionophore (or carrier) in the bimolecular lipid membrane. Its ion specificity is: $H^+ > Rb^+ > K^+ > Cs^+ >> Na^+ > Li^+$.

the resting g_K in young embryonic heart, a large hyperpolarization should occur which could be easily detected. However, we found that valinomycin slowly produced a marked reduction in the duration of the action potential of both old and young hearts without significant changes in resting potential (19). Thus, the kinetics of the K^+ activation during the action potential may be more greatly affected by valinomycin than the steady state g_K of the resting membrane.

In addition, as we obtained evidence previously that slow cation channels induced in guinea pig hearts by catecholamines and methylxanthines are dependent on metabolic energy (14, 15), we decided to test whether valinomycin would inactivate rapidly the slow channels in these hearts that are highly dependent on aerobic metabolism and oxidative phosphorylation. We found that valinomycin inactivates slow channels rapidly (2–5 min) probably through ATP depletion, and thereby blocks contraction through depression of the slow inward current. A preliminary account of some of these findings has been published (15).

MATERIALS AND METHODS

Chick hearts

Embryonic chick hearts were studied at young (3 day) and old (15–19 day) stages. In the 3-day-old embryos, the entire heart was removed; therefore, the ventricular cells were activated spontaneously. Only the ventricles of the old hearts were removed. Electrical stimulation was applied with platinum bipolar electrodes to the old hearts and to those young hearts in which automaticity failed after valinomycin addition. The hearts were bathed in Puck's medium of the following composition: 15% horse serum, 40% N-16 nutrient medium, and 45% Hank's balance salt solution (21). The bath was maintained at 37°C (±1°C) during the experiments.

Conventional intracellular recording was done using glass capillary microelectrodes filled with 3 M KCl. The microelectrode resistances were usually 30–50 megohms. The reversible half-cells were Ag:AgCl. A WP Instruments Co. model 4A d.c. preamplifier with a high input-impedance electrometer probe and negative capacitance was used, and the signal was led to a Tektronix 565 dual-beam oscilloscope. The maximum rate of rise of the action potential ($+\dot{V}_{max}$) and dV/dt were measured using a Tektronix type O operational amplifier for differentiation. The dV/dt channel was purposely not aligned with the V-t channel so as not to obscure dV/dt within the rising phase of the action potential.

Guinea pig hearts

Guinea pig hearts were removed quickly and perfusion through the coronary arteries was begun within 15 sec. The perfusing solution (37°C) was either normal physiological salt solution (PSS) or a modified PSS containing 27 mM

K^+. The composition of the normal PSS was (in millimolar concentration): 134.3 Na^+, 2.7 K^+, 2.0 Ca^{++}, 1.0 Mg^{++}, 143 Cl^-, 5 Tris•HCl (pH 7.4). The high K^+ (27 mM) solution was made by substituting KCl for NaCl on an equimolar basis. The high K^+ solution was used to depolarize the myocardial cells to about −40 mV to inactivate the fast Na^+ channels; 1-isoproterenol bitartrate (Sigma) (10^{-7} M) or caffeine (3 mM) was added to the high K^+ solution to induce the slow responses with accompanying contractions. The solutions were bubbled vigorously with 100 percent O_2. The perfusion bottles were suspended 180 cm above the heart, and gave a constant pressure perfusion and a flow rate of 10–20 ml/min.

The hearts perfused with normal PSS contracted spontaneously at about 180/min. The partially depolarized hearts were electrically paced (at 1/sec) with stimulating electrodes sutured to the ventricle. Intracellular recordings were made using hanging microelectrodes. The glass capillary micropipettes were filled with 3 M KCl and had resistances of 20–40 megaohms. A WP Instruments Co. VF-1 high input-impedance preamplifier and a Tektronix 502A oscilloscope were used for recording. dV/dt and $+\dot{V}_{max}$ were obtained from an RC circuit and displayed on the oscilloscope. Contractile force was obtained from a suture placed through the apex of the heart, and was recorded by a force transducer and penwriter; a resting tension of 10 gm was applied.

ATP analyses

ATP measurements were done by rapidly freezing the guinea pig ventricles between metal clamps which had been precooled to −70°C in liquid nitrogen. The frozen tissues were stored in liquid nitrogen until assayed for ATP. ATP was assayed by the firefly method using Sigma FLE-50 kits.

Valinomycin

Crystalline valinomycin (Sigma) was added to the bath or perfusing solutions, usually at a final concentration of 1×10^{-6} M, after a control period. The valinomycin (molecular weight of 1111) was dissolved in DMSO at a concentration of 2 mg/ml. The final concentration was usually 1×10^{-6} M for the guinea pig hearts, and was usually 1×10^{-6} M (old) or 20×10^{-6} M (young) for the embryonic chick hearts. At 1×10^{-6} M, the final DMSO concentration was 0.05 percent; this concentration of DMSO by itself in the absence of valinomycin had no effect. Other agents used included 2,4-dinitrophenol (DNP) and verapamil (Knoll Pharmaceuticals).

RESULTS

Chick hearts

Valinomycin shortened the plateau phase of the action potential in both young and old embryonic chick hearts. This effect was produced slowly, but the time course was accelerated substantially at higher concentrations of valinomycin (fig. 2*A–C*). At long times and high doses, excitability sometimes failed completely. The degree of shortening and the effective concentration of valinomycin were different in young hearts and old hearts.

In old hearts, the reduction of the plateau was very pronounced, and the action potential in the presence of valinomycin often seemed to be composed of

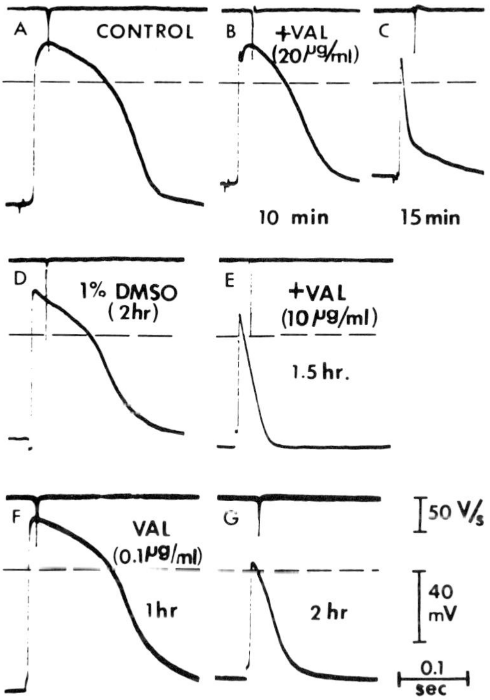

Figure 2 Effect of valinomycin (*VAL*) on old (15–19-day-old) embryonic chick hearts (ventricules), *A–C* Rapid effect of a high dose of valinomycin; *A*, control intracellularly-recorded action potential; *B–C*, 10 min (*B*) and 15 min (*C*) after addition of valinomycin, 20 μg/ml (note the pronounced shortening of the action potential); *D–E*, demonstration that the vehicle for the valinomycin was not responsible for the shortening; *D*, heart incubated for 2 hr in 1 percent DMSO without valinomycin (note normal-appearing action potential); *E*, another heart incubated for 1.5 hr in 0.5 percent DMSO plus valinomycin, 10 μg/ml (the action potentials were greatly abbreviated); *F–G*, demonstration that very low concentrations of valinomycin were effective after longer exposure periods; *F*, normal-appearing action potential recorded 1 hr after exposure to valinomycin at 0.1 μg/ml; *G*, greatly shortened and depressed action potential recorded in the same heart after 2 hr. Voltage, time, and dV/dt calibrations pertain throughout. *Upper trace* gives dV/dt (arbitrarily shifted to the right); the maximum rate of rise of the action potential ($+\dot{V}_{max}$) is given by the peak downward deflection of dV/dt. *Broken horizontal line* in each panel gives the zero potential level. (Reproduced from Shigenobu and Sperelakis (19), by permission of *American Journal of Physiology*.)

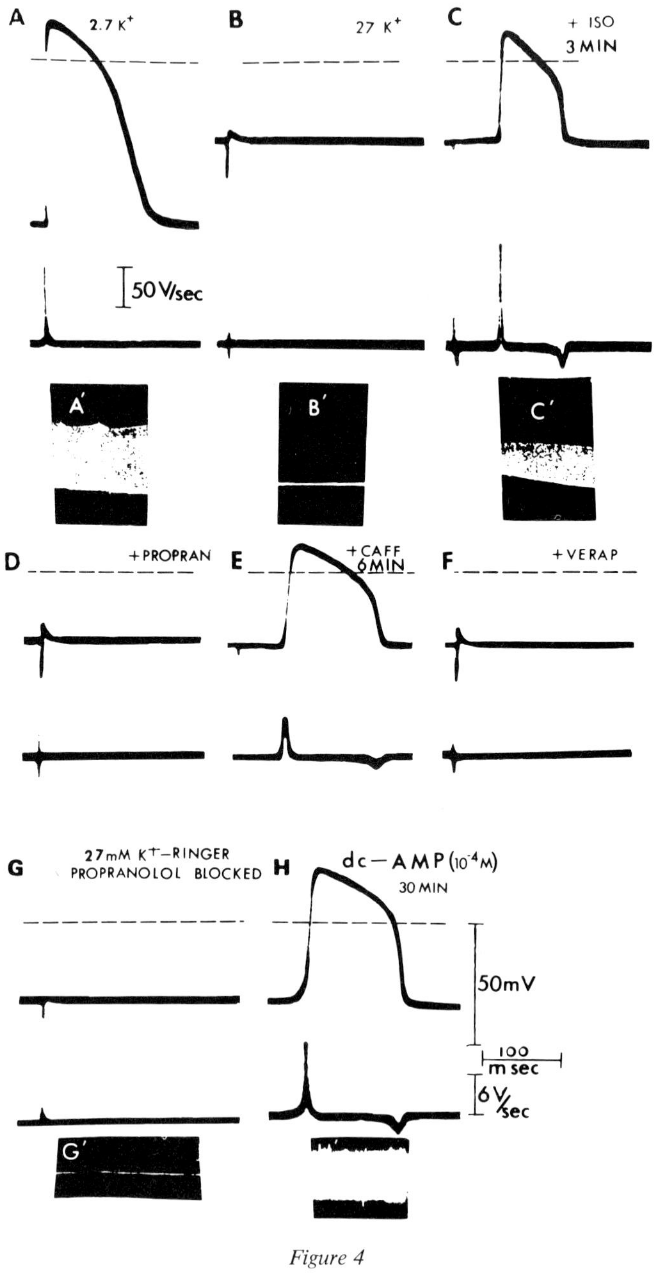
A
2.7 K+
B
27 K+
C
+ ISO
3 MIN
50 V/sec
A′
B′
C′
D
+PROPRAN
E
+CAFF
6 MIN
F
+VERAP
G
27mM K+–RINGER
PROPRANOLOL BLOCKED
H
dc–AMP (10-4M)
30 MIN
50mV
100
m sec
6 V/sec
G′

Figure 4

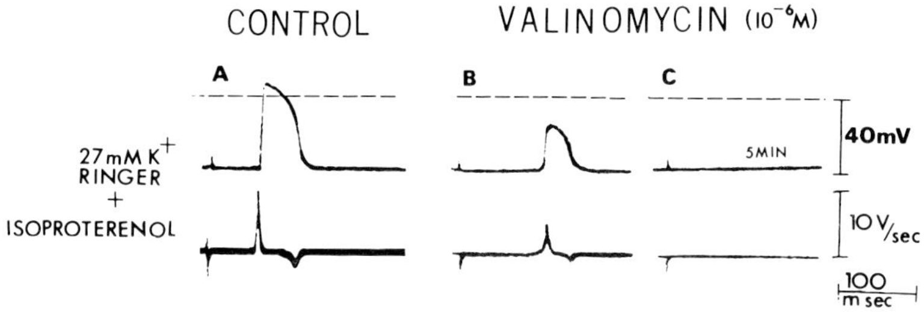

Figure 5 Slow responses from a heart perfused with 27 mM K^+ plus 10^{-7} M isoproterenol to induce slow responses. *A,* Control slow response; *B–C,* valinomycin (10^{-6} M) rapidly suppressed the slow response and accompanying contractions (not illustrated). Pacing artefacts are visible at the left of each panel. (Modified from Schneider and Sperelakis, (16) by permission of *European Journal of Pharmacology.*)

course of the decrease in force of the contractions accompanying the slow response is summarized in figure 6. The effect of valinomycin on the slow response was not readily reversible by perfusion with valinomycin-free solution.

ATP analyses. A total of 53 guinea pig hearts were analyzed for ATP. These hearts were perfused with solution containing 27 mM K^+ plus 10^{-7} M isoproterenol. After a preperfusion period of 5 min, various agents were added to the perfusion fluid. The control hearts (Group 2 of table I) were not exposed to drugs. Ten of the hearts were frozen immediately upon removal from the animals (Group 1) for comparison with the control perfused hearts. Most hearts were frozen after a timed experimental perfusion period of 10 min, but some

---

Figure 4 (p. 40) Characterization of the slow responses induced by catecholamines, methylxanthines, or cyclic AMP in isolated perfused guinea pig hearts, the fast Na^+ channels of which were inactivated in elevated K^+. *Upper tracings:* intracellular potentials (V versus t); *lower tracings* give dV/dt; maximum upward deflection of dV/dt gives $+\dot{V}_{max}$, the maximum rate of rise of the action potential. *A,* Normal cardiac action potential (2.7 mM K^+-Ringer perfusate), $+\dot{V}_{max}$ of 86 V/sec; *B,* perfusion with 27 mM K^+-Ringer depolarized the cells to –40 mV and inactivated the fast Na^+ channels (heart was unresponsive to stimulation 10 times greater than threshold); *C,* isoproterenol (10^{-7} M) restored electrical activity in the form of a slow response ($+\dot{V}_{max}$ of 15 V/sec). Mechanical contractions were always associated with this response. *A', B', C'* are the mechanical recordings on a penwriter which correspond to *A, B,* and *C. D,* Propranolol (2 mg/l), a β-adrenergic blocker, abolished the isoproterenol effect; *E,* caffeine (3 mM) induced the slow responses with accompanying contractions in the presence of propranolol; *F,* verapamil (2 mg/l) blocked the caffeine-induced slow response; *G,* inexcitability of another heart perfused with a solution containing 27 mM K^+ and propranolol; *H,* addition of dibutyryl cyclic AMP (10^{-4} M) induced the slow responses with accompanying contractions (*inset, H'*) within 30 min. Electrical stimulation was applied in all panels except *A.* Voltage and time calibrations apply to all panels; dV/dt calibration given in *H* applies to all panels except *A.* (Modified from Schneider and Sperelakis (14, 15) by permission of *Journal of Surgical Research* and *Journal of Molecular and Cellular Cardiology.*)

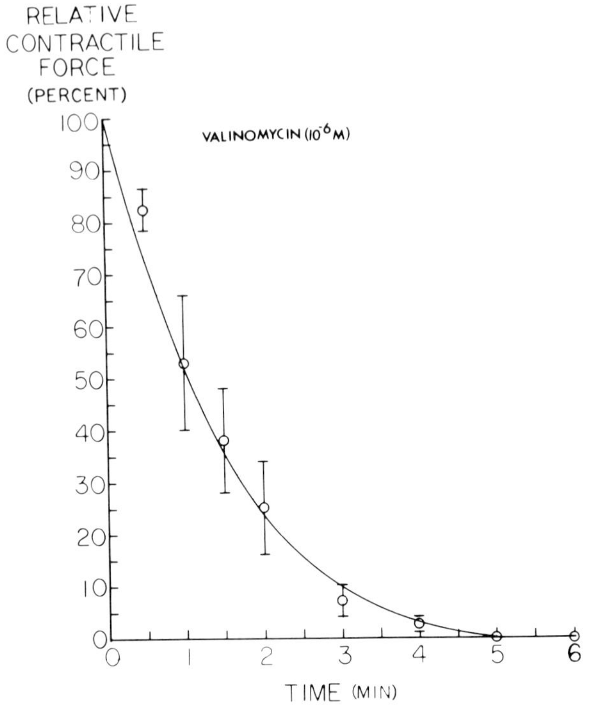

Figure 6 Time course of decay of contractile force after addition of valinomycin (10^{-6} M) to guinea pig hearts perfused with 27 mM K^+ plus isoproterenol (10^{-7} M). The values at zero time represent the twitch tensions accompanying the induced slow response. The ordinate gives the relative contractile force as a percentage of the control force before valinomycin addition (at zero time). The contractions were completely abolished within 5 min. *Vertical bars* give ± 1 SEM; the means represent data from 12 hearts. (Reproduced from Schneider and Sperelakis (16) by permission of *European Journal of Pharmacology*.)

hearts (Groups 3*A* and 4*A*) were frozen for ATP analysis immediately after the suppression of contractions (about 2 min). The results tabulated in table I demonstrate that valinomycin caused marked ATP depletion, as did the powerful inhibitor of metabolism, DNP. Further, the ATP was depleted at the time at which the slow responses and contractions were suppressed. In contrast, verapamil, which also blocks the slow responses (15, 17), did not cause ATP depletion. The isolated hearts perfused for a total of 15 min had significantly less (about one-half) ATP than did the fresh nonperfused hearts.

Effect of other metabolic poisons on slow response. For comparison with the effect of valinomycin, figures 7 and 8 illustrate the effects of 10^{-4} M DNP (fig. 7*A–D*) and 10^{-3} M cyanide (fig. 8*A–D*) on the catecholamine-induced slow responses (14). It can be seen that both DNP and cyanide abolished the slow responses rapidly, without alteration of the resting potential, in a manner similar

Table 1 Summary of data on ATP analyses in perfused guinea pig hearts*

Category	ATP (μmoles/g tissue), means ± S.E.	(n)
1. Fresh nonperfused	3.56 ± 0.42	(10)†
2. Control perfused (10 min)	1.76 ± 0.23	(10)
2A. Control perfused (2 min)	1.30	(2)
3. DNP (10^{-5} M) (10 min)	0.08 ± 0.03	(4)†
3A. DNP (10^{-5} M) ‡ (~ 2 min)	0.01	(2)
4. Valinomycin (10^{-6} M) (10 min)	0.04 ± 0.01	(12)†
4A. Valinomycin (10^{-6} M) § (~ 2 min)	0.06	(2)
5. Verapamil (2 mg/) (10 min)	1.08 ± 0.27	(8)
5A. Verapamil (50 μg/l) § (10 min)	1.98 ± 0.33	(3)

*All hearts were preperfused for 5 min before the start of the experimental perfusion period (given in parentheses with Category). All hearts were perfused with 27 mM K^+, and the slow responses were induced by isoproterenol (10^{-7} M).

†The hearts were removed and frozen for analysis within 30 sec after the contractions completely disappeared, usually about 2 min after exposure to the drug.

‡This dose of verapamil was sufficient to completely block contraction.

§Statisically significant (p = <0.001) difference from the control perfused hearts (Category 2).

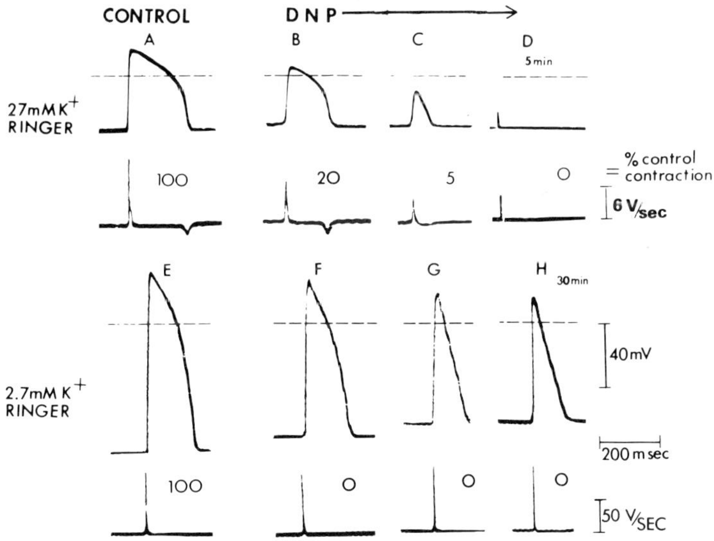

Figure 7 Inhibition by DNP (10^{-4} M) of the slow responses induced by catecholamines (*upper row*) in guinea pig hearts perfused with 27 mM K^+. Percentage of control contractile force is numerically indicated in each panel. *Top row,* the isoproterenol (10^{-7} M)-induced slow response (*A*) was abolished within 5 min after addition of DNP to the perfusion solution (*B–D*); contractions simultaneously disappeared. *Bottom row,* The action potential generated in normal Ringer (2.7 mM K^+) (*E*) continued throughout a 30-min period of DNP perfusion (*F–H*). However, the action potentials decreased in duration and magnitude. Contractions disappeared after 10 min (*F*), suggesting loss of the Ca^{++} current during the action potential. Voltage calibration applies throughout; the 200-msec time calibration in *H* represents 100 msec in *A–D*. (Reproduced from Schneider and Sperelakis (14) by permission of *Journal of Surgical Research.*)

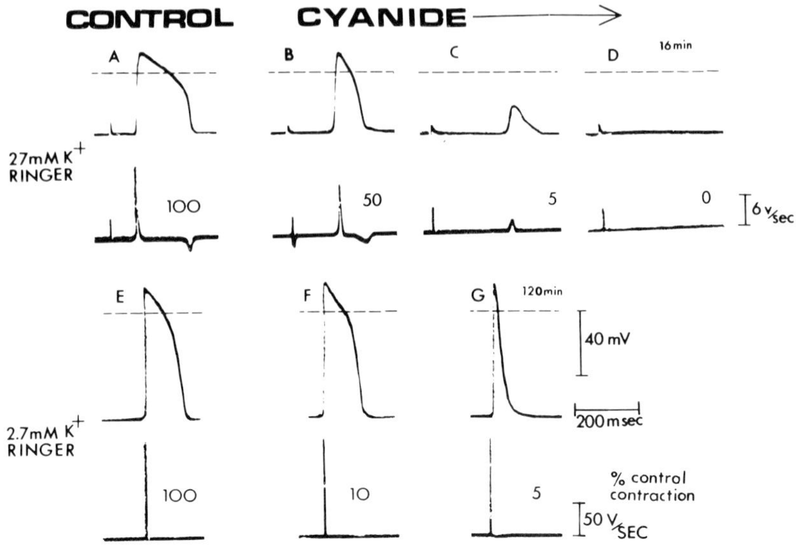

Figure 8 Inhibition by cyanide (10^{-3} M) of the slow responses induced by catecholamines (*upper row*) in guinea pig hearts perfused with 27 mM K^+. Percentage of control contractile force is numerically indicated in each panel. *Top row,* control slow response induced by isoproterenol (10^{-7} M) (*A*) was depressed (*B–C*) and finally abolished (*D*, 16 min) after addition of KCN to the perfusate; depression and loss of contractions followed a parallel time course. *Bottom row,* action potentials in normal Ringer (2.7 mM K^+) (*E*), and 60 min (*F*) and 120 min (*G*) after KCN addition. Spontaneous action potentials, with a shortened plateau at 120 min, remained, but mechanical contractions were greatly reduced by 12 min. Voltage calibration applies throughout; the 200-msec time calibration in *G* represents 100 msec in *A–D*. (Reproduced from Schneider and Sperelakis (14) by permission of *Journal of Surgical Research.*)

to that of valinomycin. DNP and cyanide had identical effect on the methylxanthine-induced slow responses.

In contrast, on the normal action potential (2.7 mM K^+), the actions of DNP (fig. 7*E–H*) and cyanide (fig. 8*E–G*) were much slower than in the case of valinomycin. That is, valinomycin markedly shortened the action potential within 5 min, whereas DNP and cyanide took considerably longer (*e.g.*, 15–90 min). However, electromechanical uncoupling was about equally rapidly produced by DNP, cyanide, and valinomycin.

Ischemia and hypoxia (not illustrated) produced effects on the slow responses and action potentials similar to those of the metabolic poisons (14). A summary of much of the results from blocking agents and metabolic poisons is given in figure 9.

Ineffectiveness of adenosine. Because adenosine has been identified as a physiological vasodilator in ischemic heart muscle (12), the possibility was tested that adenosine escaping from the myocardial cells in the presence of the metabolic poisons was responsible for the suppression of the slow responses. The slow responses induced in ventricular cells by isoproterenol in four guinea pig hearts were essentially unaltered after addition of adenosine (10^{-4} M) to the perfusate (fig. 10). The force of contractions also was not affected (fig. 10*A′*–*B′*).

DISCUSSION

Valinomycin action

The results demonstrated that valinomycin (10^{-6}) markedly shortened the action potential duration of embryonic chick and guinea pig hearts with concomitant

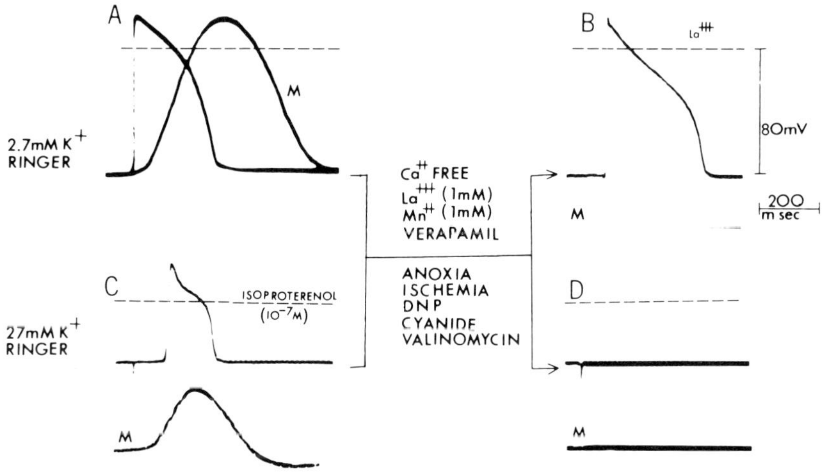

Figure 9 Electrophysiological aspects of mechanical uncoupling. Summary of results on the isolated perfused guinea pig hearts (ventricular cells). Intracellular microelectrode recordings (*upper traces*), and the associated contractions recorded with a force transducer (*lower tracings* labeled *M*). *A*, normal action potential and mechanogram recorded from a spontaneously beating heart perfused with normal Ringer solution (K^+ = 2.7 mM); *B*, although action potential generation persisted in hearts treated with any of the agents indicated in the figure, mechanical activity was lost within 5–10 min, *i.e.*, electromechanical uncoupling occurred. The agents *listed above the arrow* act directly to abolish Ca^{++} influx either by blockade of the Ca^{++} channels (La^{+++}, Mn^{++}, verapamil) or by removal of the charge-carrying ion (Ca^{++} free); the agents *listed below the arrow* all act indirectly presumably by interfering with ATP availability to the slow channels. The record illustrated was taken from a heart exposed to 1 mM La^{+++}. *C*, Intracellular recording of a slow response from a heart whose fast Na^+ channels were blocked by partial depolarization in 27 mM K^+-Ringer perfusate. The slow responses were produced by catecholamines or methylxanthines, presumably by induction of slow Ca^{++}-Na^+ channels in the sarcolemma. The slow responses were always accompanied by mechanical contraction; *D*, the slow electrical response (unlike the normal action potential) and contractions were abolished within 5–10 min by any of the agents indicated.

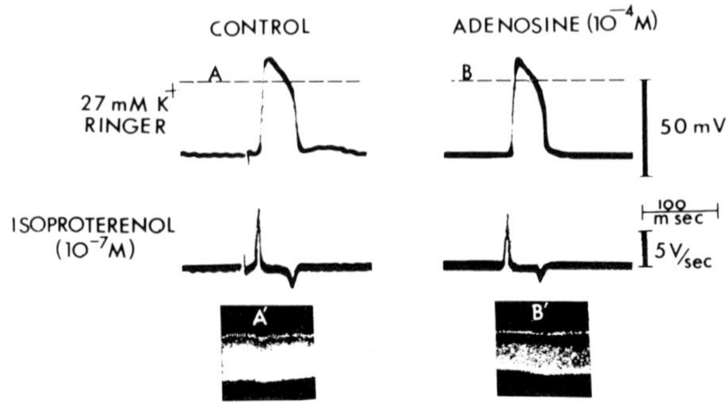

Figure 10 Illustration of the lack of effect of adenosine on the slow response induced in guinea pig ventricular cells. *A*, Typical slow response induced by isoproterenol (10^{-7} M) in 27 mM K^+; *B*, 15 min after addition of adenosine (10^{-4} M) to the perfusing solution, the slow responses and accompanying contractions (B′) were not blocked. Even $+\dot{V}_{max}$ was usually not affected. A′ and B′ are the mechanograms recorded on a penwriter corresponding to *A* and *B*.

abolition of contractions. That is, the hearts became electromechanically uncoupled. Action potential abbreviation occurred much more rapidly in the guinea pig heart (within 2–10 min) than in the chick heart (up to 60 min required). The action potential shortening was invariably followed in guinea pig hearts by arrhythmias consisting of a high frequency discharge of abbreviated action potentials. In guinea pig hearts perfused with 27 mM K^+, valinomycin (10^{-6} M) abolished the slow response and accompanying contractions rapidly, whether induced by isoproterenol or caffeine, and caused the ATP levels to be depleted nearly completely within 2–10 min. Action of valinomycin as a metabolic poison can account for the latter two findings, whereas the action responsible for the loss of the plateau of the normal action potential may include a direct effect on the sarcolemma.

The lack of hyperpolarization of any of the preparations, including the young embryonic chick hearts, suggests that valinomycin does not significantly increase resting g_K of the sarcolemma. This is consistent with the finding that the valinomycin-induced K^+ transport and accumulation found in bacterial membrane vesicles was related to respiration-linked transport and not to an increase in passive K^+ permeability (8). However, it is still possible that valinomycin could act to shorten the plateau duration in guinea pig heart by increasing resting g_K. Alternatively, valinomycin could affect the time course or magnitude of the g_K changes during the action potential. For example, a speed-up of K^+ activation, so that it occurred more quickly after the spike component, would act to terminate the action potential earlier. Carmeliet (personal communication) has found that valinomycin (10^{-6} M) produced a 2-fold increase in the rate coefficient for K^+ efflux over 1–3 hr in 7–10-day chick hearts (in K^+-free solution). Similarly, Tosteson *et al.* (24) found that valinomycin (10^{-6} M) produced a reduction of K^+ content of sheep red blood cells after 2 hr and produced a large increase in the rate constant for inward leakage of K^+. Nonactin, which has an effect on lipid bilayer membranes similar to that of valinomycin, produced a rapid (<10 min to peak) loss of myocardial K^+ in perfused dog hearts (3).

Valinomycin may lower cellular ATP levels in guinea pig hearts by acting on the mitochondria to cause uncoupling of oxidative phosphorylation. Valinomycin in low concentration also produced marked ATP depletion within 30 min in Ehrlich ascites tumor cells (7). Thus, valinomycin may readily penetrate through the cell membrane and affect the mitochondria directly. The uncoupling of oxidative phosphorylation, respiratory stimulation, increased K^+ uptake, and mitochondrial ATPase activation produced by valinomycin appear to be due to the selective increase in K^+ and H^+ permeability of the mitochondrial membrane (see Reference 6). The effect on the mitochondria may be to diminish the H^+ gradient and hence decrease ATP synthesis (6, 11). A mitochondrial site of valinomycin action could account for the greater resistance of young embryonic hearts and cultured heart cells to valinomycin, since these myocardial cells were probably less dependent on oxidative metabolism (19, 20).

Our finding that valinomycin caused a rapid (2–10 min) block of the slow responses and accompanying contractions produced by catecholamines or methylxanthines in guinea pig hearts, and depleted the ATP content of the hearts over a similar time period, is consistent with our view that ATP is necessary for production of slow channels. This action of valinomycin was similar to that produced by DNP, cyanide, or anoxia (14). These results suggest that the induced slow Na^+–Ca^{++} channels require energy for their production and/or maintenance.

The cardiac action potential is composed of spike and plateau components (2, 5, 19), and metabolic poisons, like DNP, shorten and abolish the plateau component and lead to electromechanical uncoupling (9, 14, 15). The contractions disappear quickly (2–10 min), but the plateau component disappears completely only after longer periods (*e.g.*, 20–60 min). In contrast, valinomycin quickly suppressed the plateau of the normal action potential (in 2.7 mM K^+) along the same time course that the induced slow response disappeared. This might suggest that alterations in shape of the plateau mirror changes solely in the slow inward current. We have found, however, that this is not always the case, because at a time when the slow response is blocked by anoxia, ischemia, DNP, or cyanide, only small changes are evident in the shape of the normal action potential (figs. 7, 8). Therefore, although energy-dependent slow Ca^{++}–Na^+ channels undoubtedly are open during the plateau phase, inactivation of these channels sometimes leads to only slight alteration in shape of the plateau. Further, an inward Ca^{++} current is not sufficient by itself to account for the plateau because Mn^{++} or verapamil, known blockers of Ca^{++} influx, only slightly depress the plateau (13, 15).

The electromechanical uncoupling produced by valinomycin corresponded to the time at which the induced slow channels were inactivated. The inward current during the slow response in the guinea pig ventricle is carried by both Ca^{++} and Na^{+} (15). Thus, mechanical failure of hearts (which retain electrical activity) seems to be related to loss of transmembrane Ca^{++} current. We propose that inactivation of energy-dependent slow Ca^{++} channels is responsible for electromechanical uncoupling. Thus, the electromechanical uncoupling produced by valinomycin could be caused by ATP depletion which leads to inactivation of the slow channels.

In summary, we believe that valinomycin acts to shorten the cardiac ventricular action potential, produce electromechanical uncoupling, and block the slow response, all without associated hyperpolarization, by either one or both of the following methods: 1) valinomycin may become incorporated into the sarcolemma and act to increase g_K of the resting or excited membrane, thus terminating the action potential prematurely or 2) valinomycin may act on the mitochondria to cause cellular ATP depletion, which leads to inactivation of the slow Na^{+}-Ca^{++} channels.

Control of Ca^{++} influx

Because catecholamines (through stimulation of the β-adrenergic receptor and adenyl cyclase) and methylxanthines (through inhibition of the phosphodiesterase) and cyclic AMP (or dibutyryl cyclic AMP) itself are effective in induction of the slow channels (present results and References 14, 15, and 18), it would appear that both cyclic AMP and ATP are required for the production and maintenance of the slow Ca^{++}-Na^{+} channels. Increases in cyclic AMP levels occur within seconds after catecholamine or methylxanthine addition in a variety of tissues. Tsien (25) also presented evidence that the ionic currents associated with the cardiac action potential plateau are sensitive to changes in cyclic AMP levels. Brooker (1) and Wollenberger *et al.* (26) demonstrated oscillations in cyclic AMP levels during the cardiac cycle, the peak level occurring about midway during the rise in tension. These observations suggest that cyclic AMP levels may normally fluctuate during the action potential, and perhaps serve as a control for Ca^{++} influx, which in turn would control contractile force.

The role of cyclic AMP in production of the slow channel could be as a cofactor in phosphorylation of a membrane protein by ATP (fig. 11*A*), analogous to the cyclic AMP-dependent phosphorylation of a sarcoplasmic reticulum protein (23). Phosphorylation of a protein constituent of the slow channel may make it become available for voltage-dependent activation. These channels may become dephosphorylated after a short period by the action of a phosphatase, *i.e.*, energy would be required more or less continuously in order to maintain the slow channels in an available form.

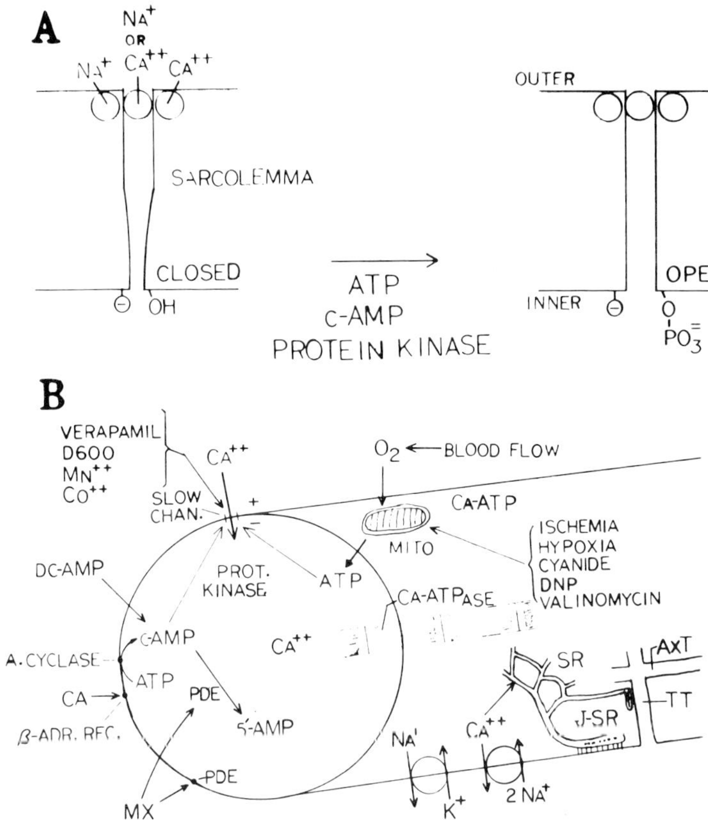

Figure 11 A proposed model is schematized for the membrane slow Ca^{++}-Na^{+} channel induced by catecholamines, methylxanthines, or cyclic AMP. The model takes into account the facts that both Na^{+} and Ca^{++} ions are required for activation of the channel, and that both Na^{+} and Ca^{++} ions participate in carrying the inward current, perhaps with competition between Na^{+} and Ca^{++} for entry. Both energy (ATP) and cyclic AMP appear essential to the production of the slow Ca^{++}-Na^{+} channels. A cyclic AMP-dependent protein kinase may phosphorylate a channel protein which produces a conformational change and makes the channel available for voltage activation. Conformational change of the phosphorylated channel may be produced by electrostatic repulsion. *B,* Schematic diagram of a myocardial cell which summarizes the known information and plausible assumptions concerning the production of membrane slow Ca^{++}-Na^{+} channels. The slow channels can be blocked directly by agents such as verapamil, D-600, Mn^{++} (1 mM), Co^{++} (1 mM), or La^{+++} (1 mM). The action of catecholamines (*CA*) on the β-adrenergic receptor (*β-ADR. REC.*) rapidly increase levels of cyclic AMP (*C-AMP*) through stimulation of adrenyl cyclase. Methylxanthines (*MX*) increase cyclic AMP levels through inhibition of phosphodiesterase (*PDE*) located in the myoplasm and/or sarcolemma. Perfusion with dibutyryl cyclic AMP or cyclic AMP (0.1 mM) also leads to the induction of slow channels, but much more slowly. The slow channel may be "induced", *i.e.,* made available for voltage activation, by phosphorylation with a cyclic AMP-dependent protein kinase. Metabolic poisons, hypoxia, or ischemia, which interfere with ATP production, thus indirectly depress transmembrane Ca^{++} influx, and therefore myocardial contraction. A mechanism is proposed by which the ATP level controls the density of slow Ca^{++}-Na^{+} channels and thereby the force of contraction. Inactivation of these channels would conserve that ATP which otherwise would be utilized in producing contractile work. This control mechanism would regulate the Ca^{++} influx per beat and would act to prevent depletion of ATP below a critical level.

We propose that there is a control mechanism normally operating between ATP level and Ca^{++} influx which serves to protect the myocardial cell (fig. 11*B*). If the ATP level is lowered, *e.g.,* because of regional ischemia or hypoxia, then the density of Ca^{++} channels available for voltage activation becomes lowered because ATP is required for maintenance of the slow channels. (Since ATP is consumed in the production of cyclic AMP, lowered ATP levels should also lead to lowered cyclic AMP levels.) The decreased number of Ca^{++} channels would lead to a smaller inward Ca^{++} current, hence a smaller elevation of $[Ca^{++}]_i$ and a smaller contractile force. The smaller activation of the myofibrillar ATPase gives a sparing action on the ATP expended for mechanical work. Thus, this would tend to compensate partially for the lowered ATP production. If $[ATP]_i$ is reduced below some critical level, all Ca^{++} channels may be inactivated, shutting off all Ca^{++} influx, and the cell would stop contracting, *i.e.*, it would become electromechanically uncoupled. ($[ATP]_i$ will fall only after the phosphocreatine stores have been nearly exhausted.) Partial recoupling would occur when ATP production is restored sufficiently to raise $[ATP]_i$ above the critical level. In this control system, it is the $[ATP]_i$ which is being sensed by the membrane Ca^{++} channels (fig. 11). If ATP production falls because of the metabolic impairment, then this control mechanism, which limits the rate of ATP expenditure by controlling Ca^{++} influx, comes into play as a possible protective mechanism.

Protection would come about by this control mechanism in that it would prevent the cellular ATP level from going so low as to produce irreversible injury or death of the cell. It is better for survival of the animal if a partially ischemic portion of the heart were to stop contracting temporarily (as actually occurs during episodes of angina pectoris) than to suffer permanent loss of this region by death of the cells. It is interesting to note that agents which increase Ca^{++} influx, such as catacholamines increase the size of experimental myocardial infarcts (2), and agents which act to depress Ca^{++} influx, such as verapamil, tend to decrease the degree of catecholamine-associated necrosis (2). The "stone-heart" syndrome (or rigor) follows prolonged ischemia of the heart during cardiac surgery (27). A metabolically crippled cell would not be able to pump out Ca^{++}, and therefore Ca^{++} would accumulate in the myoplasm, producing first contracture and then supercontraction of the myofibrils. Therefore, the protective mechanism described above would serve to prevent irreversible damage to the myocardial cell from sustained elevated $[Ca^{++}]_i$ levels.

REFERENCES

1. BROOKER, G. 1973. Oscillation of cyclic adenosine monophosphate concentration during the myocardial contraction cycle. Science 182: 933–934.
2. FLECKENSTEIN, A. 1971. Specific inhibitors and promoters of calcium action in the excitation-contraction coupling of heart muscle and their role in the prevention or production of myocardial lesions, pp. 135–188. *In* P. Harris and L. H. Opie, (eds.), Calcium and the Heart. Academic Press, New York.

3. GILMORE, J. P., GERLINGS, E. D., and MILLER, D. T. 1969. Influence of nonactin on myocardial performance, O_2 consumption, and K^+ balance. Amer. J. Physiol. 217: 131–135.
4. HERMSMEYER, K., RULON, R., and SPERELAKIS, N. 1972. Loss of the plateau of the cardiac action potential in hypertonic solutions. J. Gen. Physiol. 59: 779–793.
5. HOSHIKO, T., and SPERELAKIS, M. 1962. Components of the cardiac action potential. Amer. J. Physiol. 203: 258–260.
6. JAIN, M. K. 1972. The Bimolecular Lipid Membrane. Van Nostrand Reinhold Co., New York.
7. LEVINSON, C. 1967. Effect of valinomycin on net sodium and potassium transport in Ehrlich ascites tumour cells. Nature 216: 74–75.
8. LOMBARDI, F. J., KABACK, H. R., and REEVES, J. P. 1972. Valinomycin-induced Rb^+ transport by bacterial membrane vesicles. Fed. Proc. 31: 457.
9. McDONALD, T. F., and MacLEOD, D. P. 1973. Metabolism and the electrical activity of anoxic ventricular muscle. J. Physiol. London 229: 559–582.
10. MUELLER, P., and RUDIN, D. O. 1967. Development of K^+-Na^+ discrimination in experimental bimolecular lipid membranes by macrocyclic antibiotics. Biochem. Biophys. Res. Commun. 26: 398–404.
11. PRESSMAN, B. C., HARRIS, E. J., JAGGER, W. S., and JOHNSON, J. H. 1967. Antibiotic-mediated transport of alkali ions across lipid barriers. Proc. Nat. Acad. Sci. U.S.A. 58: 1949–1956.
12. RUBIO, R., WIEDMEIER, T., and BERNE, R. M. 1974. Relationship between coronary flow and adenosine production and release. J. Molec. Cell. Cardiol. 6: 561–566.
13. SAWANOBORI, T., LIEBERMAN, M., VEREECKE, J., and JOHNSON, E. A. 1973. Effects of valinomycin on the electrical properties of embryonic chick heart muscle [Abstract]. Physiologist 16: 441.
14. SCHNEIDER, J. A., and SPERELAKIS, N. 1974. The demonstration of energy dependence of the isoproterenol induced transcellular Ca^{++} current in isolated perfused guinea pig hearts–An explanation for mechanical failure of ischemic myocardium. J. Surg. Res. 16: 389–403.
15. SCHNEIDER, J. A., and SPERELAKIS, N. 1975. Slow Ca^{++} and Na^+ current responses induced by isoproterenol and methlxanthines in isolated perfused guinea pig hearts exposed to elevated K^+. J. Molec. Cell. Cardiol.
16. SCHNEIDER, J. A., and SPERELAKIS, M. 1974. Valinomycin blockade of slow channels in guinea pig hearts perfused with elevated K^+ and isoproterenol. Eur. J. Pharmacol. 27: 349–354.
17. SHIGENOBU, K., SCHNEIDER, J. A., and SPERELAKIS, N. 1974. Blockade of slow Na^+ and Ca^{++} currents in myocardial cells by verapamil. J. Pharmacol. Exp. Therap. 190: 280–288.
18. SHIGENOBU, K., and SPERELAKIS, N. 1972. Ca^{++} current channels induced by catecholamines in chick embryonic hearts whose fast Na^+ channels are blocked by tetrodotoxin or elevated K^+. Circ. Res. 31: 932–952.
19. SHIGENOBU, K., and SPERELAKIS, N. 1975. Valinomycin shortening of the action potential of embryonic chick hearts. Amer. J. Physiol. 228: 1113–1117.
20. SPERELAKIS, N. 1972. The electrical properties of embryonic heart cells, pp. 1–61. *In* W. C. De Mello (ed.), Electrical Phenomena in the Heart. Academic Press, New York.
21. SPERELAKIS, N., HOSHIKO, T., KELLER, R. F., Jr., and BERNE, R. M. 1960. Intracellular and external recordings from frog ventricular fibers during hypertonic perfusion. Amer. J. Physiol. 198: 135–140.
22. SPERELAKIS, N., and SHIGENOBU, K. 1972. Changes in membrane properties of chick embryonic hearts during development. J. Gen. Physiol. 60: 430–453.
23. TADA, M., KIRCHBERGER, M. A., IORIO, A. M., and KATZ, A. M. 1973. Phosphorylation of a low molecular weight component (phospholamban) in cardiac sarcoplasmic reticulum catalyzed by a cyclic AMP-dependent protein kinase [Abstract No. 96]. Circulation 48: 25.

24. TOSTESON, D. C., COOK, P., ANDREOLI, T., and TIEFFENBERG, M. 1967. The effect of valinomycin on potassium and sodium permeability of HK and LK sheep red cells. J. Gen. Physiol. 50: 2513–2525.
25. TSIEN, R. W. 1973. Adrenaline-like effects of intracellular iontophoresis of cyclic AMP in cardiac Purkinje fibres. Nature New Biol. 245: 120–122.
26. WOLLENBERGER, A., BABSKII, E. B., KRAUSE, E. G., GENZ, S., BLOHM, D., and BOGDANOVA, E. V. 1973. Cyclic changes in levels of cyclic AMP and cyclic GMP in frog myocardium during the cardiac cycle. Biochem. Biophys. Res. Commun. 55: 446–452.
27. WUKASCH, D. C., REUL, G. J., MILAM, J. D., HALLMAN, G. L., and COOLEY, D. A. 1972. The "stone heart" syndrome. Surgery 72: 1071–1080.

Mailing address: Dr. Nick Sperelakis,
Department of Physiology,
School of Medicine, University of Virginia,
Charlottesville, Virginia 22903 (USA).

Recent Advances in Studies on Cardiac Structure and Metabolism
Volume 9
The Sarcolemma
Edited by Paul-Emile Roy and Naranjan S. Dhalla

DRUG-INDUCED CHANGES IN THE SUPERFICIALLY LOCATED STORES OF CALCIUM IN HEART SARCOLEMMA[1]

W. G. NAYLER, J. DUNNETT, and A. SULLIVAN

Cardiothoracic Institute, London, England

SUMMARY

Excitation-contraction coupling in heart muscle cells depends upon an inward displacement of Ca^{2+} from the extra- to the intracellular phase. Some of the Ca^{2+} which is displaced inwards is probably derived from superficially located binding sites. Drugs which increase the peak tension developed during contraction, *e.g.,* catecholamines and the cardiac glycosides, may act by increasing the amount of Ca^{2+} which becomes available for participation in the events associated with excitation-contraction coupling. The catecholamines increase the amount of Ca^{2+} which is displaced inward from the extracellular phase, whereas the cardiac glycosides apparently act by increasing the exchangeability of the Ca^{2+} which is stored at binding sites located at the plasma membrane. Other drugs, *e.g.,* verapamil, act in the reverse manner, decreasing the amount of Ca^{2+} which is available for displacement from the superficially located binding sites.

These observations will be discussed in terms of an hypothesis that the plasma membrane represents an important site of drug action, and that many drugs which alter the contractile performance of the heart do so because of their ability to modify the ability of the plasma membrane to accumulate and exchange Ca^{2+}.

It is now generally agreed that the transition from the resting to the active state in heart muscle cells is triggered by an increase in the intracellular availability of Ca^{2+} due, in part at least, to an enhanced influx of Ca^{2+} from the extra- to the intracellular phase (3). This inward displacement of Ca^{2+} starts during the rising phase of the action potential (3, 20, 33, 34) and continues (fig. 1) throughout the plateau phase. In those cardiac muscle cells which have a T system (38–40) it is unlikely that the influx of these ions is restricted to the extreme periphery of the cell, because ultrastructural studies have shown consistently (fig. 2) that the walls which line the T tubules are continuous with the sarcolemma and, like the sarcolemma, are covered with basement coat material. Because these tubular invaginations of the cell membrane branch and ramify (12, 28), they provide a

[1] This work was carried out by a grant in aid from the British Heart Foundation and the Medical Research Council of Great Britain.

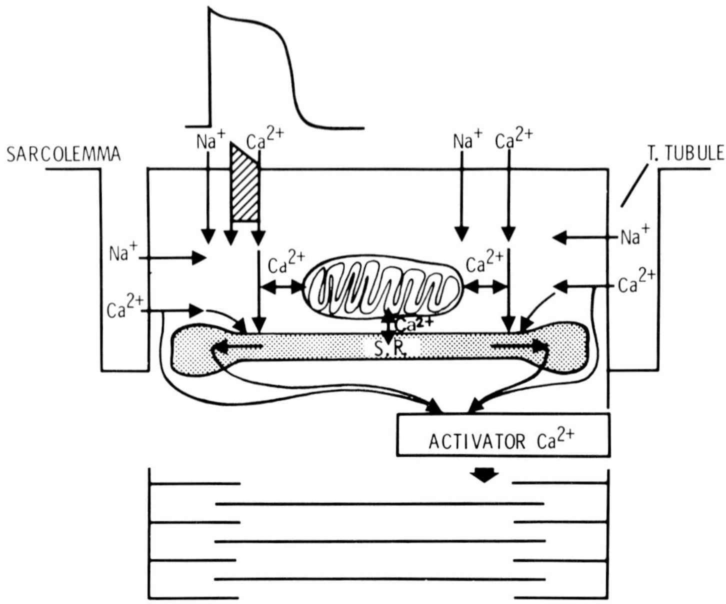

Figure 1 Schematic representation of the excitation-induced displacement of Ca^{2+} into the myoplasm of a cardiac muscle cell. *S.R.*, sarcoplasmic reticulum; *T. tubule*, transverse tubule.

vast increase in the surface area per unit cell volume, which is probably available for ionic exchange.

Although depolarization of the cell membrane results in an inward displacement of Ca^{2+} from the extra- to the intracellular phase (29), not all of the Ca^{2+} which is involved in this inward displacement will remain available for interaction with the myofibrillar proteins (13, 16). Instead, and as is shown schematically in figure 1, some of this Ca^{2+} may be sequestered by intracellular organelles which are not involved directly in the cascade of events (fig. 3) which is triggered off by the arrival of the excitatory stimulus and terminated by the onset of the mechanical response. Thus, some of this Ca^{2+} may be sequestered by the mitochondria (4) and some of it probably accumulates within the sarcoplasmic reticulum (fig. 2), where it may be stored for release in response to subsequent depolarizations (8, 11). Some of this Ca^{2+} may also function as a "trigger" to effect the release of some of the Ca^{2+} which has already been stored within the sarcoplasmic reticulum (9, 11). It is possible that the Ca^{2+} which is stored within the reticulum, and which is available for release in response to subsequent depolarizations of the cell membrane, accumulates within specialized areas of the reticulum, for example, within the subsarcolemmal cisternae (fig. 2). These cisternae represent specialized dilations of the sarcoplasmic reticulum, and they

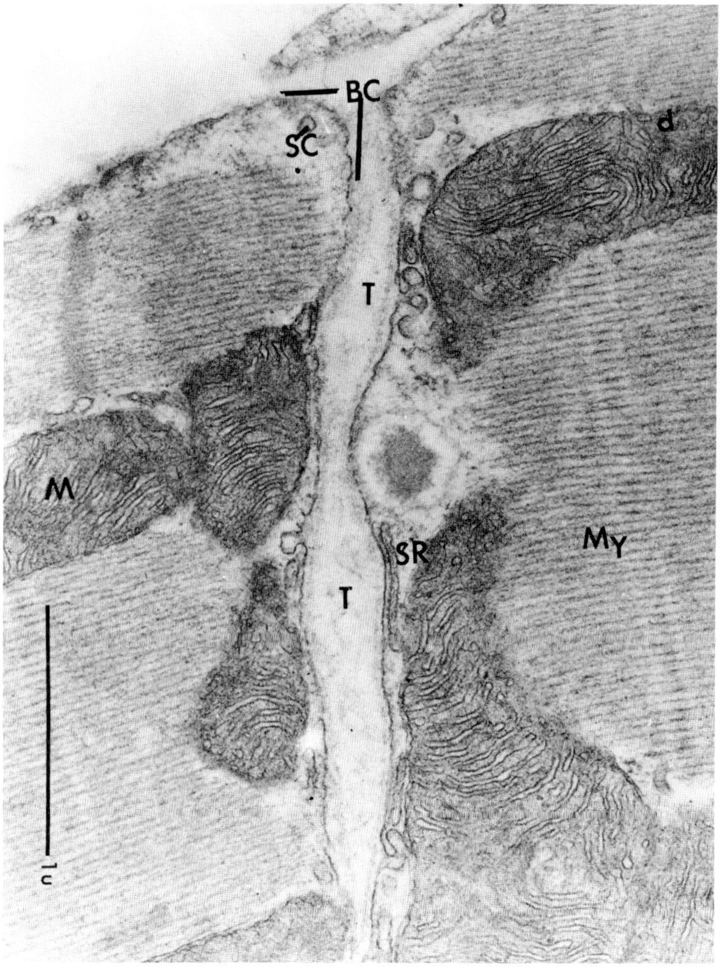

Figure 2 Longitudinal section of rabbit heart muscle showing a T tubule perfusion-fixed in glutaraldehyde and postfixed in osmium tetroxide. Note evidence of pinocytotic activity along the lining walls of the T tubule. *M*, mitochondria; *SR*, sarcoplasmic reticulum; *BC*, basement coat; *My*, myofilaments; *SC*, subsarcolemmal cisternae (magnification, × 48,000).

develop whenever the sarcoplasmic reticulum approaches either the sarcolemma or its intracellular invaginations (28).

It follows from the above discussion that the Ca^{2+} which finally interacts with the contractile proteins (13) and which, for convenience, is referred to here as "activator Ca^{2+}," must be heterogenous in origin (43), some being derived from extracellular sources and some from cellular storage depots. That the Ca^{2+} which

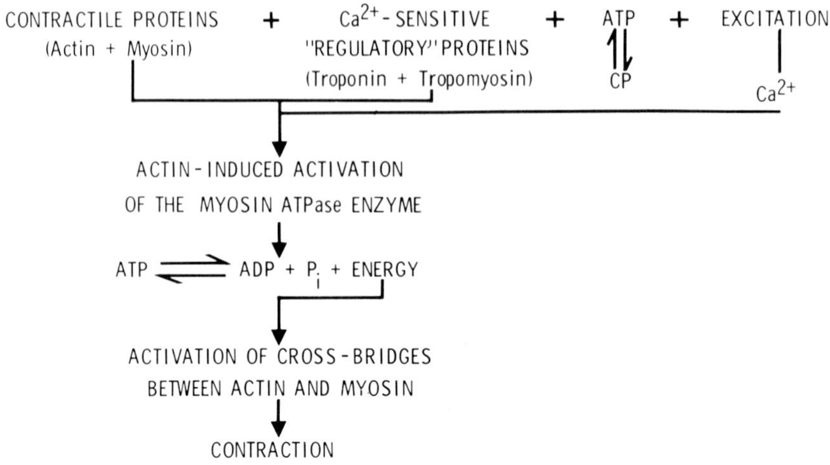

Figure 3 Schematic representation of the sequence of events intermediate between the arrival of the excitatory stimulus and the onset of the mechanical response in heart muscle cells.

is involved in excitation-contraction coupling (fig. 3) in cardiac muscle does originate from several sources is now well documented (1, 2, 5, 6, 23, 31). Recent isotope and other studies (26, 35) have shown consistently that part of this heterogenous supply of Ca^{2+} originates from superficially located depots which are closely associated with the cell membrane, and whose Ca^{2+} content is determined, in part at least, by the concentration of Ca^{2+} in the extracellular space. This raises the possibility that some cardioactive drugs may act either by altering the amount of Ca^{2+} which is available for displacement from these superficially located storage sites or, alternatively, by interfering with the displacement process itself. In the studies reported here three different techniques have been used to investigate this hypothesis.

ACTION OF INOTROPICALLY ACTIVE DRUGS ON HEART MUSCLE RENDERED HIGHLY PERMEABLE TO Ca^{2+}

If the inotropic activity of certain drugs is to be explained in terms of the ability of the drugs to alter the amount of Ca^{2+} which is available for displacement from the superficially located binding sites, then rendering the cell membranes freely permeable to Ca^{2+}, so that the Ca^{2+} which is present in the extracellular phase can diffuse freely into the cell, should interfere with their inotropic activity. The technique used to render cardiac cell membranes highly permeable to Ca^{2+} has already been described in detail in the literature (24, 46) and need not be reiterated here in detail. Briefly the technique consists of bathing the muscle in a buffered ethylenediaminetetraacetic acid (EDTA) solution to render the cell membranes highly permeable to Ca^{2+}, and then exposing it to different concentrations of Ca^{2+} in an Ca-EGTA-buffered medium. Under these conditions (24, 46), the amount of tension which develops is directly proportional to the concentration of Ca^{2+} in the extracellular phase. The typical result of such an experiment, in which EDTA-treated rabbit papillary muscle was exposed to varying concentrations of Ca^{2+} in the EGTA-buffered solution, is shown in figure 4. These results show that the developed peak tension is linearly related to the concentration of Ca^{2+} in the extracellular phase, despite the fact that the mechanical response was initiated simply by adding Ca^{2+} to the bathing fluid. When these experiments were repeated in the presence of 2.0 μg/ml of verapamil, a dose of verapamil which reduces ($p < 0.001$) the peak tension developed by electrically driven strips of heart muscle bathed in Krebs solution (fig. 5), then we find that rendering the muscle highly permeable to Ca^{2+} has abolished (fig. 6) the well documented (10, 27) cardiodepressant activity of this compound. Such a result is compatible with the theory that this drug acts as a Ca antagonist at the level of the cell membrane (10), and that it reduces the amount of Ca^{2+} which is available for displacement from superficially located sites (27).

A similar experimental approach can be used to demonstrate the fact that ouabain fails to exert a positive inotropic effect on heart muscle in which the cell membranes have been rendered highly permeable to Ca^{2+}. The experiments which are alluded to here were performed on thin strips of guinea pig papillary muscle, which were equilibrated in Krebs solution and stimulated to contract at a frequency of 12 beats/min, using supramaximal rectangular pulses of 10-msec duration delivered from a Tektronix square wave generator assembly. Initially

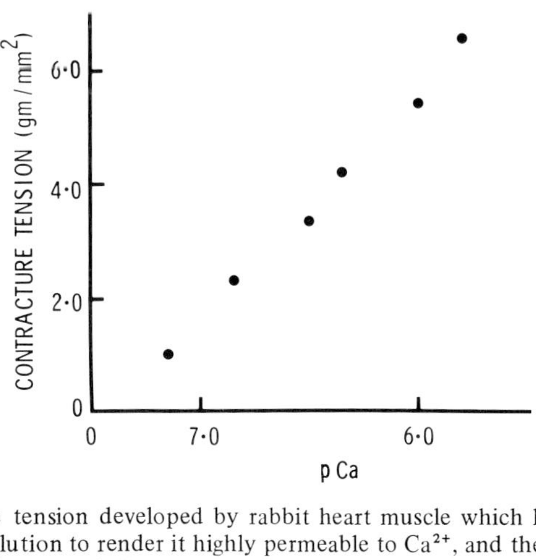

Figure 4 Contracture tension developed by rabbit heart muscle which had been bathed in an EDTA-buffered solution to render it highly permeable to Ca^{2+}, and then exposed to Ca^{2+}. Note that when Ca^{2+} was added to the bathing fluid it evoked a dose-dependent positive inotropic response.

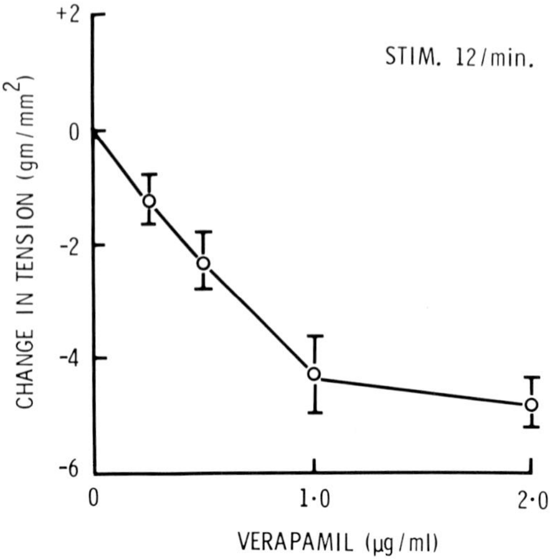

Figure 5 Dose-response curve relating to the effect verapamil exerts on the peak tension developed by isolated, electrically stimulated (*STIM.*) (12/min) thin strips of rabbit papillary muscle. Each point represents mean ± S.E.M. of six separate experiments performed in duplicate. The muscle preparations were equilibrated in aerated (95 percent O_{2+} + 5 percent CO_2) Krebs solution at 37°C.

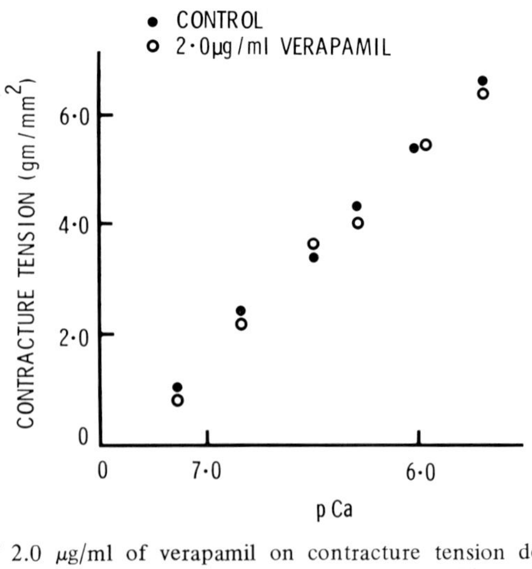

Figure 6 Effect of 2.0 μg/ml of verapamil on contracture tension developed by rabbit papillary muscle treated with EDTA to render it highly permeable to Ca^{2+}. Note that the amount of tension which was developed was not altered by this dose of verapamil.

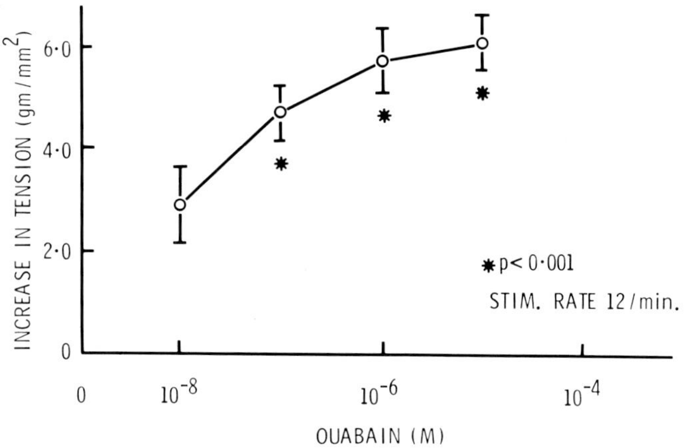

Figure 7 Dose-response curve for the positive inotropic effect of ouabain on freshly equilibrated guinea pig papillary muscle bathed in Krebs solution oxygenated with 95 percent O_2 + 5 percent CO_2 and stimulated (*STIM*) to contract at a rate of 12/min. Each point represents mean ± S.E.M. of six separate experiments. The calculated *p* values relate to the significance of the ouabain-induced change in peak developed tension.

experiments were performed to establish a dose-response curve for the normal positive inotropic effect of ouabain, and it is the mean results of these experiments which are summarized in figure 7. Other guinea pig papillary muscles were treated with a buffered EDTA solution to render them highly permeable to Ca^{2+} and then exposed to Ca^{2+} in a buffered EGTA solution, with and without ouabain (10^{-8}–10^{-6} M). Under these conditions, and as we have reported previously for dog heart muscle (24), the positive inotropic effect of ouabain is abolished (fig. 8).

A result qualitatively similar to that which has been described above for verapamil and ouabain was obtained in other experiments in which concentrations of dibutyryl-3′, 5′-AMP equal to or in excess of those (22) required to evoke a positive inotropic response from isolated segments of electrically driven strips of rabbit heart muscle (fig. 9) were added to papillary muscles (fig. 10) which had been rendered highly permeable to Ca^{2+}. The results of these experiments are summarized in figures 9 and 10. These results show that the positive inotropic effect of the dibutyryl derivative of 3′,5′-AMP is abolished under conditions which render the cell membranes highly permeable to Ca^{2+}.

Throughout these experiments care was taken to ensure that the tension which the EDTA-treated muscles developed in the absence of added drugs was not already maximal. This was done simply by adding extra Ca^{2+} to the bathing fluid and recording the resultant increase in developed tension. In general, therefore, these experiments establish that by rendering the cell membrane highly permeable to Ca^{2+}, so that the myoplasmic concentration of Ca^{2+} can be altered

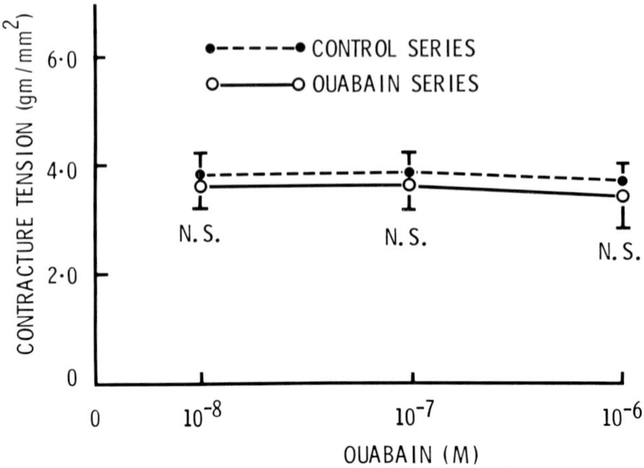

Figure 8 Contracture tension developed by EDTA-treated guinea pig heart muscle when exposed to Ca^{2+}, p_{Ca} 6.2, with (○) and without (•) 10^{-8}–10^{-6} M ouabain. Each point represents mean ± S.E.M. of six experiments performed in duplicate. *N.S.*, not significant at a level of $p = 0.05$. Test of significance relates to the change in contracture tension induced by ouabain.

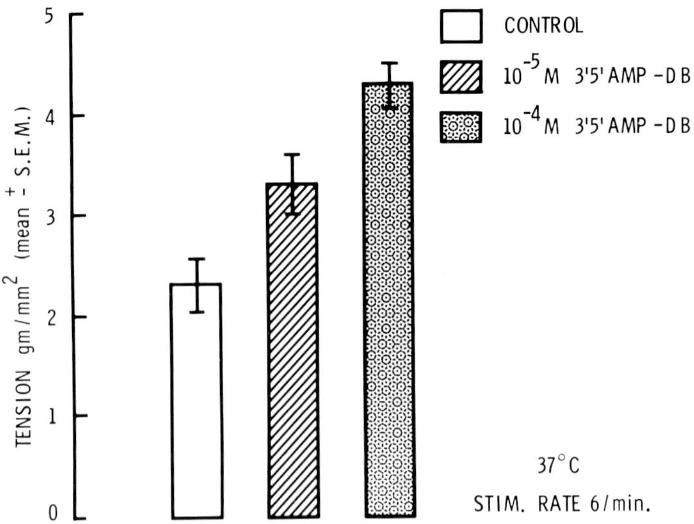

Figure 9 Effect of 10^{-5} and 10^{-4} M dibutyryl-3′,5′-AMP on the peak tension developed by isolated, isometrically suspended strips of rabbit papillary muscle, stimulated to contract 6/min. Each bar represents mean ± S.E.M. for six experiments. The control muscles were bathed in oxygenated Krebs solution.

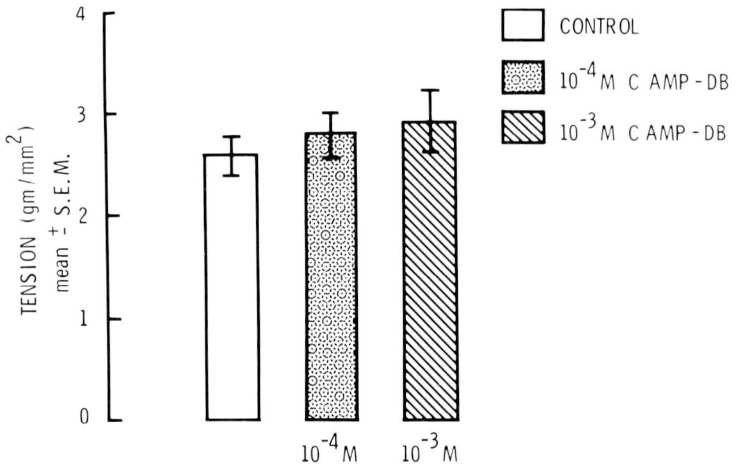

Figure 10 Effect of 10^{-4}–10^{-3} M dibutyryl 3′,5′-AMP (*C AMP-DB*) on tension development in rabbit heart muscle rendered highly permeable to Ca^{2+} (see text). In these preparations a mechanical response was initiated by adding Ca^{2+} to the bathing fluid. Each bar represents mean ± S.E.M. for six experiments.

simply by changing the extracellular concentration of Ca^{2+}, the inotropic effect which certain cardioactive compounds (including ouabain, dibutyryl-3′,5′-AMP, and verapamil) normally exert is abolished. It is therefore possible that these inotropic responses depend either on the ability of these drugs to alter the permeability of cell membranes to Ca^{2+}, to displace Ca^{2+} from binding sites located within or close to the surface membrane, or to alter the amount of Ca^{2+} which is displaced inward from these sites during the depolarization phase of the action potential. Alternatively, exposure of the tissue to EDTA may alter the properties of the cell membrane in such a way that the various drugs are no longer able to interact with the membrane, to evoke their various responses.

If the inotropic activity of certain drugs is to be accounted for, not in terms of their ability to interact with either the myofibrillar proteins or the energy yielding pathways, but in terms of their ability to react with the cell membrane in such a way that the amount of Ca^{2+} which is available (fig. 11) for displacement in response to subsequent depolarizations is altered, then it is necessary for us to establish that drugs can alter the capacity of cardiac cell membranes, or structures which are closely associated with these membranes, to accumulate Ca^{2+}. This possibility has been investigated by studying the ability of isolated fragments of the cell membrane to accumulate Ca^{2+}. Although such preparations are referred to as "plasma membrane fractions" it is almost certain that they contain other membrane-bound organelles, including some subsarcolemmal cisternae.

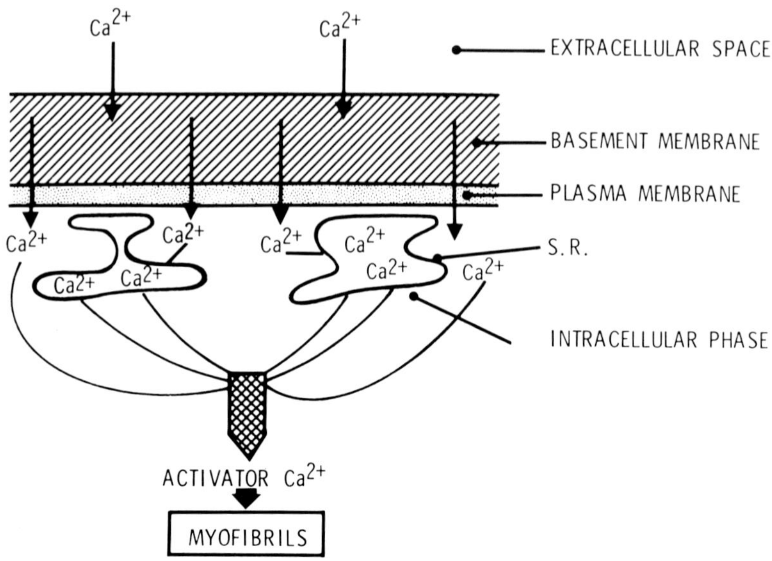

Figure 11 Schematic representation of the inward displacement of Ca^{2+} from binding sites within the cell membrane in response to 75 mV membrane depolarization. *S.R.*, sarcoplasmic reticulum.

CALCIUM BINDING BY A "PLASMA MEMBRANE" FRACTION ISOLATED FROM HOMOGENIZED HEART MUSCLE

"Plasma membrane" fractions have been isolated from a variety of tissues, including rat submandibular glands (30), liver cells (36), skeletal muscle (41), and cardiac muscle (7, 21). These isolated fractions are usually characterized in terms of the enzymatic profiles which they display, *e.g.*, the activity of the Na^+, K^+-activated, ouabain-inhibited ATPase enzyme is higher in plasma membrane than in sarcoplasmic reticulum preparations. Both fractions contain an active adenyl cyclase enzyme (17, 42).

In recent studies we have isolated fragments of the plasma membrane from guinea pig heart muscle cells by homogenization and differential centrifugation as described by Matsui and Schwartz (21), and after characterizing these fragments in terms of their Na^+,K^+-activated, ouabain-inhibited ATPase and adenyl cyclase activity, we have begun a study of their Ca^{2+} binding properties. The incubation medium used for these experiments contained 50 mM Tris-HCl buffer, pH 7.4; 100 mM KCl; 10 mM $MgCl_2$; 30–60 μM Ca^{2+}, 1 mM ATP, and 500 μg of plasma membrane protein per ml final volume. Incubation was carried out of 25°C, and Ca^{2+} binding was monitored continuously using a dual-beam spectrophotometer. The results which are summarized in figures 12–15 show

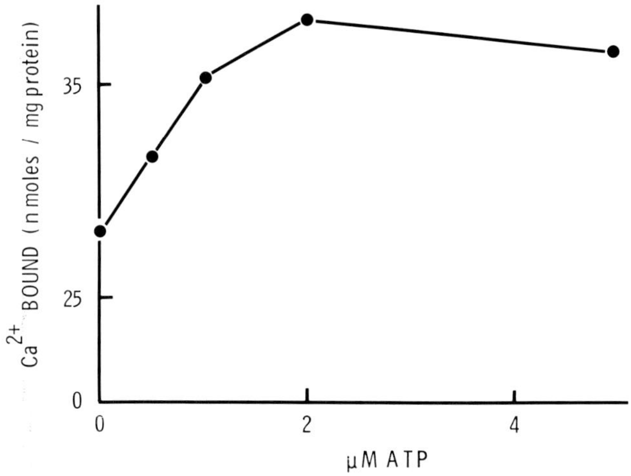

Figure 12 Ca^{2+} bound by freshly prepared guinea pig plasma membrane. The reaction was performed at 25°C for 2 min and the incubation medium contained 50 mM Tris-HCl buffer, pH 7.4, 100 mM KCl, 10 mM $MgCl_2$, 60 μM Ca^{2+}, and 500 μg of plasma membrane protein per ml and ATP as indicated by •.

that these isolated fragments of the plasma membrane bind Ca^{2+}. Part of this binding (fig. 12) proceeded in the absence of exogenously added ATP, and part of it appeared to be ATP dependent. The binding process progresses comparatively rapidly (figs. 13, 14*A*) relative to that exhibited by microsomal fractions which consisted predominantly of sarcoplasmic reticulum (figs. 13, 14*B*).

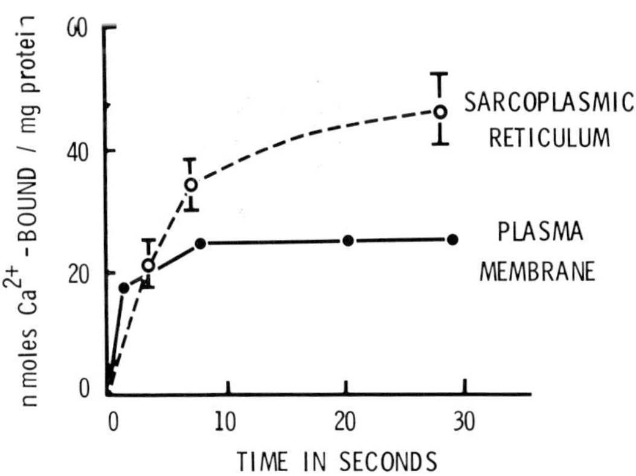

Figure 13 Time course for Ca^{2+} binding by fragments of plasma membrane and sarcoplasmic reticulum prepared from guinea pig heart muscle. Binding was monitored in a dual-beam spectrophotometer using murexide as an indicator.

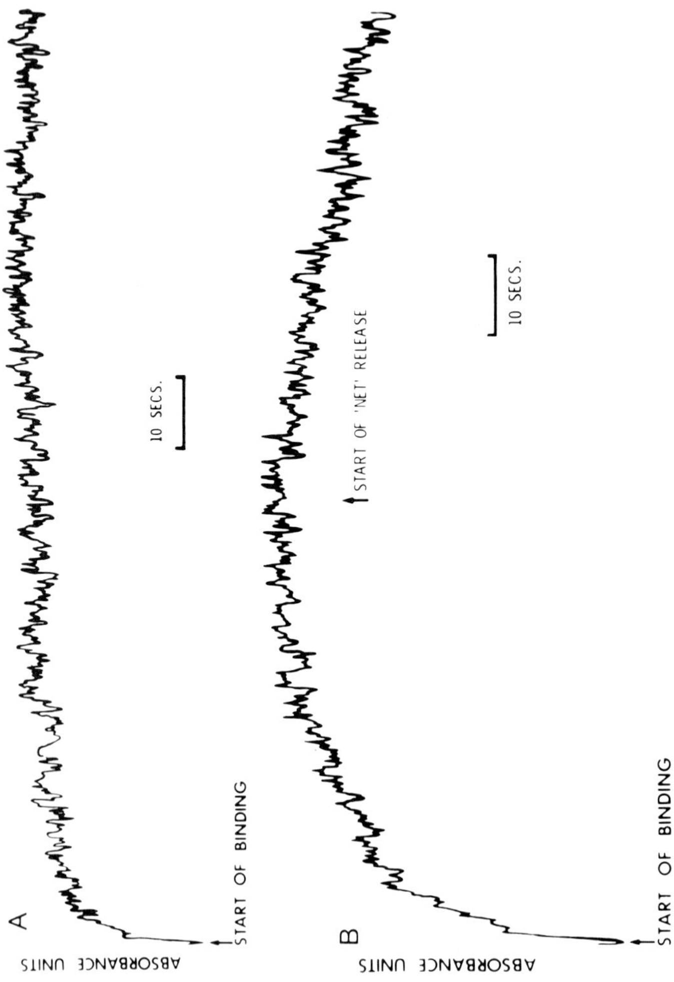

Figure 14 Record obtained during the monitoring of Ca^{2+} binding by (*A*) plasma membrane and (*B*) sareoplasmic reticulum absence of net Ca^{2+} release from the plasma membrane preparation, and the onset of net release from calcium binding was monitored in an Aminco-DW-2 dual-wave length spectrophotometer with stopped flow mixing. Wavelengths μnm 507 and 545 nm.

In marked contrast to the results obtained by Sulakhe *et al.* (41) for skeletal muscle plasma membrane preparations, but in agreement with the results described by Nijjar and Pritchard (30) for plasma membrane fractions prepared from rat submandibular glands, epinephrine (fig. 15) usually stimulated the binding of Ca^{2+} by these plasma membrane preparations. Why this epinephrine-induced stimulation of Ca^{2+} binding was not always observed is not known, nor is it known whether the simulatory effect of epinephrine is accompanied by an activation of the plasma membrane located Ca^{2+}-activated ATPase enzyme (7). Possibly this effect of epinephrine is mediated via a 3′,5′-AMP-protein kinase-dependent phosphorylation of binding sites within the plasma membrane, shown schematically in figure 16, and resembling that described in the literature (16, 25) for the sarcoplasmic reticulum. That cardiac plasma membrane preparations do display cyclic AMP-dependent protein kinase activity and that stimulation of this activity results in an enhanced binding of Ca^{2+} has already been reported in the literature (18, 45).

Irrespective of the mechanisms which are involved in this epinephrine-induced facilitation of Ca^{2+} binding the results described here are in agreement with others described in the literature (10, 14, 32, 37) and which implicate Ca^{2+} in the positive inotropic effect of the catecholamines. It is possible that a common

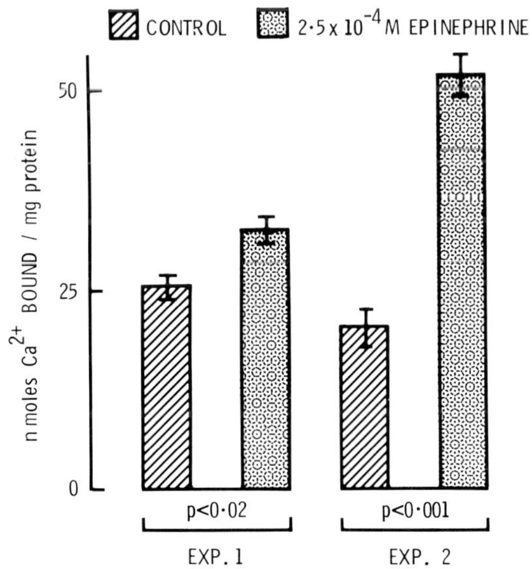

Figure 15 Increase in Ca^{2+} binding induced in guinea pig plasma membrane preparations by 2.4 × 10^4 M epinephrine (as epinephrine bitartrate). The results of experiments on two different preparations are shown, and each bar represents mean ± S.E.M. of five determinations.

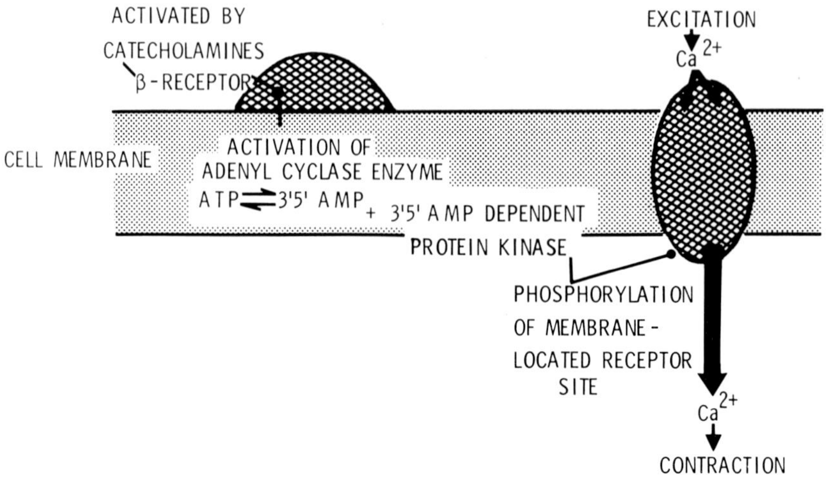

Figure 16 Schematic representation of the involvement of the 3′, 5′-AMP-dependent protein kinase in drug-induced facilitation of Ca^{2+} binding by isolated fragments of cardiac plasma membrane.

mechanism underlies the processes whereby epinephrine enhances twitch tension and abbreviates systole. Thus, by interacting with the adenyl cyclase system at the level of the cell membrane, epinephrine may increase the amount of Ca^{2+} which is available for displacement during excitation-contraction coupling, thereby evoking a positive inotropic response. By interacting with this same system at the level of the sarcoplasmic reticulum, epinephrine enhances the rate at which Ca^{2+} is retrieved from the myoplasm, resulting in an abbreviated systole. Such an argument presupposes, however, that there is in fact an epinephrine-sensitive adenyl cyclase present in the sarcoplasmic reticulum membrane. Despite recent reports in the literature (17, 42), the possibility exists that microsomal preparations claimed to contain fragments of the sarcoplasmic reticulum are heavily contaminated with plasma membrane vesicles.

LANTHANUM-INDUCED DISPLACEMENT OF Ca^{2+} FROM SUPERFICIALLY LOCATED STORAGE SITES IN HEART MUSCLE CELLS

The hypothesis that at least some of the Ca^{2+} which participates in the events associated with excitation-contraction coupling in heart muscle cells originates from storage sites which are superficially located, either within or immediately adjacent to the plasma membrane, originates from experiments in which the rare earth element lanthanum has been used to displace this Ca^{2+} into the extracellular phase (26–28, 35). Although under certain conditions La^{3+} can penetrate the surface membranes of cardiac muscle cells (44), it is usually (28) confined to the extracellular phase, and electron micrographs which purport to show an accumulation of La^{3+} in subsarcolemmal cisternae probably only provide evidence of an accumulation of La^{3+} within pinocytotic vesicles (fig. 2), because when filled with an electron dense marker such as La^{3+} it is often difficult to distinguish the subsarcolemmal cisternae from the pinocytotic vesicles, particularly in transverse section.

When added to the extracellular phase, La^{3+} uncouples excitation-contraction coupling (35), presumably by displacing the Ca^{2+} which is normally involved in this process from its superficially located storage sites. Hence, by adding La^{3+} under the appropriate conditions, it is possible to gain some information about the amount of Ca^{2+} which is stored at these superficially located sites. The data which is summarized in figure 17 shows that doses of ouabain which increase the peak tension developed during contraction but which do not cause contracture

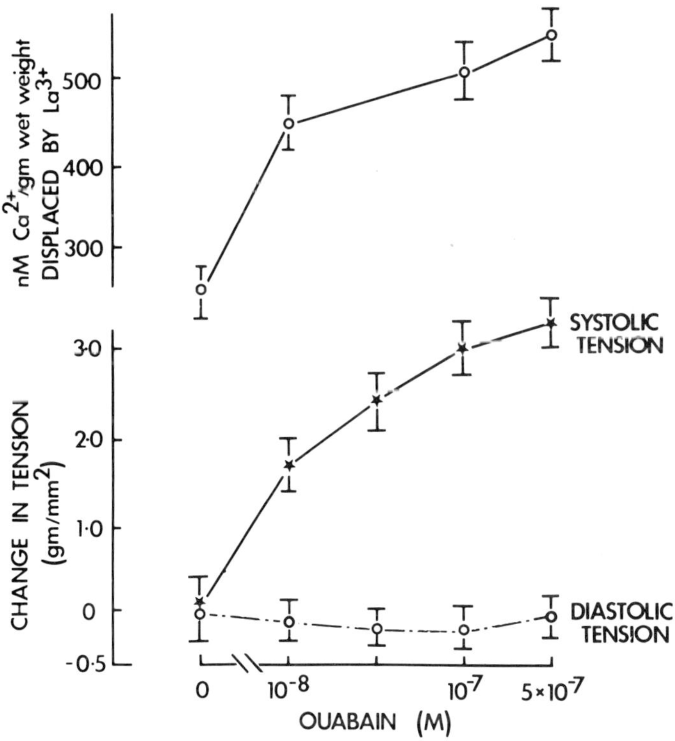

Figure 17 Effect of inotropically active doses of ouabain on the amount of $^{45}Ca^{2+}$ displaceable by La^{3+} from $^{45}Ca^{2+}$-labeled dog trabecular muscle, and on the tension developed during systole and diastole. The preparations were immersed in an HEPES-buffered Tyrode's solution. Each point represents mean ± S.E.M. of six experiments. The cross-sectional area of the muscles was <1.2 mm^2.

increase the amount of Ca^{2+} displaceable by La^{3+}. Conversely, concentrations of verapamil which reduce the peak tension developed by electrically driven strips of papillary muscle reduce the amount of Ca^{2+} which is displaceable by La^{3+} (27). For example, 1.0 μg/ml of verapamil, a dose which reduced by 4.2 gm/mm^2, the peak tension developed by electrically driven strips of rabbit papillary muscle (fig. 6), reduces the amount of Ca^{2+} displaceable by La^{3+} by 29 percent ($p < 0.01$). This effect of verapamil on the amount of Ca^{2+} which is displaceable by La^{3+} is dose dependent and significant ($p < 0.001$).

In conclusion, the three experimental approaches used here have provided evidence of the fact that, in addition to the Ca^{2+} which is derived from storage sites within the sarcoplasmic reticulum, Ca^{2+} which originates from other cellular storage sites plays an important role in regulating the development of tension in heart muscle. Some of these cellular storage sites are closely associated with the cell membrane, and some drugs which alter the peak tension developed by heart muscle may do so by altering the amount of Ca^{2+} which is available for displacement from these sites. The plasma membrane contains an adenyl cyclase and a 3′,5′-AMP-dependent protein kinase, and it is possible that the ability of certain drugs to alter the amount of Ca^{2+} which is available for displacement from the superficially located sites involves the activation of this system.

REFERENCES

1. BAILEY, L. E., ONG, S. D., and QUEEN, G. M. 1972. Calcium movement during contraction in the cat heart. J. Molec. Cell. Cardiol. 4: 121–138.
2. BAILEY, L. E., and ONG, S. D. 1974. Calcium pools in excitation-contraction coupling in the cat heart, pp. 225–271. *In* N. S. Dhalla and G. Rona (eds.), Recent Advances in Studies on Cardiac Structure and Metabolism, Vol. 4. University Park Press, Baltimore.
3. BASSINGTHWAITE, J. B., and REUTER, H. 1972. Calcium movements and excitation-contraction coupling in cardiac cells, pp. 353–395. *In* Electrical Phenomena in the Heart, Chap. 13. Academic Press, New York.
4. CARAFOLI, E. 1974. Mitochondria in the contraction and relaxation of heart, pp. 393–406. *In* N. S. Dhalla and G. Rona (eds.) Recent Advances in Studies on Cardiac Structure and Metabolism, Vol. 4. University Park Press, Baltimore.
5. CHAPMAN, R. A., and NIEDERGERKE, R. 1970. Effects of calcium on the contraction of the hypodynamic frog heart. J. Physiol. (London) 211: 389–422.
6. CHAPMAN, R. A., and NIEDERGERKE, R. 1970. Interaction between heart rate and calcium concentration in the control of contractile strength of the frog heart. J. Physiol. (London) 211: 423–443.
7. DIETZE, G., and HEPP, K. D. 1971. Calcium stimulated ATPase of cardiac sarcolemma. Biochem. Biophys. Res. Commun. 44: 1041–1045.
8. ENDO, M., and NAKAJIMA, Y. 1973. Release of calcium induced by 'depolarization' of the sarcoplasmic reticulum membrane. Nature New Biol. 246: 216–218.
9. ENDO, M., TANAKA, M., and OGAWA, Y. 1970. Calcium-induced release of calcium from the sarcoplasmic reticulum of skinned skeletal muscle fibres. Nature 228: 34–36.
10. FLECKENSTEIN, A. 1971. Specific inhibitors and promoters of calcium action in the excitation-contraction coupling of heart muscle and their role in the prevention or

production of myocardial lesions, pp. 135–188. *In* P. Harris and L. H. Opie (eds.), Calcium and the Heart. Academic Press, New York.
11. FORD, L. E., and PODOLSKY, R. J. 1970. Regenerative calcium release within muscle cells. Science 167: 58–59.
12. FORSSMANN, W. G., and GIRARDIER, L. 1970. A study of the T system in rat heart. J. Cell Biol. 44: 1–19.
13. GERGELY, J., GREASER, M., NAGY, B., and POTTER, J. 1974. Troponin: The regulatory protein complex of muscle, pp. 281–290. *In* N. S. Dhalla and G. Rona (eds.), Recent Advances in Studies on Cardiac Structure and Metabolism, Vol. 4. University Park Press, Baltimore.
14. GROSSMAN, A., and FURCHGOTT, R. F. 1964. Effects of various drugs on calcium exchange in isolated guinea-pig left auricle. J. Pharmacol. Exp. Therap. 145: 162–172.
15. KATZ, A. M. 1970. Contractile proteins of the heart. Physiol. Rev. 50: 63–158.
16. KATZ, A. M., and REPKE, D. 1973. Calcium-membrane interactions in the myocardium. Amer. J. Cardiol. 31: 193–201.
17. KATZ, A. M., TADA, M., REPKE, D. I., IORIO, J. M., and KIRCHBERGER, M. A. 1974. Adenylate cyclase: Its probable localization in sarcoplasmic reticulum as well as sarcolemma of the canine heart. J. Molec. Cell. Cardiol. 6: 73–78.
18. KRAUSE, E. G., WILL, H., PELOUCH, V., and WOLLENBERGER, A. 1973. Cyclic AMP-dependent protein kinase activity in a cell membrane-enriched subcellular fraction of pig myocardium. Acta Biol. Med. Germ. 31: 37–43.
19. LANGER, G. A., and FRANK, J. S. 1972. Lanthanum in heart cell culture: effect on calcium exchange correlated with its localization. J. Cell Biol. 54: 441–445.
20. MASCHER, D., and PEPER, K. 1969. Two components of inward membrane current in myocardial muscle fibers. Pflüger's Arch. ges. Physiol. 307: 190–203.
21. MATSUI, H., and SCHWARTZ, A. 1966. Purification and properties of a highly active, ouabain sensitive Na^+,K^+-dependent adenosinetriphosphatase from cardiac tissue. Biochim. Biophys. Acta 128: 380–390.
22. MEINERTZ, T., NAWRATH, H., and SCHOLZ, H. 1973. Stimulatory effects of DB-c-AMP and adrenaline on myocardial contraction and ^{45}Ca exchange. Experiments at reduced calcium concentration and low frequencies of stimulation. Naunyn-Schmiedeberg's Arch. Pharmacol. 279: 327–338.
23. NAYLER, W. G. 1973. An effect of ouabain on the superficially-located stores of calcium in cardiac muscle cells. J. Molec. Cell. Cardiol. 5: 101–110.
24. NAYLER, W. G. 1973. Effect of inotropic agents on canine trabecular muscle rendered highly permeable to calcium. Amer. J. Physiol. 225: 918–924.
25. NAYLER, W. G., and BERRY, D. 1975. Effect of drugs on the cyclic adenosine 3′,5′ monophosphate-dependent protein kinase induced stimulation of calcium uptake by cardiac microsomal fractions. J. Molec. Cell Cardiol. 7:387–395.
26. NAYLER, W. G., and DUNNETT, J. 1974. Regulation of myocardial contraction. Advan. Cardiol. 12: 45–58.
27. NAYLER, W. G., and SZETO, J. 1972. Effect of verapamil on contractility, oxygen utilization, and calcium exchangeability in mammalian heart muscle. Cardiovasc. Res. 6: 120–128.
28. NAYLER, W. G., and MERRILLEES, N. C. R. 1971. Cellular exchange of calcium, pp. 24–65. *In* P. Harris and L. H. Opie, (eds.), Calcium and the Heart. Academic Press, New York.
29. NIEDERGERKE, R. 1963. Movements of Ca in frog heart ventricles at rest and during contractures. J. Physiol. (London) 167: 515–550.
30. NIJJAR, M. S., and PRITCHARD, E. T. 1973. Calcium binding by a plasma membrane fraction isolated from rat submandibular glands. Biochim. Biophys. Acta 323: 391–395.
31. ONG, S. D., and BAILEY, L. E. 1972. Two functional distinct calcium pools in the excitation-contraction coupling process. Experientia 28: 1446–1447.

32. REUTER, H. 1965. Ueber die Wirkung von Adrenalin auf den cellularen Ca-Umsatz des Meerschweinchenvorhofs. Naunyn-Schmiedeberg's Arch. Pharmakol. 251: 401–412.
33. REUTER, H., and BEELER, G. W., JR. 1969. Calcium current and activation of contraction in ventricular myocardial fibers. Science 163: 399–401.
34. ROUGIER, O., VASSORT, G., GARNIER, D., GARGOUIL, Y. M., and CORABOEUF, E. 1969. Existence and role of a slow inward current during the frog atrial action potential. Pflüger's Arch. ges. Physiol. 308: 91–110.
35. SANBORN, W. G., and LANGER, G. A. 1970. Specific uncoupling of excitation and contraction in mammalian cardiac tissue by lanthanum. J. Gen. Physiol. 56: 191–217.
36. SCHLATZ, L., and MARINETTI, G. V. 1972. Calcium binding to the rat liver plasma membrane. Biochim. Biophys. Acta 290: 70–83.
37. SHIGENOBU, K., and SPERELAKIS, N. 1972. Calcium current channels induced by catecholamines in chick embryonic hearts whose fast sodium channels are blocked by tetrodotoxin or elevated potassium. Circ. Res. 31: 932–952.
38. SIMPSON, F. O. 1965. The transverse tubular system in mamallian myocardial cells. Amer. J. Anat. 117: 1–18.
39. SOMMER, J. R., and JOHNSON, E. A. 1969. Cardiac muscle: Comparative ultrastructural study with special reference to frog and chicken hearts. Z. Zellforsch. Mikrosk. Anat. 98: 437–468.
40. SPERELAKIS, N., FORBES, M. S., and RUBIO, R. 1974. The tubular systems of myocardial cells: Ultrastructure and possible function. *In* N. S. Dhalla and G. Rona (eds.), Recent Advances in Studies on Cardiac Structure and Metabolism, Vol. 4. University Park Press, Baltimore.
41. SULAKHE, P. V., DRUMMOND, G. I., and Ng, D. C. 1973. Calcium binding by skeletal muscle sarcolemma. J. Biol. Chem. 248: 4150–4157.
42. TADA, M., FINNEY, J. O., JR., SWARTZ, M. H., and KATZ, A. M. 1972. Preparations and properties of plasma membranes from guinea pig heart. J. Molec. Cell. Cardiol. 4: 417–426.
43. VASSORT, G. 1974. Evidence for two components in frog cardiac mechanical activity, pp. 67–75. *In* N. S. Dhalla and G. Rona (eds.), Advances in Studies on Cardiac Structure and Metabolism, Vol. 4, University Park Press, Baltimore.
44. WAUGH, R. A., SPRAY, T. L., and SOMMER, J. R. 1973. Fenestrations of sarcoplasmic reticulum: Delineation by lanthanum acting as a fortuitous tracer in situ negative stain. J. Cell Biol. 59: 254–260.
45. WILL, H., SCHIRPKE, B., and WOLLENBERGER, A. 1973. Binding of calcium to a cell membrane-enriched preparation from pig myocardium: Increase in calcium affinity upon membrane protein phosphorylation enhanced by a membrane-bound cyclic AMP-dependent protein kinase. Acta Biol. Med. Germ. 31: 45–52.
46. WINEGRAD, S. 1971. Studies of cardiac muscle with a high permeability to calcium produced by treatment with ethylenediaminetetraacetic acid. J. Gen. Physiol. 58: 71–93.

Mailing address: Dr. W. G. Nayler,
Cardiothoracic Institute, 2 Beaumont Street,
London, W1N 2DX (England).

Recent Advances in Studies on Cardiac Structure and Metabolism
Volume 9
The Sarcolemma
Edited by Paul-Emile Roy and Naranjan S. Dhalla

TECHNIQUES OF SKINNED CARDIAC CELLS AND OF ISOLATED CARDIAC FIBERS WITH DISRUPTED SARCOLEMMAS[1]

WITH REFERENCE TO THE EFFECTS OF CATECHOLAMINES AND OF CAFFEINE

A. FABIATO[2] and F. FABIATO

Cardiovascular Research Laboratory,
Harvard Medical School and Peter Bent Brigham Hospital,
Boston, Massachusetts, USA

SUMMARY

Single cardiac cells with disrupted sarcolemmas were obtained by homogenization in low free [Ca^{2+}] medium. Skinned cardiac cells were obtained by microdissection of remaining pieces of sarcolemmas. In both preparations the buffers bathing the cells were in contact with the structures controlling activation: myofilaments, sarcoplasmic reticulum, and mitochondria. Tension was recorded with a highly sensitive force transducer. The use of either high or low total [EGTA] permitted the study of the effect of free [Ca^{2+}] on either the myofilaments or the internal stores. At physiological levels of myoplasmic free [Ca^{2+}], the internal stores of Ca^{2+} were located mostly in the sarcoplasmic reticulum, whereas mitochondria were involved at higher free [Ca^{2+}]. Single cardiac cells with simply disrupted sarcolemmas responded to variations of perfusion media in a manner similar to the skinned cardiac cells. However, the remaining pieces of sarcolemma represented a limit to the intracellular diffusion of externally applied media and might contain receptor sites for catecholamines. In multicellular fragments, evidence of surface membrane activity (resting and action potentials) were found during certain ionic modifications.

Single cells of skeletal muscle from which the sarcolemma was peeled off (skinned fibers) permitted important insight into the mechanism of activation of skeletal muscle (3, 6–8, 11–13, 16, 17, 21, 28, 30, 36, 37). The removal of the sarcolemma allowed direct application of solutions with buffered free calcium concentration (free [Ca^{2+}]) onto the intracellular structures that control excitation-contraction coupling: myofilaments, sarcoplasmic reticulum, and possibly, mitochondria. Although cardiac cells are much smaller than skeletal muscle cells,

[1] This research was supported by Grant HL-11306 from the National Institutes of Health.
[2] This work was done during the tenure of an Established Investigatorship of the American Heart Association.

high resolution optical microscopy with phase or interference contrast as well as techniques of micromanipulation have permitted the development of a "sarcolemma-free" preparation of cardiac cells. This preparation has been developed in two steps: cardiac fibers with disrupted sarcolemmas (9) and skinned cardiac cells (10). Cardiac fibers with disrupted sarcolemmas are simple to obtain and may be useful for physiological studies under the condition that the sarcolemma is largely disrupted. However, remaining pieces of sarcolemma limit the diffusion of externally applied media and may represent receptor sites for drugs such as catecholamines. This has justified the development of a preparation in which the sarcolemma is almost completely removed from a single cardiac cell (skinned cardiac cell).

SOLUTIONS

In all cases, the buffers contained dextrose, 7 mM, and Tris-maleate, 18 mM; pH was 7.0 at 22°C. Imidazole was used initially as a pH buffer, but was deleted because it appeared to depress the cyclic contractions that were attributed to Ca^{2+} release and binding by the sarcoplasmic reticulum. This effect may be due to a stimulation of the sarcolemmal phosphodiesterase, which could inhibit Ca^{2+} binding by the sarcoplasmic reticulum (10). The concentrations of ethyleneglycol-bis(β-aminoethyl ether)-N,N′-tetraacetic acid (EGTA), $CaCl_2$, Na_2ATP, and $MgCl_2$ were varied. A computer program, developed by Reuben *et al.* (36) and adapted for our studies by A. Gordon, M. Orentlicher, and M. Kawai, was used to define the concentrations of those components necessary for obtaining the desired free [Ca^{2+}], free [Mg^{2+}], free [ATP^{4-}], [$MgATP^{2-}$], and total [EGTA]. The following apparent stability constants were used: CaEGTA, 4.9×10^6 M^{-1} (38, 40); MgEGTA, 40 M^{-1} (35); CaATP, 5×10^3 M^{-1} (29); MgATP, 11.4×10^3 M^{-1} (29). Since increase of ionic strength depresses the tension developed by the myofilaments (16, 36), the ionic strength was kept constant at 0.16 M + 0.01 by addition of an appropriate amount of KCl. In some experiments, the nature of monovalent ions was varied without modification of the ionic strength. These variations included the replacement of Cl^- by propionate and variations of the [Na^+]/[K^+] ratio, including total absence of Na^+. Some preparations were treated for 30 min to 4 days with the nonionic detergent, Brij 58 (32), obtained from Ruger Chemical Company (Irvington, N.Y.), at a concentration of 1 percent. Caffeine and isoproterenol (1-isoproterenol-D-bitartrate) were obtained from Sigma. Total calcium and magnesium concentrations were measured by atomic absorbtion spectrophotometry (Perkin Elmer 303).

CARDIAC FIBERS WITH DISRUPTED SARCOLEMMA

In some experiments, the heart was perfused for 1 hr by the Langendorff technique with a medium containing 0.2 mM total [EGTA], $10^{-8.4}$ M free

$[Ca^{2+}]$, $10^{-3.5}$ M free $[Mg^{2+}]$, $10^{-2.5}$ M $[MgATP^{2-}]$, and $10^{-3.0}$ M free $[ATP^{4-}]$. In all cases, the tissue was minced in cubic fragments of approximately 2 mm in each dimension, and the fragments were washed several times in the same medium at 2°C. Tissue (500–700 mg) was homogenized into fragments slightly smaller than a single cell in 25 ml of the same buffer using a Virtis 45 blender. A fraction of the homogenate was diluted to 1:200 in the same buffer, and placed in a 0.05-ml perfusion chamber on the stage of a Biovert inverted microscope (Reichert, Vienna, Austria) with anoptral phase contrast or Nomarski interference. This microscope had a thermoelectrically controlled heating and cooling stage (Cambion, Cambridge, Mass.). A temperature of 22 ± 0.1°C was used in most of the studies.

Ventricular cells of mammalian species have relatively large dimensions (40–100 μm in length and 10–20 μm in diameter). They are joined end to end by specialized intercellular junctions (the intercalated discs) although there is no or infrequent lateral connection between the sarcolemmas of adjacent cells (23).

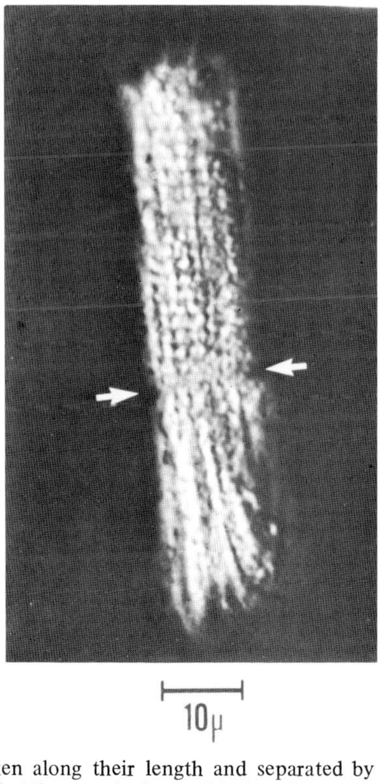

Figure 1 Two cells broken along their length and separated by an intercalated disc (indicated by the *two arrows*). Micrograph obtained in interference (type Nomarski). Disruption of the sarcolemma is particularly obvious at the end of the lower cell, where myofibrils are diverging.

During homogenization, the cells generally broke along their length (direction parallel to the long axis of the myofibrils). Sometimes two broken cells were separated by an intercalated disc, which was easily visible under interference microscopy (fig. 1). In other cases, the two ends of a single cell were broken (fig. 2), so that the disrupted cell had the same diameter as the intact cell (less than 15 μm in a 200-g rat) but had a slightly shorter length (less than 60 μm). In both instances, disruption of the sarcolemma was visible at the ends of the cell where myofibrils were partly dissociated. Generally, the lateral limits of the cell were regular, suggesting the presence of sarcolemma, which was confirmed by electron microscopy (fig. 3). Single disrupted cells (30 μm in length and 10 μm in diameter) or groups of two broken cells (about 100 μm in length and 10 μm in diameter) were used to demonstrate the existence of Ca^{2+}-triggered release of Ca^{2+} from the sarcoplasmic reticulum of cardiac cells. In both cases, the

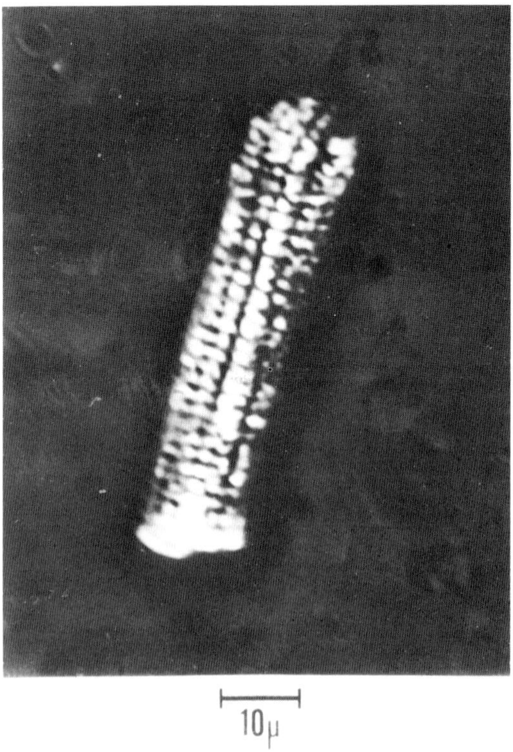

Figure 2 Single broken cell (disrupted at its two ends). Micrograph in anoptral phase contrast (negative phase contrast). Notice that the A bands of the sarcomeres are clear and the I bands are dark with this type of negative phase contrast. Mitochondria (*clear*) are irregularly superimposed on the striations of the sarcomeres.

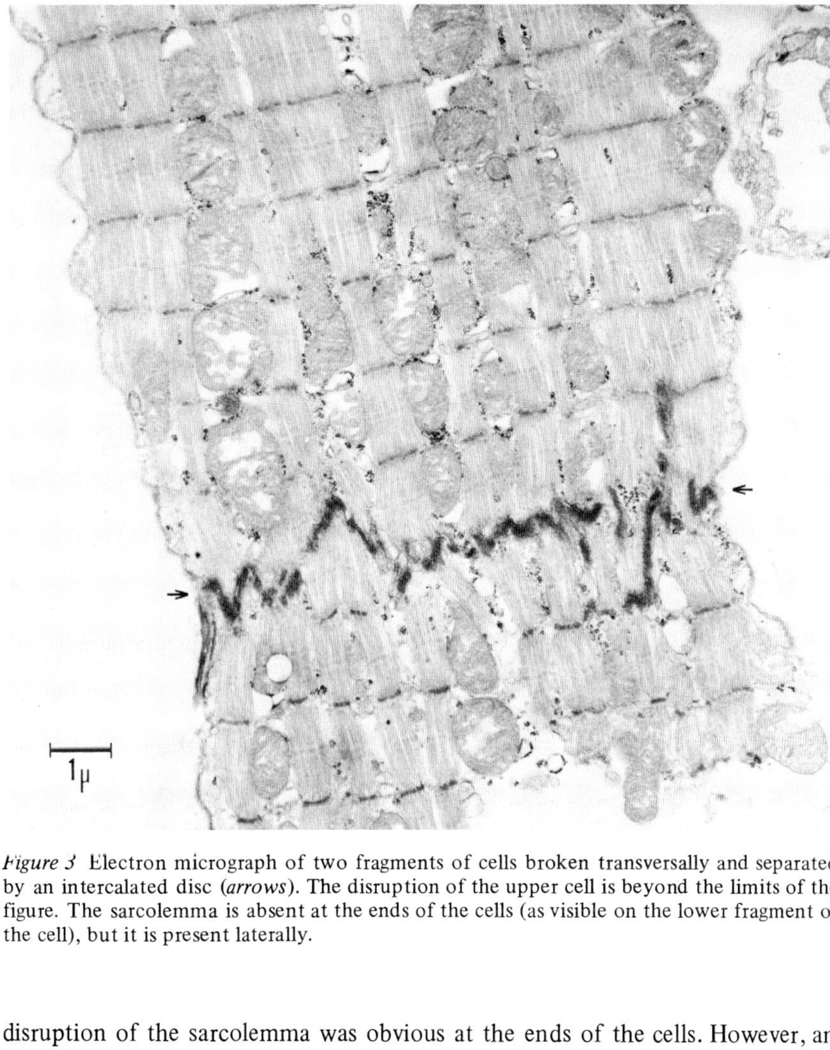

Figure 3 Electron micrograph of two fragments of cells broken transversally and separated by an intercalated disc (*arrows*). The disruption of the upper cell is beyond the limits of the figure. The sarcolemma is absent at the ends of the cells (as visible on the lower fragment of the cell), but it is present laterally.

disruption of the sarcolemma was obvious at the ends of the cells. However, an unhomogenous contraction was observed in the preparations made of two broken cells (fig. 2*B* of Reference 9), whereas contraction was homogenous in the fragments that were smaller than a single cell (figs. 6 and 7 of Reference 9).

SKINNED CARDIAC CELLS

In this preparation, remaining pieces of sarcolemma were removed by microdissection from a cell with disrupted sarcolemma. Fragments made of two broken cells separated by an intercalated disc were discarded. One end of the broken cell was immobilized with a glass microtool (heat-occluded microelectrode). A 2 μm tip glass microneedle (prepared as a microelectrode) was used to pull sarcolemma and superficial myofibrils longitudinally along the entire length of the cell, thereby stripping the sarcolemma from the cell (fig. 4). During this process superficial myofibrils and mitochondria were removed with the sarcolemma. Before the skinning, the sarcolemma was not visible under optical microscopy. However, the presence of the sarcolemma was indirectly demonstrated by a construction around the circumference of the broken cell,

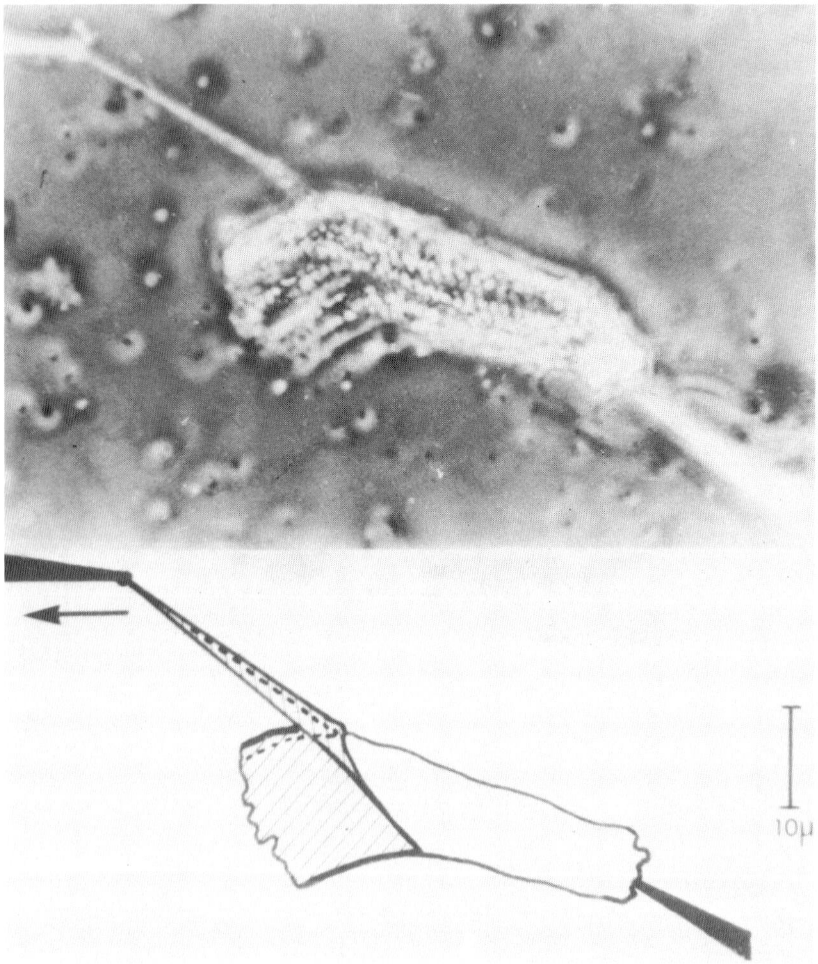

Figure 4 Skinning of a single broken cell. (Reproduced from Fabiato and Fabiato (12) by permission of the *European Journal of Cardiology*.)

produced by the stretching of the sarcolemma. During the skinning, care was taken not to stretch the broken cell beyond 30 percent of its initial length. Complete skinning without overstretch was obtained in about 1 cell of 10. The skinning was carried out in the medium used for homogenization, which contained 0.2 mM total [EGTA] and $10^{-8.4}$ M free [Ca^{2+}]. In this medium the cell was permanently in relaxation. Larger total [EGTA] would have resulted in adherence of the myofibrils between themselves and to sarcolemma, preventing the skinning.

After "complete" longitudinal skinning, fragments of surface membrane remained on the cell at the invaginations of the T system. In skinned fibers of skeletal muscle, the transverse tubules are long, narrow, and healed over into closed vesicles. Calcium sequestered within these vesicles may play a nonphysiological role in activation of this preparation (28). In cardiac muscle the T tubules have a larger diameter, and we have not as yet found evidence of their healing over into closed vesicles.

Atrial cells of mammalian species are smaller (20–60 μm in length, 6–10 μm in diameter) than ventricular cells. Furthermore, they present frequent lateral membrane junctions (23). For this reason, the fragments obtained by homogenization had a complex geometry and were not made of one single broken cell or two cells separated by an intercalated disc, as in ventricular tissue. Consequently, longitudinal skinning in accordance with the technique described previously was impossible. However, it was possible to microdissect small bundles of 5 – 15 myofibrils that were surrounded by sarcoplasmic reticulum and mitochondria and were free of any large pieces of sarcolemma. Skinned bundles of cardiac myofibrils from atrial cells are advantageous because these cells do not generally have a T system (15). Therefore, this preparation permitted control experiments to eliminate the role of Ca^{2+} hypothetically stored in healed-over transverse tubules of ventricular skinned cells. Frog ventricular or atrial cells are so thin (80–150 μm in length, 3–5 μm in width) that their mechanical skinning seems impossible.

In presence of a low free [Ca^{2+}], the myofibrils of the skinned cells developed a strong adherence to the glass. In a previous stage of the technique's development, the skinned cell was attached to a glass microtool and was transferred to a perfusion chamber that was not contaminated by other cellular elements. After transfer, it was possible to detach the cell by gentle pulling (10). However, this detachment appeared to damage some myofibrils by overstretching sarcomeres. Furthermore, when the cell was re-attached to glass microneedles for tension recording, the adherence of the myofibrils to the glass was not as strong as it was for the first attachment. For these reasons, the transfer of the skinned cell was abandoned. In contrast, the other broken cells and the large bundles of myofibrils present in the homogenate were attached to a glass microtool and were removed from the perfusion chamber. Consequently, the skinned cell was the only major piece of tissue present in the perfusion chamber.

RECORDING OF CONTRACTION

Each end of the skinned cell was attached to a glass microneedle. The cell was impaled by the microneedles, and in a few seconds the myofibrils developed such a strong adherence to the glass that excessive stretching caused the cell to break along its length rather than detach itself from the microneedles. One microneedle was immobilized, and the other was directly connected to the light 10-cm brass lever of the transducer. The mobile microneedle impaled the entire cross-section of the cell, and its direction was kept strictly perpendicular to the main axis of the cell in order to record the full tension developed by the myofilaments (fig. 5). In some experiments, the fixed end of the cell was pressed against the bottom of the chamber rather than suspended in the liquid. The microneedles were made of 2 mm external diameter Corning glass tubing, pulled with a David Kopf micropipette puller. The tips of the microneedles were

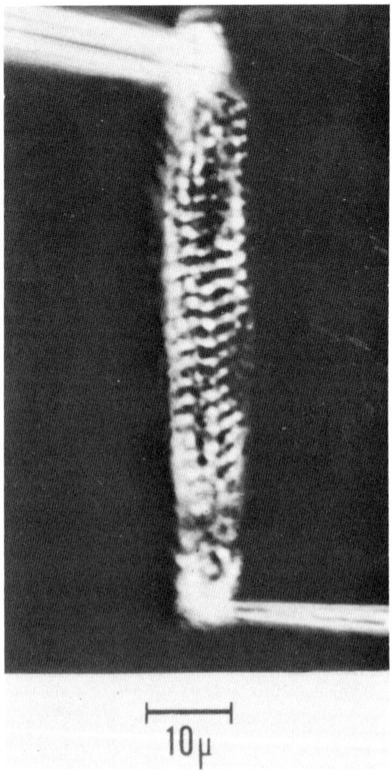

Figure 5 Attachment of the glass microneedles for tension recording. Micrographs obtained in anoptral phase contrast. The lower microneedle was attached to the transducer. Sarcomere length was 2.1 μm.

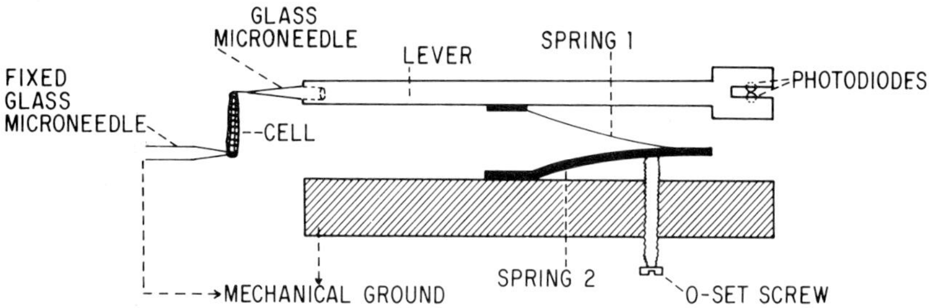

Figure 6 Transducer for tension recording, built according to plans given by R. A. Meiss.

heat-occluded with a De Fonbrune microforge. The microneedle attached to the transducer had a 1-μm tip. The microneedle attached to the fixed end of the cell had a 3-μm tip.

The free end of the transducer lever was enlarged in an opaque vane that had a 1-mm horizontal slit in its center. The light given by an infrared light-emitting diode (General Electric type SSL-5C) passed through the slit and illuminated two photodiodes (Texas Instruments, H-38). The diameter of each photodiode was 1 mm. Therefore, each of them was half-illuminated when the slit was in the central position. Any force applied to the lever resulted in a slight movement of the vane. Consequently, the illumination of one of the photodiodes was increased whereas that of the other was decreased. A Wheatstone bridge permitted the differential change in illumination to be reflected as change in electrical resistance. To increase sensitivity, the center of the lever was attached to a brass spring that served as a pivot and a restoring force. The sensitivity was 500 mV/mg, the linearity was 99 percent for a 10-mg range, the drift was 0.002 mg/min, and the natural frequency was 50 Hz. The compliance of the transducer was 2 μm/mg. The microtools were responsible for an additional compliance, and the total compliance was 4–8 μm/mg. The maximum force developed by a 10.5 μm wide and 50 μm long cell was 0.57 ± 0.05 mg (S.D.). Consequently, the shortening observed during contraction was smaller than 10 percent of the initial length of the cell. This highly sensitive isometric transducer was made in accordance with plans given by Meiss (27) (fig. 6).

Before and after each experiment, the transducer was calibrated in exactly the same position as for tension recording. The long arm of a lever with 1:40 ratio was placed in contact with the glass microneedle to which the cell was attached. The fulcrum of the lever was made from the main movement of a watch. A weight of 20 mg was applied to the short arm of the lever. Consequently, a calibration of 0.5 mg was obtained. Micromanipulations for the skinning and attachment of the glass microneedles for tension recording were done with three Leitz micromanipulators. The transducer was attached to one of these micromanipulators. The microscope and all the micromanipulators were placed on a vibration-free table (Vibrostat, Milo, Freeport, N.Y.) in a copper Faraday cage to avoid electrical interference and in a Plexiglas cage to avoid mechanical interference caused by air movement. The force transducer was protected by an additional shielding against air currents and temperature drift. When fully activated by a free [Ca^{2+}] $10^{-5.0}$ M at a sarcomere length of 2.3 μm, a single cardiac cell of rat ventricle (10.5 μm width and 7.8 μm thickness) developed a tension of about 0.57 mg.

Cyclic contractions observed in slight EGTA buffering could be localized or could propagate within the cell (9, 10). Accordingly, they might not be studied accurately from the simple tension measured from the end of the cell. Two additional techniques of contraction recording were used: 1) propagation velocity of the contraction was measured with a closed circuit television; 2) localized contractions were recorded with a photomuliplier (A. O. Reichert spectrophotometer) that was attached to the microscope. The photomultiplier recorded the variation of light transmittance occurring during the contraction of the 2-μm diameter area encompassed by the photomultiplier diaphragm. The anoptral phase contrast was necessary to obtain a good photomultiplier recording of the cell contraction. With this type of contrast, the image of the cell was brighter than the background, and the light transmittance increased during contraction. In a small single cell, the amplitude variations of the localized photomultiplier recording followed the tension variations recorded from the ends of the cell (fig. 3 of Reference 10). The electrical outputs of both the photomultiplier and the transducer were filtered above 50 Hz and were displayed on a d.c. linear penwriter (Hewlett-Packard, Inc., Palo Alto, Calif.) as well as on an oscilloscope. Additionally, the first derivative of the tension was recorded.

A high resolution television camera (GPL 990) was attached to the microscope and recorded the image of the contracting cell. This camera had a silicon tube, which was selected for its high sensitivity and short persistence of the image. Furthermore, the contrast was increased electronically by an image enhancer (Dynascience 832). The magnification of this image was adjusted by a specially developed electronic zoom. An Audiotronic television camera with an ordinary vidicon tube viewed the oscilloscope screen. A special effects generator (Shintron, Boston, Mass.) mixed the images of the two television cameras. This composite image was recorded on a Panasonic television monitor and stored by a 1-inch videotape recorder (Panasonic Videomaster). The optical system is diagrammatically represented in figure 7.

Sarcomere striations were easily visible in anoptral phase contrast (fig. 5). An appropriate optical sectioning permitted the elimination of the superimposition of the mitochondria on the sarcomeres. Distortion of the television image renders quantitative measurement of sarcomere length unreliable. For this reason, such measurements were made directly with a Unitron Filar eyepiece

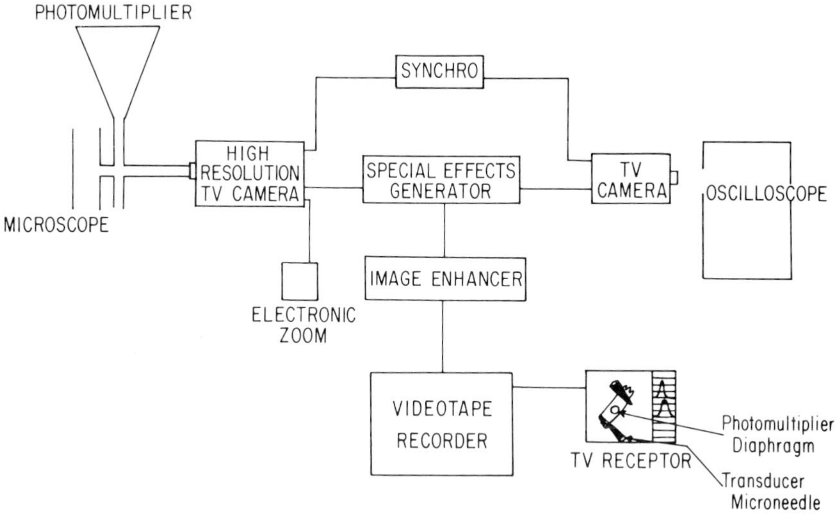

Figure 7 Closed circuit television system.

micrometer. This apparatus comprised a drum which was turned to move a scale with reference to a fixed line. This line was superimposed on a clear striation. By turning the drum, the zero of the scale was superimposed on the 11th striation. The distance between these two striations was given by division of the drum reading by the objective power. Dividing this measurement by 10 gave the sarcomere length. Furthermore, photographs of the cell were taken to document the exact sarcomere lengths.

PERFUSION TECHNIQUE

Perfusion was done continuously with several inlets and one suction outlet (fig. 8). Each medium was brought by a different inlet under pressure with a screw syringe. The inlets and outlet were placed at about 50 μm from the lateral sides of the cell. All of the inlets were placed on one side and the outlet on the other. The pipettes used for perfusion were held by Narshighe micromanipulators. Under the microscope, their openings were placed in focus with the cell. The perfusion medium was not recirculated. Complete change of the liquid contained in the 0.05-ml chamber was achieved in 3 sec. However, the cell was placed in the stream of the perfusion, and the full effect of a new medium on the tension developed by a skinned cell could be observed less than 1 sec after the introduction of this medium.

In the experiment represented in figure 9, the delay necessary for complete change of the medium bathing the cell was measured by injecting a dye.

Figure 8 Perfusion system.

Simultaneously, a Ca^{2+}-containing solution was injected, and the pattern of tension development was recorded with the force transducer from a 5 μm wide, 30 μm long fragment of skinned cell. Initially the cell was perfused through one inlet with a relaxing solution containing 4.0 mM total [EGTA], $10^{-9.0}$ M free [Ca^{2+}]. At the time indicated in figure 9 by the arrow, this perfusion was interrupted, and another was begun through another inlet with a contracting buffer containing 4.0 mM total [EGTA] and $10^{-5.8}$ M free [Ca^{2+}] plus 0.1 mg/ml of phenol red. The microscope was set in bright field, and a photomultiplier was adjusted to a wave length of 500 nm with a 30-μm diaphragm encompassing the entire cell. With these microscope and photomultiplier settings, the photomultiplier recorded the opacification resulting from dilution of the phenol red. The variation of light transmittance due to the contraction of the cell was negligible. Opacification by the dye began less than 0.01 sec after the beginning of the perfusion and reached its plateau in 1 sec. Tension development started after 0.1 sec and was maximal in 1 sec. A free [Ca^{2+}] of $10^{-5.8}$ M corresponded to incomplete activation of the myofilaments (maximum activation would have required a free [Ca^{2+}] of $10^{-5.25}$ M). Therefore, the plateau of tension observed with $10^{-5.8}$ M free [Ca^{2+}] was not due to a saturation of the myofilaments with Ca^{2+}, but corresponded to the complete equilibration between free [Ca^{2+}] in the myofilament space and in the medium. In skinned cells of larger diameter, a more important delay was observed before the plateau of tension. This additional delay may be explained by binding and release of Ca^{2+} by the sarcoplasmic reticulum (12, 13, 17) and by Ca^{2+}-binding sites within the contractile proteins themselves (1, 32).

PARAMETERS MEASURING THE FUNCTION OF EACH ORGANELLE CONTRIBUTING TO ACTIVATION OF SKINNED CARDIAC CELLS

The use of high and low [EGTA] buffering, respectively, permitted separate study of the function of myofilaments and internal stores of Ca^{2+} in skinned cardiac cells.

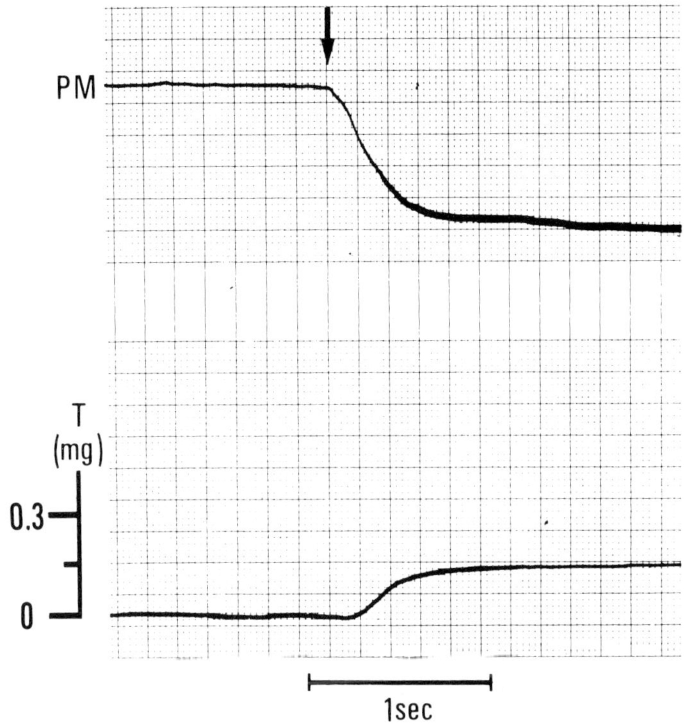

Figure 9 Measure of the delay necessary for changing the medium bathing the skinned cell. *Upper tracing:* photomultiplier recording of the opacification due to phenol red; *lower tracing:* tension recording from a 5 μm width skinned cell of rat ventricle.

Myofilaments

The function of the myofilaments was studied by the curve of tension as a function of free [Ca^{2+}] in the presence of 4.0 mM total [EGTA]. In the presence of $10^{-3.5}$ M free [Mg^{2+}] and $10^{-2.5}$ M [$MgATP^{2-}$] no tension was developed at $10^{-6.75}$ M free [Ca^{2+}], 10 percent of the maximum tension was developed at $10^{-6.5}$ M, 20 percent at $10^{-6.25}$ M, 50 percent at $10^{-6.0}$ M, 80 percent at $10^{-5.75}$ M, 95 percent at $10^{-5.5}$ M, and 100 percent at $10^{-5.25}$ M free [Ca^{2+}]. The Ca^{2+} binding capacity of the sarcoplasmic reticulum is likely to be small when compared with 4.0 mM EGTA (41). Any amount of Ca^{2+} released or sequestered by the sarcoplasmic reticulum would be small when compared with 4.0 mM total [EGTA] and would not modify the free [Ca^{2+}] bathing the myofilaments. Therefore, the tension developed at a given free [Ca^{2+}] set in the buffer represented the direct effect of this free [Ca^{2+}] on the myofilaments. The curve of the tension versus the free [Ca^{2+}] was also obtained in skinned cells that were treated for ½–1 hr in a relaxing solution containing 1 percent Brij 58, a total [EGTA] of 4.0 mM, and a free [Ca^{2+}] lower than 10^{-9} M. According to studies done on skeletal muscle, ½ hr of treatment with 1 percent Brij 58 should be sufficient to destroy the sarcoplasmic reticulum almost completely (32). In skinned cardiac cells, treatment for 1 hr with 1 percent Brij 58 resulted in no significant modification of the curve of tension as a function of free [Ca^{2+}] obtained in the presence of 4.0 mM total [EGTA]. This result supports the assumption that internal stores of Ca^{2+} play no role in the presence of 4.0 mM total [EGTA] and confirms that the force-free [Ca^{2+}] curve obtained under these conditions represents the direct effect of Ca^{2+} on the myofilaments.

Internal stores

In presence of a total [EGTA] of less than 0.2 mM in rat ventricle, or less than 0.05 mM in rabbit ventricle, cyclic contractions were observed. The inhibition of these contractions by a total [EGTA] larger than these values suggested that they were attributable to the presence of an intracellular sink in competition with EGTA for Ca^{2+}. In rat ventricle with a total [EGTA] of 0.05 mM, free [Mg^{2+}] of $10^{-3.5}$ M, and [$MgATP^{2-}$] of $10^{-2.5}$ M, cyclic contractions were observed for a free [Ca^{2+}] of $10^{-7.40}$ M, which is much lower than the threshold for direct activation of myofilaments. Thus, the cyclic contractions corresponded to release and binding of Ca^{2+} by the intracellular stores. A slight variation of free [Ca^{2+}] in the solution, from $10^{-7.65}$ M to $10^{-7.40}$ M, was sufficient to trigger a contraction with an amplitude equal to the tension produced by the direct effect of $10^{-6.1}$ M free [Ca^{2+}] on myofilaments (*i.e.*, a 20 times larger free [Ca^{2+}]). On this basis, it was concluded that this contraction was induced by a Ca^{2+}-induced release of Ca^{2+} (8–13).

In skinned cells of rat ventricle, the internal stores were unloaded by perfusion for 5 min with 4.0 mM total [EGTA] and varied free [Ca^{2+}] from $10^{-9.0}$ to $10^{-7.0}$. The cell was then perfused with 0.05 mM total [EGTA] and $10^{-7.40}$ M free [Ca^{2+}]. The delay between the beginning of the second perfusion and the first phasic contraction was shorter when the initial perfusion contained a higher free [Ca^{2+}] (data reported in detail in a manuscript in preparation). It was concluded that the Ca^{2+} was released when the internal stores were filled to a critical free [Ca^{2+}]. Accordingly, when the internal stores were unloaded by perfusion in high total [EGTA] and lower free [Ca^{2+}], the time required to reach this critical level of free [Ca^{2+}] was longer than with the delay observed after perfusion with higher free [Ca^{2+}].

The cyclic contractions observed in low total [EGTA] were inhibited by 10 mM caffeine, which inhibits Ca^{2+} sequestration by the sarcoplasmic reticulum (43), and by the surfactant deoxycholate (17) or by the detergent Brij 58, both of which destroy the sarcoplasmic reticulum. It was concluded that the sarcoplasmic reticulum played a major role in the cyclic binding and release of Ca^{2+}.

Furthermore, the use of specific inhibitors or uncouplers of the oxidative phosphorylation and inhibitors of the passive Ca^{2+} binding by mitochondria permitted the demonstration that mitochondria played no role in this process (10). Neither azide (1–10 mM), nor dinitrophenol (0.5–5 mM), nor low concentrations of ruthenium red (0.1–1 μM) modified either the cyclic contractions or the resting tension when free [Ca^{2+}] was lower than $10^{-6.5}$ M in the presence of 0.025–0.05 mM total [EGTA].

In the absence of EGTA and for a free [Ca^{2+}] higher than 10^{-6} M, the three drugs listed above produced a shortening of the latency after unloading of the Ca^{2+} stores by perfusion in high total [EGTA], a shortening of subsequent cycles, a marked decrease in the rate of relaxation of cyclic contractions, and an increase in resting tension. These effects were similar to those produced by an increase of free [Ca^{2+}]. It was concluded that in the presence of high free [Ca^{2+}], mitochondria were capable of sequestering Ca^{2+} (10). However, these concentrations of free [Ca^{2+}] clearly were supraphysiological for the diastolic free [Ca^{2+}]. During the systole, free [Ca^{2+}] may reach this level after Ca^{2+} release from the sarcoplasmic reticulum, but it stays at this level for too short a time to permit the slow binding of Ca^{2+} by mitochondria (39) to be effective. The nonphysiological character of the Ca^{2+} binding by mitochondria under these conditions was confirmed by their massive swelling and frequent disruption under electron microscopy. The swelling was so important that it was even apparent under phase microscopy, where some mitochondria were circular with a diameter of more than 2 μm. Under electron microscopy, they presented large vacuoles and their membrane system was frequently broken (in about one out of four mitochondria after 30 min of perfusion in a medium containing no EGTA at free [Ca^{2+}] higher than 10^{-6} M).

Absence of sarcolemma

In skinned cardiac cells, no "resting potential" larger than −10 mV was observed. This small negative potential may be a junction potential or may correspond to the charges associated with the myofilaments that are responsible for the Donnan osmotic phenomena observed in skinned skeletal muscle fibers (Maughan and Godt, manuscript in preparation). Cyclic contractions of skinned cells were insensitive to the ratio between [Na^+] and [K^+] in the perfusion medium. Even after a strong preload with Ca^{2+}, no cyclic contractions were induced by change of a K^+-containing, Na^+-free solution for a Na^+-containing solution (9, 10). Similarly, cyclic contractions were not induced or modified by changing propionate for choloride or methylsulfate for chloride (9). Cyclic contractions of skinned cardiac cells were insensitive to tetrodotoxin and to concentrations of verapamil, which inhibits Ca^{2+} current across the surface membrane and contraction in intact cardiac cells (10).

In summary

The function of the myofilaments was studied by the curve of tension as a function of the free [Ca^{2+}] obtained in presence of a high total [EGTA] (4.0 mM).

The following parameters were used to study the function of the sarcoplasmic reticulum in presence of low [EGTA] (0.05 mM): 1) the lowest free [Ca^{2+}] resulting in phasic contractions in 4.0 mM total [EGTA]; 2) the latency in appearance of the first contraction after the transfer of Ca^{2+}-unloaded cells from high into low [EGTA]; 3) the duration of the subsequent cycles (inverse of frequency of the cyclic contractions); 4) the rate of contraction development and relaxation (amplitude of the first derivative of contraction).

DIFFERENCES BETWEEN SKINNED CARDIAC CELLS AND CARDIAC FIBERS WITH DISRUPTED SARCOLEMMAS

The following differences were observed between skinned cardiac cells and isolated cardiac fibers with disrupted sarcolemmas.

1) The contraction was asynchronous with propagation along the fiber on the television recording and with several components in each beat on the photomultiplier recording when the disrupted fiber was made of two broken cells separated by an intercalated disc.

2) The delay between change of media and resulting effects on contraction was longer in cardiac fibers with disrupted sarcolemma even when they were made of a fragment of a single cell. This was interpreted as caused by restriction of the diffusion of externally applied media because of the presence of the remaining pieces of sarcolemma.

3) Isoproterenol, 10^{-6} M, had nearly no effect on the cyclic contractions of skinned cardiac cells. In contrast, it did have an effect on larger fibers with disrupted sarcolemmas (4–6 broken or nonbroken cells), in which no resting potential larger than −35 mV was recorded. This effect consisted of a decrease in the frequency of cyclic contractions with an increase in amplitude of their relaxation rate (10). It was concluded that remaining pieces of sarcolemma could play a role as drug receptor sites even if they were not generating action potential.

4) Cardiac fibers with disrupted sarcolemmas made of a single broken cell presented the same low "resting" potential and the same insensitivity to drugs and monovalent ions as skinned cardiac cells. However, fragments made of several cells (*i.e,* more than 120 μm in length and 20 μm in diameter) presented quite different properties. A resting potential of up to −35 mV was observed, and was increased in high [Na^{+}] and high [Ca^{2+}] and decreased in high [K^{+}]. It was concluded that these pluricellular fragments were not mechanically skinned, and they were not used (see Discussion in Reference 9).

In summary, isolated cardiac fibers with disrupted sarcolemmas are much simpler to obtain than skinned cardiac cells, and may be useful if they are smaller than 15-μm in diameter and 60-μm in length and if they do not include an intercalated disc. The partially skinned cell preparation that was used for the initial studies (9) may be compared to the preparation of longitudinally cut and split skinned skeletal muscle fibers recently developed by Nakajima and Endo (28). In contrast, multicellular cardiac fragments cannot be used in the absence of a chemical skinning (45).

OTHER PREPARATIONS OF PARTIALLY SKINNED MYOCARDIUM

Winegrad (45) developed a preparation of chemically skinned myocardium using the high Ca^{2+} permeability produced by treatment with ethylenediaminetetraacetic acid (EDTA). Winegrad defined critical conditions of buffer composition and demonstrated that the high membrane permeability was progressively reversed in presence of high free [Ca^{2+}]. With this preparation, the sensitivity of the myofilaments to Ca^{2+} and a curve of tension as a function of the free [Ca^{2+}] was obtained (45). Using this preparation, Solaro *et al.* (42) demonstrated that the curve of tension as a function of the free [Ca^{2+}] was similar to the curve of ATPase activity of isolated myofibrils as a function of free [Ca^{2+}], except for the low levels of free [Ca^{2+}]. This preparation was also used for pharmacological studies by Nayler (31), who demonstrated that the curve of tension as a function of the free [Ca^{2+}] was not modified by ouabain. In contrast, caffeine increased the contracture tension developed at a given feee [Ca^{2+}]. Nayler concluded that 1) the positive inotropic effect of ouabain, which was observed in intact tissue, was related to the surface membrane; and 2) the increase of tension induced by caffeine may reflect an inhibition of Ca^{2+} uptake in the sarcoplasmic reticulum. However, Nayler stressed that this interpretation depended upon the unproven assumption that the sarcoplasmic reticulum retains its ability to accumulate and exchange Ca^{2+} under these conditions. Our results suggest that, even if the sarcoplasmic reticulum were capable of exchanging Ca^{2+} in high [EGTA] or [EDTA] (46), the amount of Ca^{2+} exchanged would be small as compared with the 3 mM total [EGTA] present in the buffer used by Nayler and would not modify the tension. In single skinned cardiac cells, ouabain had no effect on any of the previously defined parameters of the cyclic contractions observed in low total [EGTA], thus demonstrating its lack of effect on Ca^{2+} binding and release by the sarcoplasmic reticulum (10). Furthermore, ouabain had no effect on the curve of tension as a function of free [Ca^{2+}] obtained in high total [EGTA] and therefore, had no direct effect on the myofilaments. Caffeine, 1–2 mM, produced a decrease in rate of relaxation and of tension development, and an increase in resting tension. Caffeine, 10 mM, inhibited cyclic contractions. It was concluded that caffeine inhibited Ca^{2+}

binding and induced Ca^{2+} release from the sarcoplasmic reticulum (10). Caffeine also produced a shift to the left in the curve of tension as a function of the free [Ca^{2+}] obtained in presence of 4.0 mM EGTA so that a higher tension was developed at a given free [Ca^{2+}], as shown by Nayler (31). In cells treated for 1 hr with 1 percent Brij 58, the same shift of the tension-free [Ca^{2+}] curve was observed as in the absence of Brij 58 (data to be published in detail). Since this detergent destroyed the sarcoplasmic reticulum, the effect of caffeine on the contracture tension that was observed in presence of a high total [EGTA] may not be related to its effect on Ca^{2+} binding and release by the sarcoplasmic reticulum. Caffeine could increase the sensitivity of the myofilaments to Ca^{2+}, or it could release Ca^{2+} from binding sites that are different than those generating the cyclic contractions. These binding sites might be associated with the contractile proteins (14, 32). It seems to us that the results reported by Nayler (31) demonstrate this effect of caffeine on the myofilaments rather than the effect of caffeine on the sarcoplasmic reticulum. In addition, a release of Ca^{2+} from the mitochondria might be induced by caffeine under the conditions used by Nayler (31). The low affinity binding sites of mitochondria may have a large enough capacity to permit competition with 3 mM EGTA (46).

In a recent report, Kerrick and Best (20) used the technique of cardiac fibers with disrupted sarcolemmas. However, they used multicellular muscles, which cannot be considered to be composed of skinned cells. The tension shown in the figure of this report is about 25 times larger than that obtained in a single skinned cardiac cell. Furthermore, it was indicated that the force per cross-section area was about 20–50 percent of that developed by skinned skeletal muscle fibers, while single skinned cardiac cells developed 64 percent of the tension developed by skinned skeletal muscle fibers. Thus, the diameter of the preparations used by Kerrick and Best (20) could be 50–100 μm. The attachment of the muscle between the tips of two forceps (17) may require a total length of more than 1 mm. The preparation could, therefore, be made of one or several hundred cells. It is interesting to note that other authors (24, 25) used cardiac preparations of even smaller dimensions than those used by Kerrick and Best, and that they do not consider these preparations as skinned. They even use them to study transsarcolemmal ionic currents with a voltage-clamp technique (24, 25). Therefore, the preparation used by Kerrick and Best (20) cannot be considered as skinned unless the homogenization somehow destroys the membranes of the cells located in the center of the preparation. Under electron microscopy we never found any evidence of surface membrane disruptions in nonperipheral cells of pluricellular fragments that were obtained by homogenization. We concluded that, in the presence of low free [Ca^{2+}], the preparation used by Kerrick and Best was "chemically" skinned, although not as properly by EGTA as it would be with EDTA (45). When the cell was bathed in high free [Ca^{2+}] ($10^{-4.5}$ M), the increased permeability to Ca^{2+} was progressively reversed.

Any new medium was applied on a surface membrane having recuperated part of its selective permeabilities.

Kerrick and Best transferred the muscle from 1) a relaxing solution containing propionate as major anion and K^+ as major cation, into 2) a contracting solution in high total [EGTA] (7.0 mM), with a free $[Ca^{2+}]$ of $10^{-4.5}$ M (the stability constant for CaEGTA was different from that used in our studies) with propionate and K^+, then into 3) a low [EGTA] (0.05 mM) solution at free $[Ca^{2+}]$ of $10^{-6.4}$ M, with propionate and K^+ in which no contraction was observed, and finally into 4) a low [EGTA] (0.05 mM) solution at free $[Ca^{2+}]$ of $10^{-6.4}$ M with either Cl^- as principal anion or Na^+ as principal cation, where a phasic contraction was observed. They concluded that this phasic contraction was attributable to a Ca^{2+} release from the sarcoplasmic reticulum. We tried to reproduce these experiments using the same buffers in single skinned cardiac cells. We took into account the stability constants and added the ATP regenerating system that they used. In contrast to the experiments of Kerrick and Best, we always observed cyclic contractions in the low [EGTA] potassium propionate medium, and these cyclic contractions were only slightly increased in amplitude by subsequent perfusion in Cl^- or Na^+ medium. The small increased amplitude of the contractions following the replacement of propionate by Cl^- in the perfusion cannot be attributed to a release of Ca^{2+} from the sarcoplasmic reticulum, since we observed that this ionic substitution causes the myofilaments to be more sensitive to Ca^{2+} in the presence of a low free $[Ca^{2+}]$. The increased amplitude of the contraction following the replacement of K^+ by Na^+ was not statistically significant. Therefore, our results obtained in skinned cardiac cells do not confirm those obtained in multicellular preparations by Kerrick and Best and do not demonstrate a chloride- or sodium-induced release of Ca^{2+} from the sarcoplasmic reticulum.

Our negative results contrast with the observations reported on skinned fibers of skeletal muscle. A Cl^--induced release of Ca^{2+} has been described in skinned fibers of skeletal muscle (6, 7, 28). The release of Ca^{2+} was attributed to a depolarization of the membrane of the sarcoplasmic reticulum (28). The development by Nakajima and Endo (28) of a new technique to obtain partially skinned fibers without healing over of the transverse tubules permitted the elimination of previous hypotheses (6) and the conclusion that the Ca^{2+} was released from the sarcoplasmic reticulum. This was confirmed with the use of the Ca^{2+}-bioluminescent protein aequorin (7). A large load of the internal Ca^{2+} stores was required to observe this phenomenon, and the resulting contraction was smaller when the internal stores were less completely filled (28). The sarcoplasmic reticulum of cardiac muscle has a smaller binding capacity than does skeletal muscle. Previously reported data indicated that the Ca^{2+} binding and release in skinned cardiac muscle cells were similar to the Ca^{2+} binding and release of skeletal muscle fibers when the latter were treated with 10 mM caffeine (9). It was impossible for us to clearly demonstrate contractions by Cl^--induced release of Ca^{2+} in skinned fibers of skeletal muscle treated with 10 mM caffeine. It was not surprising, therefore, that Cl^--induced release of Ca^{2+} could not be demonstrated in skinned cardiac cells. It was concluded that the cardiac sarcoplasmic reticulum was not capable of maintaining its Ca^{2+} load in K^+-propionate, as demonstrated by the cyclic contractions observed in this medium. This negative result cannot be attributed to the damage of the sarcoplasmic reticulum in skinned cardiac cells, since we also observed cyclic contractions in improperly chemically skinned trabeculae of 50–100 μm diameter. Applying the same series of perfusion media used by Kerrick and Best (22) to 50–100 μm diameter trabeculae of rat ventricle, we also observed localized cyclic contractions under the microscope in the low total [EGTA] potassium propionate medium. These localized and asynchronous contractions may be overlooked on the tension recording obtained from the ends of these large preparations. It must be stressed that this negative result does not mean that a Cl^--induced release of Ca^{2+} does not exist in skinned cardiac cells, but simply that it was impossible for us to demonstrate it under the conditions described by Kerrick and Best (20). For instance, change of methylsulphate for chloride might be more efficient than the change of propionate for chloride as shown in skeletal muscle by Nakajima and Endo (28).

Our failure to demonstrate a Na^+-induced release of Ca^{2+} from the sarcoplasmic reticulum of skinned cardiac cells is not very surprising. As pointed out by Kerrick and Best (20), a Na^+-induced release of Ca^{2+} was described in isolated cardiac microsomes (33), but this result has not been confirmed by other studies using purified preparations (19).

It remains now to explain the positive results obtained by Kerrick and Best (1974) on multicellular preparations and interpreted by these authors as a demonstration of a Cl^-- or Na^+-induced release of Ca^{2+} from the sarcoplasmic reticulum. In skinned cardiac cells the amplitude of the phasic contractions caused by a release of Ca^{2+} from the sarcoplasmic reticulum was always much smaller than the tonic tension elicited by the full activation of the myofilaments. The large phasic contractions observed by Kerrick and Best (20) represent, therefore, an intriguing phenomenon. Preliminary results suggest that they may be triggered by action potentials. In 50–100 μm diameter trabeculae of rat ventricle bathed in 10^{-7} M free [Ca^{2+}] and 0.05 mM total [EGTA] with potassium propionate, we observed a resting potential of about −20 mV. A perfusion with the same medium with Cl^- replacing the propionate resulted in an increase of this resting potential to −40 mV, sometimes followed by the development of a slow diastolic depolarization and a small action potential. This action potential was associated with a contraction of large amplitude similar to that observed by Kerrick and Best (20). The increase of Cl^- in the extracellular space may result in an increase of resting potential. This "repriming" of the membrane may permit the development of an automatism. Cyclic contractions of small amplitude not related to any fluctuation of the membrane potential were superimposed on this large contraction, which was triggered by an action potential. The small cyclic contractions, which were similar to those previously described, corresponded to cyclic releases of Ca^{2+} from the sarcoplasmic reticulum.

The large phasic contractions observed by Kerrick and Best (20) after replacement of K^+ by Na^+ also may be related to the sarcolemma. In the presence of the surface membrane, an increase of extracellular $[Na^+]$ produces a decrease in tension (26). However, this antagonism between Na^+ and Ca^{2+} might be limited to a definite range of Na^+ concentrations, and inverse effects might be produced when the initial medium is Na^+ free (5). In multicellular preparations, the interstitium and the basement membrane could represent a major Ca^{2+} pool, as it is under physiological conditions (22). During the bathing in high total [EGTA] and high free $[Ca^{2+}]$, as used by Kerrick and Best (20), these superficial binding sites may be loaded and may represent a larger Ca^{2+} pool than the sarcoplasmic reticulum, even if they have a lower affinity for Ca^{2+}. More experiments are needed in order to understand if and how Na^+ and Cl^- can produce the motion of Ca^{2+} from these sites into the myofilaments.

Bloom *et al.* (2) also used a pluricellular preparation, the dimensions of which corresponded to about 50 cells. They did not claim that the sarcolemmas of all the cells were mechanically disrupted. Since the number of cells is smaller than in the study of Kerrick and Best (20), the preparation may be intermediary between the disrupted cells and the muscles used by these authors. If one cell is disrupted, the extracellular medium will diffuse from the myoplasm of this cell into neighboring cells at the level of the intercalated discs, which present a high permeability (34, 44). However, this diffusion is limited to a small number of cells and is in competition with a healing-over process, which reduces the permeability of the intercalated discs. In low free $[Ca^{2+}]$, the preparation of Bloom *et al.* (2) was highly permeable to Ca^{2+}. When free $[Ca^{2+}]$ was increased to the high level used by these authors, a complex situation resulted, dominated by the Ca^{2+} overload of both sarcoplasmic reticulum and mitochondria. Overload of the sarcoplasmic reticulum was already achieved by homogenization in a medium which was free of EGTA and contained oxalate that penetrated into the sarcoplasmic reticulum and attracted Ca^{2+}. Cyclic contractions, not triggered by action potentials, similar to those observed by Bloom *et al.* (2), occurred also in intact muscles overloaded with Ca^{2+} (see Reference 9, for review) and can be attributed to movements of Ca^{2+} between the sarcoplasmic reticulum and the myofilaments. Their existence indicates that the sarcoplasmic reticulum is competing with the myofilaments for the excess of Ca^{2+} present in the myoplasm, and does not implicate a triggering by the Ca^{2+} influx from the extracellular space as suggested by Bloom *et al.* (2). The free $[Ca^{2+}]$ used by Bloom *et al.* (2) was generally much higher than required to activate the myofilaments. The phenomena to be explained are not the cyclic contractions but the cyclic relaxations. It is likely that Ca^{2+} binding by the sarcoplasmic reticulum was responsible for the cyclic relaxations that were observed. This competition between sarcoplasmic reticulum and myofilaments for Ca^{2+} in the presence of a low [EGTA], or in total absence of EGTA, prevents any measure of the sensitivity of the myofilaments to Ca^{2+}. At a free [Ca^{2+}], $10^{-4.0}$ M, cyclic contractions disappeared and a contracture without oscillations was obtained. This free [Ca^{2+}] is in fact 20 times greater than that required to fully activate the myofilaments in the presence of a high total [EGTA]. It would correspond to the overtaxing of the binding capabilities of the sarcoplasmic reticulum. The observation of Bloom *et al.* (2) that $10^{-3.0}$ M of lanthanum had the same effect in the presence of 10^{-5} M of free [Ca^{2+}] may simply indicate that lanthanum inhibits Ca^{2+} binding by the sarcoplasmic reticulum and does not warrant that La^{3+} has a 10-fold lower affinity for troponin than Ca^{2+}. This study demonstrates an increase of resting tension in the presence of oligomycin. However, part of this effect may be explained by the specific action of the ethanol which is used as a solvent of oligomycin. The absence of modifications of the cyclic contractions by oligomycin in the study of Bloom *et al.* (2) cannot be explained on the basis of our results obtained in skinned cardiac cells.

It seems that the use of a "chemical" skinning, in accordance with the technique described by Winegrad (45) on these very thin but pluricellular muscle fragments, could permit a precise control of the intracellular free [Ca^{2+}] in the myofilament space. However, mechanically skinned cardiac cells or single cells with disrupted sarcolemmas represent a simpler preparation to study the function of the sarcoplasmic reticulum in cardiac activation. This study of the sarcoplasmic reticulum function requires the use of very low concentrations of EGTA which do not seem to be compatible with a chemical skinning even when detergents are added to the "skinning solution." In summary, the techniques of "chemical" skinning permit valuable insight into the activation process of cardiac cells but have some limitations, as defined by Winegrad (45). It seems that these limitations have to be considered when interpreting the results obtained with preparations that are "partially mechanically" and "partially chemically" skinned, such as those used by Bloom *et al.* (2) and Kerrick and Best (20).

Acknowledgments The authors thank Dr. R. A. Meiss (Indiana University, Indianapolis, Ind.) for his help in the development of the tension transducer.

REFERENCES

1. ASHLEY, C. C., and MOISESCU, D. G. 1973. Tension changes in isolated bundles of frog and barnacle myofibrils in response to sudden changes in the external free calcium concentration [Abstract]. J. Physiol. (London) 233: 8P–9P.

2. BLOOM, S., BRADY, A. J., and LANGER, G. A. 1974. Calcium metabolism and active tension in mechanically disaggregated heart muscle. J. Molec. Cell Cardiol. 6: 137–148.
3. BRANDT, P. W., REUBEN, J. P., and GRUNDFEST, H. 1972. Regulation of tension in the skinned crayfish muscle fiber. II. Role of calcium. J. Gen. Physiol. 59: 305–317.
4. BREMEL, R. D., and WEBER, A. 1972. Cooperation within actin filaments in vertebrate skeletal muscle. Nature New Biol. 238: 97–101.
5. BUSSELEN, P., and CARMELIET, E. 1973. Protagonistic effects of Na and Ca on tension development in cardiac muscle in low extracellular Na concentrations. Nature New Biol. 243: 57–59.
6. COSTANTIN, L. L., and PODOLSKY, R. J. 1967. Depolarization of the internal membrane system in the activation of frog skeletal muscle. J. Gen. Physiol. 50: 1101–1124.
7. ENDO, M., and BLINKS, J. R. 1973. Inconstant association of aequorin luminescence with tension during calcium release in skinned muscle fibers. Nature New Biol. 246: 218–220.
8. ENDO, M., TANAKA, M., and OGAWA, Y. 1970. Calcium induced release of calcium from the sarcoplasmic reticulum of skinned skeletal muscle fibers. Nature (London) 228: 34–36.
9. FABIATO, A., and FABIATO, F. 1972. Excitation-contraction coupling of isolated cardiac fibers with disrupted or closed sarcolemmas: Calcium dependent cyclic and tonic contractions. Circ. Res. 31: 293–307.
10. FABIATO, A., and FABIATO, F. 1973. Activation of skinned cardiac cells. Subcellular effects of cardioactive drugs. Europ. J. Cardiol. 1: 143–155.
11. FORD, L. E., and PODOLSKY, R. J. 1970. Regenerative calcium release within muscle cells. Science 167: 58–59.
12. FORD, L. E., and PODOLSKY, R. J. 1972. Calcium uptake and force development by skinned muscle fibers in EGTA buffered solutions. J. Physiol. (London) 223: 1–19.
13. FORD, L. E., and PODOLSKY, R. J. 1972. Intracellular calcium movements in skinned muscle fibers. J. Physiol. (London) 223: 21–33.
14. FUCHS, F. 1970. Striated muscle. Ann. Rev. Physiol. 36: 461–502.
15. GIRARDIER, L., and FORSSMANN, W. G. 1970. Study of the T system in rat heart. J. Cell Biol. 44: 1–19.
16. GORDON, A. M., GODT, R. E., DONALDSON, S. K. B., and HARRIS, G. E. 1973. Tension of skinned frog muscle fibers in solution of varying ionic strength and neutral salt concentration. J. Gen. Physiol. 62: 550–574.
17. HELLAM, D. C., and PODOLSKY, R. J. 1969. Force measurements in skinned muscle fibers. J. Physiol. (London) 200: 807–819, 1969.
18. HENDERSON, A. H., BRUTSAERT, D. L., PARMLEY, W. W., and SONNENBLICK, E. H. 1969. Myocardial mechanics in papillary muscle of the rat and cat. Amer. J. Physiol. 217 (5): 1273–1279.
19. KATZ, A. M., and REPKE, D. I. 1967. Quantitative aspects of dog cardiac microsomal calcium binding and calcium uptake. Circ. Res. 21: 153–162.
20. KERRICK, W. G. L., and BEST, P. M. 1974. Calcium ion release in disrupted heart cells. Science 183: 435–437.
21. KERRICK, W. G. L., and DONALDSON, S. K. B. 1972. The effects of Mg^{2+} on submaximum Ca^{2+}-activated tension in skinned fibers of frog skeletal muscle. Biochem. Biophys. Acta 275: 117–122.
22. LANGER, G. A. 1973. Heart: Excitation-contraction coupling. Ann. Rev. Physiol. 35: 55–86.
23. LEGATO, M. J. 1973. Ultrastructure of the atrial, ventricular, and Purkinje Cell with special reference to the genesis of arrhythmias. Circulation 47:178–204.
24. LEOTY, C. 1974. Membrane current and activation of contraction in rat ventricular fibres. J. Physiol. (London) 239: 237–249.

25. LEOTY, C., and RAYMOND, G. 1972. Mechanical activity and ionic currents in frog atrial trabeculae. Pfluger's Arch. ges. Physiol 334: 114–129.
26. LUTTGAU, H. C., and NIEDERGERKE, R. 1959. The antagonism between Na and Ca ions on the frog's heart. J. Physiol. (London) 143: 486–505.
27. MEISS, R. A. 1974. A versatile transducer system for mechanical studies of muscle. J. Appl. Physiol. 37: 459–463.
28. NAKAJIMA, Y., and ENDO, M. 1973. Release of calcium induced by depolarization of the sarcoplasmic reticulum membrane. Nature New Biol. 246: 217–218.
29. NANNINGA, L. B. 1961. Association constant of the complexes of adenosine triphosphate with magnesium, calcium, strontium, and barium ions. Biochim. Biophys. Acta 54: 330–338.
30. NATORI, R. 1954. Property and contraction process of isolated myofibrils. Jikei Kai Med. J. I: 119–126.
31. NAYLER, W. G. 1973. Effect of inotropic agents on canine trabecular muscle rendered highly permeable to calcium. Amer. J. Physiol. 225: 918–924.
32. ORENTLICHER, M., REUBEN, J. P., GRUNDFEST, H., and BRANDT, P. W. 1974. Calcium binding and tension development in detergent-treated muscle fibers. J. Gen. Physiol. 63: 168–186.
33. PALMER, R. F., and POSEY, V. A. 1967. Ions effects on calcium accumulation by cardiac sarcoplasmic reticulum. J. Gen. Physiol. 50: 2085–2096.
34. POLLACK, G. H., and HUNTSMAN, L. L. 1973. Intracellular pathways in the heart: Direct evidence of low resistance channels. Experientia, 29: 1501–1502.
35. PORTZEHL, H., CALDWELL, P. C., and RUEGG, J. C. 1964. Dependence of contraction and relaxation of muscle fibers from the crab *Maia squinado* on the internal concentration of free calcium ions. Biochim. Biophys. Acta 79: 581–591.
36. REUBEN, J. P., BRANDT, P. W., BERMAN, M., and GRUNDFEST, H. 1971. Regulation of tension in the skinned crayfish muscle fiber. I. Contraction and relaxation in the absence of Ca^{2+} (pCa > 9). J. Gen. Physiol. 57: 385–407.
37. REUBEN, J. P., BRANDT, P. W., and GRUNDFEST, H. 1967. Tension evoked in skinned crayfish muscle fibers by anions, pH and drugs [Abstract]. J. Gen. Physiol. 50: 2501.
38. RINGBOM, A. 1963. Complexation in Analytical Chemistry, Vol. XVI. Interscience Publishers, Division of John Wiley & Sons, Inc, New York.
39. SCARPA, A., and GRAZIOTTI, P. 1973. Mechanisms of intracellular calcium regulation in heart. I. Stopped-flow measurements of Ca^{2+} uptake by cardiac mitochondria. J. Gen. Physiol. 62: 756–772.
40. SCHWARTZENBACH, G., SENN, H., and ANDEREGG, G. 1957. Komplexone. XXIX. Ein grösser chelateffekt besonderer Art. Helv. Chim. Acta 40: 1186–1900.
41. SOLARO, J., and BRIGGS, F. N. 1974. Estimating the functional capabilities of sarcoplasmic reticulum in cardiac muscle: Calcium binding. Circ. Res. 39: 531–540.
42. SOLARO, J., WISE, R. M., SHINER, J. S., and BRIGGS, F. N. 1974. Calcium requirements for cardiac myofibrillar activation. Circ. Res. 39: 525–530.
43. WEBER, A., and HERZ, R. 1968. The relationship between caffeine contracture of intact muscle and the effect of caffeine on reticulum. J. Gen. Physiol. 52: 750–759.
44. WEIDMAN, S. 1966. The diffusion of radiopotassium across intercalated disks of mammalian cardiac muscle. J. Physiol. (London) 187: 323–342.
45. WINEGRAD, S. 1971. Studies of cardiac muscle with a high permeability to calcium produced by treatment with ethylenediaminetetraacetic acid. J. Gen. Physiol. 58: 71–93.
46. WINEGRAD, S. 1973. Intracellular calcium binding and release in frog heart. J. Gen. Physiol. 62: 693–706.

Mailing Address: Drs. Alexandre and Francoise Fabioto,
Cardiovascular Research Laboratory,
Harvard Medical School,
Boston, Massachusetts (USA).

Recent Advances in Studies on Cardiac Structure and Metabolism
Volume 9
The Sarcolemma
Edited by Paul-Emile Roy and Naranjan S. Dhalla

EFFECTS OF CALCIUM-FREE PERFUSION ON EXCITATION-CONTRACTION COUPLING IN HEART AND SKELETAL MUSCLE[1]

T. L. RICH and G. A. LANGER

Department of Physiology and the Los Angeles County
Cardiovascular Research Laboratory,
University of California at Los Angeles,
Los Angeles, California, USA

SUMMARY

Interventricular septa and semimembranosus accessorius muscles (SMA) from the guinea pig and frog gastrocnemius (FG) muscles were vascularly perfused and isolated for mechanical and radioisotope flux measurements. Since calcium deprivation causes progressive electrical inexcitability, 100 μM La was included in all calcium-free solutions (0-Ca, 100-La) to stabilize the membrane. Perfusion of ventricular septa with such solutions resulted in tension (P) and dP/dt declines with half-times ($t_{1/2}$) = 11 sec, similar to the most rapid phase of ^{45}Ca efflux. In contrast, the major component of tension decline during 0-Ca, 100-La perfusion of the SMA and FG muscles was much slower ($t_{1/2}$ = 33 min). The rate of the slow component of ^{45}Ca efflux from the SMA and FG muscles was essentially identical with the rate of the major tension decline. Significant tension (10–16 percent) was still present in these perfused skeletal muscles after 1–2 hr of 0-Ca, 100-La perfusion. Subsequent addition of La (200 μM total La) did not change the slope of P decline significantly for an additional 30–45 min. Addition of up to 2 mM EGTA after 30 min 0-Ca perfusion (calculated extracellular [Ca^{2+}] <3 × 10^{-9} M) altered but slightly the rate of P decline. These experiments emphasize the difference in the source of contractile Ca in heart and skeletal muscle and make it unlikely that there is a physiological "trigger" Ca for excitation-contraction (E-C) coupling in these muscles.

The concept of "trigger Ca" or an internal regenerative release of Ca has recently been proposed as an integral phase of excitation E-C coupling (5–7). It has been proposed that contractile Ca is released from internal sites in a regenerative fashion by the entry, during each excitation, of a small amount of external Ca which acts as a "trigger." Data from skinned frog skeletal muscle fibers (5, 7), homogenized rat hearts (3, 6), and chemically treated frog ventricles (13) has been interpreted to support the hypothesis of a trigger Ca. However, the validity

[1] This study was supported by Grant HL-11351-06 and HL-11351-07 from the United States Public Health Service and a grant from the Bear Foundation.

of the extrapolation from these anatomically abnormal preparations to more physiological, intact muscles has not yet been adequately tested.

A strong challenge to the physiological regenerative release theory came from data (2) which showed that tension in single skeletal muscle fibers could be maintained for 20–25 min in EGTA-buffered 0-Ca solutions in which the calculated extracellular Ca concentrations were less than 2×10^{-9} M. Tension failure was attributed to the loss of excitability at such low Ca concentrations. This progressive electrical inexcitability is a major problem in examining the trigger Ca hypothesis in intact tissues (4, 8). However, it has been reported recently (1) that La would maintain the membrane resting potential and excitability during 0-Ca exposure. La has other advantages over other similar acting cations. La does not penetrate the intact cell membrane (10), it displaces Ca bound to the membrane (11) and it inhibits transmembrane movements of Ca (10, 12). The hypothesis which proposes the regenerative release of Ca was examined in intact, vascularly perfused muscles using 100 μM La to maintain excitability during 0-Ca (0-Ca, 100-La) exposure.

Three arterially perfused experimental preparations of different fiber type were used: the perfused interventricular septum and the perfused SMA muscle from the guinea pig, and the perfused FG muscle. The two skeletal muscle preparations were isolated by cannulating the femoral arteries of excised hind limbs of guinea pigs and frogs and tying all arterial vessels not supplying the respective muscle. The skeletal muscles were stimulated directly and transversely with short tetani via two sets of four parallel electrodes positioned on opposing sides of each muscle. Mechanical and radioisotope flux measurements could be simultaneously monitored. All perfusates contained 1×10^{-7} gm/ml pancuronium bromide to provide constant neuromuscular blockade.

Figure 1 summarizes the results of 0-Ca, 100-La perfusion in heart, SMA, and FG muscles. Changes in developed tension were normalized by plotting the tension as a percentage of control tension. Perfusion of guinea pig septa with 0-Ca, 100-La solutions resulted in a very rapid decline of tension with a $t_{1/2}$ = 11 sec. This very rapid decline of tension indicates a strong dependence on either a rapidly exchangeable extracellular source of contractile Ca or a superficial source in very rapid equilibrium with the extracellular space (ECS). Two possibilities are suggested with regard to this rapidly exchangeable Ca: 1) sufficient Ca enters the cell with each contraction to activate the myofilaments, or 2) a component of Ca from the rapidly exchangeable sites enters the cell and triggers the release of contractile Ca from intracellular stores. If the latter possibility were true, *i.e.*, if trigger Ca were operative, then perfusion with 0-Ca, 100-La would prevent the trigger Ca from crossing the membrane and contraction should be inhibited. If this explanation is used to explain the results seen in the perfused heart, then one might reasonably expect to see similar results in perfused, intact skeletal muscles.

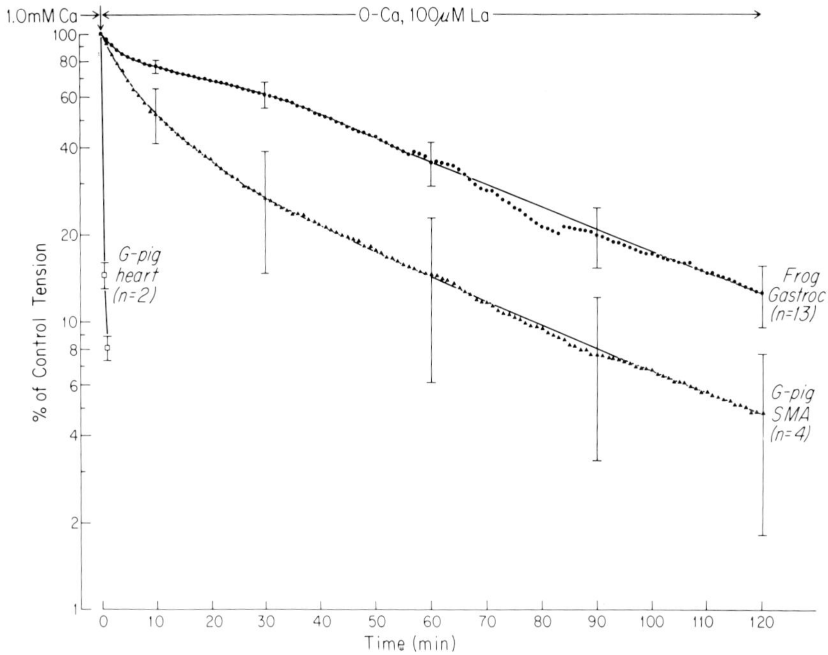

Figure 1 Rate of tension decline normalized as a percentage of control tension in vascularly perfused hearts and SMA muscles of the guinea pig (*G.-pig*) and in perfused frog gastrocnemius (*Frog Gastroc*) muscles during 0-Ca, 100-La infusions. Each point is the mean of *n* observations with 1 S.E. of the mean shown in *vertical brackets*. Note that the rate of tension decline in SMA and FG muscles is virtually identical from 30 to 120 min, whereas tension in the heart fell with $t_{1/2}$ = 11 sec.

Tension decline during 0-Ca, 100-La perfusion in SMA muscles was also immediate in onset but was much slower and biphasic in nature with approximate $t_{1/2}$ = 5 and 33 min (fig. 1). From 25–120 min the rate of fall was monophasic. The significant remnant of tension (16.5 percent) after 1 hr 0-Ca, 100-La perfusion and the much slower rate of tension decline question whether extracellular Ca serves as a trigger in these vascularly perfused SMA muscles.

The decline of tension in FG perfused with 0-Ca, 100-La solutions was virtually identical with SMA muscles for 3 min (fig. 1). The rate of decline then slowed over the next 30 min with an apparent $t_{1/2}$ = 65 min. From 33–120 min the tension decline was again very similar to that seen in SMA muscles ($t_{1/2}$ = 33 min). The common rate of tension decline might indicate the rate of intracellular Ca exchange after intracellular Ca reached a certain critical or threshold value. If the volume of the source of contractile-related Ca in the two skeletal muscles were different, the onset of the common rate of tension decline would be delayed, depending on the difference in volume. Electron micrographs of the perfused SMA and FG muscles confirm the difference in fractional volume ratios of the lateral cisternal component of the SR of these different fiber types (9). In skeletal muscles the lateral sac has been largely accepted as the source of contractile Ca. Hence the larger lateral sac volume in FG might account for the 30-min delay with slower tension decline.

It was assumed that Ca would clear the ECS during 0-Ca perfusion at least as rapidly as the much larger molecules [^{14}C] mannitol and [^{51}Cr] EGTA used as ECS markers. Less than 3 percent of the mannitol or EGTA spaces remained labeled after a 30-min 0-Ca washout after a 1-hr labeling period. A 3 percent residual of the equilibrated 1.0 mM Ca would indicate an interstitial Ca = 30 μM. In the presence of La which displaces Ca bound to the membrane and which inhibits transmembrane flux, and after 60–120 min 0-Ca, 100-La perfusion, the actual concentration of Ca in the ECS would be expected to be much lower, nearer the contaminant level of Ca in the 0-Ca perfusate. This was found by routine atomic absorption analysis to be less than 5 μM. The addition of La (200 μM total La), which would be expected to further uncouple excitation from contraction if trigger Ca is operative, did not have a significant effect on tension decline after 60–120 min 0-Ca, 100-La perfusion. These data support the

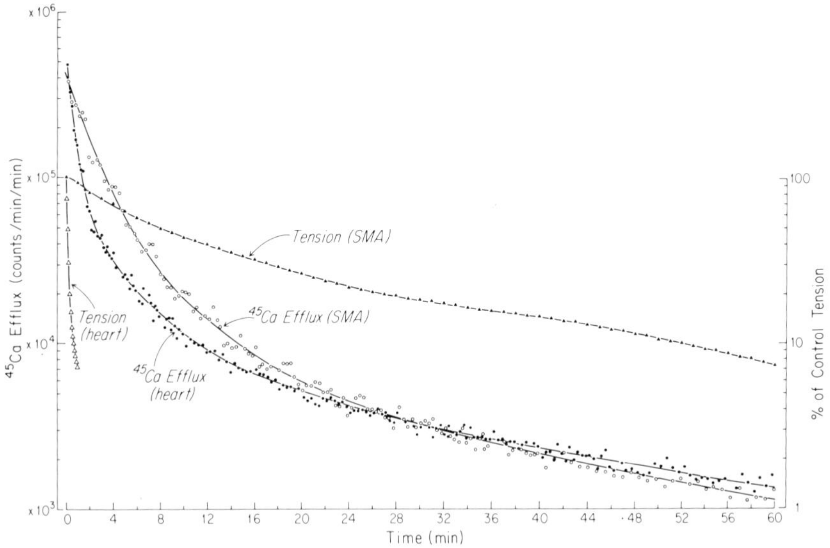

Figure 2 Effect of 0-Ca, 100-La perfusion on simultaneously monitored ^{45}Ca efflux and tension decline in a representative guinea pig heart and SMA muscle. Tension decline in the heart matched most closely the most rapid rate of ^{45}Ca exchange in contrast to the SMA muscle where tension decline paralleled the ^{45}Ca efflux from the muscle after the ECS had been cleared.

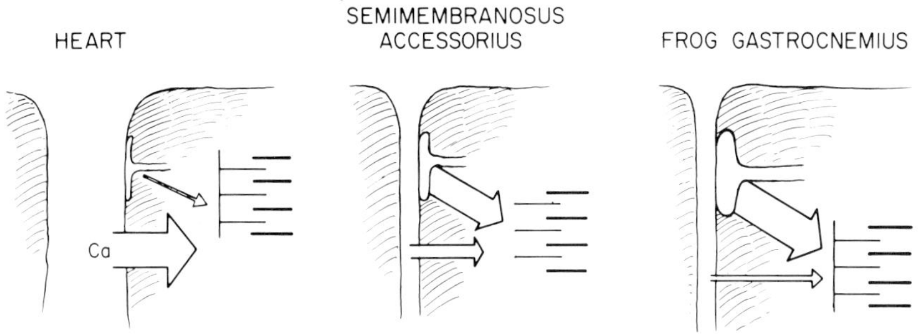

Figure 3 Model of the source of contractile Ca in heart, SMA, and FG muscles. The T tubule of the heart is represented approximately 5 times larger than the skeletal muscle T tubules. Lateral cisternal volume is in the order FG > SMA > heart. The arrows represent contractile Ca, with total amounts in each muscle approximately equal.

conclusion that external Ca does not serve as an activator or trigger Ca in these preparations.

The effect of buffering extracellular Ca during 0-Ca perfusion with 2 mM was approximated at 30 μM or less, the addition of 2 mM EGTA to the perfusate resulted in only a slight increase in the rate of tension decline. Calculated [Ca^{2+}] in the ECS during EGTA infusion was less than 3×10^{-9} M. Of far greater significance than the break in tension decline, however, was the remnant of 30 percent of original tension after 45 min 0-Ca exposure, 15 min of which were at 3×10^{-9} M Ca concentrations.

Representative results of monitoring ^{45}Ca efflux and tension decline simultaneously during 0-Ca, 100-La perfusion in the heart and SMA are seen in figure 2. It is clear that the rate of tension decline in the guinea pig heart is most closely related to the most rapid rate of ^{45}Ca efflux from the muscle. This emphasizes again the strong dependence on an external source of contractile Ca in myocardium. In contrast with the heart, the major component of tension decline and ^{45}Ca efflux in the perfused guinea pig SMA skeletal muscle are very similar after the period during which most of the ECS had been cleared. The rates of tension decline ($t_{1/2}$ = 25.3 min) and ^{45}Ca efflux ($t_{1/2}$ = 24.7 min) were virtually identical from 30–60 min. 0-Ca, 100-La washouts in perfused FG skeletal muscle were quite similar to 0-Ca, 100-La washouts in SMA muscles, with the rate of ^{45}Ca efflux and the rate of tension decline similar after the ECS had been cleared.

Figure 3 summarizes the conclusions in model form. These experiments emphasize the difference in the source of contractile Ca in the heart and skeletal muscle: In the heart (fig. 3) sufficient Ca enters the cell with each contraction to activate the myofilaments. The major source of contractile Ca in SMA is intracellular, with a small component dependent on a rapidly exchangeable extracellular source or on a superficial source in rapid equilibrium with the ECS, similar to the heart. Contractile-related Ca in FG comes mainly from intracellular stores. The conclusion is that it is unlikely that there is a physiological trigger Ca for E-C coupling in these muscles.

Acknowledgments We are pleased to acknowledge the technical assistance of Beatrice Hileman and Ray Tsuyuki.

REFERENCES

1. ANDERSSON, K.-E., and EDMAN, K. A. P. 1973. Effects of lanthanum on the coupling between membrane excitation and contraction of isolated frog muscle fibres. Acta Physiol. Scand. 90: 113–123.
2. ARMSTRONG, C. M., BEZANILLA, F. M., and HOROWICZ, P. 1972. Twitches in the presence of ethylene glycol bis(β-aminoethylether)-N,N′-tetraacetic acid. Biochim. Biophys. Acta 267: 605–608.
3. BLOOM, S., BRADY, A. J., and LANGER, G. A. 1974. Calcium metabolism and active tension in mechanically disaggregated heart muscle. J. Molec. Cell. Cardiol. 6: 137–147.
4. CAPUTO, C., and GIMENEZ, M. 1967. Effects of external calcium deprivation on single muscle fibers. J. Gen. Physiol. 50: 2177–2195.
5. ENDO, M., TANAKA, M., and OGAWA, Y. 1970. Calcium induced release of calcium from the sarcoplasmic reticulum of skinned skeletal muscle fibres. Nature 228: 34–36.
6. FABIATO, A., and FABIATO, F. 1972. Excitation-contraction coupling of isolated cardiac fibers with disrupted or closed sarcolemmas. Circ. Res. 31: 293–307.
7. FORD, L. E., and PODOLSKY, R. J. 1970. Regenerative calcium release within muscle cells. Science 167: 58–59.
8. JENDEN, D. J., and REGER, J. F. 1963. The role of resting potential changes in the contractile failure of frog sartorius muscles during calcium deprivation. J. Physiol. (London) 169: 889–901.
9. LANGER, G. A. 1973. Heart: Excitation-contraction coupling. Ann. Rev. Physiol. 35: 55–86.
10. LANGER, G. A., and FRANK, J. 1972. Lanthanum in heart cell culture: Effect on calcium exchange correlated with its localization. J. Cell Biol. 54: 441–445.
11. LANGER, G. A., SERENA, S. D., and NUDD, L. M. 1974. Cation exchange in heart cell culture: Correlation with effects on contractile force. J. Molec. Cell. Cardiol. 6: 149–161.
12. WEISS, G. B. 1970. On the site of action of lanthanum in frog sartorius muscle. J. Pharmacol. Exp. Therap. 174: 517–526.
13. WINEGRAD, S. 1971. Studies of cardiac muscle with a high permeability to calcium produced by treatment with ethylenediaminetetraacetic acid. J. Gen. Physiol. 58: 71–93.

Mailing address: Dr. Terrell L. Rich,
A. N. Richards Building,
Johnson Research Foundation,
University of Pennsylvania,
Philadelphia, Pa. 19104 (USA).

Recent Advances in Studies on Cardiac Structure and Metabolism
Volume 9
The Sarcolemma
Edited by Paul-Emile Roy and Naranjan S. Dhalla

CYTOCHEMICAL STUDIES ON SARCOLEMMA: Na^+, K^+-ADENOSINE TRIPHOSPHATASE AND ADENYLATE CYCLASE

A. WOLLENBERGER and W. SCHULZE

Division of Cellular and Molecular Cardiology,
Central Institute of Heart and Circulatory Regulation Research,
Academy of Sciences of the DDR,
Berlin-Buch, German Democratic Republic

SUMMARY

Two highly lead-sensitive ATPases, Na^+,K^+-ATPase and adenylate cyclase, can be demonstrated cytochemically by the lead precipitation technique in briefly prefixed tissue, provided that the free Pb^{2+} concentration in the incubation medium is kept below 0.1 mM by a heavy metal chelator. Under conditions suitable for Na^+,K^+-ATPase activity precipitation of final reaction product (lead phosphate) at the sarcolemma of cardiac muscle is abolished by 0.1–1mM ouabain. In contrast, reaction product deposition at the intramuscular part of the plasma membrane and at intracellular sites is not noticeably affected by the glycoside. These findings indicate either that the sarcolemma is the exclusive location of Na^+, K^+-ATPase in cardiac muscle or that the presence of the enzyme at other loci is masked by active Na^+,K^+-independent, ouabain-resistant ATPases. Under conditions favoring adenylate cyclase activity, precipitation by Pb^{2+} of orthophosphate derived, with the help of added cyclic nucleotide phosphodiesterase and 5′-nucleotidase, from cyclic AMP formed from adenylyl imidodiphosphate (AMP-PNP) is seen after prolonged incubation in myocardial cells along the entire course of the plasma membrane and also at the transverse tubules and is particularly intense at the tight junction regions of the intercalated disks. Ouabain has no effect on these reactions. Reaction product deposition is also observed at the sarcolemma in red skeletal muscle and at the terminal cisternae of the sarcoplasmic reticulum in white skeletal muscle, where the reaction is intensified by adrenaline. Sarcoplasmic reticulum of cardiac and of red skeletal muscle exhibits only relatively weak staining attributable to cyclic AMP formation. These observations are in agreement with the results of tissue fractionation studies according to which the plasma membrane is the chief site of adenylate cyclase in heart and in red, but not white skeletal muscle.

Quantitative biochemical studies on subcellular components obtained by mass isolation procedures have contributed greatly to the understanding of the mechanisms of many enzyme-dependent processes. But as is very often the case in experiments performed on isolated, fractionated, and largely broken cell components, the extrapolation of the results to the situation obtaining in the intact cell is fraught with uncertainties. A cell, such as the cardiac muscle cell, is a complex polymorphic entity containing various types of membranes and other structures, each characterized by a specific enzyme activity pattern. In most cases it has not been possible to isolate in reasonable yield a given membrane structure from heart muscle uncontaminated by other cell components. Nor has a successful attempt been made to separate and isolate, in quantities and purity sufficient for biochemical characterization, the various cell types of which heart muscle is made up, that is, muscle cells, endothelial cells, fibroblasts, and others. Cytochemistry, especially when adapted to the use of the electron microscope, provides the possibility of assigning a given enzymatic activity to a defined cellular locus and structure. In conjunction with quantitative biochemical procedures, the cytochemical method can be a powerful tool in the precise mapping of intracellular enzyme activity and can add substantially to our knowledge of cell function. The cytochemical study of ATPase activities in cardiac and skeletal muscle is a case in point.

The present paper focuses attention on two ATP-splitting enzymes that are currently of great interest in cell physiology and biochemistry, namely Na^+,K^+-ATPase and adenylate cyclase. In most of the cell types investigated both enzymes appear to be confined to the plasma membrane. Na^+,K^+-ATPase is perhaps the most commonly used enzymatic marker of the plasma membrane, including the sarcolemma in muscle. The finding that adenylate cyclase was located in the cell membrane of nucleated erythrocytes became one of the foundation stones of the two-messenger concept of hormone action (25).

Before turning to the subject proper of this report it will be useful to consider briefly the potentialities and limitations of the electron microscopic demonstration of ATPase activities.

METHODOLOGICAL CONSIDERATIONS

The cytochemical demonstration of membrane-bound ATPase is based on a modification of Gomori's (7) metal salt technique. The most frequently used method is the precipitation of liberated inorganic phosphate by lead salts, as described by Wachstein and Meisel (26). Preparatory steps include prefixation of the tissue to be examined. This treatment, as well as the lead ions needed for the precipitation of the phosphate formed, may reduce enzyme activity markedly. Hence it is desirable, and may even be necessary, to make parallel quantitative measurements of the enzyme activity in order to be able to evaluate the cytochemical results correctly.

It was shown in quantitative biochemical investigations on isolated tissue fractions and samples that prefixation with aldehydes usually reduced total ATPase activity. The individual cell organelles, however, were affected to a differing extent. In experiments on subcellular fractions of heart muscle the

ATPase of the microsomal fraction proved to be most resistant to prefixation, whereas Na^+,K^+-ATPase in the cell membrane fraction is strongly inhibited. Still, the residual activity after prefixation with either formaldehyde or glutaraldehyde is sufficient for a cytochemical demonstration of total ATPase and of specific Na^+,K^+-ATPase activities (9, 12, 21). The glutaraldehyde and formaldehyde solutions used for prefixation abolished adenylate cyclase activity in cell-free particle preparations of heart muscle, but active enzyme was preserved largely in briefly prefixed heart muscle tissue sections (18). According to these quantitative determinations the percentage activation of the adenylate cyclase by sodium fluoride was less impeded by prefixation than was activation by adrenaline.

The influence on ATPase activity of lead ions, used in the Wachstein-Meisel medium as precipitating reagent for phosphate, has likewise been examined (20). These studies can be summarized by the statement that a free lead ion concentration of less than 0.01 mM has no major inhibitory effect on ATPase activity. Employing heavy metal chelating agents, such as maleate and Tirone (catechol-3,5-disulfonic acid), the free lead ion concentration can be kept constant in the medium at a value not exceeding 0.01–0.05 mM (20). According to these preliminary quantitative investigations experimental conditions can be chosen for cytochemical studies that leave a substantial portion of ATPase activity unaffected, including the activity of Na^+,K^+-ATPase and adenylate cyclase.

LOCALIZATION OF Na^+,K^+-ATPase

The localization of various ATPases in muscle membranes of mammalian hearts has been described at length in previous publications from this laboratory (17, 21, 22). The present report will deal with the results of analogous investigations carried out in the myocardium of the chick.

Methods

The left ventricles of 1-year-old chicken were cut into 60–100-μm thick cryostat sections, prefixed at 0°C in 2.5 percent buffered glutaraldehyde or 4 percent formaldehyde (freshly produced from paraformaldehyde), and incubated in a medium containing 12.5 mM Tris-HCl (pH 7.4), 2.5 mM ATP, 2.5 mM $MgSO_4$, 2.5 mM lead acetate, and usually 3 mM Tirone. Na and K salts were added to yield concentrations varying for each salt to up to 100 mM; usually the Na:K ratio was 5:1 (100 mM Na^+, 20 mM K^+). For suppression of Na^+,K^+-ATPase the sections were pretreated for 10 min at room temperature with 10^{-3} M ouabain dissolved in incubation medium, with Pb omitted; they were then incubated with ouabain in complete medium at 37°C for up to 30 min.

Results

The designation of the various parts of the cardiac cell membrane differs greatly in the literature. This paper adheres to the terminology of Fawcett and McNutt (5) in which "sarcolemma" is the term for the plasma membrane of the muscle cell, *i.e.*, exclusive of the basement membrane and irrespective of its neighborhood. The nomenclature of the individual intramuscular membranes corresponds to Fawcett's suggestions (4).

After incubation in the medium described above a more or less intense accumulation of lead phosphate precipitate could be found at all parts of the sarcolemma (fig. 1). ATPase activity was also demonstrated at the membranes of the subsarcolemmal cisternae and at special terminal branches of the sarcoplasmic reticulum in the interior of the muscle cells (fig. 2). These reactions

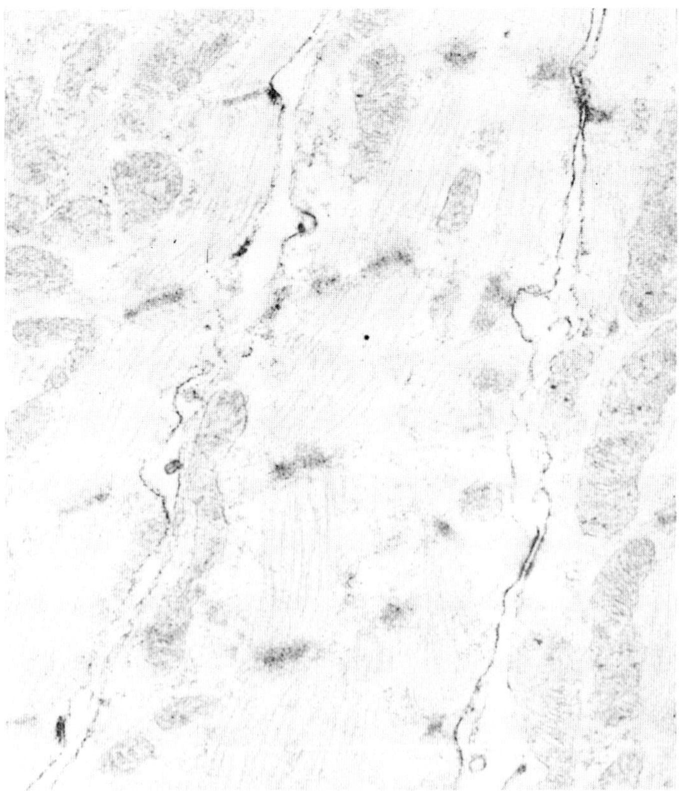

Figure 1 Electron micrograph of a section from 3-week-old chicken myocardium after 30 min of incubation. Fine deposits of electron-dense reaction product outline the sarcolemma of three cells and subsarcolemmal cisternae (× 36,000).

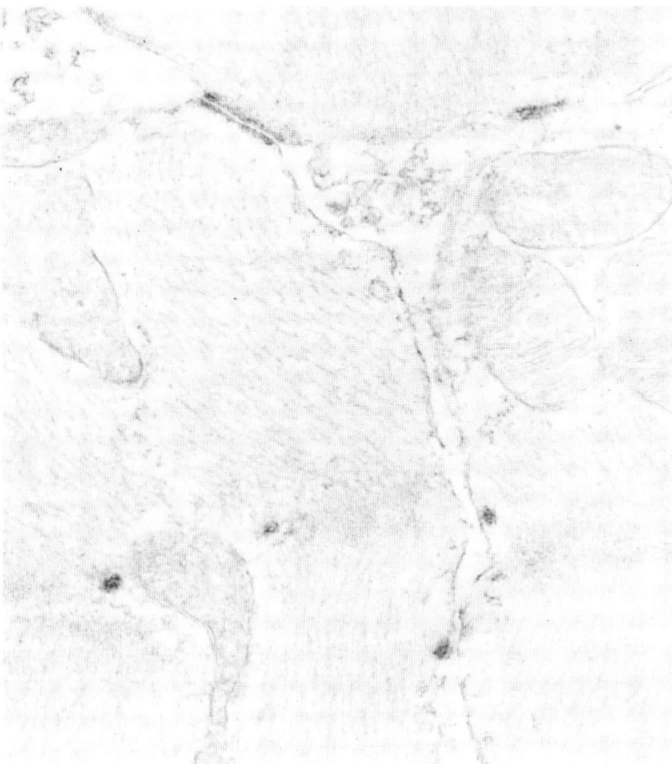

Figure 2 Electron micrograph of a section from 2-week-old chicken myocardium. Same incubation as in figure 1. Deposits of lead phosphate are localized at the sarcolemma, in the subsarcolemmal cisternae, and at special terminal branches of the sarcoplasmic reticulum (× 40,000).

occur also when the incubation medium contains 2.5 mM $CaCl_2$, with Na^+ and K^+ being present. However, no visible reaction product is formed in incubation medium lacking Mg^{2+} ions. When 10^{-3} M ouabain was added to incubation medium containing Na^+ and K^+, lead phosphate deposition at those parts of the sarcolemma facing the interstitial space was abolished (fig. 3) or greatly weakened. The corresponding reaction of the subsarcolemmal cisternae and other elements of the sarcoplasmic reticulum, however, was not influenced by ouabain, and these structures were still marked clearly by lead deposits (see fig. 3). Ouabain had no effect on the intensity of the reaction in those experiments in which 2.5 mM Ca^{2+} was present and Na^+ and K^+ were omitted. In confirmation of our earlier observations on mammalian heart tissue (21), ATPase activity at the different parts of the intercalated discs of the chick heart cells was little affected by the cardiac glycoside. In particular, the enzyme activity at the tight

Figure 3 Electron micrograph of a section from 2-week-old chicken myocardium. Incubation with Mg^{++}, Na^{+}, K^{+}, and ouabain in our medium. Lead deposits are localized at the subsarcolemmal cisternae but not on the sarcolemma (X 36,000).

junction region (nexus, zonula occludens) appeared to be unchanged after glycoside treatment.

Discussion

According to the results just described the Na^{+},K^{+}-ATPase in chick heart muscle cells is localized at those parts of the sarcolemma that are facing the interstitial space. Its presence in the sections of the cell membrane belonging to the intercalated disc is uncertain, because relatively strong basal Mg^{2+}-ATPase activity could well mask the Na^{+},K^{+}-ATPase that may perhaps be present there. Contrary to the data of Forbes and Sperelakis (6), we were unable to detect ouabain inhibition of ATPase activity at the subsarcolemmal cisternae and at other parts of the sarcoplasmic reticulum.

LOCALIZATION OF ADENYLATE CYCLASE

Methods

Left ventricular muscle, musculus psoas, and musculus soleus were removed from male guinea pigs after decapitation. The heart tissue was cut with a tissue sectioner into 100-μm thick sections, whereas the skeletal muscles were divided into small strips 0.6 mm in thickness. The material was prefixed at 0° for 10 min in 4 percent buffered formaldehyde (pH 7.4), washed thoroughly in the cold, and then treated either according to Method A or Method B. It was postfixed for several hours in cold buffered osmium tetroxide.

Method A. The incubation medium 2.5 ml, contains 80 mM Tris-maleate of pH 7.4, 10 mM $MgSO_4$, and 0.6 mM AMP-PNP (obtained from Boehringer-Mannheim), 0.25 M sucrose, 5 mM theophylline, and 2 mM Pb $(NO_3)_2$. The tissue, weighing a few milligrams, was incubated first at room temperature for about 10 min and then at 37°C for up to 60 min. Similar media and procedures have been employed for the detection of adenylate cyclase by Wagner *et al.* (27) in isolated fat cells and by Howell and Whitfield (8) in pancreas tissue.

Method B (19). After prefixation and washing the muscle tissue sections and strips were incubated in Medium A to which 0.01 unit/ml of bovine heart cyclic nucleotide phosphodiesterase (PDE), obtained from Boehringer-Mannheim, and 0.01 mg/ml of a 5′-nucleotidase preparation from crotalus atrox venum, from which theophylline was omitted, obtained from Sigma Chemical Company or from Serva Fein-Biochemica, were added. In control incubations the PDE was omitted and 5 mM theophylline was added.

In some experiments, using both methods, tissue samples were treated with 10 mM NaF or 0.2 mM adrenaline. In a few runs DL-propranolol, in a concentration of 0.02 mM, was used as β-adrenoreceptor blocker.

Results

The points of localization of the final reaction products, presumably lead imidodiphosphate (Method A) and lead phosphate (Method B), were the same in both methods. There is thus no need to describe the results separately.

The sarcolemma was the chief locus of precipitation of the two reaction products in heart muscle. Staining was in general rather weak, even after longer incubation periods. Fine granular lead precipitates were also distributed in close proximity to the membrane (fig. 4). The highest concentration of the reaction product is found at the tight junction region of the intercalated discs (fig. 5). In some micrographs lead deposits are seen at the surface of the T tubules and in a few cases also at the subsarcolemmal cisternae.

All of these reactions were either absent or of lesser intensity in Method B, when PDE was left out of the medium and theophylline was added to suppress

Figure 4 Guinea pig myocardium incubated for 60 min according to Method B. The fine granular lead precipitates are distributed at the sarcolemma and the subsarcolemmal cisternae (× 40,000).

endogenous phosphodiesterase activity. Figure 6 is an electron micrograph of a cardiac tissue section that was incubated in Medium B with cyclic AMP added. Heavy patches of reaction product are distributed at random all over the section. This random precipitation indicates that the added enzymes, phosphodiesterase and 5′-nucleotidase, were likewise evenly distributed throughout the tissue section and that the specific localization of reaction product in figure 4 could hardly have been due to an accumulation of the hydrolyzing enzyme at the membrane sites marked by reaction product deposition.

In contrast to heart muscle the reaction in white skeletal muscle (M. psoas) is primarily localized in the membranes of the sarcoplasmic reticulum. The lateral

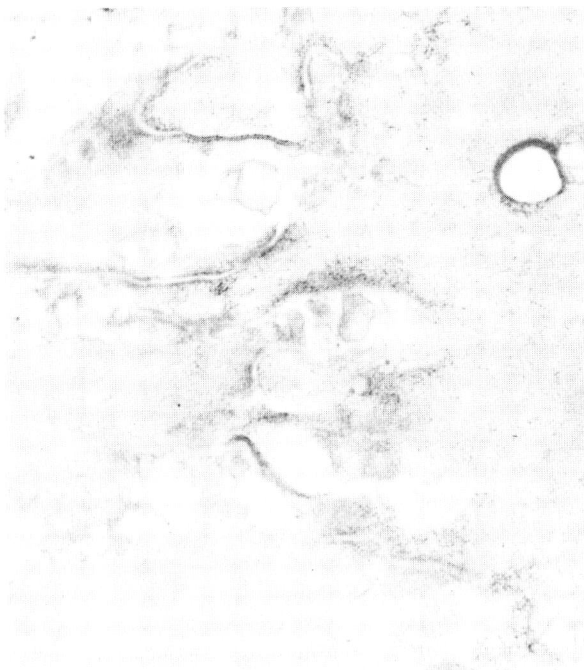

Figure 5 Ventricular heart muscle of a guinea pig incubated according to Method A. Relatively heavy deposits of reaction product were seen at the intercalated discs, especially at the tight junction (× 40,000).

cisternae, in particular, are favored loci of localization of reaction product (fig. 7). This reaction is intensified by NaF and by adrenaline (fig. 8). The enhancement of reaction product deposition by adrenaline could be prevented by the β-blocker, propranolol.

The distribution of adenylate cyclase activity in red skeletal muscle (M. soleus) is similar to that in heart muscle, with the plasma membrane and its immediate surroundings being the chief sites of lead precipitation (fig. 9).

Discussion

AMP-PNP is accepted in place of ATP as a substrate by adenylate cyclase. This has been shown to be the case in liver (16) and appears to be the case also in heart and skeletal muscle (E.-G. Krause and present authors, unpublished), where AMP-PNP, as compared with ATP, is hydrolyzed only to a negligible extent by other membrane-bound enzymes. The two products formed by adenylate cyclase from AMP-PNP are cyclic AMP and imidodiphosphate (PNP). PNP can be precipitated by Pb^{2+} as a fairly insoluble lead salt, the solubility of which

Figure 6 Ventricular heart muscle of a guinea pig incubated in medium B with cyclic AMP, final concentration 2.5 mM, added. Random distribution of rather heavy patches of precipitated reaction product (X 30,000).

at pH 7.4 and 25°C is 1.1×10^{-5} M. The solubility determinations were made in collaboration with Dr. F. Scheller by amperometric titration and the salt was found to be Pb_2PNP, so that we can calculate the solubility product, which at 25°C and pH 7.4 has a value of 5.3×10^{-15}. The incubation medium contained, as in the original Wachstein-Meisel (26) and Wagner-Kreiner-Barrnett-Bitensky (27) procedures, 2 mM Pb $(NO_3)_2$ and 80 mM maleate, a rather weak metal chelator. From the stability constant of lead maleate, determined by Yasada *et al.* (28), we calculated the free Pb^{2+} concentration in the incubation medium, correcting for the formation of magnesium maleate and for the influence of ionic strength, to be about 4×10^{-5} M. At this concentration, which is not very toxic to adenylate cyclase in tissue sections (E.-G. Krause and present authors, unpublished), Pb_2PNP precipitates, if there is no supersaturation, when the concentration of PNP reaches 3.3×10^{-6} M, or 3.3×10^{-9} moles/ml. This is a rather high concentration in face of the comparatively low velocity of the adenylate cyclase reaction which in guinea pig heart particles has been measured (2) to be 3×10^{-9} moles of cyclic AMP formed per minute and per milliter of

Figure 7 Psoas muscle of guinea pig. Incubation by Method B. The electron-dense reaction product is concentrated in the lateral sacs of the sarcoplasmic reticulum system in close contact to the T tubules with which the lateral sacs form the triads (× 40,000).

tissue. However, since the formation of this product and of inorganic pyrophosphate (and probably also PNP) is confined to, and thus concentrated in a tiny portion of the tissue (in heart muscle chiefly an inner layer of the plasma membrane), the ratio between the solubility and the rate of formation of Pb_2PNP may be sufficiently low to permit a cytochemical demonstration of adenylate cyclase.

At the above named Pb^{2+} concentration of 4×10^{-5} M, lead orthophosphate ($Pb_3(PO_4)_2$), the solubility product of which is 1.6×10^{-32}, precipitates when the orthophosphate concentration rises to 4.85×10^{-11} M, or 4.85×10^{-14} moles/ml, *i.e.*, 5 orders of magnitude lower than the corresponding saturating concentration of PNP. Since the diffusion constant of $Pb_3(PO_4)_2$ can be assumed to be not much different from that of Pb_2PNP, the great difference in solubility makes $Pb_3(PO_4)_2$ a more suitable final reaction product in the histochemical localization of enzyme activity than Pb_2PNP. For this reason recourse was taken to Method B in which one of the products of the adenylate cyclase reaction, cyclic AMP, is degraded by phosphodiesterase and 5′-nucleotidase via 5′-AMP to adenosine and orthophosphate, which is immediately

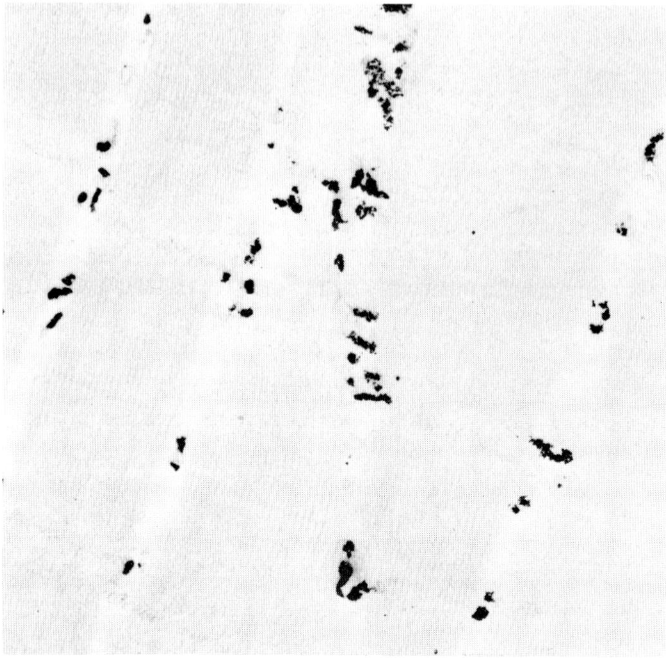

Figure 8 Psoas muscle of guinea pig. Incubation by Method B. Incubation with 2×10^{-4} M adrenaline. The lead deposits are localized in the same parts as in figure 6, but heavier than following incubation in standard medium (× 40,000).

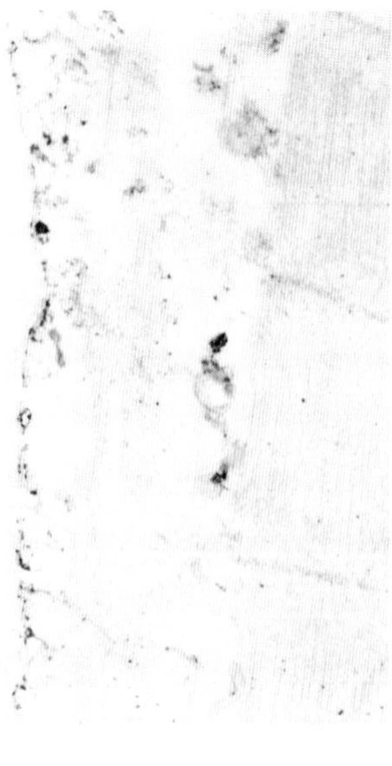

Figure 9 Soleus muscle of guinea pig after incubation in Medium A. Lead precipitates are localized at the plasma membrane and the T tubules. The reaction is similar to that seen in cardiac muscle (× 40,000).

precipitated by Pb^{2+} as the highly insoluble $Pb_3(PO_4)_2$. This is a multistep procedure with all of the possible shortcomings inherent in it and one must be careful to ensure that substances of large molecular weight, such as the enzyme proteins, penetrate well into the tissue specimen before the adenylate cyclase reaction is set in motion. This seemed to be the case in the present experiments, as shown in figure 6. On the other hand, omission of one of the enzymes provides the possibility of additional control.

Our cytochemical data on heart muscle agree reasonably well with the results of cardiac tissue fractionation studies reported from several laboratories (3, 10, 22) in which hormone-sensitive adenylate cyclase was found to be present in heart muscle not only in the sarcolemma, but also in membranes of the sarcotubules. Sulakhe and Dhalla (24) estimated that about 12 percent of the total adenylate cyclase is associated with these latter membrane structures. Our findings with AMP-PNP as substrate point in the same direction and suggest that the plasma membrane is the principal seat of the enzyme in the myocardial cells of guinea pig ventricular muscle. To judge from our still preliminary data, the tight junction region of the intramuscular part of the plasma membrane appears to be a region particularly rich in adenylate cyclase. This finding, however, needs further experimental verification and substantiation. At the moment it is difficult to think of a reason why a hormone-sensitive enzyme should be present at a site which is relatively inaccessible to exogenous chemical stimulants.

When it was first found that adenylate cyclase is present in cardiac microsomal fractions that largely consist of fragments of sarcotubules, it was potulated (3) that cyclic AMP may play a role in the hormonal and neurohumoral regulation of sarcotubular function in excitation-contraction coupling and muscular relaxation, *i.e.*, in the intracellular release of Ca^{2+} ions from the Ca^{2+} storing sites of the sarcoplasmic reticulum to troponin and their recapture by the reticulum. This idea received support from the finding (3) that added cyclic AMP can increase Ca^{2+} binding and uptake by isolated cardiac microsomes, an effect that in part is thought to be a consequence of the phosphrylation of a microsomal protein by a cyclic AMP-dependent protein kinase present in microsomal membranes (11). All this does not necessarily mean that cyclic AMP has to be produced by adenylate cyclase right at its cellular site of action, although such close proximity would probably be of advantage if the action has to be a quick one.

There is also good agreement between our results with white and red skeletal muscle (psoas and soleus, respectively) and the results of tissue fractionation experiments carried out by Robison *et al.* (15) on these two muscle types. In white fast contracting muscle adenylate cyclase was present in the subcellular fraction derived predominantly from the sarcoplasmic reticulum (see also References 14 and 23), whereas, in red slow contracting muscle, most of the cyclase was sedimented in the fraction-containing fragments of the cell membrane (15).

The same difference in the distribution of the enzyme is evident from some of the electron micrographs presented above.

So far as white skeletal muscle is concerned, there are no indications for a physiological regulation of contractility by instantaneously acting hormones capable of stimulating adenylate cyclase, such as the catecholamines (1). One may then ask what an adrenaline-sensitive adenylate cyclase, demonstrated in our cytochemical experiments, is doing in the sarcotubular membranes of white skeletal muscle. The best answer probably is (13) that it may be there for the purpose of assisting in the activation of the phosphorylase system and the resulting acceleration of the breakdown of the glycogen particles that are distributed throughout the sarcoplasm. Skeletal muscle fibers have a large diameter and a correspondingly low surface to volume ratio, and in fast contracting white muscles that in emergency situations requiring sudden great outbursts of activity are subject to large temporary anaerobic energy demands, these demands can be met, with the help of adrenaline released from adrenal medulla into the blood stream, by rapid mobilization and utilization of the cellular glycogen stores. The presence of β-adrenoreceptor-linked adenylate cyclase molecules in the internal membrane system of the muscle fibers would then serve a useful purpose.

REFERENCES

1. BOWMAN, W. C., and NOTT, M. W. 1969. Actions of sympathomimetic amines and their antagonists on skeletal muscle. Pharmacol. Rev. 21: 27–72.
2. DRUMMOND, G. I., and DUNCAN, L. Adenylate cyclase in cardiac tissue. J. Biol. Chem. 245: 976–983.
3. ENTMAN, M. L., LEVEY, G. S., and EPSTEIN, S. E. 1969. Mechanism of action of epinephrine and glucagon on the heart: Evidence of a cyclic 3′,5′-AMP mediated increase in cardiac sarcotubular calcium stores. Circ. Res. 25: 429–438.
4. FAWCETT, D. W. 1966. Atlas of Fine Structure. Saunders, Philadelphia.
5. FAWCETT, D. W., and McNUTT, D. I. 1969. Ultrastructure of myocardium. J. Cell Biol. 42: 1–45.
6. FORBES, M. S., and SPERELAKIS, N. 1972. (Na^+,K^+)-ATPase activity in tubular systems of mouse cardiac and skeletal muscles. Z. Zellforsch. 134: 1–11.
7. GOMORI, G. 1939. Microtechnical demonstration of phosphatase in tissue sections. Proc. Soc. Exp. Biol. Med. 42: 23–26.
8. HOWELL, S. L., and WHITFIELD, M. 1972. Cytochemical localization of adenyl cyclase activity in rat islets of Langerhans. J. Histochem. Cytochem. 20: 873–879.
9. JACOBSON, N. O., and JØRGENSEN, P. L., 1969. A quantitative biochemical and histochemical study of the lead method for localization of adenosine triphosphate-hydrolyzing enzymes. J. Histochem. Cytochem. 17: 443–453.
10. KATZ, A. M., TADA, M., REPKE, D. I., IORIO, J. M., and KIRCHBERGER, M. A. 1974. Adenylate cyclase: Its probable localization in sarcoplasmic reticulum as well as sarcolemma of the canine heart. J. Molec. Cell. Cardiol. 6: 73–78.
11. KIRCHBERGER, M. A., TADA, M., REPKE, D. I., and KATZ, A. M. 1972. Cyclic adenosine 3′:5′-monophosphate-dependent protein kinase stimulation of calcium uptake by canine cardiac microsomes. J. Molec. Cell. Cardiol. 4: 673–680.
12. MARCHESI, V. T., and PALADE, G. E. 1967. The localization of Mg-Na-K activated adenosine triphosphatase on red cell ghost membranes. J. Cell Biol. 35: 385–404.

13. MAYER, S. E., and STULL, J. T. 1971. Cyclic AMP in skeletal muscle. Ann. N. Y. Acad. Sci. 185: 433–438.
14. RABINOWITZ, M., DESALLES, L., MEISLER, J., and LORAND, L. 1965. Distribution of adenyl-cyclase activity in rabbit skeletal muscle fractions. Biochim. Biophys. Acta 97: 29–36.
15. ROBISON, G. A., BUTCHER, R. W., and SUTHERLAND, E. W. 1971. Cyclic AMP. Academic Press, New York.
16. RODBELL, M., BIRNBAUMER, L., POHL, S. L., and KRAUS, M. J. 1971. The glucagon-sensitive adenyl cyclase system in plasma membranes of rat liver. V. An obligatory role of guanyl nucleotides in glucagon action. J. Biol. Chem. 246: 1877–1882.
17. SCHULZE, W. 1971. Differenzietung und Ozganisierung von Membranen in embryonalen Herzmuskelzellen. Acta Histochem. Suppl. 11: 185–191.
18. SCHULZE, W., KRAUSE, E.-G., and WOLLENBERGER, A. 1972. Cytochemical demonstration and localization of adenylate cyclase in skeletal and cardiac muscle. Advan. Cyclic. Nucl. Res. 1: 249–260.
19. SCHULZE, W., KRAUSE, E.-G., and WOLLENBERGER, A. To be published.
20. SCHULZE, W., PORTIUS, H. J., SCHELLER, F., and WOLLENBERGER, A. 1973. Über Grundlagen der Anwendungsmöglichkeiten des Komplexbildners Tiron beim histochemischen Nachweis der NaK-ATPase-Aktivität. Acta Histochem. 47: 376–379.
21. SCHULZE, W., and WOLLENBERGER, A. 1969. Zytochemische Lokalisation und Differenzierung von Na^{+}-K^{+}- und anderer membranständiger ATPase-Aktivität im Herzmuskel. Histochemie 19: 302–318.
22. SCHULZE, W., and WOLLENBERGER, A. 1971. Cytochemistry of membrane-bound ATPase systems of cardiac muscle. Methods Achiey. Exp. Pathol. 5: 347–383.
23. SERAYDARIAN, K., and MOMMAERTS, W. F. H. M. 1965. Density gradient separation of sarcotubular vesicles and other particulate constituents of rabbit muscle. J. Cell Biol. 26: 641–656.
24. SULAKHE, P. V., and DHALLA, N. S. 1973. Adenylate cyclase of heart sarcotubular membranes. Biochim. Biophys. Acta 293: 379–386.
25. SUTHERLAND, E. W., ØYE, I., and BUTCHER, R. W. The action of epinephrine and the role of the adenyl cyclase system in hormone action. Recent Progr. Hormone Res. 21: 623–642.
26. WACHSTEIN, M., and MEISEL, F. 1957. Hitochemistry of hepatic phosphatase at a physiologic pH. (With special reference to the demonstration of bile canaliculi.) Amer. J. Clin. Pathol. 27: 13–23.
27. WAGNER, R. C., KREINER, P., BARRNETT, R. J., and BITENSKY, M. W. 1972. Biochemical characterization and cytochemical localization of a catecholamine-sensitive adenylate cyclase in isolated capillary endothelium. Proc. Nat. Acad. Sci. U.S.A. 69: 3175–3179.
28. YASADA, M., YAMASAKI, K., and OHTAKI, H. 1960. Stability of complexes of several carboxylic acids formed with bivalent metals. Bull. Chem. Soc. Jap. 33: 1067–1070.

Mailing address: Dr. A. Wollenberger,
Division of Cellular and Molecular Cardiology,
Central Institute of Heart and Circulatory Regulation Research,
Academy of Sciences of the DDR, Berlin-Buch (German Democratie Republic).

Recent Advances in Studies on Cardiac Structure and Metabolism
Volume 9
The Sarcolemma
Edited by Paul-Emile Roy and Naranjan S. Dhalla

BIOCHEMICAL PROPERTIES OF CARDIAC SARCOLEMMA: ADENYLATE CYCLASE AND (Na^+ + K^+)-ACTIVATED ATPase[1]

M. TADA,[2] M. A. KIRCHBERGER,[3] and A. M. KATZ[4]

Division of Cardiology,
Department of Medicine and Department of Physiology,
Mount Sinai School of Medicine of the City University of New York,
New York, New York, USA

SUMMARY

Morphologically intact plasma membranes from guinea pig ventricles were obtained by exposing isolated cell segments to osmotic shock, followed by extraction of actomyosin in 1 M KCl. These preparations contained approximately 1/6 of the protein and 5–10 percent of the mitochondrial markers present in the original cell preparation. Both adenylate cyclase and (Na^+ + K^+)-activated ATPase activities were enriched 3–4-fold. The receptor for epinephrine stimulation of adenylate cyclase was retained.

The "basal" ATPase activity of 5–6 μmoles of Pi/mg/hr, measured in 120 mM NaCl or KCl, was approximately doubled in 100 mM NaCl + 20 mM KCl. This increment, the (Na^+ + K^+)-activated ATPase, was abolished by 10^{-5} M ouabain, the K_i for ouabain being approximately 3×10^{-7} M. Adenylate cyclase, which had a basal activity of approximately 0.33 nmole of cyclic AMP produced/min/mg of protein, was significantly stimulated by both l-epinephrine and NaF. Half-maximal stimulation was seen at approximately 5×10^{-6} M l-epinephrine.

Increasing Ca^{2+} in the range between 10^{-7} and 10^{-3} M inhibited basal, l-epinephrine-, and NaF-stimulated adenylate cyclase activities. Basal rates of cyclic AMP production were more sensitive to Ca^{2+} than was l-epinephrine-stimulated adenylate cyclase activity, so that l-epinephrine stimulation was increased from approximately 60 percent in 0.5 mM ethylene glycol bis(β-aminoethyl ether)-N,N′-tetraacetic acid to approximately 150 percent in 10^{-7} M Ca^{2+} and 400 percent in 10^{-5} M Ca^{2+}. The inhibitory effect of Ca^{2+} on adenylate cyclase activity may represent a negative feedback mechanism by which elevation of intracellular Ca^{2+} concentration lowers cellular levels of cyclic AMP and thus reduces Ca^{2+} influx into the myocardium.

[1] This research was supported by the United States Public Health Service, contract NIH-NHLI-72-2973-M; grants HL-13191 and HL-15764 from the National Heart and Lung Institute; grants AA-00316 from the National Institute on Alcoholic Abuse and Alcoholism; a Grant-in-Aid from the New York Heart Association; and gifts from the Jack Martin Fund.

[2] Present address: The First Department of Internal Medicine, Osaka University Medical School, Fusushima-ku, Osaka 553, Japan.

[3] A portion of this work was carried out during the tenure of a New York Heart Association Research Fellowship. Recipient of a Research Career Development Award from the U.S. Public Health Service, HL-00053.

[4] Philip J. and Harriet L. Goodhart Professor of Medicine (Cardiology).

In recent years plasma membranes of many tissues have been investigated intensively because they contain a number of proven and postulated transport systems and serve as the initial site of interaction between cells and a number of drugs and hormones. In the heart, in which contractile performance is regulated by systems intrinsic to the muscle, a number of stimuli impinge upon the plasma membranes that can, ultimately, modify the interactions between the contractile apparatus of the heart and metabolic processes. Therefore, *in vitro* preparations of morphologically intact and biochemically distinct heart plasma membranes are of considerable value in defining the characteristics of active ion transport systems and cell receptors. In view of evidence that in most cells adenylate cyclase activity is found on the plasma membrane (31, 33, 34), such preparations are also useful in defining drug and ion interactions with this important enzyme. Although many procedures, such as those of Kono and Colowick (24), McCollester (28), Westort and Hultin (51), and Carvalho *et al.* (8), are available for preparation of skeletal muscle membranes, and the procedures of Matsui and Schwartz (26) for cardiac microsomal membranes enriched in plasma membrane marker enzyme activities, there is currently no standard method for isolating morphologically intact cardiac plasma membranes. Portius and Repke (32) and Stam *et al.* (38) described cardiac plasma membrane preparations in which high concentration of KBr and NaI, respectively, were used to remove the contractile proteins, but biochemical purity is difficult to define because salts of this type inactivate most membrane markers (19).

We have attempted to prepare morphologically recognizable plasma membranes by means that would preserve most plasma membrane markers including hormone-sensitive adenylate cyclase (43, 45). The present communication will review our findings on the biochemical characteristics of these plasma membranes.

PREPARATION AND CHARACTERIZATION OF MYOCARDIAL PLASMA MEMBRANES

Our preparation employs brief and repeated homogenization of guinea pig ventricle slices in a Virtis Omnimixer, osmotic shock of the cell segments through a narrow gauge needle into a large volume of hypotonic solution, and extraction of the contractile proteins with hypertonic KCl, in some cases followed by exposure to changes in ambient pressure (fig. 1). The preparation described in this report is that obtained after the extraction step, unless otherwise stated. When observed by phase-contrast microscopy, the initial homogenate consists of cell segments, 20–30 μm in diameter (fig. 2). In this figure, cross-striations are seen within two cells joined by an intercalated disc. After extraction of the contractile proteins with KCl, the cross-striations disappear almost completely, but a small amount of particulate material is still present

VENTRICLE SLICES

↓ Homogenize briefly in 0.25 mM EDTA
Collect cells from supernatant of the homogenate

Initial homogenate preparation

↓ Extrude through No. 23 gauge needle into 20 volumes of 0.25 mM EDTA

After shock preparation

↓ Incubate in 1 M KCl - 30 percent sucrose for 15 hrs
Wash cells with 10 percent sucrose

After extraction preparation

↓ Expose to N_2 at 600 p.s.i.
Eject gently through No. 18 gauge needle

After bomb preparation

Figure 1 Outline of the procedures to prepare myocardial plasma membranes. See Tada *et al.* (43) for detail.

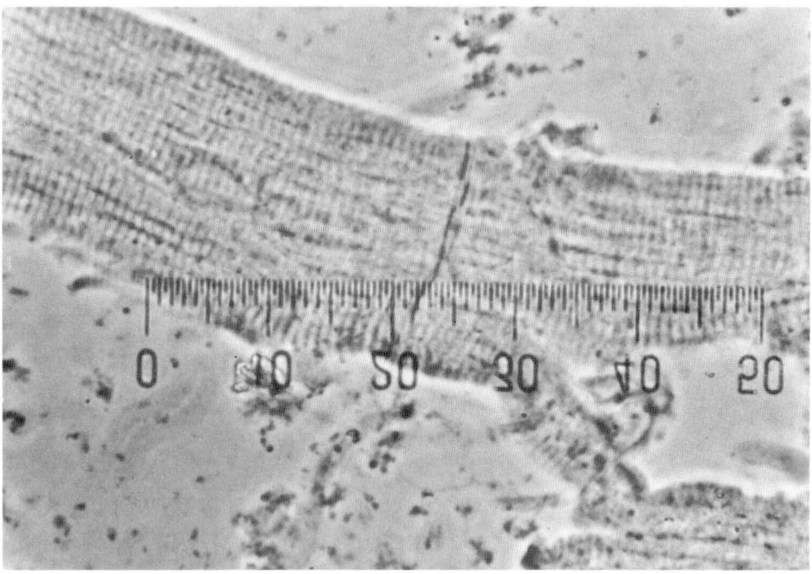

Figure 2 Phase contrast microscopic appearance of the initial homogenate under high power (10 scale divisions = 12.5 μm). (Reproduced from Tada *et al.* (43) by permission of *Journal of Molecular and Cellular Cardiology.*)

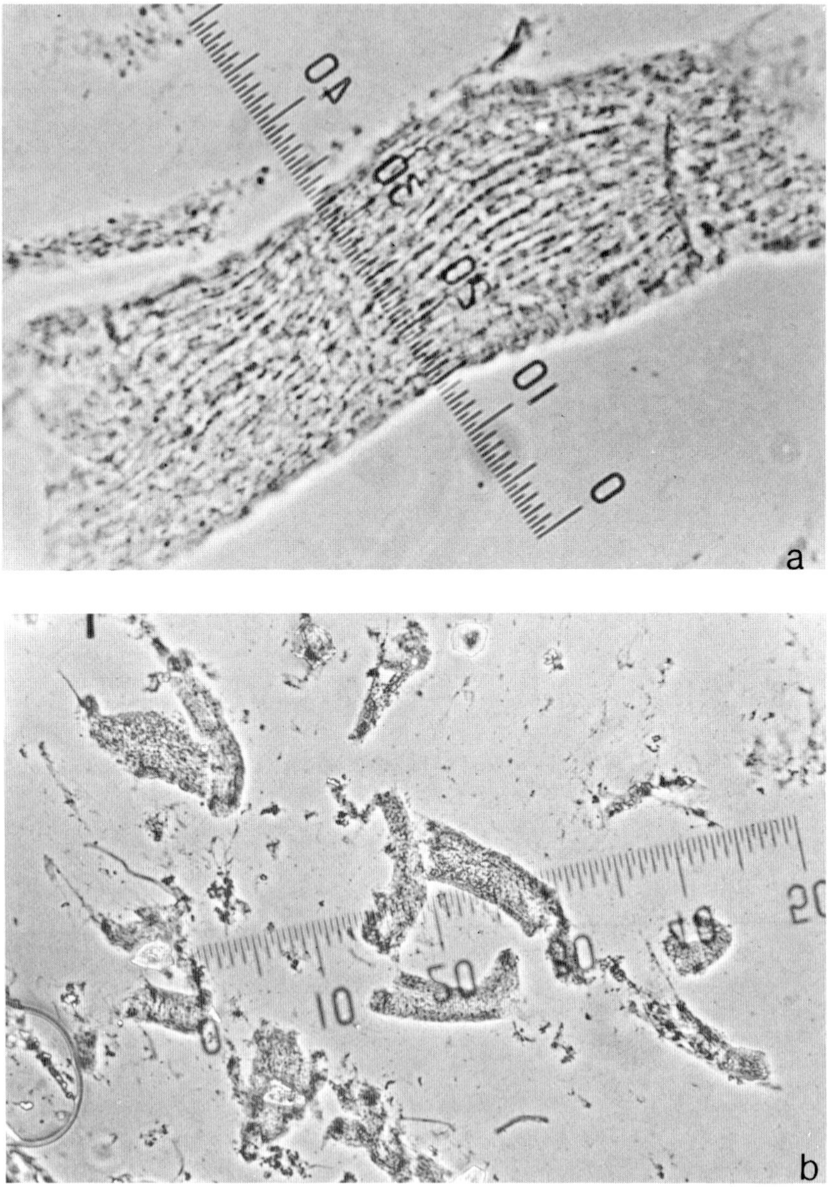

Figure 3 Phase contrast microscopic appearance of the after extraction preparation under high power (*a:* 10 scale division = 12.5 μm) and low power (*b:* 10 scale divisions = 50 μm). (Reproduced from Tada *et al.* (43) by permission of *Journal of Molecular and Cellular Cardiology*.)

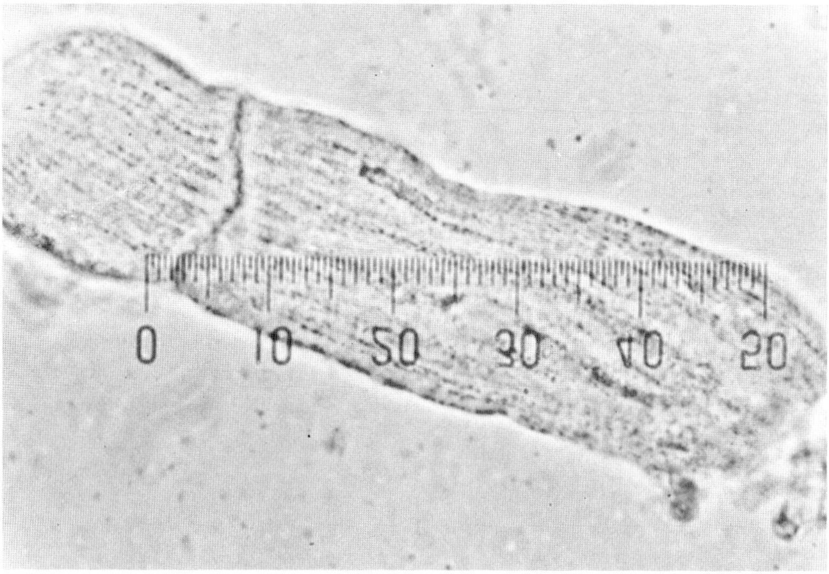

Figure 4 Phase contrast microscopic appearance of the after bomb preparation under high power (10 scale division = 12.5 μm). (Reproduced from Tada *et al.* (43) by permission of *Journal of Molecular and Cellular Cardiology.*)

within the plasma membranes (fig. *3a, b.*) After the nitrogen bomb step, the material within the membrane is significantly reduced (fig. 4), but the shape of the membranes tends to be destroyed.

By these procedures, mitochondrial marker activities, namely azide-sensitive ATPase and succinate cytochrome *c* reductase, are more than 90 percent eliminated (table I). At the same time, the specific activities fall to within a few percent that of a purified mitochondrial preparation. Enrichment of plasma

Table I Mitochondrial markers during preparation of cardiac sarcolemma[1]

Preparation	Recovery of azide-sensitive ATPase (μmoles/min/gm ventricle)	Recovery of succinate cytochrome *c* reductase (nmoles/min/gm ventricle)
Initial homogenate	5.68	1,880
After shock	1.28	730
After extraction	0.30	161
After bomb	0.22	

[1] Data reprinted from Tada *et al.* (43) by permission of *Journal of Molecular and Cellular Cardiology.*

Table II ($Na^+ + K^+$)-activated ATPase and adenylate cyclase activities during preparation of cardiac sarcolemma[1]

Preparation	($Na^+ + K^+$)-activated ATPase		Adenylate cyclase	
	Specific activity (μmoles/hr/mg)	Total recovery (μmoles/hr/gm ventricle)	Specific activity (pmoles/min/mg)	Total recovery (pmoles/min/gm ventricle)
Initial homogenate	2.15	63.2	62.6	2,279
After shock	2.83	60.9	83.0	1,120
After extraction	4.43	42.0	236.0	1,321
After bomb	5.42	32.6	246.0	763

[1] Data reprinted from Tada *et al.* (43) by permission of *Journal of Molecular and Cellular Cardiology*.

Table III Ca^{2+}-activated ATPase activity during preparation of cardiac sarcolemma[1]

Preparation	ATPase activity (μmoles/mg/min)		
	10^{-5} M Ca^{2+}	10^{-8} M Ca^{2+}	(10^{-5}–10^{-8} M)
Initial homogenate	0.226	0.208	0.018
After shock	0.170	0.132	0.038
After extraction	0.110	0.097	0.013
After bomb	0.116	0.104	0.012

[1] Data reprinted from Tada *et al.* (43) by permission of *Journal of Molecular and Cellular Cardiology.*

membranes are indicated by the increased specific activities and relatively small loss of the total activity of ($Na^+ + K^+$)-activated ATPase and adenylate cyclase (table II). The Ca^{2+}-activated ATPase activity is somewhat reduced after osmotic shock, but the proportion of ATPase activity lost when Ca^{2+} is lowered to 10^{-8} M increases after osmotic shock (table III). It is unlikely that this Ca^{2+}-sensitive ATPase represents an actomyosin contamination because the specific activity of the cardiac contractile proteins (18, 42) is almost 1 order of magnitude lower than the levels seen here. This ATPase activity is unlikely to be a mitochondrial ATPase because of the depletion of other mitochondrial markers. One explanation for the high levels of Ca^{2+} activation noted here might be the persistence of subsarcolemmal cisternae, which have been described in electron microscopic studies of the mammalian myocardium (13, 35). Such structures are probably analogous to the sarcoplasmic reticulum which has a highly active Ca^{2+}-activated ATPase. On the other hand, the cardiac plasma membrane itself, like its skeletal muscle counterpart (40, 41), may contain Ca-transport sites that would be manifest biochemically as a Ca^{2+}-activated ATPase. Interestingly, however, a preparation of the cardiac sarcolemma, having basic characteristics similar to our preparation, binds appreciable amounts of Ca^{2+} in the absence of ATP or oxalate (52). These Ca-binding properties of cardiac sarcolemma are similar to those of liver plasma membranes (36), but strikingly different from those of skeletal muscle sarcolemma (40, 41). Until each of these membrane properties is defined, and its Ca^{2+} sensitivity determined, full interpretation of these data in the Ca-binding properties of cardiac sarcolemma is not possible.

($Na^+ + K^+$)-ACTIVATED ATPase

ATPase activity of this preparation shows the pattern of activation typical of ($Na^+ + K^+$)-activated ATPase. Thus, as shown in figure 5, ATPase activity is stimulated when Na^+ and K^+ are present together, maximal activation being

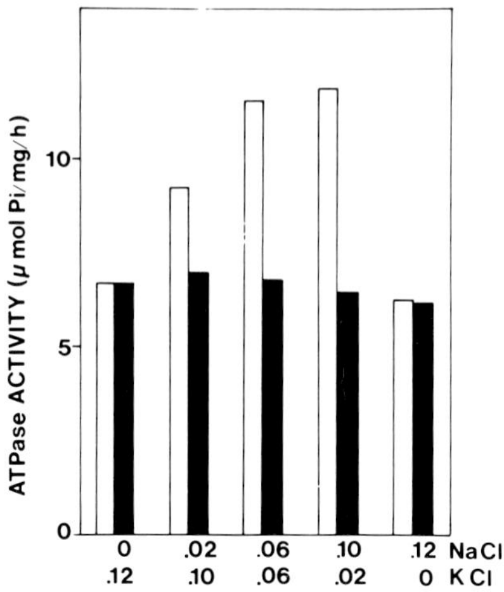

Figure 5 Effects of different combinations of Na^+ and K^+ on ATPase activity of cardiac sarcolemma. Plasma membranes (0.15 mg/ml) were incubated for 10 min at 37°C in 5 mM MgATP, Ca-EGTA buffer (10^{-8} M Ca^{2+}), and 40 mM imidazole buffer (pH 6.8) in the presence of different combinations of NaCl and KCl without (*open bars*) and with (*solid bars*) 10^{-5} M ouabain. The amount of Pi released was determined by the procedures of Taussky and Shorr (49). Each bar represents mean of triplicate determinations. (Reproduced from Tada *et al.* (45) by permission of *Circulation Research.*)

attained at 0.02 M KCl and 0.1 M NaCl. The specific activity of ($Na^+ + K^+$)-activated ATPase, estimated by subtracting the ATPase in 0.12 M NaCl alone from that in 0.02 M KCl + 0.1 M NaCl, is usually within the range of 5–6 μmoles of Pi/mg/hr. This value is comparable with those of other myocardial (Na^+ + K^+)-activated ATPases (38, 52), but is about one-fifth that of a heart microsomal preparation designed to yield the highly active enzyme (26). To some extent, these low activities probably reflect the persistence of nonplasma membrane protein. On the other hand, it appears that the ($Na^+ + K^+$)-activated ATPase is a latent enzyme which is activated by a variety of structure-disrupting agents. Thus, the low activities noted here may reflect the more "native" state of these membranes, attributable to efforts to preserve the more labile hormone-responsiveness of adenylate cyclase, rather than the presence of large quantities of contaminating protein or extensive enzyme denaturation by the extraction procedure. Cardiac glycosides, like ouabain at 10^{-5} M, abolish activation by K^+ + Na^+ completely, reducing ATPase activity to the basal level seen with Na^+ or K^+ alone (fig. 5). Half-maximal inhibition of the ($Na^+ + K^+$)-activated ATPase is seen at approximately 3×10^{-7} M ouabain, which is similar to the previous reports by Matsui and Schwartz (26) and Stam *et al.* (38). Unlike that of brain homogenate (54), the ($Na^+ + K^+$)-activated ATPase activity of the present preparation is not stimulated by epinephrine.

ADENYLATE CYCLASE

Plasma membrane adenylate cyclase, the basal activity of which is enriched several fold and is over 0.2 nmole/min/mg of protein, was stimulated by NaF and l-epinephrine, maximal stimulation being within the range of 100–150 percent for 8 mM NaF and 30–50 percent for 10^{-4} M l-epinephrine. Maximal stimulation by l-epinephrine was obtained at 10^{-4} M, half-maximal stimulation being seen at approximately 5×10^{-6} M. When stimulatory effects of l-epinephrine and NaF on the adenylate cyclase activity were examined as functions of ATP concentrations at the fixed $MgCl_2$ concentration of 10 mM (fig. 6), Lineweaver-Burk plots show that the maximal velocities are 0.33, 0.42, and 0.63 nmole of cyclic AMP/min/mg of protein for the basal, l-epinephrine- and NaF-stimulated activities, respectively, whereas the Michaelis constant, K_m, for ATP is not significantly altered by either l-epinephrine or NaF. The maximal rate of basal adenylate cyclase, 0.33 nmole of cyclic AMP/min/mg of protein, is comparable with the value obtained by Williamson *et al.* (52) and is much higher than those reported previously for myocardial adenylate cyclase in other preparations (11, 12, 27, 30). The high specific activity of this preparation may reflect the extraction of contractile proteins during the preparation, and the removal of Ca^{2+}, an inhibitor of myocardial adenylate cyclase, by inclusion of EGTA during the preparation and in the assay media.

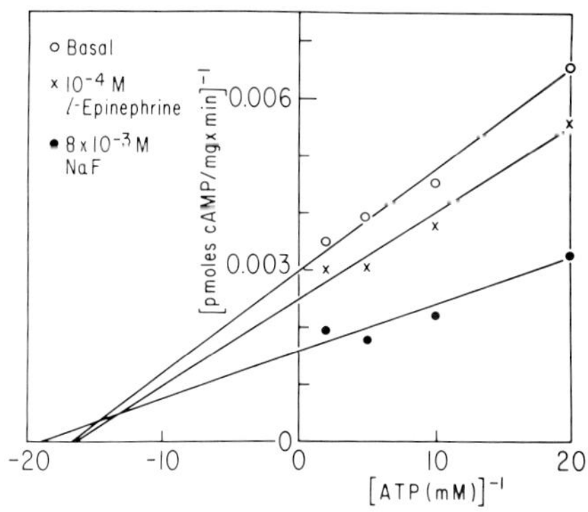

Figure 6 Lineweaver-Burk plot of the ATP concentration dependence of basal, l-epinephrine- and NaF-stimulated adenylate cyclase activities of cardiac sarcolemma. Plasma membranes (0.4 mg/ml) were incubated at 37°C for 10 min in 50 mM Tris-HCl (pH 7.5), 10 mM $MgCl_2$, 1 mM EGTA, 0.1 percent bovine serum albumin, 0.5 mM unlabeled cyclic AMP, 25 mM di-Tris phosphocreatine, 1 mg/ml creatine phosphokinase, and different concentrations of [α-^{32}P] ATP (about 50 μCi/μmole) in the absence (○) and presence of l-epinephrine (×) and 8 mM NaF (•). The amount of cyclic AMP released was determined by thin layer chromatography in 0.3 M LiCl on polyethyleneimine-impregnated plastic sheets by the procedures described by Bär and Hechter (2). (Reproduced from Tada *et al.* (45) by permission of *Circulation Research.*)

CONTROL OF CYCLIC AMP LEVELS BY Ca^{2+}

The activities of basal, l-epinephrine and NaF-stimulated adenylate cyclase depend on the ionized Ca^{2+} concentration. The absolute levels of basal and l-epinephrine-stimulated adenylate cyclase activities decrease as Ca^{2+} concentration is increased, but the extent of stimulation by l-epinephrine is greater at higher Ca^{2+} concentrations (fig. 7). The l-epinephrine sensitivity of adenylate cyclase activity was examined in the presence of 10^{-7} M and 10^{-5} M Ca^{2+} (fig. 8). Increasing the Ca^{2+} concentration causes very little shift in the l-epinephrine concentration needed for half-maximal stimulation: from 2 μM in 10^{-7} M Ca^{2+} to 5 μM in 10^{-5} M Ca^{2+} (both of which are similar to the value of 5 μM measured in 0.5 mM EGTA under slightly different conditions), suggesting that Ca^{2+} inhibits primarily an l-epinephrine-insensitive component of adenylate cyclase. It should be noted that adenylate cyclase activity is much less when reactions are carried out at pH 6.8 and 25°C, in contrast to the other studies reported here, in which the pH is 7.4 and temperature is 37°C. However, the pH requirements for use of Ca-EGTA buffers makes it necessary to work away from the optimal conditions. Ca^{2+} has been generally, but not always, considered to be an inhibitor of

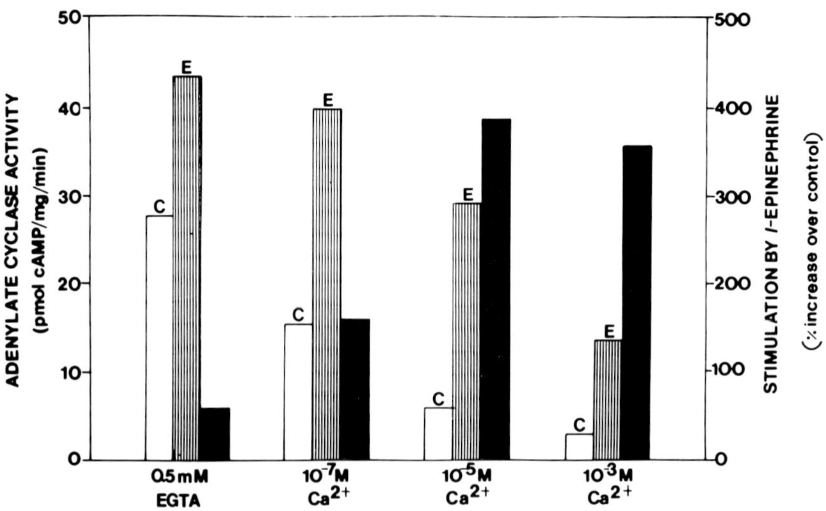

Figure 7 Effects of Ca^{2+} on adenylate cyclase of cardiac sarcolemma. Plasma membranes (0.8 mg/ml) were incubated at 25°C for 20 min in 5 mM [α-^{32}P] ATP (~100 μCi/μmole), 5 mM $MgCl_2$, 0.5 mM unlabeled cyclic AMP, various concentrations of ionized Ca^{2+}, and 100 mM imidazole buffer (pH 6.8). In studies at low Ca^{2+} concentrations (below 10^{-5} M), levels of ionized Ca^{2+} were maintained by Ca-EGTA buffers. The amount of cyclic AMP produced was determined as described in figure 5. (Reproduced from Tada *et al.* (45) by permission of *Circulation Research.*)

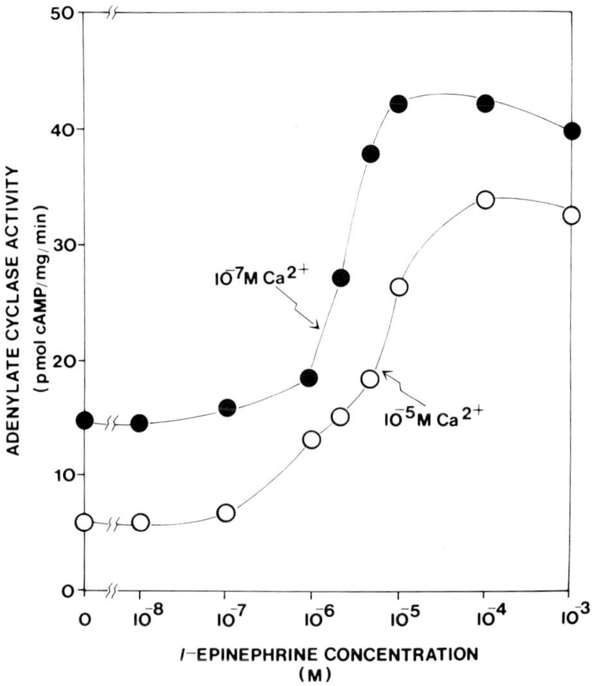

Figure 8 Depencence on Ca^{2+} concentration of l-epinephrine stimulation of cardiac sarcolemmal adenylate cyclase activity. Experimental conditions were the same as those described in figure 6. Ca^{2+} concentrations were maintained by Ca-EGTA buffers at 10^{-5} M (○) and 10^{-7} M (●). Each point represents an average of four determinations. (Reproduced from Tada *et al.* (45) by permission of *Circulation Research.*)

adenylate cyclase from many tissues, including myocardium (10, 39). In contrast, however, other studies with noncardiac preparations indicate that EGTA can inhibit basal and/or polypeptide hormone-stimulated adenylate cyclase (1, 4, 5, 7, 25). These studies also show that Ca^{2+} has a stimulatory action at low concentrations while higher Ca^{2+} concentrations inhibit these activities.

The functional significance of the apparent Ca^{2+} dependence of cardiac plasma membrane adenylate cyclase is not clear. The present findings indicate that profound changes in basal levels of cyclic AMP production can occur in the range of Ca^{2+} concentrations that is found within the myocardial cell. Thus, at 10^{-7} M Ca^{2+}, which approximates the Ca^{2+} level during diastole (16), basal adenylate cyclase activity is almost twice that in 10^{-5} M Ca^{2+}, the systolic level of Ca^{2+}. These findings stand in contrast to the fact that the opposing enzyme, cyclic nucleotide phosphodiesterase, from many tissues including myocardium is stimulated by Ca^{2+} ions (9, 14, 15, 29, 50). Cyclic AMP levels within the myocardial cells have been reported to fluctuate during the contraction-relaxation cycle (6, 51). These findings might reflect the fluctuations of the activities of adenylate cyclase and phosphodiesterase in response to varying Ca^{2+} concentrations, which fluctuate within the cells by more than 100-fold during each contraction-relaxation cycle (16). If these findings are relevant to the physiological behavior of the myocardium, they might represent a negative feedback mechanism by which excessive increases in contractility could be avoided under conditions where Ca^{2+} influx into the myocardial cell is increased (45). Thus, high levels of intracellular Ca^{2+}, by inhibiting adenylate cyclase (fig. 7) and stimulating phosphodiesterase, would reduce the levels of cyclic AMP within the cell, thereby lessening the retention of Ca^{2+} within the sarcoplasmic reticulum, the mechanism of which we have also been studying (20–23, 44, 46–48). Such an inhibitory effect of Ca^{2+} can occur while the response to epinephrine is retained (fig. 8), so that Ca^{2+} appears not to impair the ability of the heart to respond to the inotropic influences of β-adrenergic agonists. This negative feedback mechanism might serve to protect the ischemic heart from the detrimental effects of a continued high rate of energy utilization by the contraction-relaxation system and thereby delay the advent of rigor (3, 17).

REFERENCES

1. BÄR, H. P., and HECHTER, O. 1969. Adenyl cyclase and hormone action. III. Calcium requirement of ACTH stimulation of adenyl cyclase. Biochem. Biophys. Res. Commun. 35: 681–686.
2. BÄR, H. P., and HECHTER, O. 1969. Adenyl cyclase in fat cell ghosts. Anal. Biochem. 29: 476–489.
3. BING, O. H. L., BROOKS, W. W., and MESSER, J. V. 1973. Heart muscle viability following hypoxia: Protective effect of acidosis. Science 180: 1297–1298.
4. BOCKAERT, J., ROY, C., and JARD, S. 1972. Oxytocin-sensitive adenylate cyclase in frog bladder epithelial cells: Role of calcium, nucleotides and other factors in hormonal stimulation. J. Biol. Chem. 247: 7073–7081.
5. BRADHAM, L. S., HOLT, D. A., and SIMS, M. 1970. The effect of Ca^{2+} on the adenyl cyclase of calf brain. Biochim. Biophys. Acta 201: 250-260.
6. BROOKER, G. 1973 Oscillation of cyclic adenosine monophosphate concentration during the myocardial contraction cycle. Science 182: 933–934.
7. CAMPBELL, B. J., WOODWARD, G., and BORBERG, V. 1972. Calcium-mediated interactions between the antidiuretic hormone and renal plasma membranes. J. Biol. Chem. 247: 6167–6175.
8. CARVALHO, A. P., MADEIRA, V. M. C., and ANTUNES-MADEIRA, M. C. 1971. Effect of potentiators of muscular contraction on contractile and enzymatic activities of sarcolemma. Biochim. Biophys. Acta 234: 210–221.
9. CHEUNG, W. Y. 1971. Cyclic 3′,5′-nucleotide phosphodiesterase: Evidence for and properties of a protein activator. J. Biol. Chem. 246: 2859–2869.
10. DRUMMOND, G. I., and DUNCAN, L. 1970. Adenyl cyclase in cardiac tissue. J. Biol. Chem. 245: 976–983.
11. DRUMMOND, G. I., SEVERSON, D. L., and DUNCAN, L. 1971. Adenylate cyclase: Kinetic properties and nature of fluoride and hormone stimulation. J. Biol. Chem. 246: 4166–4173.

12. EPSTEIN, S. E., LEVEY, G. S., and SKELTON, C. L. 1971. Adenyl cyclase and cyclic AMP: Biochemical links in the regulation of myocardial contractility. Circulation 43: 437–450.
13. FAWCETT, D. W., and McNUTT, N. S. 1969. The structure of cat myocardium. I. Ventricular papillary muscle. J. Cell Biol. 42: 1–45.
14. KAKIUCHI, S., and YAMAZAKI, R. 1970. Calcium dependent phosphodiesterase activity and its activating factor (PAF) from brain. Biochem. Biophys. Res. Commun. 41: 1104–1110.
15. KAKIUCHI, S., YAMAZAKI, R., TESHIMA, Y., and UENISHI, K. 1973. Regulation of nucleoside cyclic 3′,5′-monophosphate phosphodiesterase activity from rat brain by a modulator and Ca^{2+}. Proc. Nat. Acad. Sci. U. S. A. 70: 3526–3530.
16. KATZ, A. M. 1970. Contractile proteins of the heart. Physiol. Rev. 50: 63–158.
17. KATZ, A. M. 1973. Effects of ischemia on the contractile processes of heart muscle. Amer. J. Cardiol. 32: 456–460.
18. KATZ, A. M., REPKE, D. I., and COHEN, B. R. 1966. Control of the activity of highly purified cardiac actomyosin by Ca^{2+}, Na^+ and K^+. Circ. Res. 19: 1062–1070.
19. KATZ, A. M., REPKE, D. I., UPSHAW, J. E., and POLASCIK, M. A. 1970. Characterization of dog cardiac microsomes. Use of zonal centrifugation to fractionate fragmented sarcoplasmic reticulum, $(Na^+ + K^+)$-activated ATPase and mitochondrial fragments. Biochim. Biophys. Acta 205: 473–490.
20. KATZ, A. M., TADA, M., and KIRCHBERGER, M. A. 1974. Control of calcium transport in the myocardium by the cyclic AMP-protein kinase system, pp. 453–472. *In* G. I. Drummond, P. Greengard and G. A. Robison (eds.), Advances in Cyclic Nucleotide Research, Vol. 5. Raven Press, New York.
21. KIRCHBERGER, M. A., TADA, M., and KATZ, A. M. 1974. Adenosine 3′:5′-monophosphate-dependent protein kinase-catalyzed phosphorylation reaction and its relationship to calcium transport in cardiac sarcoplasmic reticulum. J. Biol. Chem. 249: 6166–6173.
22. KIRCHBERGER, M. A., TADA, M., and KATZ, A. M. 1974. Phospholamban: A regulatory protein of the cardiac sarcoplasmic reticulum, pp. 103–115. *In* A. Fleckenstein and N. S. Dhalla (eds.), Recent Advances in Studies on Cardiac Structure and Metabolism, Vol. 5. University Park Press, Baltimore.
23. KIRCHBERGER, M. A., TADA, M., REPKE, D. I., and KATZ, A. M. 1972. Cyclic adenosine 3′,5′-monophosphate-dependent protein kinase stimulation of calcium uptake by canine cardiac microsomes. J. Molec. Cell. Cardiol. 4: 673–680.
24. KONO, T., and COLOWICK, S. P. 1961. Isolation of skeletal muscle cell membrane and some of its properties. Arch. Biochem. Biophys. 93: 520–533.
25. LEFKOWITZ, R. J., ROTH, J., and PASTAN, I. 1970. Effects of calcium on ACTH stimulation of the adrenal: Separation of hormone binding from adenyl cyclase activation. Nature 228: 864–866.
26. MATSUI, H., and SCHWARTZ, A. 1966. Purification and properties of a highly active ouabain sensitive Na^+,K^+-dependent adenosinetriphosphatase from cardiac tissue. Biochim. Biophys. Acta 128: 380–390.
27. MAYER, S. E. 1972. Effects of adrenergic agonists and the antagonists on adenylate cyclase activity of dog heart and liver. J. Pharmacol. Exp. Therap. 181: 116-125.
28. McCOLLESTER, D. L. 1962. A method for isolating skeletal-muscle cell-membrane components. Biochim. Biophys. Acta 57: 427–437.
29. MIKI, N., and YOSHIDA, H. 1972. Purification and properties of cyclic AMP phosphodiesterase from rat brain. Biochim. Biophys. Acta 268: 166–174.
30. MURAD, F., CHI, Y. M., RALL, T. W., and SUTHERLAND, E. W. 1962. Adenyl cyclase III. The effect of catecholamines and choline esters on the formation of adenosine 3′,5′-phosphate by preparations from cardiac muscle and liver. J. Biol. Chem. 237: 1233–1238.
31. PERKINS, J. P. 1973. Adenyl cyclase, pp. 1–63. *In* P. Greengard and G. A. Robison (eds.), Advances in Cyclic Nucleotide Research Vol. 3. Raven Press, New York.

32. PORTIUS, H. J., and REPKE, K. R. H. 1967. Darstellung des Na^+ + K^+-aktivierten, Mg^{++}-abhängigen Adenosintriphosphat Phosphohydrolase-Systems des Herzmuskels durch Isolierung der Zellmembran. Acta Biol. Med. Germ. 19: 876–906.
33. ROBISON, G. A., BUTCHER, R. W., and SUTHERLAND, E. W. 1967. Adenyl cyclase as an adrenergic receptor. Ann. N. Y. Acad. Sci. 139: 703–723.
34. ROBISON, G. A., BUTCHER, R. W., and SUTHERLAND, E. W. 1971. Cyclic AMP. Academic Press, New York.
35. ROSTGAARD, J., and BEHNKE, O. 1965. Fine structural localization of adenine nucleotide phosphatase activity in the sarcoplasmic reticulum and the T-system of rat myocardium. J. Ultrastruct. Res. 12: 579–591.
36. SCHLATZ, L., and MARINETTI, G. V. 1972. Calcium binding to the rat liver plasma membrane. Biochim. Biophys. Acta 290: 70–83.
37. SEVERSON, D. I., DRUMMOND, G. I., and SULAKHE, P. V. 1972. Adenylate cyclase in skeletal muscle: Kinetic properties and hormone stimulation. J. Biol. Chem. 247: 2949–2958.
38. STAM, A. C., JR., WEGLICKI, W. B., JR., FELDMAN, D., SHELBURNE, J. C., and SONNENBLICK, E. H. 1970. Canine myocardial sarcolemma - its preparation and enzymatic activity. J. Molec. Cell. Cardiol. 1: 117–130.
39. SULAKHE, P. V., and Dhalla, N. S. 1973. Adenylate cyclase of heart sarcotubular membranes. Biochim. Biophys. Acta 293: 379–396.
40. SULAKHE, P. V., DRUMMOND, G. I., and NG, D. C. 1973. Calcium binding by skeletal muscle sarcolemma. J. Biol. Chem. 248: 4150–4157.
41. SULAKHE, P. V., DRUMMOND, G. I., and NG, D. C. 1973. Adenosine triphosphatase activities of muscle sarcolemma. J. Biol. Chem. 248: 4158–4162.
42. TADA, M. 1967. Adenosine triphosphatase activity and superprecipitation of canine cardiac myosin B. J. Biochem. 62: 658–672.
43. TADA, M., FINNEY, J. O., JR., SWARTZ, M. H., and KATZ, A. M. 1972. Preparation and properties of plasma membranes from guinea pig hearts. J. Molec. Cell. Cardiol. 4: 417–426.
44. TADA, M., and KIRCHBERGER, M. A. 1975. Regulation of calcium transport by cyclic AMP: A proposed mechanism for the beta-adrenergic control of myocardial contractility. Acta Cardiologica, in press.
45. TADA, M., KIRCHBERGER, M. A., IORIO, J. A., and KATZ, A. M. 1975. Control of cardiac sarcolemmal adenylate cyclase and sodium, potassium-activated adenosinetriphosphatase activities. Circ. Res. 36: 8–17.
46. TADA, M., KIRCHBERGER, M. A., and KATZ, A. M. 1975. Phosphorylation of a 22,000-dalton component of the cardiac sarcoplasmic reticulum by adenosine 3′:5′-monophosphate-dependent protein kinase. J. Biol. Chem. 250: 2640–2647.
47. TADA, M., KIRCHBERGER, M. A., and KATZ, A. M. 1975. Regulation of calcium transport in cardiac sarcoplasmic reticulum by cyclic AMP-dependent protein kinase. *In* P. E. Roy and N. S. Dhalla (eds.), Recent Advances in Studies on Cardiac Structure and Metabolism, Vol. 9, University Park Press, Baltimore.
48. TADA, M., KIRCHBERGER, M. A., REPKE, D. I., and KATZ, A. M. 1974. Stimulation of calcium transport in cardiac sarcoplasmic reticulum by adenosine 3′:5′-monophosphate-dependent protein kinase. J. Biol. Chem. 249: 6174–6180.
49. TAUSSKY, H. H., and SHORR, E. 1953. A microcolorimetric method for the determination of inorganic phosphorus. J. Biol. Chem. 202: 675–685.
50. TEO, T. S., and WANG, J. H. 1973. Mechanism of activation of a cyclic adenosine 3′:5′-monophosphate phosphodiesterase from bovine heart by calcium ions. J. Biol. Chem. 248: 5950–5955.
51. WESTORT, C., and HULTIN, H. O. 1966. A procedure for the preparation of empty cell segments from skeletal muscle using solutions of low ionic strength. Anal. Biochem. 16: 314–319.
52. WILLIAMSON, J. R., WOODROW, M. L., and SCARPA, A. 1975. Calcium binding to cardiac sarcolemma. *In* A. Fleckenstein and N. S. Dhalla (eds.), Recent Advances in

Studies on Cardiac Structure and Metabolism, Vol. 5, pp. 61–71. University Park Press, Baltimore.

53. WOLLENBERGER, A., BABSKII, E. B., KRAUSE, E.-G., GENZ, S., BLOHM, D., and BOGDANOVA, E. V. 1973. Cyclic changes in levels of cyclic AMP and cyclic GMP in frog myocardium during the cardiac cycle. Biochem. Biophys. Res. Commun. 55: 446–452.
54. YOSHIMURA, K. 1973. Activation of Na-K activated ATPase in rat brain by catecholamine. J. Biochem. 74: 389–391.

Mailing address: Dr. M. Tada,
The First Department of Internal Medicine,
Osaka University Medical School,
Fukushima-ku, Osaka 553, Japan.

Recent Advances in Studies on Cardiac Structure and Metabolism
Volume 9
The Sarcolemma
Edited by Paul-Emile Roy and Naranjan S. Dhalla

CHARACTERIZATION OF SARCOLEMMA FROM RABBIT SKELETAL MUSCLE[1]

W. DRABIKOWSKI and E. ZUBRZYCKA

Department of Biochemistry of Nervous System and Muscle,
Nencki Institute of Experimental Biology,
Warsaw, Poland

SUMMARY

Sodium dodecyl sulfate polyacrylamide gel electrophoresis reveals in the rabbit skeletal muscle sarcolemma the presence of four major protein bands corresponding to molecular weight 216,000, 110,000, 44,000, 15,000, and smaller amounts of 148,000, 78,000, 68,000, 37,000, 27,000 proteins. (Na^+-K^+)-ATPase isolated from rabbit skeletal muscle contains 102,000, 44,000, and 32,000 protein bands. Sarcolemma is resistant to the solubilizing effect of Triton X-100, Lubrol, and deoxycholate. Sarcolemmal proteins, especially sodium-potassium-ATPase, are sensitive to digestion with trypsin.

Chemical composition of the skeletal muscle plasma membrane has been less studied in relation to the second membranous system specific for muscle, *i.e.*, to sarcoplasmic reticulum and to plasma membrane from other tissues, especially from liver.

The aim of this work was to characterize the protein composition of sarcolemma with the use of sodium dodecyl sulfate polyacrylamide gel electrophoresis. An attempt was also made to solubilize sarcolemma and to fractionate proteins present in this biomembrane. The amount and composition of protein-bound carbohydrates and lipids was also analyzed. Part of the results was presented at the Ninth International Congress of Biochemistry (30).

MATERIAL AND METHODS

Preparation of sarcolemma

Sarcolemma was prepared from back and leg rabbit muscles according to the method of Abood *et al.* (1) or that of Peter (23). In the latter case two

[1] This work was partly supported by National Institutes of Health Foreign Research Agreement under P.L. 480 Grant No. 05-015-1.

additional extractions with 0.6 M KCl were employed to remove actomyosin contamination (see Results). All preparations were suspended in deionized water and homogenized in a glass-Teflon homogenizer. Purity of the preparations, examined under phase contrast and electron microscopy, revealed a high degree of purity. The preparations obtained were free of contamination by fragments of sarcoplasmic reticulum, mitochondria, or myofilaments. In agreement with Abood *et al.* (1) it was found also in this work that because of prolonged treatment with 0.6 M KCl, the obtained sarcolemma preparations consisted of plasma (inner) and basement (middle) membranes, and were essentially devoid of collagen fibers.

Preparation of (Na^+-K^+)-ATPase

(Na^+-K^+)-ATPase was prepared from rabbit skeletal muscle according to the method of Matsui and Schwartz (19) or that of Pitts *et al.* (24). Activity of (Na^+-K^+)-ATPase was determined in a medium containing 60 mM imidazole buffer, pH 7.4, 5 mM ATP, 100 mM NaCl, 20 mM KCl, 5 mM $MgCl_2$. The suspension was incubated at 37°C for 10 min. Ouabain sensitivity was determined in the presence of 2 mM ouabain.

Polyacrylamide gel electrophoresis

Sodium dodecyl sulfate (SDS) polyacrylamide gel electrophoresis was performed according to the method of Weber and Osborn (29). For solubilization about 100 μg of sarcolemma were heated for 5 min at 100°C in a medium containing 1.5 percent SDS and 0.33 percent β-mercaptoethanol in phosphate buffer, pH 7.0. Five percent polyacrylamide gel was used. Current was 8 mA/tube. After electrophoresis the gels were stained with Coomassie blue for several hours and then destained in a solution containing acetic acid and methanol. Catalase, α-actinin, ovalbumin, actin, myosin, cytochrome *c*, chymotrypsin, and the TN-C component of troponin were used as standard proteins.

Analytical procedures

Protein content was determined according to the method of Lowry *et al.* (15) after solubilization of the suspension of sarcolemma in 1 N NaOH with bovine serum albumin as standard. Total lipids were extracted by the method of Folch *et al.* (10). Lipid phosphorus was determined according to the method of Chen *et al.* (4). Lipids were separated by thin layer chromatography on silica G plates using chloroform-methanol-water (65:25:4, v/v/v) as a system for phospholipids and petrol ether-diethyl ether-acetic acid (70:30:1, v/v/v) as a system for neutral lipids. The content of total sterols was determined according to the method of Searcy *et al.* (26).

RESULTS

Protein composition

During the early stage of this work various procedures for preparation of sarcolemma were used and the protein pattern of the preparations obtained was compared. Finally two procedures were used, each based on different principles. One, employing only low ionic strength solutions, was used in this work in the modification of the method of Peter (23). The other method, as described by Abood *et al.* (1), was based on extraction with high ionic strength solutions (0.4 M LiBr and 0.6 M KCl).

In our case the latter method gave a reproducible yield of sarcolemma, approximately 16 mg of protein/100 g wet weight of muscle. Somewhat more material was obtained by the method of Peter. This value is similar to the yield obtained by other authors (23, 27).

Rabbit muscle sarcolemma obtained with the use of either method revealed the same qualitative pattern of protein composition when examined in SDS-polyacrylamide gel electrophoresis (fig. 1, *gels 1* and *2*). Four major protein bands with the average molecular weights of 216,000, 110,000, 44,000, and 15,000 were present. Other proteins of molecular weight 148,000, 78,000, 68,000, 37,000, and 27,000 could be also detected, although in somewhat smaller amounts. A very similar protein pattern was obtained from crayfish sarcolemma (fig. 1, *gel 3*). Preparations of sarcolemma obtained according to the method of Peter (23) contained much larger amounts of proteins with molecular weights 216,000 and 44,000 than the preparations obtained with the use of high ionic strength solutions.

The protein pattern obtained in this work is very similar to that reported recently by Madeira and Antunes-Madeira (17). It is somewhat simpler, since in our case four major protein bands were found to contain most of the total protein, whereas some minor protein bands found in sarcolemma by Madeira and Antunes-Madeira were virtually absent in our preparations. The molecular weight of 216,000 and 44,000 proteins suggested that they may correspond to a heavy chain of myosin and actin, respectively, and hence orginate from myofibrils which can contaminate preparations of sarcolemma obtained at low ionic strength. In addition, the band in the region 216,000 can originate from collagen.

Indeed, additional extractions with 0.6 M KCl of sarcolemma obtained according to the method of Peter (23) removed part of the protein material of these two bands. However, all preparations of sarcolemma, even those obtained according to the method of Abood (1), in spite of several exaustive extractions with high ionic strength solutions, still contained 216,000 and 44,000 molecular weight proteins (fig. 2). These proteins were not released even when sarcolemma was extracted with 0.6 M KCl after disruption by sonication. As pointed out by

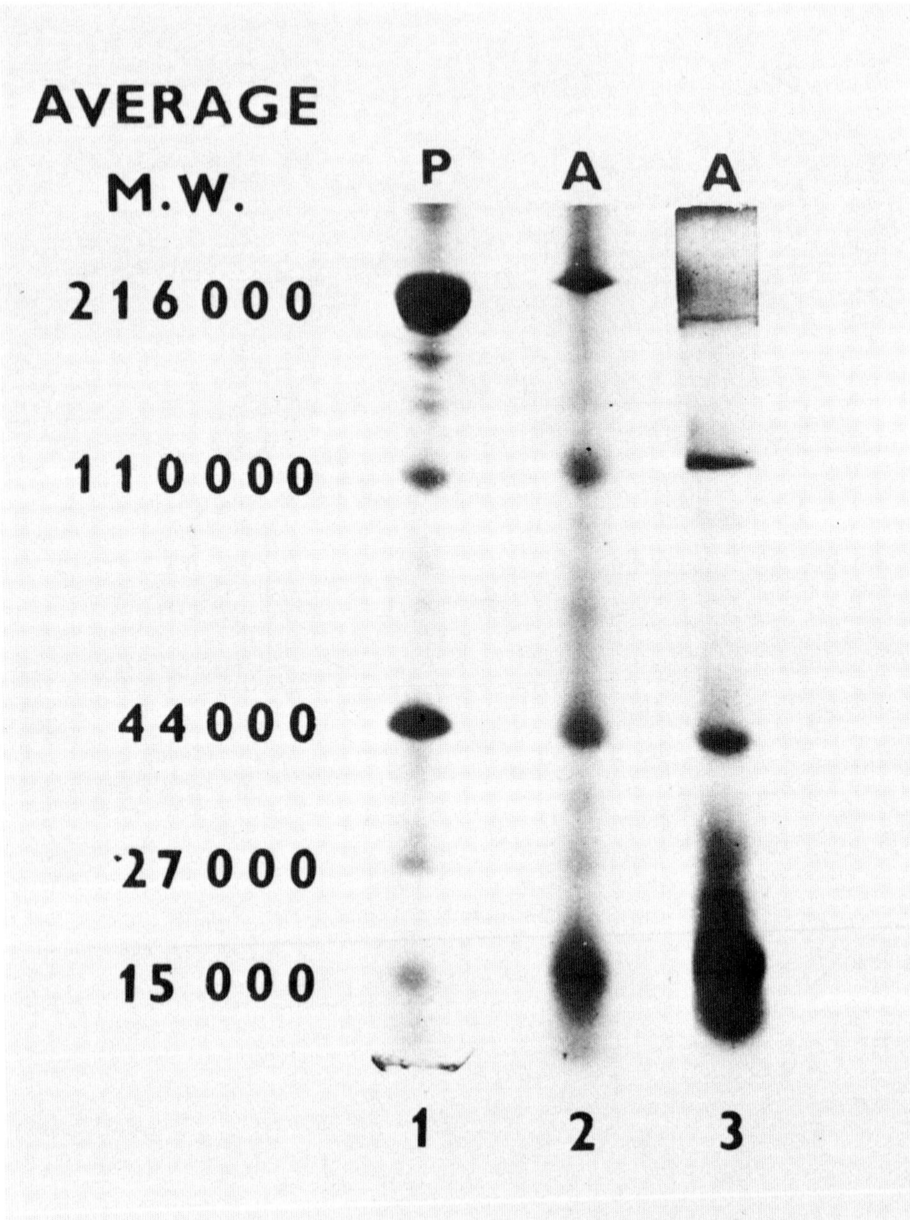

Figure 1 SDS-gel electrophoretic pattern of sarcolemma. *P:* Preparation of sarcolemma prepared according to the method of Peter (23); *A:* those obtained according to the method of Abood *et al.* (1). *Gels 1* and *2:* sarcolemma obtained from rabbit skeletal muscle; *gel 3:* from crayfish.

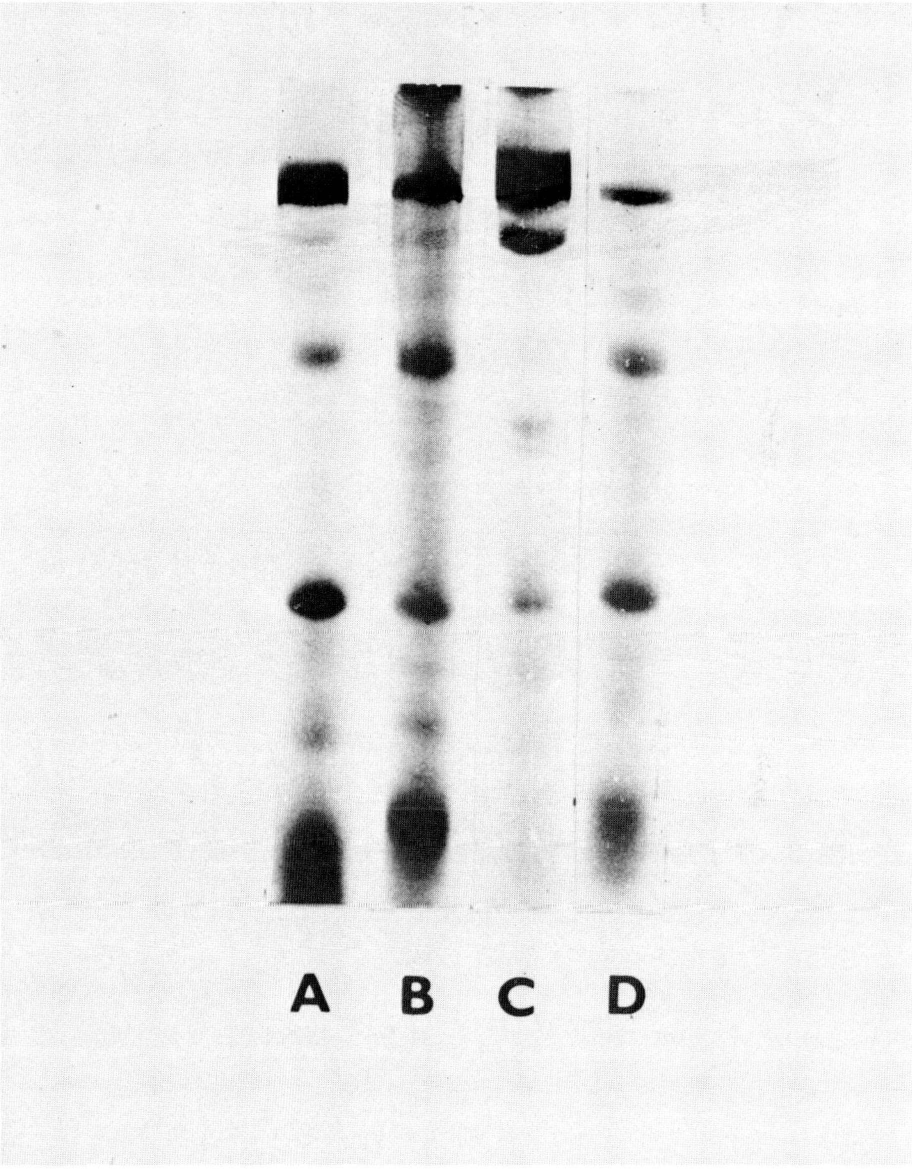

Figure 2 Extraction of sarcolemma with 0.6 M KCl. Sarcolemma prepared according to the method of Abood *et al.* (1) before extraction with 0.6 M KCl (*A*) was treated twice with 0.6 M KCl for 2 hr and subsequently centrifuged at 5,000 × *g* for 15 min. *B:* Precipitate; *C:* extract. Precipitate *B* was sonicated, treated again with 0.6 M KCl, and centrifuged at 80,000 × *g*. *D:* the precipitate obtained. In the supernatant only traces of protein were detected.

Abood *et al.* (1), extraction with high salt concentrations leads to the removal not only of actin and myosin but also of collagen. Thus, it may be concluded that proteins found in the regions of 216,000 and 44,000 are genuine constituents of sarcolemma.

Protein-bound carbohydrates

Determination of the amount of protein-bound carbohydrates performed in this laboratory (H. Walkowiak, unpublished data) showed that sarcolemma contained, on the average, 1.5 moles of hexosamines, 2.0 moles of hexoses, 0.08 mole of sialic acid, and 0.2 mole of methylpentoses per 100,000 g of protein (table I). The ratio of hexosamines to sialic acid is about 20. In addition, preparations of sarcolemma contained 0.9 mole of ribose and 1.9 moles of deoxyribose as contaminants.

The values for protein-bound sialic acid and hexoses found in this work are very similar to those reported by Abood *et al.* (1). On the other hand, Kono *et al.* (12) recorded much higher amounts of sialic acid and Madeira and Antunes-Madeira (17) higher amounts of hexoses.

With the use of gas-liquid chromatography (6), as well as paper chromatography (11), galactosamine, glucosamine, galactose, glucose, fucose, mannose,

Table I Protein-bound carbohydrate content in sarcolemma (H. Walkowiak, unpublished data)[1]

	Moles/10^5 gm protein
Sialic acid	0.05–0.12 (5)
Total hexosamines	0.80–1.70 (5)
Glucosamine	0.60–0.90 (2)
Galactosamine	0.18–0.22 (2)
Methylpentoses	0.10–0.40 (5)
Hexoses	1.70–2.60 (4)
Contaminating sugars	
Ribose	0.70–1.60 (5)
Deoxyribose	1.00–3.20 (5)

[1] Sarcolemma was prepared according to the method of Peter (23) and extracted twice with 0.6 M KCl. Carbohydrate content was determined after delipidation according to the method of Folch *et al.* (10). The individual sugars were determined by the following colorimetric methods: sialic acid (28), hexosamines (3), methylpentoses (8), hexoses (21), ribose (20), and deoxyribose (7). Glucosamine and galactosamine were determined by gas-liquid chromatography (6). The number of sarcolemma preparations used for determinations is given in parentheses.

xylose, and rhamnose were identified among the protein-bound carbohydrates present in sarcolemma.

Lipid content and composition

Determination of lipids performed in this laboratory (M. G. Sarzala and M. Pilarska, unpublished data) showed that sarcolemma prepared according to the method of Abood *et al.* (1) contained, on the average 0.24 mg of lipid/mg of protein. Phospholipids accounted for 45–50 percent of total lipids. The main lipid in the fraction of neutral lipids is cholesterol, most of which is esterified. The lipid pattern is essentially in agreement with that reported by Madeira and Antunes-Madeira (17) and Mahrla and Zachar (18).

Determinations of glycolipids in sarcolemma (B. Czartoryska, unpublished data) indicated that the amount of these lipids was very low. Thin layer chromatography showed that the lower phase of extract in the procedure of Folch *et al.* (10) consisted mainly of monohexosylceramide (cerebroside) and dihexosylceramide. In the upper phase no gangliosides were detected.

The total amount of hexoses in sarcolemma lipids was about 0.6 μmole and 0.9 μmole/gm dry weight of sarcolemma in the lower and upper phase, respectively. Sialic acid was found only in amounts of about 0.02 μmole/gm sarcolemmal dry weight.

Solubilization and fractionation

Sarcolemma is relatively resistant to the solubilizing effect of detergents. Thus, Triton X-100, even in a concentration of 10 mg/mg of protein, solubilized only about 10 15 percent of sarcolemmal proteins. Another nonionic detergent, Lubrol, used by Palatini *et al.* (22) to solubilize (Na^+-K^+)-ATPase from bovine heart, also released only a small part of protein from sarcolemma. Preliminary results showed that sarcolemma treated with 8 mg of Lubrol/mg of protein was poorer in 110,000 molecular weight protein (fig. 3).

Sarcolemma was found also resistant to solubilization by deoxycholate. This detergent in the concentration of 1 mg/mg of sarcolemma protein did not solubilize any protein from this membrane. This rules out the possibility of contamination of preparations of sarcolemma by fragments of sarcoplasmic reticulum since the latter membrane is totally soluble under these conditions. Even at much higher concentrations (10 mg of deoxycholate/mg of protein) only about 10 percent of protein material, not sedimentable at 100,000 $\times$ *g*, was released from sarcolemma obtained according to the method of Abood *et al.* (1). This fraction consisted mainly of 44,000 and 37,000 proteins (fig 4). Much more protein (up to 60 percent) was extracted with high concentrations of deoxycholate from sarcolemma obtained according to the method of Peter (23). A similar value was found by Ferdman *et al.* (9). SDS-gel electrophoresis revealed in this case in the soluble fractions the presence of considerable amounts of

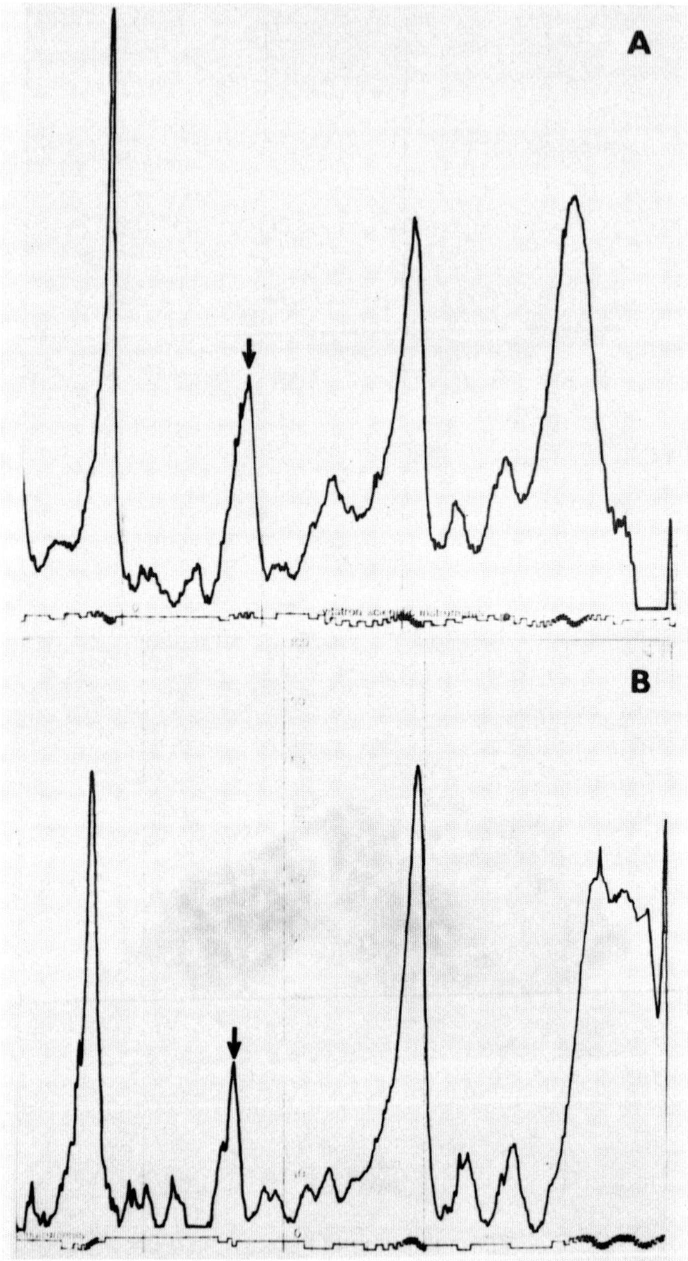

Figure 3 Extraction of sarcolemma with Lubrol. Incubation mixture contained sarcolemma (3 mg of protein/ml), 0.1 M KCl, 20 mM histidine, pH 7.2. After extraction with 8 mg of Lubrol/mg of protein the suspension was centrifuged at 80,000 × *g*. *A:* Densitometric tracing of untreated sarcolemma, *B:* tracing of residue after solubilization of sarcolemma with Lubrol. *Arrows* indicate the position of 110,000 molecular weight protein.

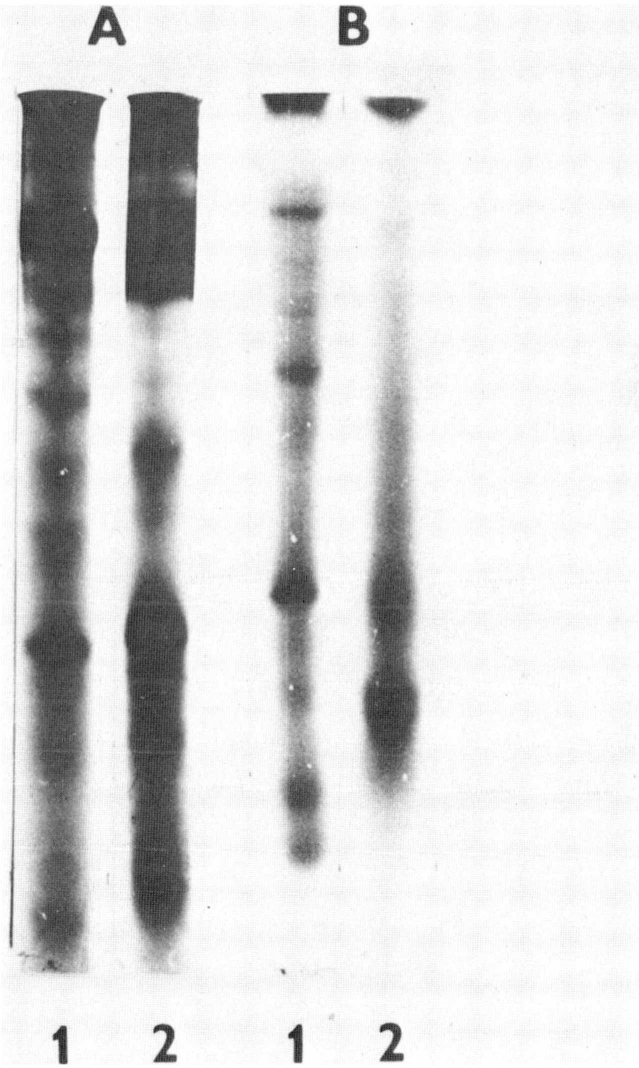

Figure 4 Treatment of sarcolemmal membranes with deoxycholate. Suspension of sarcolemma preparations obtained according to the method of Peter (*A*) or that of Abood *et al.* (*B*) were extracted with 10 mg of deoxycholate/mg of protein for 30 min at 4°C and centrifuged subsequently at 8,000 for 1 hr. *1:* Precipitate; *2:* solubilized proteins in supernatant.

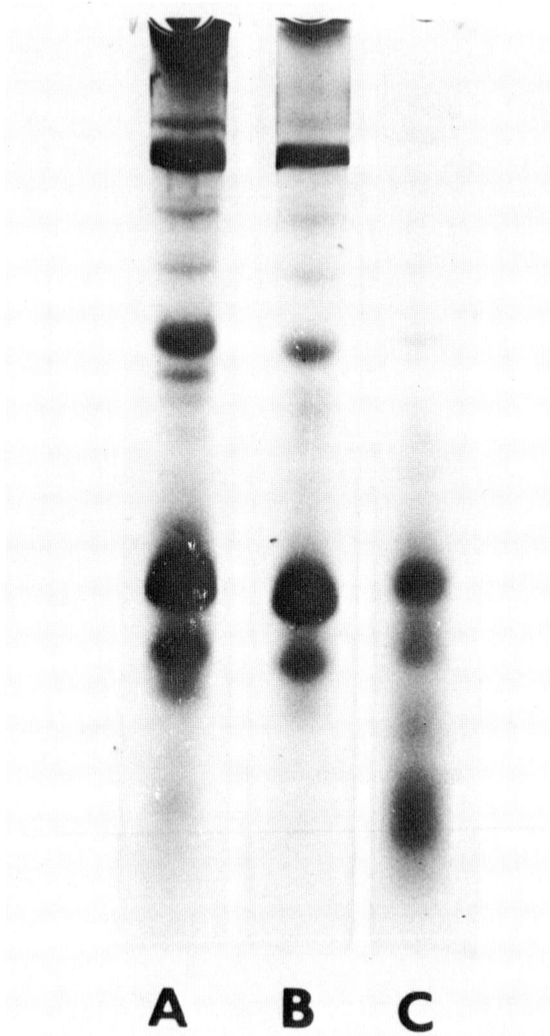

Figure 5 Treatment of sarcolemmal membranes with NaI. Suspension of sarcolemma (*gel A*) in deionized water was treated for 30 min with 3 M NaI at 4°C and centrifuged subsequently at 80,000 × *g* for 1 hr yielding precipitate (*gel B*) and extract; its protein pattern after dialysis against water is visible on *gel C*.

216,000 molecular weight protein, in addition to the 44,000 and the lower molecular weight proteins.

As an example of chaotropic agents, NaI was used for solubilization of sarcolemma. Treatment of sarcolemma with 3 M NaI for 30 min at 0°C resulted in the solubilization of a part of the protein in amounts somewhat depending on the method of preparation of sarcolemma. All proteins present in the original 3 M NaI extract precipitated after dialysis against water. That fraction consisted of a part of 44,000 and 37,000 and of the total 27,000 and 15,000 molecular weight proteins (fig. 5). Proteins of 216,000 and 110,000 were virtually not extracted by 3 M NaI.

Contrary to other biomembranes (16, 25, 31), no protein was released from sarcolemma as a result of a treatment with *n*-butanol. Neither were sarcolemma proteins degraded during several days of lasting incubation in a medium containing sodium dodecyl sulfate and β-mercaptoethanol.

Preparation of (Na^+-K^+)-ATPase

One of the functions of sarcolemma is the active transport of Na^+ and K^+ ions across the membrane. Therefore, as a further characterization of skeletal plasma membrane, an attempt was made to identify on the SDS-gel electropherogram the protein band(s) corresponding to the (Na^+-K^+)-ATPase. For this purpose this enzyme was isolated from skeletal muscle homogenate. The preparation obtained showed on the SDS-gel electropherogram protein bands corresponding to the following molecular weights: 102,000, 44,000, and 32,000 (fig. 6). (Na^+-K^+)-ATPase from various tissues other than skeletal muscle reveals in SDS-gel electrophoresis two subunits, one between 91,000 and 102,000 and the other with 57,000 molecular weight (2, 5, 13, 14). The two bands found in this work in the preparations of (Na^+-K^+)-ATPase from skeletal muscle correspond most probably to these two bands. The reason for the presence of the 32,000 protein band in our preparation is not clear at present.

The enzymatic activity of the obtained preparations of ATPase in the presence of Mg^{++}, and Na^+, and K^+ was about 11.0 μmoles Pi/mg of protein/hr and decreased to about 1.2 μmoles Pi in the presence of ouabain. Similar activity was found recently by Collins and Albers (5) in (Na^+-K^+)-ATPase prepared from cat heart.

Digestion of sarcolemma with trypsin

During 10-min digestion of sarcolemma with trypsin, at a ratio as low as 1:300 of trypsin to sarcolemmal protein, splitting of a part of 216,000, 110,000 and 44,000 molecular weight proteins was observed (fig. 7). Concomitantly, (Na^+-K^+)-ATPase activity completely disappeared, whereas Mg^{++}-ATPase activity decreased to about one-half. At higher concentrations of trypsin a more pronounced breakdown of individual proteins took place, so that, at a ratio of 1:20,

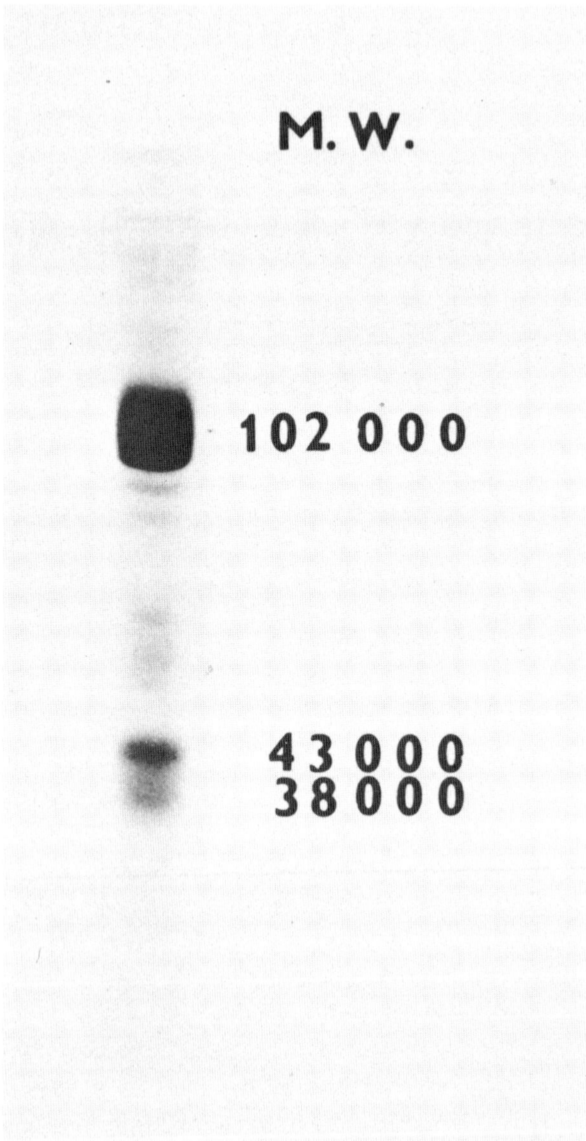

Figure 6 (Na^+-K^+)-ATPase from rabbit skeletal muscle prepared according to the method of Matsui and Schwartz (19).

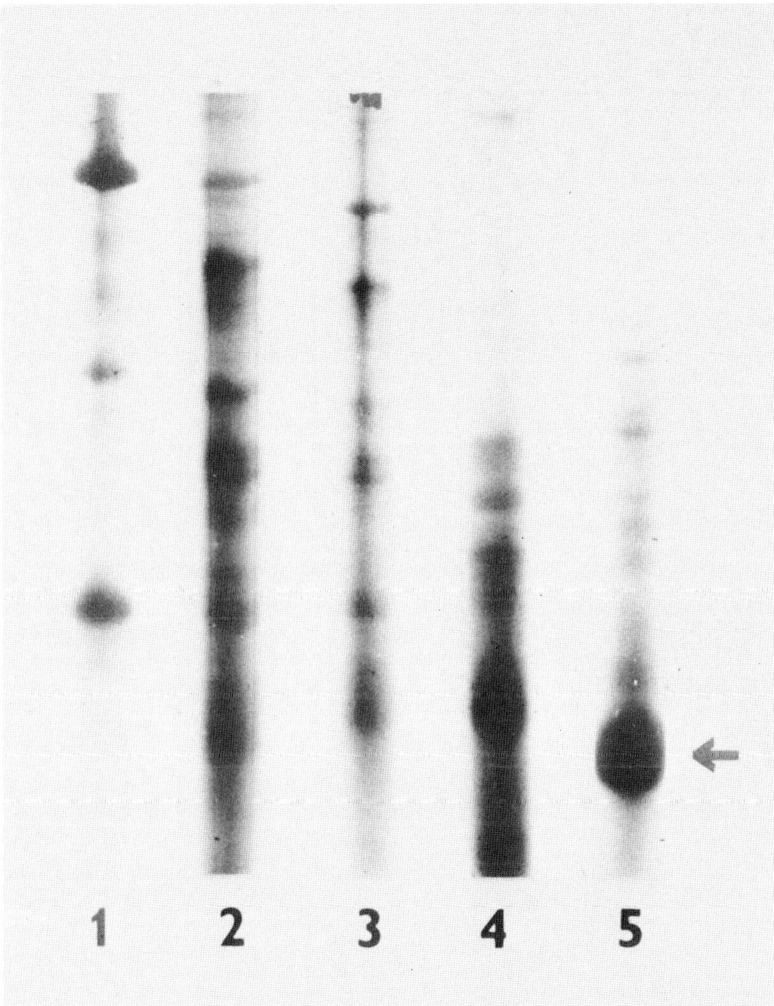

Figure 7 Effect of trypsin on sarcolemmal proteins. Sarcolemma suspended in a medium containing 20 mM histidine, pH 7.2, and 0.1 M KCl was digested with various concentrations of trypsin at room temperature for 10 min. The reaction was stopped by adding 4 mg of soya bean trypsin inhibitor per 1 mg of trypsin. The mixture was then centrifuged at 80,000 × *g* for 1 hr. The protein pattern in the pellets and supernatants was analyzed by gel electrophoresis. *Gel 1:* pellet obtained from control vesicles (trypsin and soya bean inhibitor added at zero time); *gel 2:* pellet from sarcolemma after digestion with trypsin (trypsin to sarcolemmal protein ratio 1:300); *gel 3:* 1:100; *gel 4, 1:20; gel 5,* supernatant after digestion at trypsin to protein ratio 1:20. *Arrow* indicates the position of soya bean inhibitor.

only proteins of molecular weight below 66,000 could be found. Most of the protein fragments of lower molecular weight, formed as a result of tryptic digestion, remained bound to the membrane. Only a small part of the protein fragments were released from the membrane. The amount of Lowry-positive material, which appears in the supernatant after 1 hr of centrifugation at 100,000 $\times$ g of digested sarcolemma, accounted for 7, 15, and 30 percent of the total protein for the digestion with trypsin at the ratios of 1:300, 1:100, and 1:20, respectively.

REFERENCES

1. ABOOD, L. G., KURAHASI, K., BRUNNGRABER, E., and KOKETSU, K. 1966. Biochemical analysis of isolated bullfrog sarcolemma. Biochim. Biophys. Acta 112: 330–339.
2. ALEXANDER, D. R., and RODNIGHT, R. 1970. Separation at neutral pH of a ^{32}P-labeled membrane protein associated with the sodium-plus-magnesium ion activated adenosine triphosphatase from ox brain. Biochem. J. 119: 44P.
3. BELCHER, R., NUTTEN, A., and SAMBROOK, C. M. 1954. The determination of glucosamine. Analyst 79: 201–208.
4. CHEN, P. S., TORIBARA, T. Y., and WARNER, H. 1956. Microdetermination of phosphorus. Anal. Chem. 28: 1756–1758.
5. COLLINS, R. C., and ALBERS, R. W. 1972. The phosphoryl acceptor protein of Na-K-ATPase from various tissues. J. Neurochem. 19: 1209–1213.
6. CRAVEN, D. A., and GEHRKE, C. W. 1968. Quantitative determination of N-acetylneuraminic acid by gas-liquid chromatography. J. Chromatogr. 37: 414–421.
7. DISCHE, Z. 1944. Two characteristic and sensitive color reactions between sulfhydryl compounds and thymonucleic acid. Proc. Soc. Exp. Med. 55: 217–218.
8. DISCHE, Z., and SHETTLES, L. B. 1948. A specific color reaction of methylpentoses and a spectrophotometric micromethod for their determination. J. Biol. Chem. 175: 595–603.
9. FERDMAN, D. L., HIMMELREICH, N. G., and DYADYUSHA, G. P. 1970. The enzyme activity in a detergent-treated sarcolemma of skeletal muscles. Biochim. Biophys. Acta 219: 372–378.
10. FOLCH, J., LEES, M., and SLOANE STANLEY, G. H. 1957. A simple method for the isolation and purification of total lipids from animal tissues. J. Biol. Chem. 226: 497–509.
11. HAKOMORI, S. I. 1966. Release of carbohydrates from sphingoglycolipid by osmium-catalyzed periodate oxidation followed by treatment with mild alkali. J. Lipid Res. 7: 789–792.
12. KONO, T., KAKUMA, F., HOMME, M., and FUKUDA, S. 1964. The electron-microscopic structure and chemical composition of the isolated sarcolemma of the rat skeletal muscle cell. Biochim. Biophys. Acta 8: 155–176.
13. KYTE, J. 1971. Purification of the sodium and potassium dependent adenosine triphosphatase from canine renal medulla. J. Biol. Chem. 246: 4157–4165.
14. KYTE, J. 1972. Properties of the two polypeptides of sodium- and potassium-dependent adenosine triphosphatase. J. Biol. Chem. 247: 7642–7649.
15. LOWRY, O. H., ROSEBROUGH, N. J., FARR, A. L., and RANDALL, R. J. 1951. Protein measurement with the Folin phenol reagent. J. Biol. Chem. 193: 265–275.
16. MADDY, A. H. 1966. The properties of the plasma membrane of ox erythrocytes. Biochim. Biophys. Acta 117: 193–200.
17. MADEIRA, V. M. C., and ANTUNES-MADEIRA, M. C. 1973. Chemical composition of sarcolemma isolateted from rabbit skeletal muscle. Biochim. Biophys. Acta 298: 230–238.

18. MAHRLA, Z., and ZACHAR, J. 1974. Lipid composition of isolated external and internal skeletal muscle membranes. Comp. Biochem. Physiol. 47B: 493–502.
19. MATSUI, H., and SCHWARTZ, A. 1966. Purification and properties of a highly active ouabain-sensitive Na^+,K^+-dependent adenosine-triphosphatase from cardiac tissue. Biochim. Biophys. Acta 128: 380–390.
20. MEJBAUM, W. 1939. Über die Bestimmung kleiner Pentosemengen, insbesondere in Derivaten der Adenylsäme. Z. Physiol. Chem. 258: 117–120.
21. MOKROSCH, L. C. 1954. Analysis of hexose phosphates and sugar mixtures with the anthrone reagent. J. Biol. Chem. 208: 55–59.
22. PALATINI, P., DABBENI-SALA, F., and BRUNI, A. 1972. Reactivation of a phospholipid-depleted sodium, potassium-stimulated ATPase. Biochim. Biophys. Acta 288: 413–422.
23. PETER, J. B. 1970. A (Na^+,K^+)ATPase of sarcolemma from skeletal muscle. Biochem. Biophys. Res. Commun. 40: 1362–1367.
24. PITTS, B. J. R., LANE, L. K., and SCHWARTZ, A. 1973. Nature of the transport ATPase-digitalis complex. VI. Purification of and ouabain binding to a highly active cardiac Na^+,K^+-ATPase. Biochem. Biophys. Res. Commun. 53: 1060–1066.
25. SARZALA, M. G., ZUBRZYCKA, E., and DRABIKOWSKI, W. 1974. Characterization of the constituents of sarcoplasmic reticulum membrane, pp. 315–346. *In* W. Drabikowski, H. Strzelecka-Golaszewska, and E. Carafoli (eds.), Calcium Binding Proteins. Polish Scientific Publishers Warszawa and Elsevier Co., Amsterdam.
26. SEARCY, R. L., BERGQUIST, L. M., and JUNG, R. C. 1960. Rapid ultramicro estimation of serum total cholesterol. J. Lipid Res. 1: 349–351.
27. SULAKHE, P. V., FEDELESOVA, M., McNAMARA, D. B., and DHALLA, N.S. 1971. Isolation of skeletal muscle membrane fragments containing active Na^+-K^+ stimulated ATPase: Comparison of normal and dystrophic muscle sarcolemma. Biochem. Biophys. Res. Commun. 42: 793–800.
28. WARREN, L. 1959. The thiobarbituric acid assay of sialic acids. J. Biol. Chem. 234: 1971–1975.
29. WEBER, K., and OSBORN, M. J. 1969. The reliability of molecular weight determination by dodecyl sulfate-polyacrylamide gel electrophoresis. J. Biol. Chem. 244: 4406–4412.
30. ZUBRZYCKA, E., WALKOWIAK, H., SARZALA, M. G., and DRABIKOWSKI, W. 1973. Protein and carbohydrate pattern in sarcolemma, p. 274. Ninth International Congress of Biochemistry, Stockholm, 1973. (Abstracts.)
31. ZWAAL, R. F. A., and VAN DEENEN, L. L. M. 1968. Protein patterns of red cells membranes from different mammalian species. Biochim. Biophys. Acta 163: 44–49.

Mailing address: Dr. W. Drabikowski,
3 Pasteur Street, 02-093 Warsaw (Poland).

Recent Advances in Studies on Cardiac Structure and Metabolism
Volume 9
The Sarcolemma
Edited by Paul-Emile Roy and Naranjan S. Dhalla

THE INCORPORATION OF HYDROPHOBIC PROTEIN RECEPTORS AND ARTIFICIAL LIPID MEMBRANES

T. A. READER,[2] S. FISZER DE PLAZAS,[3]
P. J. I. SALAS, and E. DE ROBERTIS[3]

Département de Pathologie, Faculté de Médecine,
Université Laval, Québec, Canada,
and Instituto de Biologia Celular, Facultad de Medicina,
Universidad de Buenos Aires,
Buenos Aires, Argentina

SUMMARY

The mechanism of chemical synaptic transmission implies: 1) the existence of a specific protein receptor at the postsynaptic membrane, and 2) the interaction between the transmitter released and the receptor, thus producing a change in ionic permeability. Previous studies from our laboratory have shown that special hydrophobic proteins extracted from postsynpatic membranes of different tissues showed a high affinity binding for the different pharmacological agents. The present paper describes experiments in which different hydrophobic proteins binding acetylcholine, noradrenaline, γ-aminobutyric acid, and glutamate were incorporated into artificial lipid membranes, similar to those first described by Mueller *et al.* (19). The effect of the different pharmacological agents was tested under experimental conditions of voltage clamp and the d.c. current changes measured. The results were then compared for the different lipid-protein membranes employed during the steady state and during transicnt conductance changes. The specificity of the responses indicate that artificial lipid membranes containing these hydrophobic proteins from electroplax, myocardium, spleen capsule and shrimp muscle can be used as a model to study pharmacologic receptors.

Previous reports have shown that special hydrophobic proteins (*i.e.*, proteolipids) extracted from the nerve ending membranes of the cerebral cortex and other gray areas of the brain had the property of interacting with dimethyl-*d*-tubocurarine (5, 6), serotonin (8), and atropine (16). Further studies were pursued with a special proteolipid extracted from the electric organ of *Torpedo* and *Electrophorus* which showed high affinity binding for acetycholine and other cholinergic agents (5). When this special cholinergic receptor protein was

[1] This work was supported by Grant 2 RIO NS 06953-07 from NEUA, National Institutes of Health, and Grant U. Neurobiol. 1824 G/71 from CONICET, Argentina.
[2] T. A. Reader is a Postdoctoral Fellow from the Medical Research Council, Canada.
[3] Career Investigators from CONICET, Argentina.

incorporated into artificial lipid membranes, similar to those first described by Mueller *et al.* (19), these membranes "reacted" to the application of acetylcholine with a transient change in conductance (21–23). In these membranes striking changes in surface fine structure, which accompanied the conductance change, were observed under the electron microscope (26, 27). Fiszer de Plazas and De Robertis (9) extracted and purified a special proteolipid from bovine spleen capsule which binds *dl*-[7-^{3}H] norepinephrine with high affinity in contrast to other proteolipid fractions of the same tissue. Ochoa *et al.* (20) isolated a proteolipid which has a high affinity to bind [^{3}H]isoproterenol from the heart ventricle of the cat, and recently Fiszer de Plazas and De Robertis (11) have isolated a special hydrophobic protein from shrimp muscle showing a high affinity binding for γ-aminobutyric acid.

We here describe experiments in which these hydrophobic proteins binding pharmacological agents were incorporated into artificial lipid membranes. The results indicate that these ultrathin membranes can be used to study the binding and interactions between membrane proteins and ligands.

MATERIALS AND METHODS

Artificial lipid membranes were made with a membrane-forming solution of chloroform-methanol-tetradecane (1.0:0.8:0.6). Type a membranes contained 10 mg/ml of synthetic dipalmitoyl-DL-α-lecithin and 10 mg/ml of synthetic cholesterol. Type b membranes contained only 20 mg/ml of cholesterol. In some cases 10 mg/ml of total phospholipids from bovine cerebral cortex extracted according to Folch *et al.* (12) were employed. The lipid-proteolipid membranes contained, in addition, 50 μg/ml of hydrophobic protein. The protein fractions used were: Peak 1 from the electric organ of *Electrophorus electricus,* which has no binding capacity for acetylcholine; Peak 3 from the electric organ of *E. electricus,* which binds nicotinic agents with high affinity (5), including α-bungarotoxin; the adrenergic hydrophobic protein from heart ventricle (20); the adrenergic hydrophobic protein from spleen capsule (9) and the γ-aminobutyric acid binding protein from shrimp muscle (11). In all cases the hydrophobic proteins were added directly into the membrane-forming solution.

The membranes were made across a 1-mm diameter hole in an horizontal Teflon septum separating two chambers containing a solution of 100 mM of NaCl and 1 mM Tris-Cl buffer, pH 7.0. In some experiments a vertical chamber was used. The pattern of membrane formation was checked with a stereomicroscope and all experiments were begun after the membranes were completely black. A voltage difference across the membrane was maintained constant by a d.c. source and it was measured, via AgCl/Cl electrodes with a Keithley d.c. voltmeter 200B (Keithley Instruments, Cleveland, Ohio). The current was determined with a Keithley 150A microammeter and recorded with a Heath EUW

Servo recorder (Heath Co., Benton Harbor, Mich.). In some experiments voltage and current were displayed on the screen of a storage oscilloscope Tektronix 51013N (Tektronix, Inc., Beaverton, Ore.).

The pharmacological agents were added with capillary tubes (Kimax 34500, size 1.5–2.0 × 100 mm) in 50–100-μl aliquots, at a distance of 2–3 mm from the membrane. In all cases the drugs were applied to the positive side of the membrane kept at a constant voltage of 100 mV. Stirring was provided by an electrically driven small helix of Teflon. All experiments were performed at 22°C (for further details see References 21, 22, 25).

RESULTS

Control membranes of cholesterol and lecithin (type a) had a basal conductance at 100 mV of about 1.5×10^{-6} mho·cm^{-2}. With the incorporation of 50 μg/ml of hydrophobic protein the conductance increased several fold. With Peak 3 fraction from electroplax it increased up to 4 times and discrete conductance steps were sometimes observed (21). These steps appear as a special type of fluctuation superimposed upon the basal current and have a magnitude of about 6×10^{-10} mho. Control membranes made only of cholesterol (type b) had a lower conductance (7.0×10^{-8} mho·cm^{-2}) which, with the addition of Peak 3 fraction (50 μg/ml) was raised about 10 times (7.0×10^{-7} mho·cm^{-2} at 100 mV).

Different pharmacological agents were added to the positive side of the artificial lipid membrane. Table I summarizes the experiments in which the reactivity to acetylcholine of artificial lipid membranes containing hydrophobic proteins was studied. Only when the membranes contained the Peak 3 protein from electroplax a conductance change was detected (table I). Since the applied voltage remains constant throughout the experiment, an increase in conductance of the membrane is reflected in the d.c. effect. This change in conductance with acetylcholine did not appear in control membranes or in those containing a noncholinergic protein from electroplax (Peak 1 protein fraction). Results were also negative in those membranes that contained the adrenergic protein from the spleen capsule, the adrenergic protein fraction from heart ventricle or the protein binding γ-aminobutyric acid from shrimp muscle. Negative results were also obtained with injections of saline solution, NaCl, KCl, Na-acetate, Tris-Cl, sucrose, and choline chloride (table II). All these agents were applied at a concentration of 10^{-1} M in the pipette, dissolved in saline solution, making a total ionic strength of 2×10^{-1} M in the pipette.

Other cholinergic agonists tested were carbamylcholine chloride, phosphorylcholine chloride, butyrylcholine chloride, acetylthiocholine chloride, and bromoacetylcholine bromide (24). Conductance changes were induced by these nicotinic agents only in artificial membranes containing the Peak 3 protein

Table I Chemical composition of membrane-forming solutions and response to acetylcholine chloride[1]

1.0	50 μg/ml	0.8	0.6		
1a. Chloroform		Methanol	Tetradecane	Cholesterol	Lecithin −
1b. Chloroform		Methanol	Tetradecane	Cholesterol	−
1c. Chloroform		Methanol	Tetradecane	Cholesterol	TBPh −
2a. Chloroform	Peak 1	Methanol	Tetradecane	Cholesterol	Lecithin −
2b. Chloroform	Peak 1	Methanol	Tetradecane	Cholesterol	−
2c. Chloroform	Peak 1	Methanol	Tetradecane	Cholesterol	TBPh −
3a. Chloroform	Peak 3	Methanol	Tetradecane	Cholesterol	Lecithin +
3b. Chloroform	Peak 3	Methanol	Tetradecane	Cholesterol	+
3c. Chloroform	Peak 3	Methanol	Tetradecane	Cholesterol	TBPh +
4a. Chloroform	ABPS	Methanol	Tetradecane	Cholesterol	Lecithin −
5a. Chloroform	ABPH	Methanol	Tetradecane	Cholesterol	Lecithin −
6a. Chloroform	GBP	Methanol	Tetradecane	Cholesterol	Lecithin −

[1] A concentration of 5×10^{-2} M in the pipette; see text for further details. Type a membranes contain 10 mg/ml of synthetic cholesterol and 10 mg/ml of synthetic dipalmitoyl-DL-α-lecithin. Type b membranes contain 20 mg/ml of synthetic cholesterol only. Type c membranes contain 10 mg/ml of synthetic cholesterol and 10 mg/ml of total phospholipids from bovine cerebral cortex (TBPh). Peak 1: noncholinergic protein fraction from the electroplax of *E. electricus;* Peak 3: cholinergic protein fraction from the electroplax of *E. electricus;* ABPS: adrenergic binding protein from spleen capsule; ABPH: adrenergic binding protein from heart ventricle; GBP: γ-aminobutyric acid binding protein from shrimp muscle.

Table II Reactivity of artificial lipid membranes to different agents

Agent and Concentration	M	Total ionic strength of the solution (× 10^{-1} M)	Control		Peak 1		Peak 3		ABPS, ABPH	
			1a	1b	2a	2b	3a	3b	4a	5a
Saline solution	10^{-1}	1.0	–	–	–	–	–	–	–	–
NaCl	10^{-1}	2.0	–	–	–	–	–	–	–	–
KCl	10^{-1}	2.0	–	–	–	–	–	–	–	–
Na-acetate	10^{-1}	2.0	–	–	–	–	–	–	–	–
Tris-Cl	10^{-1}	2.0	–	–	–	–	–	–	–	–
Sucrose	10^{-1}	2.0	–	–	–	–	–	–	–	–
Choline-Cl	10^{-1}	2.0	–	–	–	–	–	–	–	–
Acetylcholine-Cl	10^{-2}	1.1	–	–	–	–	+	+	–	–

[1] Concentrations of the agents in the pipette.

fraction from electroplax and could be repeated on the same membrane by succesive injections. The cholinergic drugs, including acetylcholine chloride, were injected at a concentration of 5×10^{-2} M in the pipette; however, as previously reported responses could be obtained with a concentration of 10^{-4} M in the pipette (25).

Four different cholinergic blocking agents were tested: *d*-tubocurarine chloride, mecamylamine chloride, PK-97, and PK-119. Both *d*-tubocurarine and mecamylamine, when injected upon the membranes containing the protein of Peak 3 from electroplax induced a conductance response but afterward they blocked a second response to an agonist (table III). The blocking agents were injected at a concentration of 10^{-3} M in the pipette dissolved in saline solution, making a total ionic strength of 1.01×10^{-1} M. A single dose of 50 μl of 10^{-3} M *d*-tubocurarine blocks the effect of 5×10^{-2} M acetylcholine or carbamylcholine. Since in the bath there is 100 times dilution the final concentration of *d*-tubocurarine is probably less than 10^{-5} M. The effect of the two nicotinic blocking agents derived from terephtalic acid (PK-97 and PK-119) were also tested. Both drugs when added to the bath in a concentration of 10^{-5} M were able to block an injection of 5×10^{-2} M acetylcholine or carbamylcholine. When these two agents were injected upon the membrane (at a concentration of 10^{-3} M in the pipette) PK-97 produced a fast transient increase in conductance, but the base line very rapidly returned to its original value. This effect was not observed with PK-119.

The injection of *dl*-isoproterenol at a concentration of 10^{-3} M in the pipette (total ionic strength 1.01×10^{-1} M) upon membranes containing the adrenergic protein from heart ventricle produced a rapid and transient increase in d.c. current intensity. This response could be much reduced or in some cases completely blocked by a previous injection of propanolol (5×10^{-4} M in the pipette).

Table III Reactivity of artificial lipid membranes to cholinergic agents

Agent employed	Concentration of the drug in the pipette ($\times 10^{-2}$ M)	Type of membrane: Control, 1a	Peak 1, 2a	Peak 3, 3a
Acetylcholine chloride	5	–	–	+
Carbamylcholine chloride	5	–	–	+
Butyrylcholine chloride	5	–	–	+
Phosphorylcholine chloride	5	–	–	+
Acetylthiocholine chloride	5	–	–	+
Bromoacetylcholine bromide	5	–	–	+

DISCUSSION

The results described here may be discussed considering artificial lipid membranes as a model to study interactions between ligands and membrane proteins. Artificial lipid membranes resemble the structure of cell membranes, that is a bimolecular leaflet separating two aqueous phases (14, 17, 19). The possibility to introduce in the bilayer hydrophobic proteins (*i.e.*, proteolipids) has enabled us to use these model membranes to further study and characterize these proteins from biophysical and pharmacological points of view.

The introduction of small quantities of hydrophobic proteins into the membrane-forming solution results in an artificial lipid membrane with a reduced resistance as compared with the control containing only phospholipids and cholesterol, or only cholesterol (21). Using special fixation techniques the membranes could be examined under the electron microscope (26, 27). Artificial membranes made only of phospholipids from the cerebral cortex and cholesterol (control membranes) had a globular fine structure with lipid micelles of about 40 Å. The introduction of small amounts of proteolipid resulted in a lower electron density and a smoother texture of the membranes. These findings suggested that the hydrophobic proteins produce a molecular reorganization of the bilayer.

When small amounts of the cholinergic protein fraction from *E. electricus* (Peak 3 fraction) were incorporated into artificial lipid membranes these membranes "reacted" to the application of acetylcholine with a transient increase in conductance (21–23). This membrane "activated" by acetylcholine had a striking change in ultrastructure which accompanied the increase in conductance (26, 27). The conductance change induced by acetylcholine, carbamylcholine, or other cholinergic agonists could be blocked by *d*-tubocurarine, mecamylamine, PK-97, and PK-119.

De Robertis (4) proposed a model for the cholinergic receptor constituted by a tetrameric arrangement of the hydrophobic protein molecules across the membrane (figure 6 of Reference 4). The transport of ions through the ionophoric portion of the receptor could be triggered by the interaction of the transmitter with the receptor site present in each of the protein molecules. The conductance changes developed by the incorporation of the cholinergic protein into lipidic bilayers is reminiscent of the bridges or channels produced by the EIM (7, 17) or by alamethicin (13, 18). However, in our material there is a special chemosensitivity of the membrane which sofar has not been observed with the other channel-forming agents. Recently Reader and De Robertis (24) have considered the effect of a cholinergic agonist as having the interactions shown in Scheme 1. The first part of Scheme 1 (ACh R_c) corresponds to the interaction of acetylcholine with the binding site, being the ionophore in the closed position.

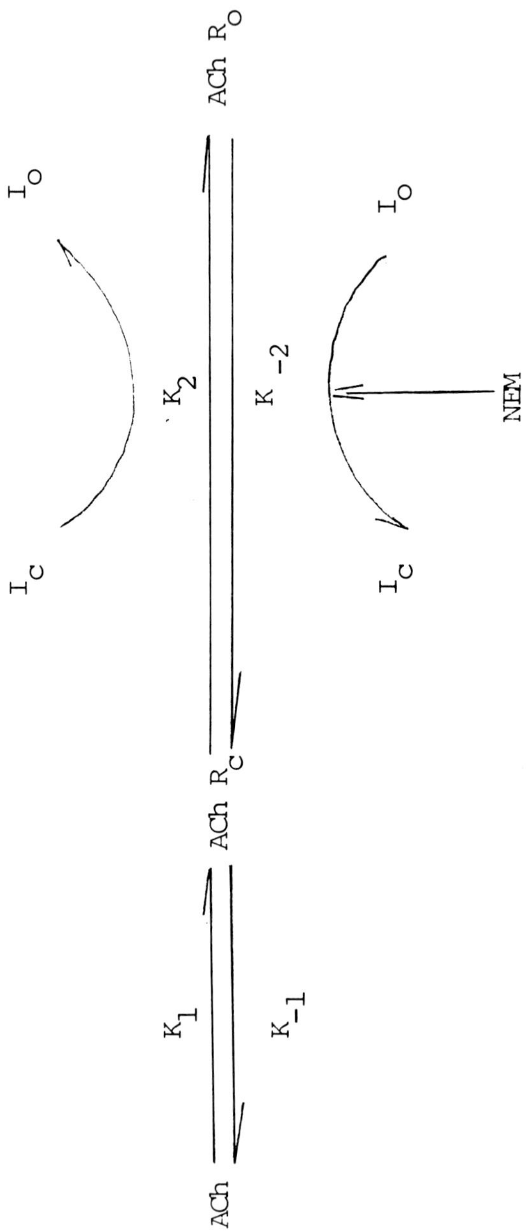

Scheme 1 Effect of cholinergic agonist. For description, see text.

The second part corresponds to the opening of the ionophore (ACh R_o) in the receptor complex. Support to this hypothesis is provided by the observation that histrionicotoxin is able to interact with the receptor ionophore, maintaining it in the closed condition (1). On the other hand, N-ethylmaleimide (NEM) inhibits the repolarization of the living electroplax after acetylcholine and carbamylcholine (15) and interacts with artificial lipid membranes containing the Peak 3 protein fraction, maintaining the conductance at a high level (24). The combined action of 1,4-dithiothreitol and N-ethylmaleimide greatly reduces the binding of cholinergic agents (5) and blocks the cholinergic response in artificial lipid membranes (24). These results are consistent with the idea put forward by Karlin (15) that disulfide groups are essential in the function of the cholinergic receptor. Del Castillo *et al.* (3) have also suggested that the process of activation of the cholinergic receptor implies a change in the free sulfhydryl groups of the receptor protein (2).

The results obtained in artificial lipid membranes containing hydrophobic proteins extracted from postsynaptic membranes in general resemble those obtained in living tissue and show specificity in their response to the different pharmacological agents tested. The experiments described represent a first approach to the study of pharmacologic receptors incorporated in model membrane systems. However, differences with the physiological situation are encountered and they refer mainly to the high concentrations of acetylcholine and isoproterenol needed to activate the receptors, the different time base of the response, and the possible ionic mechanism involved.

Acknowledgments We acknowledge the help of Professor M. J. Michelson for providing us with the drugs PK-97 and PK-119 and Dr. Karlin for supplying us with bromoacetylcholine bromide. We also acknowledge the assistance of Dr. M. Parisi with whom we discussed many of the experiments.

REFERENCES

1. ALBUQUERQUE, E. X., BARNARD, E. A., CHIU, T. H., LAPA, A. J., OLIVER, J., DOLLY, J., JANSSON, S.-E., DALY, J., and WITKOP, B. 1973. Acetylcholine receptor and ion conductance modulator sites at the murine neuromuscular junction: evidence from specific toxin reactions. Proc. Nat. Acad. Sci. U. S. A. 70: 949–953.
2. DEL CASTILLO, J. 1973. Personal communication.
3. DEL CASTILLO, J., ESCOBAR, I., and GIJON, E. 1971. Effects of the electrophoretic application of sulfhydryl reagents to the end-plate receptors. Intern. J. Neurosci. 1: 199–209.
4. DE ROBERTIS, E. 1971. Molecular biology of synaptic receptors. Science 171: 963–971.
5. DE ROBERTIS, E., and FISZER de PLAZAS, S. 1970. Acetylcholinesterase and acetylcholine Proteolipid receptor: Two different components of electroplax membranes. Biochim. Biophys. Acta 219: 388–398.

6. DE ROBERTIS, E., FISZER, S., and SOTO, E. F. 1967. Cholinergic binding capacity of proteolipids from isolated nerve-ending membranes. Science 158: 928–929.
7. EHRENSTEIN, G., LECAR, H., and NOSSAL, R. 1968. The nature of the negative resistance in bimolecular lipid membranes containing excitability-inducing material. J. Gen. Physiol. 55: 119–133.
8. FISZER, S., and DE ROBERTIS, E. 1969. Subcellular distribution and chemical nature of the receptor for 5-hydroxytryptamine in the central nervous system. J. Neurochem. 16: 1201–1209.
9. FISZER de PLAZAS, S., and DE ROBERTIS, E. 1972. Isolation of a proteolipid from spleen capsule binding (±)-(^{3}H) Norepinephrine. Biochim. Biophys. Acta 266: 246–254.
10. FISZER de PLAZAS, S., and DE ROBERTIS, E. 1972. Binding of α-bungarotoxin to the cholinergic receptor proteolipid from *Electrophorus* electroplax. Biochim. Biophys. Acta 274: 258–265.
11. FISZER de PLAZAS, S., and DE ROBERTIS, E. 1973. Hydrophobic proteins isolated from crustacean muscle having glutamante and γ-aminobutyrate receptor proteins. Fed. Eur. Biochem. Soc. Lett. 33: 45–48.
12. FOLCH, J., LEES, H., and SLOANE STANLEY, G. H. 1957. A simple method for the isolation and purification of total lipids from animal tissues. J. Biol. Chem. 226: 497–509.
13. GORDON, L. G. M., and HAYDON, D. A. 1972. The unit conductance channel of alamethicin. Biochim. Biophys. Acta 255: 1014–1018.
14. HENN, F. A., and THOMPSON, T. E. 1969. Synthetic lipid bilayer membranes. Ann. Rev. Biochem. 38: 241–288.
15. KARLIN, A., and BARTELS, E. 1966. Effects of blocking sulfhydryl groups and of reducing disulfide bonds on the acetylcholine-activated permeability system of the electroplax. Biochim. Biophys. Acta 126: 525–535.
16. LA TORRE, J. L., LUNT, G. S., and DE ROBERTIS, E. 1970. Isolation of a cholinergic proteolipid receptor from electric tissue. Proc. Nat. Acad. Sci. U. S. A. 65: 716–720.
18. MUELLER, P., and RUDIN, D. O. 1968. Action potentials induced in biomolecular lipid membranes. Nature 217: 713–719.
19. MUELLER, P., RUDIN, D. O., TIEN, T. H., and WESCOTT, W. C. 1963. Methods for the formation of single bimolecular lipid membranes in aqueous solutions. J. Phys. Chem. 67: 535–535.
20. OCHOA, E., LLORENTE de CARLIN, M. C., and DE ROBERTIS, E. 1972. Isolation of a proteolipid binding ^{3}H-isoproterenol from heart ventricle of the cat. Europ. J. Pharmacol. 18: 367–374.
21. PARISI, M., READER, T. A., and DE ROBERTIS, E. 1972. Conductance properties of artificial lipid membranes containing a proteolipid from *Electrophorus:* Response to cholinergic agents. J. Gen. Physiol. 60: 454–470.
22. PARISI, M., READER, T. A., and DE ROBERTIS, E. 1973. Reactivity to acetylcholine developed by artificial lipid membranes containing a proteolipid from *Electrophorus electricus.* Experientia 29: 1451–1452.
23. PARISI, M., RIVAS, E., and DE ROBERTIS, E. 1971. Conductance changes produced by acetylcholine in lipidic membranes containing a proteolipid from *Electrophorus.* Science 172: 56–57.
24. READER, T. A., and DE ROBERTIS, E. 1974. The response of artificial lipid membranes containing a cholinergic hydrophobic protein from *Electrophorus* electroplax. Biochim. Biophys. Acta.
25. READER, T. A., PARISI, M., and DE ROBERTIS, E. 1973. Reactivity to acetylcholine developed by artificial membranes containing a protein from *Electrophorus electricus.* The effect of uranyl ions. Biochem. Biophys. Res. Commun. 53: 10–17.
26. VASQUEZ, C., PARISI, M., and DE ROBERTIS, E. 1971. Fine structure of ultrathin

artificial membranes. I. Changes by acetylcholine addition in lipid proteolipid membranes. J. Membrane Biol. 6: 353–367.
27. VASQUEZ, C., READER, T. A., and PARISI, M. 1975. The surface structure of artificial lipid membranes.

Mailing address: Dr. T. A. Reader,
Département de Pathologie, Faculté de Médecine,
Université Laval, Québec G1K 7P4, P.Q. (Canada).

Adenylate Cyclase and Catecholamines

Recent Advances in Studies on Cardiac Structure and Metabolism
Volume 9
The Sarcolemma
Edited by Paul-Emile Roy and Naranjan S. Dhalla

ADENYLATE CYCLASE AND CATECHOLAMINE BINDING IN PLASMA MEMBRANE-ENRICHED PREPARATIONS OF CARDIAC AND SKELETAL MUSCLE[1]

G. I. DRUMMOND, J. VALLIÈRES,[2]
and M. DRUMMOND

Department of Pharmacology, Faculty of Medicine,
University of British Columbia,
Vancouver, British Columbia, Canada

SUMMARY

Binding of [^{3}H]epinephrine to plasma membrane-enriched fractions from guinea pig heart and rabbit skeletal muscle was investigated using the micropore filtration technique. [^{3}H]Epinephrine and [^{3}H]norepinephrine were found to be degraded rapidly in aqueous buffer at pH 7.6 and 37°C. Deterioration of the compounds could be prevented by low concentrations of dithiothreitol. Binding of [^{3}H]epinephrine to both membrane preparations was a slow process requiring 60 min to approach equilibrium in the case of cardiac membranes at 37°C, and 20 min for skeletal muscle membranes at 0°C. Binding was antagonized by the unlabeled β agonists, isopropylnorepinephrine, epinephrine, and norepinephrine; each was equipotent, in contrast to their different potency is eliciting β-adrenergic responses and in stimulating adenylate cyclase. A variety of catechol compounds were as effective as antagonists of binding as the catecholamines. Methylation of one ring hydroxyl abolished inhibitory activity. The β-adrenergic antagonists, propranolol, pronethalol, and dichloroisoproterenol, were not effective in inhibiting binding to either membrane preparation. D-Norepinephrine and L-norepinephrine were equieffective in antagonizing binding of [^{3}H]norepinephrine to skeletal muscle membranes, suggesting that binding was not stereospecific. These and other data led to the conclusion that binding of labeled catecholamine to isolated tissue membranes using the micropore filtration technique probably does not reflect interaction with the specific β-adrenergic receptor, but more likely reflects a less specific binding of compounds with one or more hydroxyl groups on a ring.

The mechanisms by which hormones interact with their receptor sites to initiate biochemical processes leading to a physiological response is currently of great interest. In the past few years, a direct approach to this problem has become feasible by examination of the binding of radioactive hormones to receptor sites

[1] This investigation was supported by grants from the Medical Research Council of Canada and the British Columbia Heart Foundation.
[2] Recipient of a Medical Research Council Centennial Fellowship.

on plasma membranes from various tissues. *In vitro* studies of such specific interactions have been described for several polypeptide hormones and acetylcholine. In these instances, convincing evidence has been obtained that binding occurs at the specific and functional receptor site. Radioactive catecholamines have also been used to identify the adrenergic receptor in liver (6), heart (9, 11), spleen capsule (5), erythrocytes (1, 17), and cultured myocardial cells (10). The nature of binding of labeled catecholamines to isolated cells or membrane fragments in these studies has been interpreted as characteristic of interaction with the specific adrenergic receptor. Some of the data, however, seem to reflect a more general specificity for the catechol moiety and there is a complete lack of stereospecificity for binding (1, 6, 10, 11).

In the present communication, we describe studies on the binding of [^{3}H] catecholamines to plasma membrane-enriched preparations from rabbit skeletal muscle and guinea pig heart. Our data are in agreement with recent reports of Cuatrecasas *et al.* (3) and Maguire *et al.* (14), that binding of [^{3}H] catecholamines to membrane structures using the micropore filtration technique probably does not represent interaction with the specific β receptor.

METHODS

Sarcolemma were prepared from rabbit leg muscle as described previously (19), except that the number of washings of each sediment were increased. Thus, fresh muscle was minced with scissors and homogenized in 5 volumes of 50 mM $CaCl_2$ in a Sorvall Omnimixer (40-ml vessel) for 15 sec at maximal velocity. After filtration through a nylon seive (pore size 1 mm^2) the homogenate was centrifuged at 2,000 $\times$ g for 10 min. The sediment was washed four times by suspending in 10 volumes of 10 mM Tris-HCl, pH 8.0 (based on initial tissue weight), and centrifuged at 2,000 $\times$ g for 10 min. The washed particles were suspended in 5 volumes of the above buffer (based on initial tissue weight) and 4.0 M LiBr was added to produce a final concentration of 0.4 M. The suspension was stirred magnetically (4°C) for 3 hr. The viscous suspension was diluted 4-fold with buffer and centrifuged at 3,000 $\times$ g for 10 min. The particles were then suspended in 8 volumes of 25 percent (w/w) KBr (based on initial tissue weight) containing 10 mM Tris-HCl, pH 7.6, and centrifuged at 7,500 $\times$ g for 20 min. The pellet was washed twice with 8 volumes of buffer (pH 8.0 and based on initial tissue weight) by centrifuging at 2,000 $\times$ g for 10 min. The final pellet was suspended with a motor-drivenTeflon-tipped pestle in 1 ml of 52 mM Tris-HCl, pH 7.6, per gram of initial tissue weight and used fresh for the binding assays.

Cardiac plasma membrane-enriched fractions were prepared from male guinea pigs (300–400 gm). Hearts were removed, perfused with warm Krebs-Ringer bicarbonate, and placed on ice. All subsequent procedures were carried out at

4°C. After removal of the atria and great veins, the ventricles were blotted, weighed (2–3 gm from two hearts), minced with scissors, and homogenized for 30 sec in a Polytron PT 10 at low velocity in 5 volumes of 10 mM Tris-HCl, 2 mM dithiothreitol (DTT), pH 7.5, in a 40-ml polycarbonate tube. Homogenization was then carried out for 2 sec at maximal velocity. The homogenate was diluted with an additional 5 volumes of buffer and filtered (light suction) through a nylon mesh (250-μm pore size) and centrifuged for 10 min at 600 $\times$ g. The pellet was covered with 2 ml of buffer and the tube was moved gently by a rocking motion. This caused suspension of a dense layer of mitochondria which could then be removed with a Pasteur pipette. The pellet was washed four times by suspending in 10 volumes of Tris buffer and centrifuging successively at 375, 195, and twice at 120 $\times$ g for 10 min. The last pellet was suspended in 10 volumes of 0.25 M sucrose containing 10 mM Tris-HCl, 2 mM DTT, pH 7.5, and homogenized for two 15-sec intervals in a Polytron PT 10 homogenizer at maximal velocity, with a 1-min cooling interval between. To the homogenate was added an equal volume of 2.5 M KCl in sucrose-buffer with gentle stirring for 5 min. The resulting viscous suspension was centrifuged at 100,000 $\times$ g for 1 hour; the sediment was washed with 10 volumes (based on tissue weight) of sucrose-buffer followed by centrifugation at 38,000 $\times$ g for 10 min. The particles were dispersed thoroughly in 10 ml of 0.25 M sucrose, 10 mM Tris-HCl, 2 mM DTT, pH 8.2, in a Potter-Elvehjem homogenizer with a tighly fitting Teflon pestle (12 passes). The suspension was then subjected to differential centrifugation on a discontinuous sucrose gradient. The gradient consisted of sucrose, 68.5 percent (2 ml), 65 percent (5 ml), 60 percent (5 ml), 55 percent (5 ml), and 50 percent (5 ml) in 10 mM Tris-Hcl, pH 8.2, containing 2 mM DTT, in each of three 30-ml tubes. Three-milliliter aliquots of the particle suspension were placed on each gradient and centrifugation was carried out in a Beckman SW 25.1 rotor for 1 hr at 20,000 rpm (40,000 $\times$ g average). Five distinct particle zones were obtained. The first was located at the interface between the loading medium and the 50 percent sucrose layer; the second, third, and fourth were just below the interface between the 50 and 55 percent, 55 and 60 percent and 60 and 65 percent sucrose layers, respectively; the fifth comprised a transparent pellet. When examined by phase contrast microscopy, Bands 1 and 2 contained small particles, mitochondria, and tubular structures resembling capillaries. Routinely, Bands 3 and 4 were harvested with a Pasteur pipette, diluted with 2 volumes of 10 mM Tris-HCl, pH 7.6, and centrifuged for 1 hr at 100,000 $\times$ g. The pellet was suspended in approximately 5 ml of 10 mM Tris-HCl, 1 mM DTT, pH 7.2, and stored at 4°C until analyzed for catecholamine binding. Fresh preparations were used for every experiment.

For measurement of [^{3}H]epinephrine binding to skeletal muscle plasma membranes, 175–225 μg of protein were incubated in a medium containing 40 mM Tris-HCl, pH 7.6, 10 μM DTT, in the absence or presence of unlabeled catecholamines, or other substances as indicated, in a final volume of 0.65 ml at 0°C. The reaction was initiated by adding [^{3}H]epinephrine solution in 0.002 N HCl (usually 5 nM to 50 nM final concentration) and the incubation time was 30 min. The reaction was terminated by filtering a 0.60-ml aliquot through a cellulose filter (Millipore HAWP047, 0.45 μm) which had been soaked in 1 mM pyrogallol and washed on the filter apparatus with 10 ml of 10 mM Tris-HCl, pH 7.6. Presoaking in pyrogallol greatly reduced binding of [^{3}H]epinephrine to the filters. The filter was immediately washed with 20 ml of 10 mM Tris-HCl, pH 7.6, dried at 75°C for 1 hr, placed in a scintillation vial, covered with 21 ml of scintillation fluid [4 gm of Omnifluor (New England Nuclear Corp., Boston, Mass.) per liter of toluene] and radioactivity was determined by scintillation spectrometry. Appropriate controls containing all components except membranes were included. Experimental values are corrected for binding to the filters under each experimental condition. Unless otherwise stated, results are the mean of triplicate determinations.

[^{3}H]Epinephrine binding to cardiac membranes was measured by a slightly different procedure. Membrane protein (100–150 μg) was incubated in 50 mM Tris-HCl, 1 mM DTT, pH 7.2, in a final volume of 0.65 ml. [^{3}H]Epinephrine solution (freshly prepared in 0.002 N HCl), final concentration usually 50 nM, was added to start the reaction; the incubation temperature was 37°C. At the end of the incubation interval, the entire incubation mixture was filtered through a cellulose filter (Millipore HAWPO24, 0.45 μm) and washed on the filter assembly with 5 ml of 10 mM Tris-HCl, pH 7.5. Immediately after filtering, the filter was washed with 10 ml of cold 10 mM Tris-HCl, pH 7.5, placed in a scintillation vial, dried at 80°C, dissolved in 1 ml of ethyleneglycol monomethyl ether; 16 ml of scintillation fluid (toluene-ethyleneglycol monomethyl ether, 3:1, containing 4 gm of Omnifluor/l) were added and radioactivity was determined by scintillation spectrometry. All assays were performed in triplicate, and triplicate controls containing all components except membranes were performed for each experimental condition.

Adenylate cyclase, (Na^+ + K^+)-ATPase, 5′-nucleotidase, and cytochrome *c* oxidase were assayed by methods used previously (19). Protein was determined by the method of Lowry *et al.* (13).

RESULTS

Properties of cardiac membranes

Some characteristics of the cardiac membrane preparation are given in table I. Bands 3 and 4 represented 71 percent of the adenylate cyclase activity with an increase in specific activity (F^- stimulated) of about 10-fold. Extraction with KCl greatly decreased (Na^+ + K^+)-ATPase activity; the yield at this step was only 19 percent. Seventy percent of the Na^+ + K^+-ATPase applied to the gradient was

Table I Characteristics of cardiac membrane fractions*

Fraction	Protein (%)	Adenylate cyclase (nmoles/min/mg)	Percent	$(Na^+ + K^+)$-ATPase (nmoles/min/mg)	Cytochrome oxidase (nmoles/min/mg)
Homogenate	100	0.33	100	26.2	153.6
Washed particles	39.7	1.19	150	56.2	183.0
KCl particles	10.9	2.38	79	40.8	331.2
Bands 3 and 4	7.2	3.26	71	83.0	103.5

*Values are the mean of nine preparations. Adenylate cyclase was assayed in the presence of 10mM NaF.

recovered in Bands 3 and 4. The final preparation contained 4.85 percent of the original cytochrome oxidase, indicating the presence of mitochondrial contamination. More detailed studies of the enzymic characteristics of the preparation will be published elsewhere.

Lability of catecholamines

During early experiments, we observed that catecholamines were degraded rapidly when incubated in buffers such as Tris-HCl, phosphate, glycylglycine, imidazole-histidine, or imidazole-HEPES (N-2-hydroxylethylpiperazine-N′-2-ethane sulfonic acid) at alkaline pH. This transformation could not be prevented by addition of sodium metabisulfite, sodium borohydride, and ethylenediaminetetraacetic acid at concentrations as high as 5 mM. In an attempt to solve this problem, we have subjected solutions of [^{3}H] epinephrine to thin layer chromatography after incubation in the absence and presence of DTT. When [^{3}H]epinephrine (0.5 μM) was incubated in 50 mM Tris-HCl, pH 7.6, at 30°C for 30 min and then subjected to thin layer chromatography, virtually no radioactivity was present at the R_F of authentic [^{3}H]epinephrine; radioactivity, however, appeared at a site closer to the origin (fig. 1*A*). Not only did a new radioactive spot appear, but the total recovery of radioactivity on the chromatogram was greatly diminished; only 52 percent of that applied was recovered compared with 98.4 percent recovery from the chromatogram of the zero time control. When 1 mM DTT was present in the incubation mixture (fig. 1*B*), these changes did not occur; radioactivity in the 30 min incubation mixture coincided precisely with that in the zero time control (prepared in the presence of 1 mM DTT), and the recovery of radioactivity was 98 percent. When the period of incubation was reduced to 15 min, [^{3}H]epinephrine was already degraded extensively. Although some radioactivity was present at the R_F of the zero time control (fig. 1*C*), the new peak of radioactivity with lower mobility was also present and the total recovery of counts was 57.3 percent. Again such

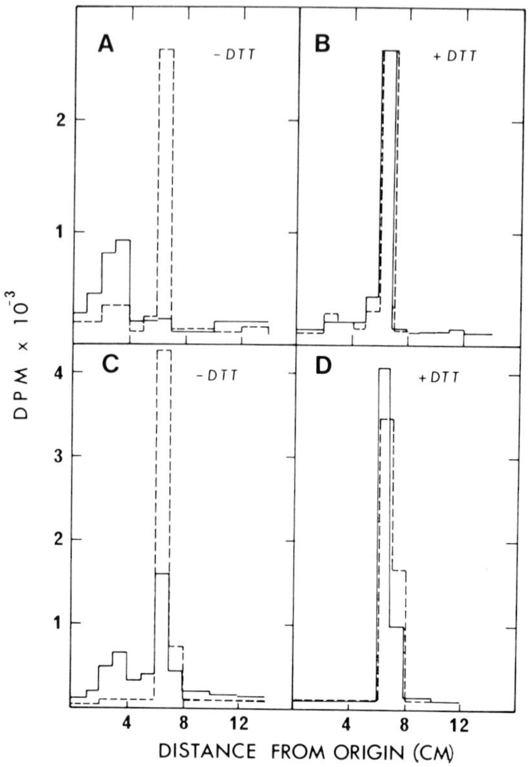

Figure 1 Lability of [^{3}H]epinephrine in solution at 37°C and pH 7.6. [^{3}H]epinephrine was incubated in a final concentration of 0.5 μM (1.83 × 10^6 d. p. m./assay) in 0.65 ml of 50 mM Tris-HCl, pH 7.6, at 37°C for 30 min (*A*) in the absence and presence (*B*) of 1 mM DTT. Fifty microliters 0.002 N HCl was then added to bring the pH to about 2.5 and duplicate 2.5-μl aliquots (5,534 d.p.m.) were spotted on thin layer cellulose plates (Eastman Kodak) and subjected to ascending chromatography using *n*-butanol saturated with 5 percent trifluoroacetic acid for 2 hr. The solvent front advanced 14.3 cm. Each lane was divided into 1-cm sections, the cellulose was scraped off, placed in scintillation vials, 1 ml of water was added followed by 15 ml of Aquasol (New England Nuclear Corp.), and radioactivity was determined. *A*:– – –, chromatogram of a zero time control solution prepared in the absence of DTT; ——, chromatogram of a solution incubated for 30 min in the absence of DTT. *B*: – – –, chromatograms of the zero time control prepared in the presence of DTT; ——, solution incubated for 30 min in the presence of DTT. *C, D*: similar experiment (2.1 × 10^6 d.p.m./assay in which the incubation time was 15 min. Each point is the mean of duplicate chromatograms.

changes were prevented completely by the presence of 1 mM DTT (fig. 1*D*). Identical changes occurred when [^{3}H]norepinephrine was incubated under similar conditions; this catecholamine appeared to be degraded extensively after 30 min at pH 7.6 and, in fact, measurable deterioration had occurred after 5 min, as revealed by thin layer chromatography (fig. 2). These findings indicate that when [^{3}H]catecholamines of high specific radioactivity are incubated in low concentration in aqueous buffer above neutral pH, chemical degradation occurs rapidly. The nature of the degradation is such that upon subsequent chromatography in *n*-butanol saturated with 5 percent trifluoroacetic acid [and also in *n*-butanol, acetic acid, water (12:3:5], tritium is lost from the chromatogram. This might indicate that, in addition to oxidative processes, deterioration through self-radiolysis (radiochemical degradation) may occur. Experiments were also performed in which [^{14}C]epinephrine was incubated and then subjected to chromatography. Figure 3 shows such a chromatogram in which the solvent was phenol (88 percent)-water (9:1). After 2 hr at 37°C, pH 7.6, less

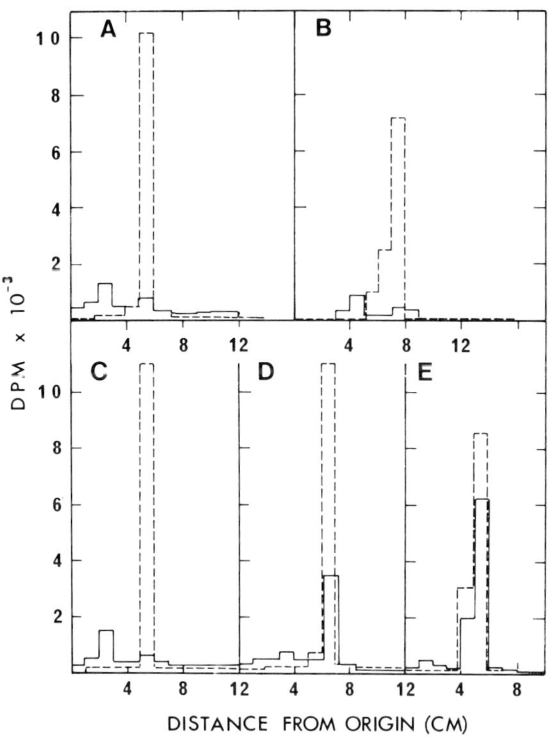

Figure 2 Lability of [^{3}H]norepinephrine in solution at pH 7.6 and 37°C. The conditions are as for figure 1, except that DL-norepinephrine bitartrate (7-^{3}H(N)) was used (3.7 × 10^6 d.p.m./assay). *A, B;* chromatograms of solutions incubated for 2 hr at 37°C; *B:* the chromatographic solvent was *n*-butanol, acetic acid, water (12:3:5); in all others solvent was as in figure 1. *C, D, E:* chromatograms of solutions incubated for 30, 15, and 5 min, respectively. Recovery of radioactivity on chromatograms of solutions incubated in the absence of DTT (——) in *A, B, C, D,* and *E* were 29, 34, 44, 48, and 74 percent, respectively. In each plate the *broken lines* (–––) represent chromatograms of zero time controls which were identical both in the presence or absence of DTT. All profiles are means of duplicate chromatograms.

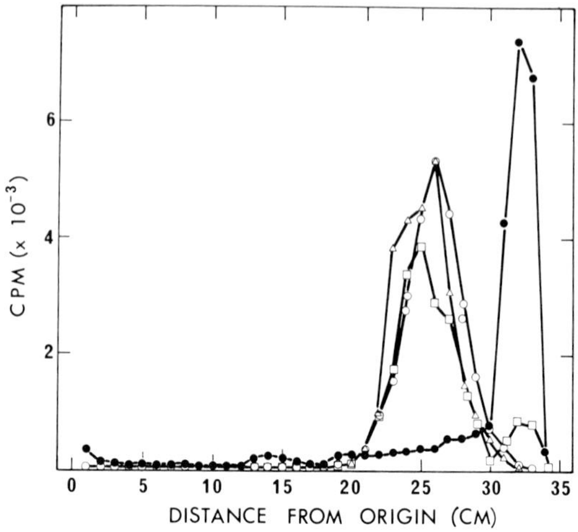

Figure 3 Lability of [^{14}C]epinephrine in solution at pH 7.6. DL-[^{14}C]Epinephrine bitartrate (50 mCi/mmole) was incubated at 54 mM Tris-HCl, pH 7.6, at a final concentration of 5 μM for 120 min in the presence and absence of 1 mM DTT at 37°C and 10 μM DTT at 0°C, final volume 0.65 ml. Fifty microliters of 0.1 N HCl were then added to bring the pH to 2.5, and duplicate 2.5-μl aliquots were spotted on No. 1 Whatman paper and subjected to descending chromatography using phenol (88 percent)-water (9:1) as the solvent system. ○, Chromatogram of control solution prepared immediately before chromatography; ●, solution incubated at 37°C in the absence of DTT; △, solution incubated at 0°C in the presence of 10 μM DTT, and □, solution incubated at 37°C in the presence of 1 mM DTT. Recovery of radioactivity ranged from 90–92 percent.

than 10 percent of the labeled material migrated with authentic [^{14}C]epinephrine. Incubation at 0°C in the presence of 10 μM DTT or at 37°C in the presence of 1 mM DTT largely prevented these changes. The recovery of applied radioactivity regardless of the incubation conditions was 90–92 percent. This experiment would indicate that oxidation of catecholamine is one of the degradative reactions occurring. Indeed, unlabeled epinephrine and "oxidized" epinephrine (epinephrine incubated at 37°C at pH 7.6 to form a dark pink solution), when chromatographed in this solvent system, were effectively separated, the "oxidized" material coinciding precisely (when iodine vapor was used for visualization) with the front running peak of radioactivity (fig. 3). Thus complex degradative reactions occur when radioactive catecholamines are incubated in aqueous buffer at pH values above neutrality. Experiments performed using these materials without a stabilizing agent must be open to serious error. Because of the ability of DTT to prevent such changes, we have employed it in all of our catecholamine binding assays.

Membranes prepared from skeletal muscle and heart, however, responded to DTT in different ways. Binding of [^{3}H]epinephrine to skeletal muscle membranes was inhibited by 1 mM DTT when the incubation temperature was 37°C (measured at incubation periods of less than 2 min). It was found that at 0°C, concentrations of DTT as low as 10 μM had no inhibitory effect on binding while preventing degradation of the catecholamine. Standard conditions finally selected for measuring binding to skeletal muscle membranes involved incubating at 0°C in the presence of 10 μM DTT. Standard conditions for measuring binding to cardiac membranes involved incubating at 37°C in the presence of 1 mM DTT.

Studies on [^{3}H]epinephrine binding

The time course of [^{3}H]epinephrine (5 nM) binding to two preparations of cardiac membranes at 37°C is shown in figure 4*A*. The binding capacity of many preparations varied only slightly and were generally similar to Preparation II (figure 4*A*). Preparation I had a somewhat higher binding capacity and this was much reduced when the incubation was carried out at 0°C. Although binding was measurable at intervals as low as 30 sec, it did not reach equilibrium until

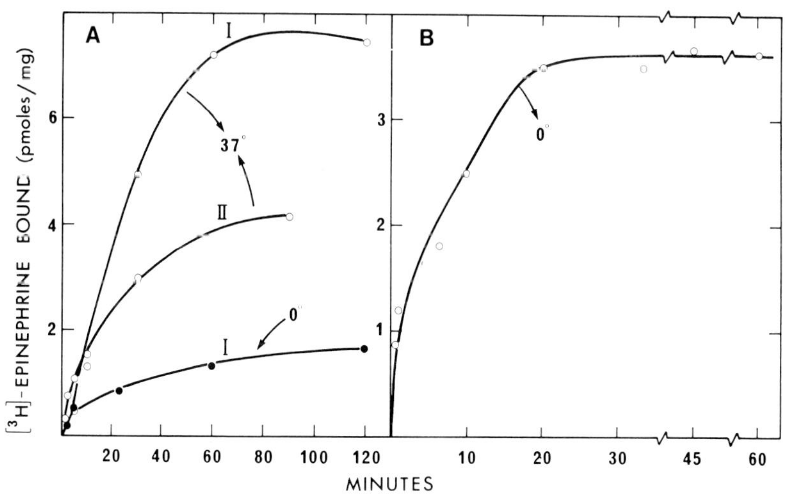

Figure 4 Association rate of [^{3}H]epinephrine with cardiac and skeletal muscle membranes. Two preparations of cardiac membranes (100 μg of protein/tube) were used (*A*). Preparation I was assayed at 37°C (○) and at 0°C (●); Preparation II was assayed at 37°C only. The concentration of [^{3}H]epinephrine was 50 nM; 6.6 × 10^5 d.p.m./assay. Values are means of triplicate determinations, corrected for binding to the filters (triplicates) which was 0.081 pmole at zero time. Binding to skeletal muscle muscle membranes (*B*) was determined under standard conditions at 0°C using 175 μg of membrane protein per assay. The concentration of [^{3}H]epinephrine was 25 nM, 3.3 × 10^5 d.p.m./assay. Before addition of plasma membranes, epinephrine had been incubated with buffer (0°C) for a period such that at the end of each assay it had been in contact with the medium for 60 min. Values are means of triplicate determinations.

after 60 min at 37°C, resembling the cardiac microsomal preparation studied by Lefkowitz *et al.* (11). Binding to skeletal muscle membranes at 0°C reached a maximum after 20 min (fig. 4*B*). With both membrane preparations, binding was proportional to membrane protein up to 400 μg in the assay.

Binding was also studied as a function of epinephrine concentration. Cardiac membranes seemed to have a very large binding capacity; the amount of [^{3}H]epinephrine bound began to level off only at concentrations above 10 μM (fig. 5*A*) when measured at 37°C. Binding was a linear function of epinephrine concentration between 5 nM and 0.5 μM. Skeletal muscle membranes were saturable at 2.5 μM epinephrine, with a half-maximum occurring at 0.55 μM (fig. 5*B*). Lefkowitz *et al.* (11) have reported two orders of binding sites of [^{3}H]nor-

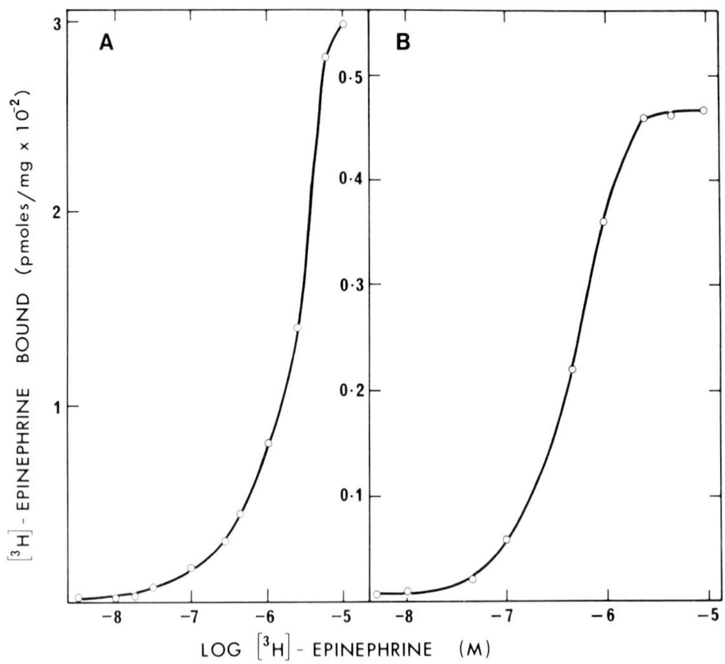

Figure 5 Concentration dependence of [^{3}H]epinephrine binding to cardiac and skeletal muscle membranes. For cardiac membranes (125 μg of protein), [^{3}H]epinephrine, specific activity 9.45 Ci/mmole was used up to 0.5 μM; above this, labeled material was appropriately diluted with unlabeled amine to produce the concentrations indicated. The time of incubation was 120 min at 37°C. Values are the mean of triplicate assays corrected for controls (binding to the filters) at each concentration also performed in triplicate. Binding to the filters never exceeded 10 percent of that bound to membranes. For skeletal muscle membranes (*B*), 240 μg of membrane protein were used; concentrations above 0.5 μM were obtained by diluting the radioactive material with unlabeled amine. Otherwise, binding was determined in triplicate (with controls in triplicate) at 0°C as described in Materials and Methods.

epinephrine to dog heart microsomal membranes, the first occurring with an association constant of 0.1 μM, and a second at 1.33 μM. [^{3}H] Isopropylnorepinephrine bound to turkey erythrocyte membranes with a half-maximal value of 4.4 μM (1).

Several authors have reported recently that binding of labeled norepinephrine to cell membranes is essentially irreversible (3, 14). Maguire *et al.* (14) have suggested that norepinephrine binding involves oxidation and covalent bond formation with macromolecules present in the membrane. The rate of dissociation of [^{3}H]epinephrine from cardiac membranes by unlabeled amine at 37°C is shown in figure 6*A*. At three points in time during the incubation, unlabeled epinephrine was added to a final concentration of 1 mM and the amount of label bound was measured at 1, 6, and 30 min thereafter. It is seen that part of the bound material exchanged rapidly and this amount appeared to be constant regardless of the time of incubation. These results are strikingly similar to those of Lefkowitz *et al.* (11) using cardiac microsomal preparations. These authors concluded that two binding compartments exist, one of which exchanges rapidly, the other of which fails to dissociate. When binding was measured at 0°C (fig. 6*A, inset*), dissociation from cardiac membranes occurred rapidly and approached completion after 30 min. This may indeed suggest that two processes are involved in binding, one which is not readily dissociable being favored at 37°C. When measured at 0°C, dissociation of [^{3}H]epinephrine from skeletal muscle membranes (fig. 6*B*) was similar to that of cardiac membranes at the same temperature. Thus, bound label was rapidly and almost completely dissociated regardless of the time of exposure of the membranes to the labeled material. In additional experiments (data not shown), less that 10 percent of label bound after 15 min at 15°C or 37°C in the absence of DTT could be dissociated by high concentrations of unlabeled epinephrine, HCl, acetic acid, high temperature (80°C), or chelating agents (5 mM).

The chemical nature of the radioactive material bound to skeletal muscle membranes was analyzed by paper chromatography. Membranes were equilibrated with [^{3}H]epinephrine, washed by filtration, and then extracted with 0.5 N acetic acid. The acetic acid solution was lyophylized and chromatographed. Virtually all (93.7 percent) of the bound radioactivity was recovered in the acid extract. Figure 7 shows that more than 96 percent of the extracted radioactivity moved with the R_F of authentic epinephrine. Virtually, no rapidly moving "oxidized" epinephrine was present. This experiment led us to conclude that, at least at 0°C and in the presence of DTT, [^{3}H]epinephrine did not undergo chemical alteration when bound to skeletal muscle plasma membranes.

Lefkowitz *et al.* (11) have concluded that β-adrenergic drugs compete with [^{3}H]norepinephrine for binding sites in cardiac microsomal fractions with a relative potency which parallels their potency in causing increased force of cardiac contraction and their ability to stimulate adenylate cyclase. We have not

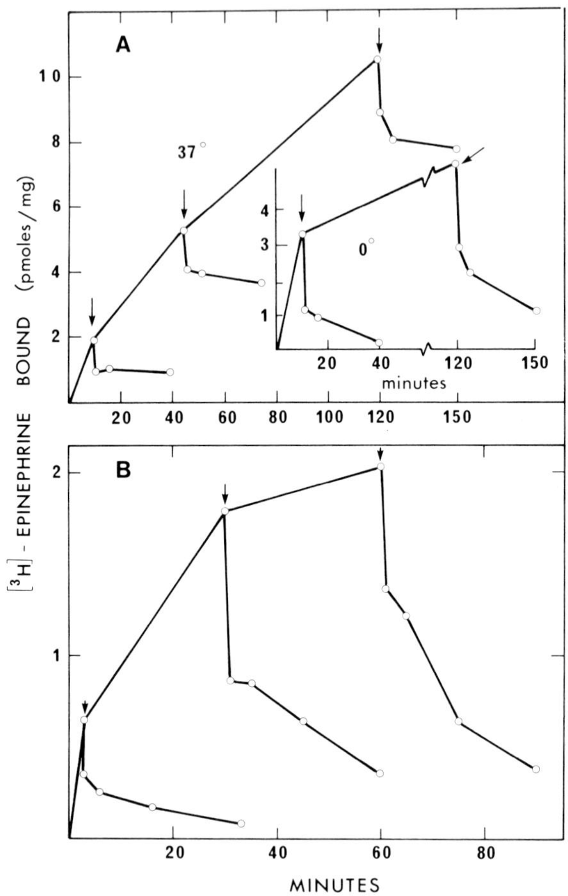

Figure 6 Dissociation of bound [^{3}H]epinephrine. In *A*, standard conditions were used for measuring binding to cardiac membranes except that the volume was increased 4-fold (625 μg of protein in 2.4 ml). The [^{3}H]epinephrine concentration was 50 nM. Each incubation as well as controls was performed in triplicate. At the end of each incubation interval, a 0.5-ml aliquot of each tube was removed, filtered and washed. Immediately thereafter, 10 μl of 0.1 M unlabeled epinephrine was added to each tube (indicated by the *arrows*), and a 0.5-ml aliquot removed at 1, 6, and 30 min thereafter. The temperature was 37°C. The *inset* in *A* represents a similar experiment in which the incubation temperature was 0°C. In *B*, tubes containing 1 mg of skeletal muscle membrane protein was incubated (in triplicate) at 0°C with [^{3}H]epinephrine (25 nM) in a final volume of 3 ml. At each time interval a 0.6-ml aliquot was removed, filtered, and washed. Immediately thereafter (indicated by the *arrows*), unlabeled epinephrine (50 μl) was added to a final concentration of 1 mM and 0.6-ml aliquots were removed 1, 5, 15, and 30 min thereafter.

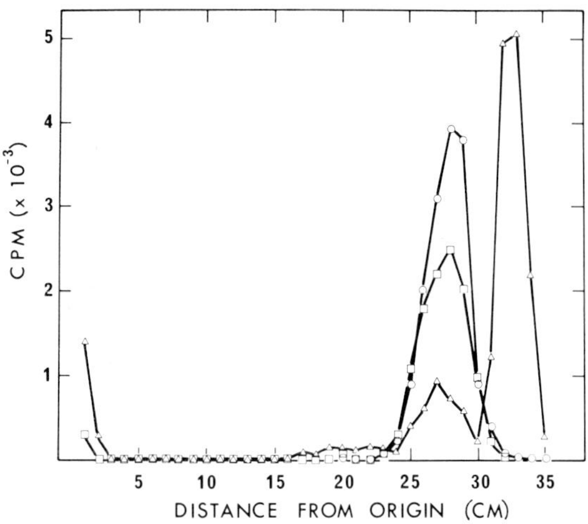

Figure 7 Chromatographic analysis of bound label extracted from skeletal muscle membranes. [^{3}H]Epinephrine (1 μM; 1.3 × 10^7 d.p.m.) was incubated under standard conditions for 60 min at 0°C. An aliquot (0.6 ml) was then filtered and washed with 20 ml of 10 mM Tris-HCl, pH 7.6. Radioactivity was then extracted from membranes on the filter with two l-ml volumes of 0.5 N acetic acid over a 5-min interval. This procedure removed 97.7 percent of the bound ^{3}H. The acetic acid solution was then lyophylized, the residue was taken up in 0.002 N HCl and subjected to descending chromatography (No. 1 paper) using phenol (88 percent)-water (9:1). Two other samples containing the same amount of radioactivity, one a solution of [^{3}H]epinephrine freshly prepared, and a similar one incubated for 1 hr at 37°C in 40 mM Tris-HCl, pH 7.6, were also chromatographed. Recovery of the authentic control radioactivity (○) was 89 percent, for epinephrine incubated at 37°C (△) 43 percent, and for the material extracted from membranes (□), 94 percent. Values are means from triplicate chromatograms.

found this to be the case for either cardiac or skeletal muscle sarcolemmal preparations. Unlabeled isopropylnorepinephrine, epinephrine, and norepinephrine competed effectively for binding of [^{3}H]epinephrine in both cardiac (fig. 8*A*) and skeletal muscle membranes (fig. 8*B*) (see also table II), but all were equally effective, bearing no resemblance to the well known gradation in potency of these agents for eliciting β receptor responses. The ability of a wide variety of agents to antagonize binding of [^{3}H]epinephrine to both cardiac and skeletal muscle membranes is shown in table II. A variety of catechol compounds were highly effective inhibitors, including dopamine, L-dopa, 6-hydroxydopamine, pyrogallol, and pyrocatechol. DL-3,4-Dihydroxymandelic acid was an effective antagonist, but compounds bearing one methoxyl group, namely DL-4-hydroxy-3-methoxymandelic acid, 4-hydroxy-3-methoxyphenylacetic acid, normetanephrine, and metanephrine were almost totally inactive. A compound with only one hydroxyl group on the ring, metaraminol, had intermediate potency. Mephentermine, a compound bearing no ring hydroxyl, was inactive. Trans-2-

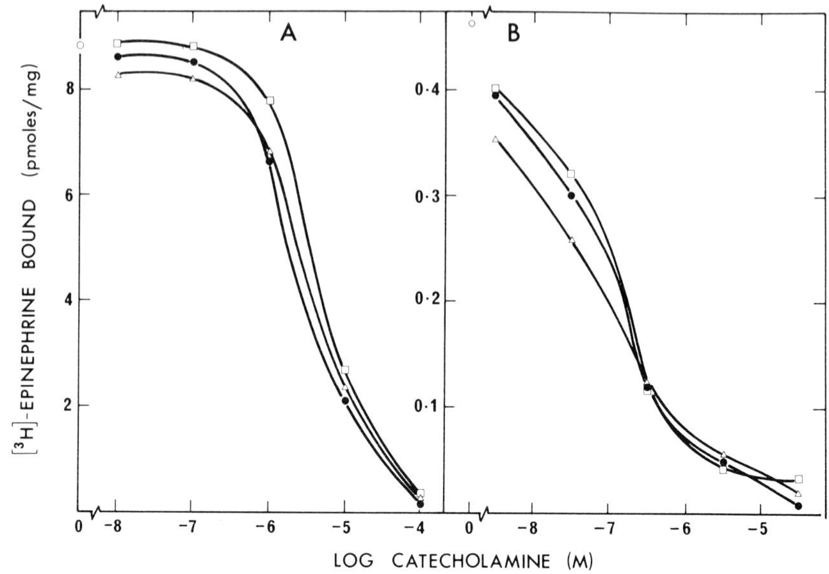

Figure 8 Inhibition of [^{3}H]epinephrine binding to cardiac and skeletal muscle membranes by catecholamines. For cardiac membranes (*A*), [^{3}H]epinephrine concentration was 0.1 μM and 150 μg of membrane protein was used. Solution of unlabeled amine (50 μl) was added to the membranes in buffer at 37°C, 30 sec before the solution of [^{3}H]epinephrine. The time of incubation was 75 min at 37°C. Skeletal muscle membranes (*B*) (208 μg of protein) were incubated for 10 min at 0°C with the indicated concentration of catecholamine before adding the [^{3}H]epinephrine (50 nM). The incubation period was 30 min at 0°C. In *A* and *B*, values are the mean of triplicate determinations; controls at each concentration of amine were also done in triplicate. •, DL-Epinephrine; □, DL-norepinephrine; and △, DL-isopropylnorepinephrine.

phenylcyclopropylamine and pargyline, two monoamine oxidase inhibitors, were ineffective. The catechol-O-methyltransferase inhibitors, pyrogallol, tropolone, and β-isopropyltropolone, displaced binding almost completely at a molar ratio of 10,000:1. The β-adrenergic blocking agents, propranolol, dichloroisoproterenol, and pronethalol, were extremely weak antagonists, quite inconsistent with their potency as β receptor antagonists. The α-adrenergic antagonist, phentolamine, in fact, was slightly more effective than propranolol. These data suggest strongly that the binding of [^{3}H]epinephrine to both of these membrane preparations is not coincident with interaction with the specific β receptor, but more likely reflects binding of compounds with one or more hydroxyl groups on a ring.

Stereospecificity for binding sites was examined using [^{3}H]norepinephrine since D and L isomers of this catecholamine were available. Figure 9 shows that unlabeled D- and L-norepinephrine were equipotent in inhibiting binding of DL-[^{3}H]norepinephrine, indicating lack of stereospecificity. Again, this is incon-

Table II Effect of various agents on [^{3}H]epinephrine binding*

Compound (M)	Cardiac membranes						Skeletal muscle membranes	
	1×10^{-8}	1×10^{-7}	1×10^{-6}	3.2×10^{-6}	1×10^{-5}	1×10^{-4}	5×10^{-7}	5×10^{-5}
DL-Epinephrine	96.4	93.4	80.6		20.3		26.1	3.1
DL-Norepinephrine	85.9	84.5	81.7		25.1		35.2	6.2
DL-Isopropylnorepinephrine	100.9	98.5	76.0				26.7	5.0
Dopamine		93.6		68.3		4.1	57.5	16.5
L-Dopa		95.3		59.1		6.9	58.1	11.7
DL-α-methyldopa		95.9		89.0		6.3		
DL-3,4-dihydroxymandelic acid		104.8		100.0		13.6	80.1	10.3
6-Hydroxydopamine		100.9		43.4			85.7	18.8
Pyrocatechol		98.2		71.5		3.7	50.0	18.0
4-Hydroxy-3-methoxyphenylacetic acid		102.8		108.9		99.0	97.1	83.3
DL-4-Hydroxy-3-methoxymandelic acid		102.4		98.1		101.4	95.0	86.4
DL-Normetanephrine		99.5		99.1		109.0	91.2	79.0
DL-Metanephrine							97.6	90.8
Metaraminol							78.5	65.0
Mephentermine		107.1		103.3		79.2	105.3	100.0
Trans-2-phenylcyclopropylamine		97.3		95.5		88.7	99.6	132.3
Pargyline		99.5		106.6		93.8	100.0	89.3
Pyrogallol		94.9		63.1		5.4	62.8	14.6
Tropolone		92.8		47.1		0.0	44.6	5.2
β-Isopropyltropolone							88.1	12.8
Phentolamine		99.4		107.5		52.3	121.4	94.6
Phenoxybenzamine							98.7	104.6
DL-Propranolol				92.2	86.8	66.7	114.8	126.4
Pronethalol		96.0		98.7		92.8	110.0	112.0
Dichloroisoproterenol		105.6		94.6		96.6	114.1	110.6

*For cardiac membranes [^{3}H]epinephrine was 10 nM. A solution of each compound in 0.002 N HCl (50 μl) was added 30 sec before the [^{3}H]epinephrine to give the final concentrations indicated. Incubation time was 75 min at 37°C. Values are the mean of triplicate determinations. Controls (without membranes) were performed in duplicate for each concentration of each compound. Control binding was 1.54 pmoles/mg of protein. For skeletal muscle membranes [^{3}H] epinephrine was 5 nM. Membranes (208 μg of protein) were incubated for 10 min at 0°C in the absence and presence of each compound in the final concentrations indicated before addition of [^{3}H]epinephrine. Incubation time was 30 min. Control binding was 0.45 pmole/mg of protein. Values are the mean of triplicate determinations. For both preparations, values are expressed as percentage of control binding.

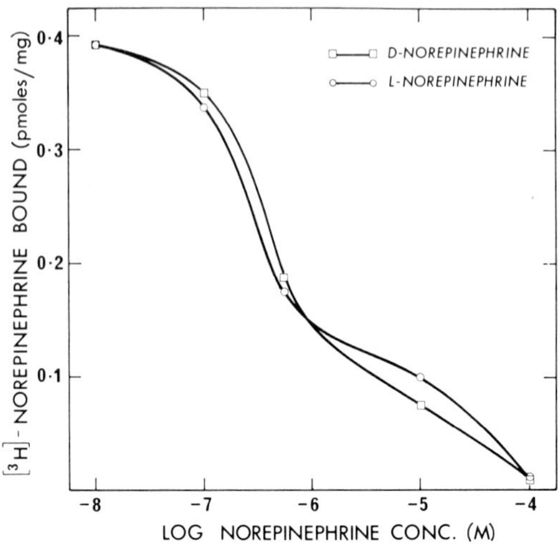

Figure 9 Inhibition of [^{3}H]norepinephrine binding by D- L-norepinephrine. Skeletal muscle membranes (180 μg of protein) were incubated for 5 min at 0°C under standard conditions with unlabeled norepinephrine at the concentrations indicated. [^{3}H]Norepinephrine was then added to a final concentration of 10 nM and the tubes were incubated for 5 min before filtering. Values are means of triplicate determinations corrected for binding to the filters at each concentration of norepinephrine. ○, L-Norepinephrine; □, D-norepinephrine.

sistent with binding to the specific β receptor since it is well known that β receptor effects are elicited almost exclusively by the L isomer.

Adenylate cyclase is present in skeletal muscle (16, 18, 22), where it is concentrated largely in the plasma membrane (19). As shown in figure 10, β-adrenergic amines stimulated the enzyme about 3-fold. Concentrations of isopropylnorepinephrine, epinephrine, and norepinephrine required for half-maximal stimulation were 0.15, 0.5, and 10 μM respectively. This is strikingly different from the data in figure 8 in which these three agonists were equipotent in inhibiting [^{3}H]epinephrine binding. In similar experiments, the adenylate cyclase stimulation produced by 10 μm epinephrine was completely blocked by 0.1 mM propranolol.

Experiments were conducted to examine whether the catecholamine binding described here could result from uptake into nerve storage vesicles such as have been described for cardiac muscle (2, 20). Animals were sympathectomized by chronic treatment with either 6-hydroxydopamine or guanethidine. The former drug is known to destroy adrenergic nerve endings selectively (7, 20). Guanethidine is concentrated *in vivo* in vesicular storage granules where it impairs their function (8, 12, 24). Table III reveals that [^{3}H]epinephrine binding to plasma membranes isolated from guanethidine or 6-hydroxydopamine-treated rabbits

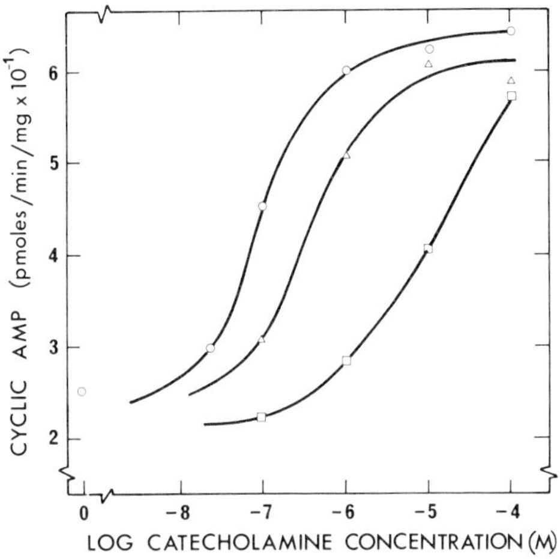

Figure 10 Action of catecholamines on adenylate cyclase activity in skeletal muscle membranes. Adenylate cyclase was assayed (19) using 0.3 mM [U-^{14}C]ATP, 6 mM Mg^{2+}, and 390 μg of membrane protein. Catecholamines were present at the concentrations indicated ○, Isopropylnorepinephrine; △, epinephrine; □, norepinephrine.

was not different from control preparations. *In vitro*, guanethidine is taken up by sympathetic nerve terminals and inhibits the catecholamine-concentrating mechanism (15, 21). Mg^{2+}ATP enhances accumulation of catecholamines by storage granules (20). To explore further the possible participation of nerve storage vesicles which might be present in our preparations, binding of [^{3}H]epinephrine was measured in the presence of guanethidine, ATP, and Mg^{2+}ATP. The

Table III Effect of treatment with guanethidine and 6-hydroxydopamine on [^{3}H]epinephrine binding to skeletal muscle membranes*

Treatment	[^{3}H]Epinephrine bound (pmoles/mg protein)
None	1.30 ± 0.02
Guanethidine	1.38 ± 0.06
6-Hydroxydopamine	1.32 ± 0.05

*Two rabbits were treated for 15 days with either 6-hydroxydopamine (20 mg/kg, twice a day, on the 1st, 6th, and 12th days), or with guanethidine (45 mg/kg, once a day), or with saline only. On the 16th day, plasma membranes were isolated and [^{3}H]epinephrine binding was measured using 10 mM by the standard assay. Results represent the mean ± S.E.M. of two preparations assayed in triplicate.

Table IV Effect of guanethidine, ATP and Mg^{2+} on [^{3}H]epinephrine binding to skeletal muscle membranes*

Additions	[^{3}H]Epinephrine bound (pmoles/mg protein)
Control	1.03 ± 0.07
Guanethidine (5 mM)	1.17 ± 0.03
ATP (1 mM)	0.43 ±0.09
ATP (5 mM)	0.16 ± 0.05
ATP (5 mM) + Mg^{2+} (5 mM)	0.21 ± 0.05

*The concentration of [^{3}H]epinephrine was 10 nM; 193 μg of membrane protein were used; time of incubation was 30 min. Values are the mean of triplicate determinations on a single preparation ± S.E.M.

data (table IV) show that binding was enhanced slightly by guanethidine but decreased markedly by ATP and Mg. Such findings would suggest that nerve ending vesicles are not involved in the binding of the [^{3}H]catecholamines studied here.

In general, the binding of [^{3}H]catecholamines to the cardiac and skeletal muscle plasma membranes in this study is similar to that described previously for liver plasma membranes (6, 23), heart microsomes (9, 11), cultured myocardial cells (10), and erythrocyte ghosts (1, 17). The inference from these studies is that binding reflects an interaction with the specific β receptor. We interpret our data as indicating that binding as measured by the micropore separation technique is not specific. During the course of these studies we observed that binding of [^{3}H]epinephrine to filters made of mixed esters of cellulose seemed to be altered by several of the agents which inhibited binding to plasma membranes. When examined in detail, it was found that those agents which inhibited binding to muscle membranes (see table II) also prevented binding of [^{3}H]epinephrine to the filters themselves (table V). Similarly, those agents which were less potent or inactive in preventing binding to membranes were equally ineffective in preventing binding to the filters. These results lend further evidence that the association of [^{3}H]catecholamines with plasma membrane preparations is likely a nonspecific process. In fact, the binding is remarkably similar to the binding of [^{14}C]epinephrine to bovine serum albumin reported by Danon and Sapira (4).

DISCUSSION

Our studies show that [^{3}H]epinephrine and norepinephrine and [^{14}C]epinephrine undergo degradation at an alarming rate at pH 7.6 and 37°C; in fact, measurable degradation is evident ater 5 min. The pattern of degradation of

Table V Effect of various agents on binding of [^{3}H]epinephrine to millipore filters*

Compound	Concentration (M)		Compound	Concentration (M)	
	10^{-6}	10^{-4}		10^{-6}	10^{-4}
DL-Epinephrine	47.1	2.4	Dichloroisoproterenol	154.9	126.5
Dopamine	36.5	0.0	Propranolol	100.1	112.6
6-Hydroxydopamine	85.8	34.3	Tranylcypromine	126.6	101.3
3,4-Dihydroxymandelic acid	66.5	10.5	Pargyline	118.8	117.4
4-Hydroxy-3-methoxymandelic acid	88.6	90.9	Pyrogallol	72.5	7.0
DL-Metanephrine	84.5	85.9	Tropolone	38.0	8.3
Metaraminol	92.5	70.6	β-Isopropyl-		
Mephentermine	101.6	103.5	Tropolone	67.1	11.3
Phentolamine	108.7	101.5	Pyrocatechol	47.3	22.8

*[^{3}H]Epinephrine (10 nM final concentration) was added to tubes containing 40 mM Tris-HCl, pH 7.6, and the compounds at the final concentration shown (volume 0.65 ml). An aliquot (0.60 ml) was then filtered through a cellulose filter (Millipore HAWPO47) and washed as usual. The filters were dried at 75°C, and dissolved in 2 ml of ethyleneglycol monomethyl ether. Eighteen milliliters of scintillation fluid (toluene-ethyleneglycol monomethyl ether (3:1) containing 4 gm of Omnifluor/l) was added and radioactivity was determined by scintillation spectrometry. The amount of unbound radioactivity retained on filters was calculated by running a parallel series of tubes containing [^{3}H]epinephrine and 50 mM pyrogallol. Control binding was 0.045 pmole/filter. Values are given as percentage of control binding.

[^{3}H]epinephrine (and [^{3}H]norepinephrine) is complicated in that not only does a new compound appear as revealed by thin layer chromatography, but also there is a large loss of tritium when such solutions are chromatographed. Chromatography of [^{14}C]epinephrine (carbinol labeled) solutions revealed that incubation in solution at pH 7.6 and 37°C led to the complete destruction of the compound after 2 hr and that the label was totally recovered in a new spot running near the solvent front and separated clearly from authentic epinephrine. It seems that the degradation of labeled epinephrine is complex, possibly involving oxidative reactions and degradation through self-radiolysis. This latter process could account in part for the loss of tritium. These degradative reactions were prevented by the presence of DTT.

As reported for other membrane preparations (1, 3, 11, 23), binding of [^{3}H]epinephrine to cardiac and skeletal muscle sarcolemma was a slow process requiring 20 min in the case of skeletal muscle membranes at 0°C, and at least 60 min for cardiac membranes to reach equilibrium. Binding was saturable for skeletal muscle sarcolemma with a maximal capacity of 46 pmoles/mg of protein. This value is very similar to that for cardiac microsomes (11) and plasma membranes from liver and turkey erythrocytes (1, 23). The amount of [^{3}H]epinephrine that could be extracted from the membranes with unlabeled epinephrine or acid was dependent on the conditions of incubation. In general, binding was readily dissociable when the incubation was at 0°C, and occurred less rapidly and only partially at 37°C. Such results might indicate the existence of two simultaneous binding processes. One of these, the readily dissociable one at 0°C, is supported by the chromatographic analysis of extracted labeled material. The second process may involve covalent bond formation as suggested by (14).

Binding to microsomal and membrane preparations from liver (3, 6, 23), heart (3, 11), and turkey erythrocytes (1, 3), and to cultured myocardial cells (10) required a catechol structure. In addition, binding was characterized by lack of stereospecificity. In the present studies, isopropylnorepinephrine, epinephrine, and norepinephrine were equally effective in displacing [^{3}H]epinephrine from binding sites, but large differences were evident in the relative concentration of each required for half-maximal stimulation of adenylate cyclase. Only the latter action truly resembled activation of β receptors. All other catechol derivatives, including pyrocatechol, pyrogallol, and 6-hydroxydopamine, in which phenolic hydroxyl groups were present, were effective competitors of binding. The catechol-O-methyl transferase inhibitors, tropolone and β-isopropyltropolone, were also potent inhibitors. Cuatrecasas *et al.* (3) have found recently that these compounds inhibit binding of catecholamines to several membrane preparations and have suggested, therefore, that binding of catecholamines to membranes (as measured by the micropore filtration technique) may reflect binding to catechol-O-methyltransferase. We would like to point out the analogy between the structure of pyrogallol, tropolone, and β-isopropyltropolone and that of catechol. Perhaps the carbonyl group adjacent to the ring hydroxyl in the latter two compounds can contribute to binding. Propranolol is a well known antagonist of catecholamine stimulation of adenylate cyclase as well as of other β-adrenergic effects such as cardiac inotropism. However, it was not an effective antagonist of [^{3}H]epinephrine binding to our membrane preparations nor were the other β antagonists, pronethalol and dichloroisoproterenol. Our studies with [^{3}H]norepinephrine show that the binding was not stereospecific.

Our results are remarkably similar to those that have been found by other investigators using a variety of membrane preparations. Our data, as well as that of others, lead us to conclude that binding of labeled catecholamines to cell membranes using the micropore filtration technique should not be interpreted as a specific process. The reaction is complex, possibly consisting of at least two processes. It seems to reflect a general specificity for compounds having one or more hydroxyl groups on a ring, and is not remarkably different from binding to cellulose esters (Millipore filters) or to bovine serum albumin (4). We tend to agree with the recent suggestion of Cuatrecasas *et al.* (3) that more sensitive methods must be developed to reveal the true nature of the β-adrenergic receptor.

REFERENCES

1. BILEZEKIAN, J., and AURBACH, G. 1973. A β-adrenergic receptor of the turkey erythrocyte. I. Binding of catecholamines and relationship to adenylate cyclase activity. J. Biol. Chem. 248: 5577–5589.
2. BURGEN, A. S. V., and IVERSEN, L. L. 1965. The inhibition of noradrenaline uptake by sympathomimetic amines in the rat isolated heart. Brit. J. Pharmacol. 25: 34–49.
3. CUATRECASAS, P., TELL, G. P. E., SICA, V., PARIKH, I., and CHANG, K. J. 1974. Noradrenaline binding and the search for catecholamine receptors. Nature 247: 92–96.
4. DANON, A., and SAPIRA, J. D. 1972. Binding of catecholamines to human serum albumin. J. Pharmacol. Exp. Therap. 182: 295–302.
5. De PLAZAS, S., and DE ROBERTIS, E. 1972. Isolation of a proteolipid from spleen capsule binding (I)-[^{3}H]norepinephrine. Biochim. Biophys. Acta 266: 246–254.
6. DUNNICK, J. K., and MARINETTI, G. V. 1971. Hormone action at the membrane level III. Epinephrine interaction with the rat liver plasma membrane. Biochim. Biophys. Acta 249: 122–134.
7. FINCH, L., HAEUSLER, G., KUHN, H., and THOENEN, H. 1973. Rapid recovery of vascular adrenergic nerves in the rat after chemical sympathectomy with 6-hyroxydopamine. Brit. J. Pharmacol. 48: 59–72.
8. KUNTZMAN, R., COSTA, E., GESSA, G. L., and BRODIE, B. B. 1962. Reserpine and guanethidine action on peripheral stores of catecholamines. Life Sci. 3: 65–74.
9. LEFKOWITZ, R. J., and HABER, E. 1971. A fraction of the ventricular myocardium that has the specificity of the cardiac beta-adrenergic receptor. Proc. Nat. Acad. Sci. U. S. A. 68: 1773–1777.
10. LEFKOWITZ, R. J., O'HARA, D., and WARSHAW, J. B. 1974. Surface interaction of [^{3}H]norepinephrine with cultured chick embryo myocardial cells. Biochim. Biophys. Acta 332: 317–328.

11. LEFKOWITZ, R. J., SHARP, G., and HABER, E. 1973. Specific binding of β-adrenergic catecholamines to a subcellular fraction from cardiac muscle. J. Biol. Chem. 248: 342–349.
12. LUNDBORG, P., and STITZEL, R. E. 1968. Studies on the dual action of guanethidine in sympathetic nerves. Acta Physiol. Scand. 72: 100–107.
13. LOWRY, O. H., ROSEBROUGH, N. J., FARR, A. L., and RANDALL, R. J. 1951. Protein measurement with the Folin phenol reagent. J. Biol. Chem. 193: 265–275.
14. MAGUIRE, M. E., GOLDMAN, P. H., and GILMAN, A. G. 1974. Covalent reaction of [^{3}H]norepinephrine with particulate fractions of cells responsive to catecholamines. Fed. Proc. 33(3): 288.
15. MAITRE, L., and STAEHELIN, M. 1971. Guanethidine uptake and noradrenaline depletion in noradrenaline storage particles of the rat heart. Biochem. Pharmacol. 20: 1233–1242.
16. RABINOWITZ, M., DESALLES, L., MEISLER, J., and LORAND, L. 1965. Distribution of adenylate cyclase activity in rabbit skeletal muscle fractions. Biochim. Biophys. Acta 97: 29–36.
17. SCHRAMM, M., FEINSTEIN, H., NAIM, E., LANG, M., and LASSER, M. 1972. Epinephrine binding to the catecholamine receptor and activation of the adenylate cyclase in erythrocyte membranes. Proc. Nat. Acad. Sci. U. S. A. 69: 523–527.
18. SERAYDARIAN, K., and MOMMAERTS, W. F. H. M. 1965. Density gradient separation of sarcotubular vesicles and other particulate constituents of rabbit muscle. J. Cell Biol. 26: 641–652.
19. SEVERSON, D. L., DRUMMOND, G. I., and SULAKHE, P. V. 1972. Adenylate cyclase in skeletal muscle. Kinetic properties and hormonal stimulation. J. Biol. Chem. 247: 2949–2958.
20. SHORE, P. A. 1972. Transport and storage of biogenic amines. Ann. Rev. Pharmacol. 12: 209–226.
21. SHORE, P. A., and GIACHETTI, A. 1966. Dual actions of guanethidine on amine uptake mechanisms in adrenergic neurones. Biochem. Pharmecol. 15: 899–903.
22. SUTHERLAND, E. W., RALL, T. W., and MENON, T. 1962. Adenyl cyclase. I. Distributing, preparation, and properties. J. Biol. Chem. 237: 1220–1227.
23. TOMASI, V., KORETZ, S., RAY, T. K., DUNNICK, J., and MARINETTI, G. V. 1970. Hormone action at the membrane level. II. The binding of epinephrine and glucagon to the rat liver plasma membrane. Biochim. Biophys. Acta 211: 31–42.
24. VALLIÈRES, J., BUREAU, M., and LEBLANC, J. 1972. Studies on the sensitization to cardiovascular effects of catecholamines. Can. J. Physiol. Pharmacol. 50: 576–583.

Mailing Address: Dr. G. I. Drummond,
Department of Pharmacology, Faculty of Medicine,
University of British Columbia,
Vancouver, British Columbia (Canada).

Recent Advances in Studies on Cardiac Structure and Metabolism
Volume 9
The Sarcolemma
Edited by Paul-Emile Roy and Naranjan S. Dhalla

ADENYLATE CYCLASE, CYCLIC NUCLEOTIDE PHOSPHODIESTERASE, AND NOREPINEPHRINE BINDING IN RAT HEART MEMBRANES[1]

F. J. MOFFET,[2] A. M. KIDWAI, and H. P. BÄR

Department of Pharmacology,
University of Alberta,
Edmonton, Alberta, Canada

SUMMARY

The method of Kidwai *et al.* (1971. Biochem. Biophys. Res. Commun. 45: 901) offers a rapid and simple technique for the preparation of membrane fractions from rat heart, using sucrose density centrifugation of a 100,000 × *g* pellet. We have investigated the distribution of adenylate cyclase, cyclic AMP phosphodiesterase, and norepinephrine binding activity in these fractions. Specific activity of adenylate cyclase was high in the plasma membrane (PM) and sarcoplasmic reticulum, as well as the nuclear (N) fractions, but 80–90 percent of the total activity was found in the N fraction. Epinephrine stimulation of adenylate cyclase was present in all fractions but did not exceed 25 percent of basal activity. The activity in the mitochondrial fraction was low and insensitive to epinephrine. About 10 percent of phosphodiesterase activity was found in the 100,000 × *g* pellet. After density gradient centrifugation, 70 percent of this portion was recovered in the N fraction, with the highest specific activity present in the PM fraction. Binding of [^{3}H] norepinephrine was measured by a membrane filtration technique. Binding activity was found in all fractions, and it paralleled roughly the distribution of adenylate cyclase. However, only about 25–30 percent of the binding was blocked by propranolol, except in the PM fraction where binding was not prevented by this drug. Further, a comparison of adenylate cyclase activities with norepinephrine binding yields a turnover number for the enzyme of the order of 10^{-2} sec^{-1}, assuming a 1:1 relationship between β-adrenergic receptor and adenylate cyclase. Since this value seems unrealistic, we suspect that either the binding activity measured is unrelated to the β-adrenergic receptor or the receptor to cyclase ratio is considerably larger than unity.

Adenylate cyclase is generally thought to be associated with plasma membranes (18). Isolation of plasma membrane fractions from various tissues (9, 16, 17, 23) has, however, resulted in only small degrees of purification of the enzyme. In the case of cardiac tissue the enzyme has been reported to be associated with both plasma membranes (22) and with sarcoplasmic reticulum (8, 21). Tada *et al.* (22)

[1] This research was supported by the Medical Research Council of Canada and by the Alberta Heart Foundation.
[2] Recipient of a Medical Scientist Fellowship of the Canadian Heart Foundation.

have separated a plasma membrane fraction in which adenylate cyclase is enriched approximately 3-fold over the homogenate. Sulakhe *et al.* (21) and Entman *et al.* (7) have reported a similarly enriched adenylate cyclase activity associated with preparations of sarcoplasmic reticulum from myocardial tissue. The methods used by these workers unfortunately did not provide other purified subcellular fractions for comparative studies.

The method of Kidwai *et al.* (10) utilizes a single-step sucrose density gradient fractionation to isolate four subcellular fractions from rat heart, thus facilitating such comparative studies on the subcellular distribution of enzyme activities. The four fractions, as characterized by enzyme markers and electron microscopy (10) represent plasma membrane (F-I), sarcoplasmic reticulum (F-II), mitochondria (F-III), and nuclei (F-IV). We have investigated the distribution of adenylate cyclase, cyclic AMP phosphodiesterase activity, and catecholamine binding activity among these fractions in order to evaluate the utility of this method for the study of subcellular distributions of enzyme activities. The system may also be useful for studies of adenylate cyclase-catecholamine interactions and as a starting point for further purification of adenylate cyclase and the associated β-adrenergic receptor.

METHODS

Hearts from female Wistar rats (200–250 gm) were used. Ventricular muscle was chopped and homogenized using a polytron homogenizer (Probe PT20; Brinkmann Instruments), and the resuspended 100,000 $\times$ *g* pellets from the homogenate were subjected to nonlinear sucrose density gradient centrifugation as described by Kidwai *et al.* (10). Assays of adenylate cyclase were carried out by the polyimine chromatography method described previously (3).

Phosphodiesterase was assayed by an adaption of the method of Gulyassy (9) using [^{3}H]-cyclic AMP (^{3}[*G*-^{3}H]-adenosine 3′,5′-cyclic phosphate, New England Nuclear Corp., 24 Ci/mmole) as a substrate. Reaction samples were spotted on polyethyleneimine cellulose thin layer chromatography plates (Macherey-Nagel/Brinkmann Instruments) and developed in 0.1 M lithium chloride. Reaction products were identified under ultraviolet light and eluted by 10 percent ammonia in 50 percent ethanol (v/v). Assays of binding of DL-[^{3}H] norepinephrine (DL-[7-^{3}H] norepinephrine (N), New England Nuclear Corp., 5 Ci/mmole) were carried out by a membrane filtration assay similar to that of Lefkowitz *et al.* (11) using 100–200 μg of protein in a total volume of 1 ml containing 100 mM Tris-HCl (pH 8.0) and 5 mM $MgCl_2$. After incubation for 2 hr at 37°C with 10^{-6} M DL-[^{3}H] norepinephrine, the reaction mixtures were filtered on 0.45 mμ membrane filters (Millipore), washed with 100 mM Tris-HCl (pH 8.0), 5 mM $MgCl_2$, and counted by liquid scintillation. Corrections for binding of DL-

[^{3}H] norepinephrine to blank filters were generally in the range of 1–2 percent of bound radioactivity.

All assays were carried out on fractions immediately after termination of the density gradient centrifugation. Results are reported as mean and S.D. of triplicate determinations.

RESULTS AND DISCUSSION

Adenylate cyclase

The total recoveries of adenylate cyclase activity and protein in the four subcellular fractions of rat heart prepared by the Kidwai method are shown in table I. The recovery of both protein and adenylate cyclase activity is very similar and constitutes approximately a third of the amounts present in the homogenate. The distribution of adenylate cyclase activity among the fractions is detailed in table II. It can be seen that the specific activity is high in fractions F-I, plasma membrane, and F-II, sarcoplasmic reticulum. The mitochondrial fraction F-III shows the lowest values, whereas an intermediate level of activity is present in the nuclear fraction (F-IV).

In general, these data show that adenylate cyclase activity is widely distributed in subcellular fractions of cardiac muscle obtained by the method of Kidwai (10). Only a rather small percentage of the total recoverable activity is associated with the plasma membrane; by far the largest portion of the activity is associated with the heavy nuclear fraction F-IV. Sulahke and Dhalla (21) found a similarly large portion of adenylate cyclase in heavy 1,000 × *g* precipitates from dog and rabbit heart. Their procedure, however, purified only the sarcoplasmic reticulum fraction, which showed high specific activity adenylate cyclase but a low portion of the total activity. They did not examine plasma membranes. Tada *et al.* (22) have reported a method of preparing only plasma membranes from guinea pig hearts which results in the recovery of up to 80 percent of initial homogenate activity in the plasma membrane fraction. The method, however, involves extraction with high concentrations of KCl and these authors suggest that their

Table I Total recovery of protein and adenylate cyclase in subcellular fractions

	Total protein		Total adenylate cyclase	
Fraction	mg	Recovery (%)	pmole/min	Recovery (%)
Homogenate	287.0	100.0	4,649	100.0
F-I–F-IV	98.9	34.4	1,812	38.9

Table II Distribution of adenylate cyclase in subcellular fractions*

Fraction	Adenylate cyclase (pmoles/min/mg)	Stimulation (%) NaF	Stimulation (%) Epinephrine	% Distribution of basal activity (F-I–F-IV)
Homogenate	16.2 ± 1.0	335	140	
100,000 × *g* pellet	26.2 ± 1.0	128	40	
F-I	31.0 ± 1.3	565	25	2.5
F-II	45.1 ± 3.7	315	20	
F-III	7.8 ± 1.2	59	0	6.4
F-IV	19.3 ± 2.4	190	25	85.4

*Enzyme assays contained in a volume of 50 μl, 40 mM Tris-HCl (pH 8) 5 mM $MgCl_2$, 10 mM sodium creatine phosphate, 0.1 mg/ml creatine phosphokinase, 1 percent bovine serum albumin, 0.5 mM sodium cyclic AMP, 0.1 mM α-[^{32}P] ATP (about 600,000 c.p.m.) and 0.1–0.5 mg/ml of enzyme protein. Incubations were for 15 min at 37°C. NaF was 10 mM and epinephrine was 0.1 mM.

high levels of recovered activity may reflect stimulation of a latent enzyme activity. Their method also does not provide other subcellular fractions for comparative studies.

The method of Kidwai does provide several purified subcellular fractions and our results using this technique suggest that high specific activity adenylate cyclase is associated with both plasma membrane and sarcoplasmic reticulum. However, despite the excellent purity of these fractions by enzyme marker studies and electron microscopy (10) it is still very difficult to exclude the cross-contamination of fractions by small membrane fragments. The strong shearing forces and ultrasonic action of the polytron homogenizer probe used in the preparation of the original homogenate may facilitate such a process.

The large amount of adenylate cyclase activity found in the F-IV nuclear fraction may be due in part to such a redistribution process. Other sources of this activity might include the possibility of a nuclear adenylate cyclase as has been reported in the liver and prostate tissue of the rat (15, 20). The inclusion of some unbroken cells in the F-IV fraction which lyse in the hypotonic adenylate cyclase assay medium might contribute to this activity, although this seems to be a minor source. In the absence of specific sarcoplasmic reticulum markers one might speculate that this fraction, in fact, remains associated with nuclei to a large extent, thus contributing to the F-IV adenylate cyclase activity. The existence of nervous, vascular, and connective tissue elements in association with myocardium may further complicate an interpretation of the activity distribution.

Sensitivity to NaF stimulation is clearly present but somewhat variable in the subfractions. The degree of stimulation by epinephrine was found to be the same as norepinephrine. The small degree of epinephrine sensitivity probably indicates

Table III Total recoveries of phosphodiesterase

Fractions	Phosphodiesterase (nmoles/min)	% Recovery*
Homogenate	580	100
Supernatant†	324	56
F-I–F-IV	745	13

*Based on activity in original homogenate.
†After separation of bulk membranes from homogenate, i.e., 100,000 × *g* supernatant.

a lability of the adrenergic receptor or of its coupling to adenylate cyclase. The use of phospholipids to stabilize hormone sensitivity according to the system developed by Levey and Klein (14) might result in greater hormone sensitivity in these preparations and will be investigated.

Cyclic AMP phosphodiesterase

The distribution of phosphodiesterase is outlined in table III. The total recovery of phosphodiesterase after 100,000 × *g* centrifugation and density gradient centrifugation of the resuspended pellet was in the range of 70–90 percent of the homogenate activity. Approximately 80 percent of this recovered activity was found in the 100,000 × *g* supernatant and the remaining 20 percent in subcellular fractions.

Table IV shows the distribution of specific activities and relative amounts of phosphodiesterase in the particulate fractions. The largest portions are associated with the F-IV (nuclear) and F-III (mitochondrial) fractions, whereas relatively

Table IV Distribution of cyclic AMP phosphodiesterase in subcellular fractions*

Fractions	Phosphodiesterase (pmoles/min/mg)	% Distribution of basal activity (F-I–F-IV)
Homogenate	2,827 ± 524	
Pellet	579 ± 115	
100,000 × *g* supernatant	8,117 ± 61	
F-I	2,338 ± 107	7.2
F-II	1,021 ± 43	3.9
F-III	1,114 ± 249	20.6
F-IV	969 ± 237	68.2

*Diesterase assays contained in a volume of 50 μl, 60 mM Tris-HCl (pH 8), 5 mM $MgCl_2$, 100 μM [^{3}H] cyclic AMP, and 5–10 μg of enzyme protein. Incubations were for 10 min at 37°C. Reactions were terminated by heating for 4 min in a boiling water bath.

small portions of the total activity are associated with the plasma membrane (F-I) and sarcoplasmic reticulum (F-II) fractions. The highest specific activity of phosphodiesterase among the subcellular fractions is found in plasma membranes (F-I), although none of these fractions is enriched over the homogenate. The specific activity found in the 100,000 × *g* supernatant is, however, 3-fold enriched relative to the homogenate.

There is not a great deal of information available in the literature concerning the subcellular distribution of phosphodiesterase. DeRobertis *et al.* (6) found that 60 percent of rat brain phosphodiesterase was in the particulate fraction of 11,500 × *g* centrifugations, and they also observed that highest specific activities were found in the supernatant. Our data show a wide subcellular distribution of phosphodiesterase activity and suggest the possibility of an enzyme of high specific activity localized in the plasma membrane. In the present study no attempts have been made to determine whether different kinetic properties are present with the diesterase activities in different fractions; such information may be valuable in the differentiation of possible isoenzymes.

DL-[^{3}H]Norepinephrine binding

All subcellular fractions were found to bind DL-[^{3}H]norepinephrine as indicated in table V. F-I, F-II, and F-VI all showed considerably higher specific binding activity than F-III. This distribution seems to correspond to that observed for specific activities of adenylate cyclase (table II). However, the degree of binding that is blocked by the β blocker propranolol is much less than the total binding in all cases, and does not correlate at all with the distribution of adenylate cyclase activity. F-III, which has the lowest specific activity of the cyclase enzyme, shows the largest percentage of propranolol-sensitive binding, whereas F-I, with relatively high specific activity adenylate cyclase, shows no

Table V Binding of DL-[^{3}H]norepinephrine to subcellular fractions*

Fraction	pmoles/mg bound	% Propranolol sensitive
Homogenate	256 ± 18	9.8
Pellet	408 ± 25	21.3
100,000 × *g* supernatant	235 ± 12	6.8
F-I	274 ± 19	0
F-II	415 ± 27	21.7
F-III	174 ± 17	32.7
F-IV	479 ± 26	26.5

*Propranolol was 0.1 mM. Reaction mixtures contained 50 mM Tris-HCl (pH 8), 5 mM $MgCl_2$, 10^{-6} M DL-[^{3}H]norepinephrine (about 3 × 10^6 c.p.m.) and 10–20 μg of subcellular fraction protein. Incubation was 120 min at 37°C. Reactions were stopped by membrane filtration at 4°C.

Table VI Turnover numbers for adenylate cyclase*

	Turnover No. (sec^{-1})	
Fraction	Total binding sites	Propranolol-sensitive sites
Homogenate	0.0011	0.0108
Pellet	0.0011	0.0050
F-I	0.0019	
F-II	0.0018	0.0084
F-III	0.0007	0.0023
F-IV	0.0007	0.0025

*Calculated by dividing specific activities of adenylate cyclase (table II) by norepinephrine binding activity (table V).

inhibition of binding at all. It is also interesting to note that the 100,000 X *g* supernatant contains significant norepinephrine binding activity, very little of which is blocked by propranol. The interpretation of catecholamine binding to both intact cells and microsomal preparations such as these must be approached cautiously. Lefkowitz *et al.* (11, 13) have examined the binding of norepinephrine to canine heart microsomes and have isolated a proposed β-adrenergic binding protein (12). Others have reported similar studies on turkey erythrocytes (4, 19). However, considerations of the lack of correlation between the rate of catecholamine activation of *in vitro* adenylate cyclase and the rate of catecholamine binding to microsomal preparations (1), as well as differences in the amounts of binding by these preparations relative to biological activity of the hormones (5), cast doubt on whether these results represent binding to a true β-adrenergic receptor. Cuatrecasas *et al.* (5) have presented evidence that such studies of catecholamine binding may actually be measuring the relatively nonspecific binding of these hormones to a particulate catechole-O-methyl transferase enzyme.

In this regard, the turnover numbers for adenylate cyclase which can be calculated from data on adenylate cyclase specific activity (table II) and hormone binding capacity (table V), assuming a 1:1 relationship between cyclase catalytic unit and hormone receptor, are of interest. These values for our data on rat heart subcellular fractions are given in table VI. The values are unreasonably low; one might expect a value of greater than 100 sec^{-1} to be more realistic. Such low values for turnover numbers can also be derived from literature data on catecholamine binding, and a number of possible explanations may account for them, as reviewed by Bär (2), including an adenylate cyclase to receptor ratio of greater than 1:1. They are, however, also consistent with the possibility that catecholamine binding studies are not, in fact, measuring the true cyclase-related β-adrenergic receptor.

Considering the above discrepancies in the relationship of catecholamine binding and adenylate cyclase activity measurements, we feel strongly that our binding studies do not reflect binding to a β-adrenergic receptor. Nevertheless, the principle of such binding studies is still of great interest and with more selective methods for β receptor binding assays, the correlation of hormone binding with adenylate cyclase may become possible.

Although the Kidwai technique provides a very rapid and simple method of obtaining subcellular fractions from myocardial tissues, and these fractions appear to be reasonably pure by enzyme marker studies and electron microscopy, our studies indicate a widespread distribution of adenylate cyclase, phosphodiesterase, and catecholamine binding activity in these fractions. The results could be interpreted to suggest that these activities may be localized at various sites in the intact cell, but we feel that such conclusions must be approached cautiously for the reasons outlined above, despite the apparent purity of the fractions by conventional standards. Furthermore, our studies indicate that the finding of adenylate cyclase activity enriched over the homogenate in a single subcellular fraction does not necessarily indicate the primary localization of the enzyme at that site in the intact cell. Other subcellular fractions should also be examined before such conclusions are drawn. In this regard we feel that the Kidwai technique offers an excellent starting point for comparative studies of such activities. We expect that detailed characterization and comparison of the adenylate cyclase, phosphodiesterase, and catecholamine binding in these subfractions will lead to more definite conclusions concerning the subcellular distribution of these activities.

REFERENCES

1. BÄR, H. P. 1974. On the kinetics and temperature dependence of adrenaline-adenylate cyclase interactions. Molec. Pharacol. 10: 597–604.
2. BÄR, H. P. 1974. Catecholamine binding and the turnover number of adenylate cyclase. Can J. Physiol. Pharm. 52: 1030–1032.
3. BÄR, H. P., and HECTER, O. 1969. Adenylate cyclase assay in fat cell ghosts. Anal. Biochem. 29: 476–483.
4. BILEZIKIAN, J. P., and AURBACH, G. D. 1973. A β-adrenergic receptor of the turkey erythrocyte. I. Binding of catecholamine and relationship to adenylate cyclase activity. J. Biol. Chem. 248: 5575–5583.
5. CUATRECASAS, P., TELL, C. P. E., PARIKH, I., and CHANG, K. J. 1974. Noradrenaline binding and the search for catecholamine receptors. Nature 247: 92–97.
6. DeROBERTIS, E., ARNAIZ, G. R. D. L., ALBERICI, M., BUTCHER, R. W., and SUTHERLAND, E. W. 1967. Subcellular distribution of adenylate cyclase and cyclic phosphodiesterase in rat brain cortex. J. Biol. Chem. 242: 3487–3493.
7. ENTMAN, M. L., LEVELY, G. S., and EPSTEIN, S. E. 1969. Demonstration of adenylate cyclase activity in canine cardiac sarcoplasmic reticulum. Biochem. Biophys. Res. Commun. 35: 728–733.
8. FORTE, L. R. 1972. Characterization of the adenylate cyclase of rat kidney plasma membranes. Biochim. Biophys. Acta 266: 524–542.

9. GULYASSAY, P. F. 1971. Inhibition of cyclic 3′,5′-nucleotide phosphodiesterase by adenine compounds. Life Sci. 10: 451–461.
10. KIDWAI, A. M., RADCLIFFE, M. A., DUCHON, G., and DANIEL, E. E. 1971. Isolation of plasma membrane from cardiac muscle. Biochem. Biophys. Res. Commun. 45: 901–910.
11. LEFKOWITZ, R. J., and HABER, E. 1971. A fraction of the ventricular myocardium that has the specificity of the cardiac β-ardrenergic receptor. Proc. Nat. Acad. Sci. U.S.A. 68: 1773–1777.
12. LEFKOWITZ, R. J., HABER, E., and O'HARA, D. 1972. Identification of the cardiac β-adrenergic receptor protein: Solubilization and purification by affinity chromatography. Proc. Nat. Acad. Sci. U. S. A. 69: 2828–2832.
13. LEFKOWITZ, R. J., SHARP, G. W. G., and HABER, E. Specific binding of β-adrenergic catecholamines to a subcellular fraction from cardiac muscle. J. Biol. Chem. 248: 342–349.
14. LEVEY, G. S., and KLEIN, I. 1972. Solubilized myocardial adenylate cyclase: Restoration of histamine responsiveness by phosphatidyl serine. J. Clin. Invest. 51: 1578–1582.
15. LIAO, S., LIN, A. H., RYMOCZKO, J. L. 1971. Adenyl cyclase of cell nuclei isolated from rat ventral prostate. Biochim. Biophys. Acta 230: 535–539.
16. NAKEEL, D. W., and JARETT, C. 1970. Preparation and characterization of a plasma membrane fraction from isolated fat cells. J. Cell Biol. 44: 417–432.
17. POHOL, S. L., BIRNBAUMER, L., and RODBELL, M. 1969. Glucagon-sensitive adenylate cyclase in plasma membrane of hepatic paranchymal cells. Science 164: 566–567.
18. ROBINSON, G. A., BUTCHER, R. W., and SUTHERLAND, E. W. 1971. Cyclic AMP, pp. 75–78. Academic Press, New York.
19. SCHRAMM, M., FEINSTEIN, H., NAIN, E., LANG, M., and LASSER, M. 1972. Epinephrine binding to the catecholamine receptor and activation of the adenylate cyclase in erythrocyte membranes. Proc. Nat. Acad. Sci. U. S. A. 69: 523–527.
20. SOIFFER, C., and HECHTER, O. 1971. Adenylate cyclase activity in rat liver nuclei. Biochim. Biophys. Acta 230: 539–542.
21. SULAKHE, P. V., and DHALLA, N. S. 1973. Adenylate cyclase of heart sarcotubular membranes. Biochim. Biophys. Acta 293: 379–396.
22. TADA, M., FINNEY, J. O., SWARTZ, M. H., and KATZ, A. M. 1972. Preparation and properties of plasma membranes from guinea pig hearts. J. Molec. Cell. Cardiol. 4: 417–426.
23. WOLFF, J., and JONES, A. B. 1971. The purification of bovine thyroid plasma membranes and the properties of membrane bound adenylate cyclase. J. Biol. Chem. 246: 3939–3947.

Mailing address: Dr. H. P. Bär,
Dept. of Pharmacology, University of Alberta,
Edmonton, Alberta (Canada).

Recent Advances in Studies on Cardiac Structure and Metabolism
Volume 9
The Sarcolemma
Edited by Paul-Emile Roy and Naranjan S. Dhalla

INFLUENCE OF THYROID STATE ON THE SPECIFIC BINDING OF NORADRENALINE TO A CARDIAC PARTICULATE FRACTION AND ON CATECHOLAMINE-SENSITIVE CARDIAC ADENYLATE CYCLASE ACTIVITY

A. WOLLENBERGER and L. WILL-SHAHAB

Division of Cellular and Molecular Cardiology,
Central Institute of Heart and Circulatory Regulation Research,
Academy of Sciences of the DDR,
Berlin-Buch, German Democratic Republic

SUMMARY

Specific binding of [^{3}H]noradrenaline, possibly to β-adrenoreceptors, and adrenaline-sensitive adenylate cyclase activity were measured in particulate preparations from the myocardium of hypothyroid, euthyroid, and hyperthyroid rats. Noradrenaline affinity to the binding sites was decreased and the dose-response curve for the stimulation of adenylate cyclase activity by adrenaline was shifted to the right in hypothyroidism. The opposite changes (an increase in noradrenaline affinity and a shift in the adrenaline-adenylate cyclase dose-response curve to the left) were associated with the hyperthyroid state. These findings correlate well with the result of numerous studies according to which thyroid hormone increases and thyroid hormone deprivation decreases the sensitivity of the myocardium to cardiac sympathetic stimulation and circulating catecholamines.

Cardiac responses to sympathetic nerve activity and to circulating sympathetic amines are known to be exaggerated in hyperthyroidism and to be attenuated in hypothyroidism. This phenomenon has for many years been the subject of extensive investigation (4, 9, 10, 17, 37), but a convincing explanation has not yet been forthcoming. Recent work on the purification and characterization of what was believed to be the cardiac β-adrenoreceptor (14–16) may have opened a new approach to the problem, although the correctness of this belief was recently seriously questioned (8, 36). Nevertheless, in an attempt to make use of this approach, the binding of [^{3}H]noradrenaline to the putative β-adrenergic receptor sites (39) and the relation of this binding to catecholamine-stimulated adenylate cyclase activity were studied in the present work in particulate preparations from hearts of hyperthyroid, euthyroid, and hypothyroid animals.

The results could help to explain some of the known differences between the cardiac responses to adrenergic stimulation in animals and persons differing in thyroid state.

MATERIALS AND METHODS

Male rats of Wistar strain were used. The hyperthyroid state was induced by a single intraperitoneal injection of 100 μCi of carrier-free ^{131}I 2–3 days after birth. Body weight and heart rate were measured every other week up to 8–10 weeks after the ^{131}I injection, when the animals were killed. Hyperthyroidism was produced by daily subcutaneous injections on 7 consecutive days of 50 μg of 3,5,3′-triiodo-L-thyronine/100 gm body weight. Thyroid status was ascertained by determination of the heart rate from electrocardiograms recorded in the conscious animals and by measurements of the thyroid hormone concentration in blood plasma, using the Res-O-Mat effective thyroxine ratio diagnostic rest of Byk-Mallinckrodt Chemische Produkte GmbH (Frankfurt/Main).

Immediately after death the heart was excised quickly and chilled and a 78,000 $\times$ *g* particulate fraction, said to contain β-receptor sites (14), was prepared from homogenates of the ventricles by the procedure described by Lefkowitz and Haber (14), employing ice-cold 0.25 M sucrose-0.001 M $MgCl_2$-0.005 M Tris-HCL (pH 7.4) solution for homogenization and dilution. Suspensions of the particulate fraction could be stored at −18°C for several months without change in noradrenaline binding activity. Aliquots of 0.25 ml of the particulate fraction, containing 150 μg of protein, were incubated at 37°C with varying concentrations of [^{3}H]noradrenaline, usually for 2 hr. Compounds to be tested for effects on [^{3}H]noradrenaline binding were included in the incubation mixture at the stated molar concentrations. Binding of [^{3}H]noradrenaline to the particulate fraction was assessed after collection on Millipore filters (type HAWP, 25 mm in diameter, and pore size of 0.45 μm) by counting in a Nuclear Chicago Mark II liquid scintillation spectrometer. The amount of [^{3}H]noradrenaline retained by the particulate fraction in the presence of 10^{-4} M unlabeled noradrenaline was taken to represent nonspecific binding. It never exceeded 5 percent of the total bound radioactivity and was subtracted from the latter to yield values for specific binding, which are the values reported below. Protein was determined by the method of Lowry *et al.* (19).

Adenylate cyclase was assayed by measuring the rate of formation of ^{32}P-labeled cyclic AMP from α-[^{32}P] ATP (28). The reaction mixture contained, in a total volume of 0.15 ml, 40 mM Tris-HCl of pH 7.4, 5.5 mM KCl, 15 mM $MgCl_2$, 0.1 percent bovine serum albumin, 8 mM theophylline, 20 mM phosphoenolpyruvate, 20 μg of pyruvate kinase, 0.4 mM α-[^{32}P]ATP (20–40 c.p.m./pmole), and 150–250 μg of protein of a particulate cardiac fraction freshly prepared according to the method of Murad (24). The incubation was carried out at 30°C for 15 min. The reaction was stopped by the addition of 250 μl of 10 mM Tris HCl, pH 7.4, containing 2 mM cyclic AMP and 0.05 μCi of [^{3}H] cyclic AMP, in order to estimate the yield of cyclic AMP recovery, and by boiling the tubes in water for 3 min. The cyclic nucleotide was separated on columns for alumina according to Ramachandran (28). ^{32}P impulses were measured by means of Cerenkov counting in a LKB-Wallac 81000 liquid scintillation counter.

RESULTS

Characterization of thyroid status

Significant bradycardia and a lowered plasma concentration of thyroid hormone characterized the hypothyroid state in the ^{131}I-treated rats (table I). The growth of the animals was retarded. Eight to 10 weeks after the injection of ^{131}I, just before killing, their body weight averaged 124 gm as compared with 298 gm for the control rats. In order to exclude effects on noradrenaline binding due to lower body weight, an additional control group was employed in which the animals were matched for weight.

The hyperthyroid state in the rats treated with triiodothyronine was made evident by increases in heart rate and in the concentration of thyroid hormone in the blood plasma (table I). The rats had lower body weights and higher heart weights than the control rats of the same age.

Table I Physiological characteristics of rats differing in thyroid status*

Group	Body weight (gm)	Heart weight (mg)	Heart rate (min^{-1})	Effective thyroxine ratio test†
Control, matched for age	298 ± 18	789 ± 6	448 ± 23	0.98
Control, matched for weight‡	120 ± 7	290 ± 2	488 ± 20	0.95
^{131}I-treated rats	124 ± 12	295 ± 1	326 ± 18	0.81
Thyroxine-treated rats	250 ± 16	968 ± 8	612 ± 28	1.23

*Ten to 20 rats per group, mean values ± S.D.
†Pooled sera of all animals in a given group.
‡Average age 6 weeks.

Binding of [^{3}H]noradrenaline to cardiac particulate fraction

Data on the binding of [^{3}H]noradrenaline to the subcellular fraction from ventricular myocardium of euthyroid rats are presented in figure 1 in form of a Scatchard plot (31). It is evident that there exist in this myocardial fraction two kinetically distinct orders of noradrenaline binding sites, with association constants of 1.26×10^7 M^{-1} and 1.60×10^5 M^{-1}. The maximum number of binding sites of the higher order of affinity was 42 pmoles of noradrenaline per mg of protein. In agreement with the results obtained by Lefkowitz *et al.* (14, 15) in dog heart ventricle muscle, catecholamines were found to compete for this class of binding sites with relative potencies corresponding to their β-adrenergic activity in the heart, *i.e.*, isoproterenol > adrenaline > noradrenaline/dopamine. As was found in dog heart muscle (14, 15), inhibition of noradrenaline binding by propranolol required much higher concentrations of this adrenergic antagonist than those necessary to achieve blockage of noradrenaline effects on myocardial contraction and metabolism. Guanosine triphosphate in concentrations of 0.01–0.1 mM had no effect on noradrenaline binding. The time course of [^{3}H]noradrenaline binding to the subcellular particulate fraction from hearts of euthyroid, hypothyroid, and hyperthyroid rats is shown in figure 2. Under the prevailing experimental conditions (37°C, pH 7.4, initial [^{3}H]noradrenaline concentration 3×10^{-8} M), equilibrium with respect to binding of the catechol-

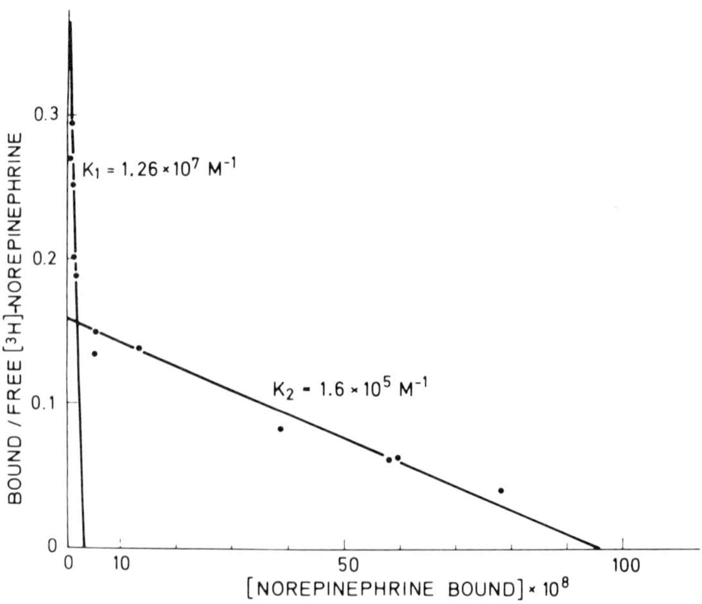

Figure 1 Scatchard plot for [^{3}H]noradrenaline binding.

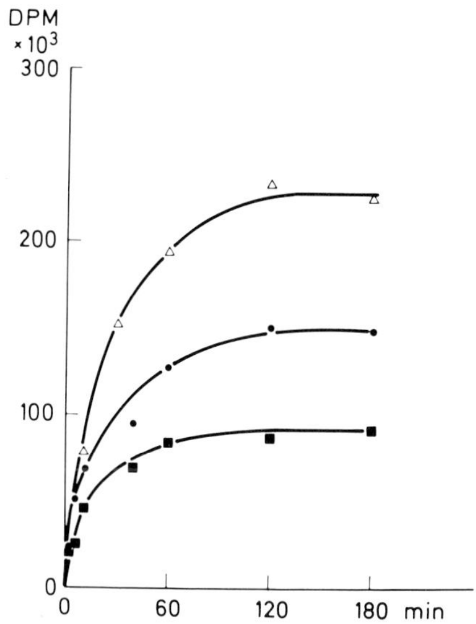

Figure 2 Time-course of [^{3}H]noradrenaline binding to particulate fraction from hearts of euthyroid, hypothyroid, and hyperthyroid rats.

amine was reached within 2 hr in all three cases, with a half-life of the process of approximately 12 min. It is seen from figure 2 that the particulate fraction from the hyperthyroid hearts bound more and that from the hypothyroid hearts bound less noradrenaline than did the fraction from the euthyroid hearts. As mentioned above, nonspecific binding [^{3}H] noradrenaline was almost negligible in all cases.

A comparision of the noradrenaline binding characteristics of the particulate cardiac fraction in the three thyroid states is presented in the Scatchard plots shown in figure 3. Whereas no major change in the number of high affinity noradrenaline binding sites was induced by alterations in thyroid status, their affinity for noradrenaline was markedly affected by lack and by excess of thyroid hormone, the binding constants being 0.54×10^7 M^{-1} in hypothyroidism and 2.00×10^7 M^{-1} in hyperthyroidism as compared with 1.26×10^7 M^{-1} in euthyroidism.

Stimulation of adenylate cyclase

The question arose at this point of our investigation of whether the observed changes in noradrenaline affinity caused by lack and excess of thyroid hormone would be associated with corresponding changes in cardiac adenylate cyclase activity. This seems to be the case. As can be seen from figure 4, the dose-response curve for the stimulation of adenylate cyclase activity by adrenaline

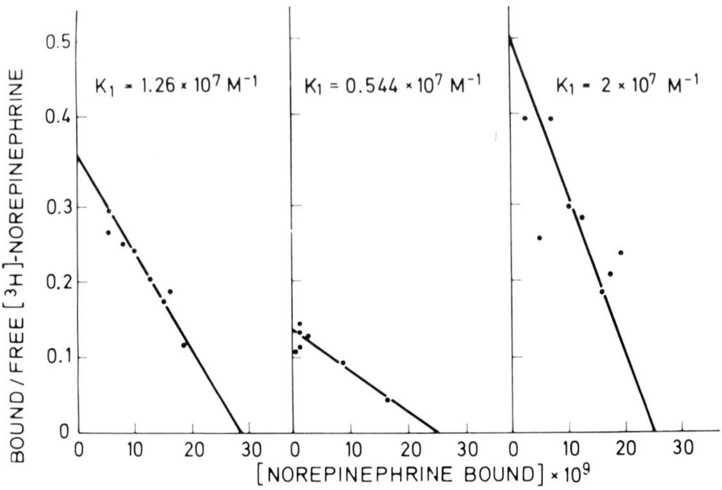

Figure 3 Scatchard plots for [^{3}H]noradrenaline binding to particulate fraction from hearts of euthyroid, hypothyroid, and hyperthyroid rats.

was shifted to the right in the heart particles from the hypothyroid rats and to the left in those from the hyperthyroid animals. The concentrations of adrenaline required to produce half-maximal stimulation of adenylate activity were 5×10^{-7} M in the heart particle preparations of the hyperthyroid rats, 9.5×10^{-6} M in the preparations of the controls, and 6×10^{-5} in the preparations of the hypothyroid animals. Thus the sensitivity of rat heart adenylate cyclase to adrenaline is increased in hyperthyroidism and decreased in hypothyroidism.

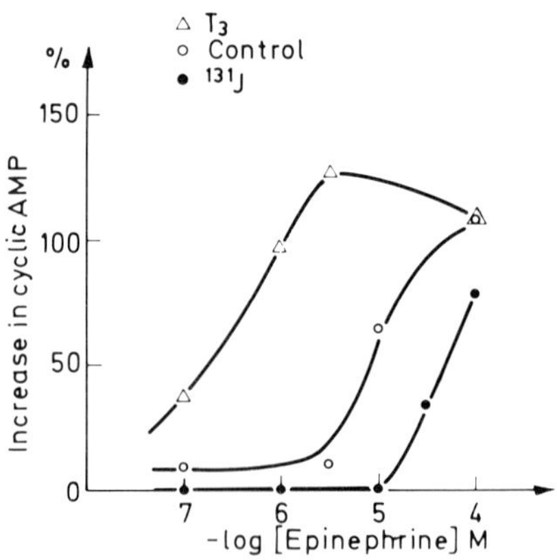

Figure 4 Effect of adrenaline on the adenylate cyclase activity of the particles from euthyroid, hypothyroid, and hyperthyroid rat hearts.

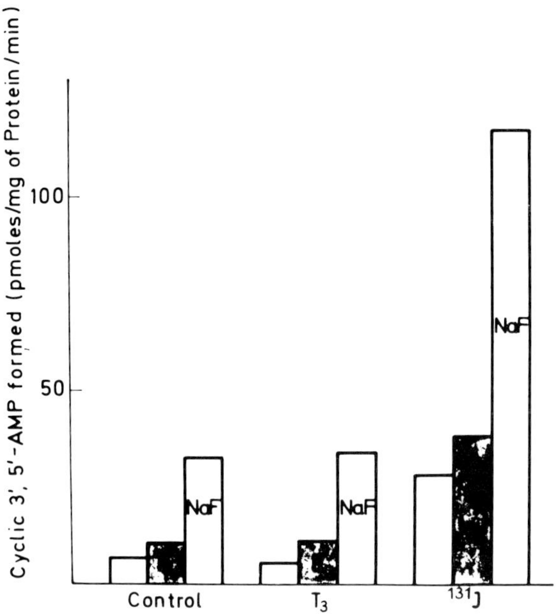

Figure 5 Basal, adrenaline-stimulated, and NaF-stimulated adenylate cyclase activities in euthyroid, hypothyroid, and hyperthyroid rat heart particles.

Hypothyroidism was associated with a striking increase in basal cardiac adenylate cyclase activity, which was elevated more than 3-fold above the activity in the heart particle preparations of the control rats (see fig. 5). This difference persisted in the presence of 12 mM sodium fluoride which produced about the same percentage of increase in cyclase activity in preparations of the control and hyperthyroid animals. As noted previously by other authors (6, 32) in cardiac adenylate cyclase preparations, basal and fluoride-stimulated activity levels were not significantly affected by hyperthyroidism (Fig. 5). Maximum stimulation of adenylate cyclase by adrenaline was 80 and 95 percent, respectively, in the preparations of the control and hyperthyroid rats and 38 percent in that of the hypothyroid rats. Yet, on account of the high basal activity level in the hypothyroid group, the absolute increment in cyclase activity achievable by adrenaline was nearly twice as great as in this group as it was in the controls.

DISCUSSION

Lefkowitz *et al.* (14, 15) described the binding of [^{3}H]noradrenaline to a 78,000 X *g* particulate fraction from dog myocardium which in their opinion exhibited, judging from the specificity and affinity of binding, properties of β-adrenoreceptors. In the present study similar adrenergic binding sites were found to exist in a 78,000 × *g* particulate fraction from rat myocardium. The association constants at these sites for noradrenaline, 1.26×10^7 M^{-1} and 1.60 $\times 10^5$ M^{-1}, are comparable with corresponding values reported for dog myocardium (14, 15). Lefkowitz and coworkers (15) also observed that exposure of the cardiac particulate fraction to thyroid hormones *in vitro* was without influence on the binding of noradrenaline. The present experiments revealed that thyroid hormone *in vivo* exerts a marked influence on the binding of noradrenaline to the 78,000 × *g* cardiac particulate fraction in that the binding constant for interaction with the catecholamine at the high affinity binding sites is reduced to one-half by radiothyroidectomy and increased 2-fold by treatment with triiodothyronine. If these binding sites belong indeed to β-adrenoreceptors or if they at least resemble these receptors with respect to catecholamine binding properties as influenced by thyroid status, this finding would correlate well with the results of studies (1, 3, 7, 20, 22, 33), according to which thyroid hormone increases and thyroid hormone deficiency decreases the sensitivity of the myocardium to cardiac sympathetic stimulation and to circulating catecholamines. If the presently observed effects of lack and excess of thyroid hormone on noradrenaline binding to what may be a β-receptor in heart muscle are representative of changes occuring at β-adrenergic receptor sites elsewhere in the body, an explanation would also be provided for the altered sensitivity to catecholamines that has been observed by numerous authors (2, 12, 23, 27, 30, 38) in other organs and tissues of thyroid hormone-deficient and thyroid hormone-treated animals and patients. On the other hand, the finding that only the affinity, but not the number of putative cardiac noradrenaline β-receptor sites is affected by the thyroid state does not readily account for results of studies (see Reference 17), indicating that the magnitude, but not the sensitivity of responses of the heart to adrenergic stimuli is reduced by deficiency and increased by excess of thyroid hormone.

The observed shifts caused by thyroid status alterations in the dose-response curves for the stimulation of cardiac adenylate cyclase by adrenaline show clearly that the sensitivity of this enzyme to the stimulatory action of the adrenergic amine is increased by thyroid hormone and decreased by deficiency of this hormone. In view of the important regulatory functions of the adenylate cyclase-cyclic AMP system in cardiac metabolism and probably also cardiac function (11, 35, 40), this observation may provide a basis for a rational explanation of the results of above cited studies (1, 3, 7, 20, 22, 33) in which the sensitivity of the heart to adrenergic stimulation was found to be increased by excess and decreased by lack of thyroid hormone. The increase in the adrenaline sensitivity of cardiac adenylate cyclase in hyperthyroidism may conceivably be due to corresponding changes in the affinity of the catecholamine to the β-adrenoreceptor which is widely regarded to be a regulatory subunit of adenylate cyclase (29). On the assumption that the noradrenaline binding characteristics of the 78,000 X *g* subcellular cardiac particles represent or resemble the binding characteristics of the cardiac β-adrenoreceptor, the presently observed changes in the noradrenaline affinity of the specific catecholamine binding sites in the 78,000 X *g* heart particles could then be held responsible for the altered adrenaline sensitivity of cardiac adenylate cyclase. At any rate, the striking correlation established by the present study between specific noradrenaline binding and adrenaline sensitivity of adenylate cyclase in cardiac particulate preparations as influenced by thyroid status may be taken as a piece of indirect evidence in support of the contention (14, 15) that the binding sites studied belong to β-adrenoreceptors or at least resemble the receptors with respect to the binding of noradrenaline. It must be mentioned that other authors (6, 18, 22) failed to notice a change in the adrenaline sensitivity of cardiac adenylate cyclase preparations after changes in thyroid state. The reason for the discrepancy between their findings and ours is not immediately apparent.

The finding of a high basal cardiac adenylate cyclase activity in the hypothyroid state is in agreement with data obtained by Broekhuysen and Ghislain (5) in heart muscle of hypothyroid rats, but contrary to the results of Levey *et al.* (18), who reported that the activity of the enzyme decreased below normal in the myocardium of hypothyroid cats. Species differences in biochemical responses to thyroid hormone administration and deprivation may account for these differing findings. Such a difference has been noted, for example, in the case of mitochondrial glycerolphosphate dehydrogenase, the activity of which is greatly enhanced by thyroid hormone in certain tissues of the rat, but not of the guinea pig and some other species (13). The fact that fluoride-stimulated adenylate cyclase activity, which probably represents the unstrained catalytic activity of the enzyme, was elevated to the same degree as was the basal activity, suggests that the activity increase in hypothyroidism of the rat may be due to an increase in the amount of the enzyme per milligram of particle protein. It is of interest in connection with this finding that particle-bound cyclic AMP phosphodiesterase has likewise been found to be much elevated in adipose tissue of hypothyroid rats (26, 34).

REFERENCES

1. ARMSTRONG, K. J. 1972. A membrane site of action for thyroxine. Fed. Proc. 31: 834.
2. BRAY, G. A. 1966. Studies on the sensitivity to catecholamines after thyroidectomy. Endocrinology 79: 554–564.
3. BRAY, G. A., and GOODMAN, H. M. 1967. Studies on cardiac glycogen in normal and thyroidectomized rats. Proc. Soc. Exp. Biol. Med. 125: 1310–1313.
4. BREWSTER, W. R., ISAACS, J. P., OSGOOD, P. F., and KING, T. L. 1911. The

hemodynamic and metabolic interrelationships in the activity of epinephrine, norepinephrine and the thyroid hormones.
5. BROEKHUYSEN, J., and GHISLAIN, M. 1972. Increased heart adenylcyclase activity in the hypothyroid rat. Biochem. Pharmacol. 21: 1493–1500.
6. BRUS, R., and HESS, M. E. 1973. Effect of norepinephrine and sodium fluoride on heart adenyl cyclase in newborn and adult rats. Endocrinology 93: 982–985.
7. COVILLE, P. F., and TELFORD, J. M. 1970. Influence of thyroid hormones on the sensitivity of cardiac and smooth muscle to biogenic amines and other drugs. Brit. J. Pharmacol. 39: 49–68.
8. CUATRECASAS, P., TELL, G. P. E., SICA, V., PARIKH, I., and CHANG, K.-J. 1974 Noradrenaline binding and the search for catecholamine receptors. Nature 247: 92–97.
9. HARRISON, T. 1964. Adrenal medullary and thyroid relationships. Physiol. Rev. 44: 161–185.
10. HESS, M. E., and SHANFELD, J. 1965. Cardiovascular and metabolic interrelationships between thyroxine and the sympathetic nervous system. J. Pharmacol. Exp. Therap. 148: 290–297.
11. HIMMS-HAGEN, J. 1972. Effects of catecholamines on metabolism, pp. 363–462. *In* H. E. Blaschko and E. Muscholl (eds.), Catecholamines, Handbuch der Experimentellen Pharmakologie. Springer Verlag, Heidelberg.
12. HIMMS-HAGEN, J. 1973. Relation of changes in metabolic effects of catecholamines to alterations in adenyl cyclase activity, p. 346. [Abstract] Ninth International Congress of Biochemistry, Stockholm, 1973.
13. KLEITKE, B., WOLLENBERGER, A., HEIER, G., and SYDOW, H. 1969. Einfluβ von Schilddrüsenhormonenauf die Aktivität der mitochondrialen Glyzerin-3-phosphat-oxydase: Vergleichende Untersuchungen an Geweben verschiedener Säugerarten, pränatale Veränderungen der Oxydaseaktivität in Rattenlebermitochondrien und Abhängigkeit des Glyzerophosphat/Dioxyazetonphosphat-Gewebsquotienten von der Aktivität der Oxydase. Acta Biol. Med. Germ. 22: 7–22.
14. LEFKOWITZ, R. J., and HABER, E. 1971. A fraction of the ventricular myocardium that has the specificity of the cardiac betaadrenergic receptor. Proc. Nat. Acad. Sci. U. S. A. 68: 1773–1777.
15. LEFKOWTIZ, R. J., SHARP, G. W. G., and HABER, E. 1973. Specific binding of β-adrenergic catecholamines to a subcellular fraction from cardiac muscle. J. Biol. Chem. 248: 342–349.
16. LEFKOWITZ, R. J., HABER, E., and O'HARA, D. 1972. Identification of the cardiac beta-adrenergic receptor protein: Solubilization and purification by affinity chromatography. Proc. Nat. Acad. Sci. U. S. A. 69: 2828–2832.
17. LEVEY, G. S. 1971. Catecholamine sensitivity, thyroid hormone and the heart. Amer. J. Med. 50: 413–420.
18. LEVEY, G. S., SKELTON, C. L., and EPSTEIN, S. E. 1969. Decreased myocardial adenyl cyclase activity in hypothyroidism. J. Clin. Invest. 48: 2244.
19. LOWRY, O. H., ROSEBROUGH, N. J., FARR, A. L., and RANDALL, R. J. 1951. Protein measurement wit the Folin phenol reagent. J. Biol. Chem. 193: 265–275.
20. MARGOLIUS, H. S. 1964. Chronotropic effects of injected and endogenously relased norepinephrine in hyperthyroid dogs. Fed. Proc. 23: 232.
21. McNEILL, J. H., and BRODY, T. M. 1968. The effect of triiodothyronine pretreatment on amine-induced rat cardiac phosphorylase activation. J. Pharmacol. Exp. Ther. 161: 40–46.
22. McNEILL, J. H., MUSCHEK, L. D., and BRODY, T. M. 1969. The effect of triiodo thyronine on cyclic AMP, phosphorylase, and adenyl cyclase in rat heart. Canad. J. Physiol. Pharmacol. 47: 913–916.
23. McNEILL, J. H., and SCHULZE, S. 1972. Thyroxine-induced supersensitivity of histamine and norepinephrine activation of cardiac phosphorylase *a*. Pharmacol. Res. Commun. 4: 25–30.

24. MURAD, F., CHI, Y.-M., RALL, T. W., and SUTHERLAND, E. W. 1962. Adenyl cyclase. III. The effect of catecholamines and choline esters on the formation of adenosine 3′,5′-phosphate by preparations from cardiac muscle and liver. J. Biol. Chem. 237: 1233–1238.
25. NELSON, T. E., and STOUFFER, J. E. 1972. Thyroxine modulation of epinephrine stimulated secretion of rat parotid α-amylase. Biochem. Biophys. Res. Commun. 48: 480–485.
26. NUNEZ, J. 1974. Personal communication.
27. PRANGE, A. J., MEEK, J. C., LIPTON, M. A. 1970. Catecholamines: Diminished rate of synthesis in rat brain and heart after thyroxine pretreatment. Life. Sci. 9: 901–907.
28. RAMACHANDRAN, J. 1971. A new simple method for separation of adenosine 3′,5′ cyclic monophosphate from other nucleotides and its use in the assay of adenyl cyclase. Anal. Biochem. 43: 227–239.
29. ROBISON, G. A., BUTCHER, R. W., and SUTHERLAND, E. W. 1967. Adenyl cyclase as an adrenergic receptor. Ann. N. Y. Acad. Sci. 139: 703–723.
30. ROSENQUIST, U. 1972. Inhibition of noradrenaline-induced lipolysis in hypothyroid subjects by increased α-adrenergic responsiveness. Acta Med. Scand. 192: 353–359.
31. SCATCHARD, G. 1949. The attraction of proteins for small molecules and ions. Ann. N. Y. Acad. Sci. 51: 660–667.
32. SOBEL, B. E., DEMPSEY, P. J., and COOPER, T. 1969. Normal myocardial adenyl cyclase activity in hyperthyroid cats. Proc. Soc. Exp. Biol. Med. 132: 6–9.
33. SPITZER, J. J., GOLD, M., MILLER, A. N., and SCOTT, J. C. 1968. Metabolic effects of norepinephrine on the heart in relation to the functional status of the thyroid. Metabolism 17: 740–746.
34. STOUFFER, J. E., ARMSTRONG, K. J., INWEGEN, R. V., THOMPSON, W. J., and ROBISON, G. A. 1970. Thyroid hormone regulation of cyclic AMP and hormone sensitive lipolysis. Fed. Proc. 33: 1391.
35. SUTHERLAND, E. W., ROBISON, G. A., and BUTCHER, R. W. 1968. Some Aspects of the biological role of adenosine 3′,5′-monophosphate (cyclic AMP). Circulation 37: 279–306.
36. TELL, G. P. E., and CUATRECASAS, P. 1974. β-Adrenergic receptors: stereospecificity and lack of affinity for catechols. Biochem. Biophys. Res. Commun. 57: 793–800.
37. WALDSTEIN, S. S. 1966. Thyroid-catecholamine interrelations. Ann. Rev. Med. 17: 123–132.
38. WILDENTHAL, K. 1972. Studies on isolated fetal mouse hearts in organ culture: Evidence for a direct effect of triiodothyronine in enhancing cardiac responsiveness to norepinephrine. J. Clin. Invest. 51: 2702–2709.
39. WILL-SHAHAB, L., and WOLLENBERGER, A. 1974. Influence of thyroid state on the binding of noradrenaline to a cardiac subcellular fraction containing the β-adrenoreceptor. Acta Biol. Med. Germ. 32: K1–K8.
40. WOLLENBERGER, 1975. The role of cyclic AMP in the adrenergic control of the heart. *In* W. G. Nayler (ed.), Contraction and Relaxation of the Myocardium. Academic Press, London and New York.

Mailing address: Dr. A. Wollenberger,
Division of Cellular and Molecular Cardiology,
Central Institute of Heart and Circulatory Regulation Research,
Academy of Sciences of the DDR,
Berlin-Buch, (Germany Democratic Republic).

Recent Advances in Studies on Cardiac Structure and Metabolism
Volume 9
The Sarcolemma
Edited by Paul-Emile Roy and Naranjan S. Dhalla

EPINEPHRINE, CYCLIC AMP, CALCIUM, AND MYOCARDIAL CONTRACTILITY[1]

J. R. WILLIAMSON and S. SCHAFFER

Johnson Research Foundation,
University of Pennsylvania
Philadelphia, Pennsylvania, USA

SUMMARY

In the first part of this paper, a short review of the literature on the effects of catecholamines, Ca^{2+}, and cyclic AMP on the biochemical, electrical, and mechanical properties of the heart is presented. It is concluded that the effects of epinephrine on activation of glycogenolysis and on the inotropic response of cardiac muscle are both mediated by the combined actions of Ca^{2+} and cyclic AMP. Since the inotropic response precedes the glycogenolytic response, it is evident that increased energy metabolism is a consequence of increased heart work and not a causative factor. The available data suggest that increased tissue cyclic AMP levels resulting from catecholamine stimulation of adenyl cyclase activity alter cardiac mechanics by modulation of the basic calcium flux cycle of the heart. These effects may be mediated by cyclic AMP-stimulated phosphorylation of specific proteins located in the sarcolemma, sarcoplasmic reticulum, and myofilaments via one or more protein kinases, although experimental verfication of the possible relationship between protein phosphorylation and altered Ca^{2+} binding properties at specific sites in the membranes is at present largely lacking. The increased rate of tension development induced by catecholamines appears to be caused by an increased rate of Ca^{2+} influx across the sarcolemma during the plateau phase of the action potential, whereas the increased rate of relaxation is probably attributable to an increased rate of sequestration of intracellular Ca^{2+} by the sarcoplasmic reticulum.

In the second part of the paper, data are presented using a working rat heart preparation to illustrate the effect of catecholamines in facilitating Ca^{2+} entry across the sarcolemma. Decreased left ventricular pressure development induced by addition of ruthenium red or verapamil to inhibit Ca^{2+} influx was relieved in a concentration-dependent manner by catecholamines. Curves of percentage change of left ventricular pressure against external Ca^{2+} concentration are presented which show that epinephrine increased the sensitivity of the heart to Ca^{2+} whereas verapamil decreased it. A half-maximal increase of left ventricular pressure was obtained with 0.6 mM Ca^{2+} in control hearts, 0.3 mM Ca^{2+} with epinephrine, and 2.9 mM Ca^{2+} with verapamil (10^{-7})-treated hearts. Verapamil combined with epinephrine at the above concentrations gave a half-maximal increase of left ventricular pressure with 1.3 mM Ca^{2+}. These results are discussed in relation to a model for the Ca^{2+} flux in the heart.

[1]This work was supported by Grant HE-14461 from the United States Public Health Service, and by Contract No. NIH-71-2494 under the Myocardia Infarction Program, National Heart and Lung Institute.

DISSOCIATION OF GLYCOGENOLYTIC AND INOTROPIC EFFECTS OF EPINEPHRINE

The effects of epinephrine on glycogen metabolism have been studied intensively in cardiac muscle, a tissue which has lent itself to measurements of the time sequence of the various activation steps between administration of epinephrine and the appearance of the products of glycogen metabolism. After the intial observation of Hess and Haugaard (26) that epinephrine addition to perfused rat hearts caused an apparently simultaneous augmentation of contractile force and increased level of phosphorylase *a,* much work has been devoted to the question of whether glycogenolysis was directly involved in the inotropic effect of epinephrine and related catecholamines (9, 24, 27). It is now clear that activation of glycogenolysis and provision of energy by glycolysis and the further oxidative metabolism of pyruvate are not required to elicit the inotropic response. Although cyclic AMP levels increased to a maximum simultaneously with (8) or even before (11, 57) the full development of the contractile response after addition of epinephrine to perfused rat hearts, a distinct lag of a few seconds was observed before phosphorylase *a* levels started to increase. Maximal activation of phosphorylase was not observed until well after the peak of the inotropic response. On the other hand, phosphroylase kinase was maximally activated within 1 sec of epinephrine addition, and its kinetics of activation could not be distinguished from the increase of tissue cyclic AMP levels. The effects of epinephrine on the contractile response, on increases of cyclic AMP, phosphorylase kinase, and phosphorylase *a* levels, and on glycogenolysis in heart muscle are all prevented by β-adrenergic blocking agents such as dichloroisoproterenol, nethalide, or propranolol; this suggests that all these events are secondary to the binding of epinephrine to specific receptor sites in the plasma membrane.

Studies with intact muscle or isolated enzyme preparations have shown that phosphorylase kinase can be activated either by a cyclic AMP-dependent process or by an independent mechanism that appears to be closely connected to increased availability of intracellular Ca^{2+} (5, 18). The question arises therefore whether the Ca^{2+}-activated process is also operative in heart muscle after epinephrine stimulation, in addition to an activation of the cyclic AMP-dependent protein kinase. Several findings suggest that both processed are involved in the overall effect. Namm *et al.* (47) observed that removal of Ca^{2+} from the perfusion medium prevented heart contractions and greatly diminished the epinephrine-induced conversion of phosphroylase *b* to *a.* However, the rise of cyclic AMP level and activation of phosphorylase kinase were not greatly affected, except for a short delay in the onset of the latter effect. On the other hand, elevation of the Ca^{2+} concentration to 3 times normal (9.6 mEq/l) increased phosphorylase *a* levels but produced only a small, statistically insignificant effect on phosphorylase kinase activity (pH 6.8/pH 8.2) and no effect on cyclic AMP levels. These findings can be interpreted on the basis of a stimulatory effect of Ca^{2+} on the catalytic activity of both the nonphosphorylated and phosphorylated forms of phosphorylase kinase and an inhibitory effect on phosphorylase phosphatase activity (75). The cyclic AMP-mediated activation of phosphorylase kinase appears to be slower than the Ca^{2+} activation effect in both heart and skeletal muscle, so that the rapid response obtained after addition of epinephrine to spontaneously contracting heart muscle can be explained only on the basis of a combined action of cyclic AMP and Ca^{2+}.

CYCLIC AMP AND CARDIAC CONTRACTILITY

The following evidence implicates cyclic AMP directly in the positive inotropic effect of epinephrine on cardiac muscle: 1) cyclic AMP levels appear to increase in parallel with or before the increased force of contraction after the addition of epinephrine to intact heart preparations; 2) the relative potencies of a series of catecholamines in stimulating cyclic AMP production *in vitro* are similar to their relative effects as positive inotropic agents in *in vivo* (isoproterenol > epinephrine > norepinephrine); 3) drugs that stimulate (*e.g.*, imidazole) or inhibit (*e.g.*, theophylline) phosphodiesterases *in vitro* have parallel effects in potentiating or inhibiting the response of mechanical activity to catecholamines; and 4) the effects of catecholamines on heart contractility and on cyclic AMP levels are prevented by β receptor blockade. Furthermore, externally added cyclic AMP or its N^6-2′-O-dibutyryl derivative will mimic the effects of catecholamines. Effects of dibutyryl cyclic AMP have been obtained on isolated ventricular papillary muscle of the cat (63) and on perfused mammalian hearts (36). A positive chronotropic response on isolated cultured heart cells has also been observed (35). Skelton *et al.* (63) showed that dibutyryl cyclic AMP caused a concentration-dependent increase in isometric tension and rate of tension development in the cat papillary muscle. With maximum concentrations of dibutyryl cyclic AMP (3×10^{-3} M), the effects were similar to those observed with maximum norepinephrine concentrations (10^{-5} M). The inotropic response to dibutyryl cyclic AMP, unlike that to norepinephrine, was insensitive to blockade by propranolol. A notable difference between the effects of these two agents was that, whereas the contractile response to norepinephrine was maximal in 2–4 min, the dibutyryl cyclic AMP effect required 30–40 min. This may reflect the slow diffusion time of dibutyryl cyclic AMP to reach its site of action. Since the dibutyryl cyclic AMP effect was insensitive to inhibition by β blockade, it is clear that it was not caused by direct stimulation of the β receptor or by a secondary release of endogenous norepinephrine.

In addition to the effect of catecholamines in raising cardiac cyclic AMP levels, recent studies have shown that the basal level fluctuates during the contraction cycle. Brokker (4) measured cyclic AMP levels in frog ventricle strips at different times during the contraction-relaxation cycle, and found that peak levels representing a 50 percent increase occurred in the early phase of the increased tension development during systole. A small increase of cyclic AMP content was observed at the peak of systole relative to diastole, and this difference was augmented along with the absolute levels after epinephrine administration. Further studies by Wollenberger *et al.* (79), also with frog hearts, confirmed that cyclic AMP levels increased to a maximum during the early phase of the contraction cycle and showed, in addition, a reciprocal fall of cyclic GMP levels. The mechanisms responsible for an apparent activation of adenyl cyclase and inhibition of guanyl cyclase are not clear at present, but would seem to be related to membrane depolarization during the action potential, possibly via displacement of sarcolemma-bound Ca^{2+} from inhibitory sites on adenyl cyclase or alternatively resulting from a conformational change of the plasma membrane.

Although the precise mechanism whereby the catecholamines can augment the rate and force of contraction of cardiac muscle remains unknown, binding of the hormone to the receptor site must be the initiating event. Various models have been proposed for the so-called β-adrenergic receptor (15, 58), but it is generally agreed that binding of the hormone to specific groups on the membrane surface must lead to some kind of conformational perturbation and alteration of the catalytic activity of adenyl cyclase. Thus adenyl cyclase in the cell membrane may consist of two types of subunits: 1) a regulatory subunit on the external surface; and 2) a catalytic subunit with its active site facing the interior of the cell and accessible to intracellular ATP. The connection between the hormone receptor site and adenyl cyclase is probably complex, involving both proteins and phospholipid (15, 73). A number of attempts have been made to study catecholamine binding to particulate fragments isolated from cardiac tissue (*e.g.*, Reference 38), but the results appear to represent primarily nonspecific binding rather than interaction with the β receptor sites (13).

Adenyl cyclase from cardiac tissue has been solubilized with the nonionic detergent, Lubrol-PX (39). However, unlike the particulate adenyl cyclase, the soluable enzyme was unresponsive to hormone activation. In further experiments, Levey (40, 41) found that addition of phosphatidylserine to soluble, detergent-free adenyl cyclase restored responsiveness to glucagon and histamine, but not to norepinephrine, whereas addition of phosphatidylinositol restored the responsiveness to norepinephrine. This activation was abolished by propranolol. These results provide further evidence in favor of the postulate that norepinephrine activates cardiac adenyl cyclase by a receptor mechanism separate and distinct from the glucagon receptor and support the arguments that membrane lipids have an important role in maintaining the functional integrity of the hormone receptor complex (50). Various cations affect the activity of adenyl cyclase, but whether these effects are related to the inotropic action of catecholamines is not clear. Thus, although elevation of K^+ concentration from 5.6 to 56 mM in perfused rat hearts blocked the increase of cyclic AMP normally elicited by epinephrine (47), K^+ was found to have a stimulatory effect on a particulate preparation of adenyl cyclase from rat heart, which suggested that the effect *in vivo* was indirect and possibly associated with depolarization of the myocardial membrane (43). Isolated particulate adenyl cyclase from cardiac muscle has a K_m for ATP of about 0.08 mM, is activated by Mg^{2+} with a K_a of 2–3 mM, and is inhibited by Ca^{2+} with a K_i of about 0.3 mM (10). From further studies, Drummond *et al.* (12) concluded that epinephrine, like fluoride ions, stimulated the cardiac enzyme (increase of V_{max}) at all Mg^{2+} (or Mn^{2+}) concentrations and had little effect on the apparent K_a for Mg^{2+} or the K_m for the MgATP substrate. Concentrations of ATP in excess of the Mg^{2+} concentration were inhibitory, presumably because of competition with MgATP at the catalytic site or removal of free Mg^{2+}. These effects could have physiological significance if the availability of Mg^{2+} to adenyl cyclase changed relative to that of ATP. Likewise, membrane-bound Ca^{2+} may have a critical role in the modulation of adenyl cyclase activity.

Tissue measurements of cyclic AMP have shown that the levels exhibit large overshoots shortly after administration of catecholamines, indicative of an initial overproduction relative to the biological response elicited. Regulation of tissue cyclic AMP levels in the steady state is achieved by a concerted action between the activities of adenyl cyclase and cyclic AMP phosphodiesterase. A number of studies have shown that a specific protein activator of cyclic AMP phosphodiesterase is present in several mammalian tissues (7, 19, 28), and both phosphodiesterase and the protein activator have been isolated from bovine heart (20, 68). Unlike the purified phosphodiesterase, the protein activator binds Ca^{2+} at high and low affinity sites (dissociation constants of 3 μM and 12 μM, respectively), and the activational effect of this protein on the phosphodiesterase is entirely dependent on the presence of Ca^{2+} over the range 1–10 μM (67). In addition, cyclic AMP enhances the interaction between the phosphodiesterase and the protein activator in a manner which is slow relative to the rate of enzyme catalysis (68). It is apparent, therefore, that a rise of either Ca^{2+} or cyclic AMP in the heart over physiological concentration ranges will stimulate the rate of removal of cyclic AMP. Since cyclic AMP levels in the frog ventricle fall during the time when tension is developing rapidly (79) and the intracellular Ca^{2+} concentration is expected to be highest, this effect may be ascribed to a Ca^{2+} activation of cyclic AMP phosphodiesterase. It remains to be seen whether cyclic GMP phosphodiesterase is regulated by Ca^{2+} in the opposite sense. The above findings implicate cyclic AMP (and possibly also cyclic GMP) as factors involved in the overall sequence of events during the cardiac contraction cycle. Because of the central role of Ca^{2+} in the cardiac contractiong cycle, cyclic AMP is likely to exert an effect on contractility via an influence on the calcium cycle.

EFFECTS OF EPINEPHRINE AND CYCLIC AMP ON CALCIUM CYCLE OF HEART

Besides an increase of heart rate, the characteristic effects of epinephrine on isometrically contracting cardiac muscle are an increase in peak tension, rate of tension development, and rate of tension decay, together with a decrease of twitch duration (or systole), time to peak tension, and relaxation time (59). In contrast to the effect of cetecholamines in increasing both the rate of tension (or pressure) development and the rate of relaxation, Ca^{2+} acting as an inotropic agent does not result in a shortening of the relaxation time although the rate of tension development is increased. Data alluded to earlier have indicated that the glycogenolytic effect of epinephrine is mediated by the two chemical transmitters, cyclic AMP and Ca^{2+}. The remarkable interplay between these two agents on a wide variety of excitatory and secretory processes has been described by Rasmussen and associates (52, 53). It is important therefore to consider carefully whether a combined action of Ca^{2+} and cyclic AMP might also be involved in determining the characteristic dynamics of the catecholamine effect on cardiac mechanics.

Several relationships become apparent from considerations of the reported changes in the electrophysiological and ionic events induced by β-adrenergic stimulation, as well as possible modulations in the normal flux of Ca^{2+} between intracellular organelles induced by cyclic AMP-dependent phosphorylation of membrane-bound proteins. Initiation of the action potential causes depolarization of the plasma membrane and increased permeability to Na^+,K^+ and Ca^{2+}. In heart, Ca^{2+} is made available by increased Ca^{2+} entry and release from internal storage sites, so that the intracellular free Ca^{2+} concentration rises possibly by as much as 2 orders of magnitude from a value in resting muscle of less than 10^{-7} M. The Ca^{2+} diffuses to the myofibrils and combines with a subunit of troponin, which causes an alteration of the configuration of the troponin-tropomysin complex with the net result that inhibition of actomyosin ATPase is relieved. This allows combination of F actin with myosin to produce shortening of the myofibril with utilization of energy from the hydrolysis of ATP. Relaxation occurs by a decrease of the free intracellular Ca^{2+} concentration and sequestration of calcium from its binding to troponin until the point is reached at which the troponin-tropomyosin complex re-exerts an inhibitory effect on the actomyosin ATPase. The sarcoplasmic reticulum is thought to be the major system responsible for removal of free Ca^{2+} from the environment of the myofibrils. This is brought about by an ATP-dependent, active reaccumulation of Ca^{2+}

followed by its redistribution in the sarcotubule system and sarcolemma membrane binding sites. The calcium balance is restored by transport of Ca^{2+} through the plasma membrane to the external medium during the interval between beats. A number of reviews are available which describe the various stages of the cardiac Ca^{2+} cycle in detail (3, 14, 23, 31, 37, 42, 44, 48, 55, 74).

In cardiac muscle the duration of the action potential (200–600 msec) is about the same as that of the contraction cycle, and contraction is totally dependent on the presence of Ca^{2+} in the external medium. Studies with Purkinje fibers, ventricular strips, or ventricular trabeculae muscles using the voltage clamp technique have established that the initial spike of the action potential is due to a rapid inward Na^+ current and that repolarization is caused by an outward K^+ flux. A slower, secondary current associated with Ca^{2+} influx has been identified as occurring during the plateau phase of the action potential by its insensitivity to replacement of external Na^+ by choline or by the presence of tetrodotoxin, and its sensitivity to alterations of the external Ca^{2+} concentrations and inhibition by Mn^{2+}. Measurements of the magnitude of the secondary current together with flux studies using $^{45}Ca^{2+}$ have indicated that the influx of Ca^{2+} during the action potential is insufficient to account for the amount of Ca^{2+} required for troponin binding and elicitation of maximum tension development. Hence it has been generally assumed by analogy with skeletal muscle that membrane depolarization or the Ca^{2+} influx itself triggers the release of additional Ca^{2+} from internal storage sites presumably associated with the sarcoplasmic reticulum tubules. Morad and Goldman (44), in a recent review, present evidence suggesting that two functionally separate sources of activator Ca^{2+} exist in mammalian myocardium, both of which summate to determine the strength of contraction, with control exerted by the action potential but via two separate mechanisms. The first is envisioned as a trigger release of stored Ca^{2+} from an internal pool which exchanges very slowly with the extracellular space when the muscle is not stimulated, and is controlled by membrane depolarization during the first 100 msec of the action potential (Pool II of fig. 1). It is suggested that the amount of releasable Ca^{2+} in this store is determined by the size and duration of several preceding action potentials and is sufficient to account for up to 60 percent of the total Ca^{2+} bound to troponin during the contraction phase. The remainder of the activator Ca^{2+} is provided by entry of Ca^{2+} across the sarcolemma or release from superficial membrane sites during the entire plateau phase of the action potential (Pool I of fig. 1). Calcium provided by both mechanisms forms a common pool of free intracellular Ca^{2+}, which therefore rises transiently (1, 2) because its rate of formation is greater than its rate of binding to troponin and the sarcoplasmic reticulum.

Direct estimates of the concentration of Ca^{2+} required to elicit tension development have been made in crab muscle fibers by injection of calcium-EGTA buffers (51). These studies provided threshold values of Ca^{2+} as low as 3×10^{-7}

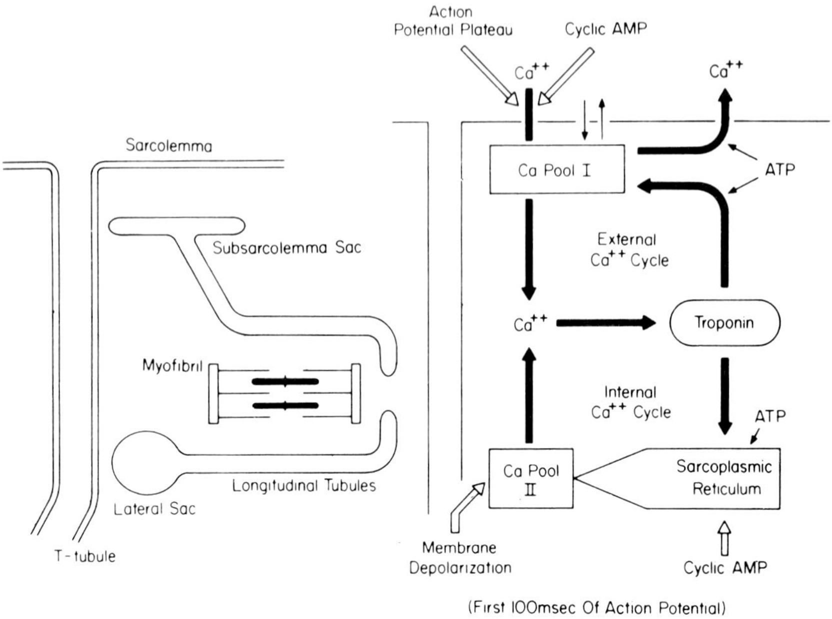

Figure 1 Diagrammatic representation of control of the Ca^{2+} cycle in cardiac muscle and its interaction with cyclic AMP-dependent processes.

M with half-maximal effects at about 10^{-6} M Ca^{2+}. Further studies with a variety of preparations from skeletal and cardiac muscle with different degrees of intactness (*e.g.*, EGTA-treated ventricular strips, skinned fibers, glycerol-extracted fibers, and myofibrils, in addition to natural or reconstituted actomyosin) provide a similar relation between tension development, ATPase activity, and the variation of the Ca^{2+} concentration (25, 78). The calcium affinity of these preparations is very similar to that reported for the TN-C subunit of troponin (21), which is responsible for the overall sensitivity of the myofibrils to Ca^{2+}. On the basis of the number of high affinity of Ca^{2+} binding sites on troponin and the troponin content per unit weight of cardiac tissue, Katz (31) estimates that about 50 μmoles of Ca^{2+}/kg of muscle are required for saturation of the troponin binding sites in the actomyosin complex, whereas an extra 10 μmoles of Ca^{2+}/kg of muscle are needed to raise the free intracellular Ca^{2+} from 10^{-7} to 10^{-5} M.

Muscle relaxation occurs by the active uptake of Ca^{2+} from the myoplasm by the sarcoplasmic reticulum, which thereby displaces the equilibrium of Ca^{2+} binding to the troponin sites. The ATPase of isolated sarcoplasmic reticulum has a sufficiently high affinity for Ca^{2+}, but at present a discrepancy exists between measurements of the rate of Ca^{2+} uptake by isolated cardiac sarcoplasmic reticulum vesicles at Ca^{2+} concentrations below 10 μM and the calculated rates required to cause relaxation (60). According to the model proposed by Morad and Goldman (44), a fixed fraction of the Ca^{2+} sequestered by the sarcoplasmic reticulum is returned to the rapid release store (Pool II) after some delay, whereas the remainder is sequestered by the surface membranes or pumped out of the cell during each beat (fig. 1). The left-hand side of figure 1 provides a diagrammatic representation of the anatomical structures possibly associated with the different Ca^{2+} pools. Electron microscopic studies (17, 49, 62) show that the sarcolemma membrane of mammalian cardiac muscle is a multilayered structure 300–425 Å in thickness. The thin plasma membrane (75 Å) is separated from the basement membrane (75–200 Å thick) by a space of approximately 150 Å. At approximately 2–3-μm intervals along the membrane in the region of the Z lines, the sarcolemma invaginates to form a tubular system (T tubules) that extends deep into the myoplasm. The lumen of the tubules communicates directly with the extracellular space, and depolarization of these membranes acts as a conducting mechanism for the spread of electrical activity from the surface to the interior of the cell. Additional fine intracellular tubules (200–500 Å in diameter), continuous across the Z lines and in close apposition to the myoplasmic surface of the sarcolemma have also been observed anatomically and may provide a locus for superficial rapid-release storage sites for Ca^{2+}. The junctional sarcoplasmic reticulum (lateral sac), which is closely apposed to the T tubules, is probably responsive to membrane depolarization (16, 29) and may account for most of the triggered release of activator Ca^{2+}.

A number of studies have shown that epinephrine (10^{-6} M) increases the amplitude of the plateau phase of the action potential in cardiac muscle and increases the inward Ca^{2+} current during the plateau (6, 30, 45, 54, 72). This effect was independent of the initial fast depolarization due to increased Na^+ conductance and was relatively independent of external Na^+ concentration, but was critically dependent on the external Ca^{2+} concentration. That influx of Ca^{2+} increased by epinephrine is shown by the sensitivity of its effects on the plateau phase of the action potential to alterations of the external Ca^{2+} concentration and to inhibition by Mn^{2+}. Recent studies with calf Purkinje fibers exposed to norepinephrine, or N^6-monobutyryl cyclic AMP (70) or after iontophoretic injection of cyclic AMP (69) show directly that cyclic AMP closely mimics the electrical effects of catecholamines on the heart. In these studies catecholamines and cyclic AMP increased the initial plateau level of the action potential, indicating an increase in the magnitude of the slow inward current, but also shortened the duration of the plateau. This latter effect is attributed to activation of an outward K^+ current which is responsible for terminating the plateau. Both effects appear to be mediated by cyclic AMP, and although increased Ca^{2+} influx can account for the positive inotropic effect of catecholamine, the early repolarization may account for the chronotropic action. A stimulation of the exchange of Ca^{2+} across the plasma membrane induced by catecholamines is in accordance with earlier studies showing an increased uptake of $^{45}Ca^{2+}$ by contracting guinea pig atria after addition of norepinephrine (22, 56). The above studies clearly suggest that the increased force of contraction and rate of tension development induced by epinephrine is caused by an increased rate of availability of free Ca^{2+} to the myofibrils. This is probably achieved in part by increased Ca^{2+} entry across the sarcolemma and in part by an increased rate of delivery of activator Ca^{2+} to internal Ca^{2+} storage sites followed by a more complete filling of these sites and an augmentation of the speed and amount of Ca^{2+} released by the trigger of membrane depolarization. The interval for release of activator calcium appears to be primarily in the early phase of the action potential plateau, since membrane repolarization by the voltage clamp technique before termination of the normal action potential does not diminish the augmented rate of tension development induced by epinephrine (45). Postulation of the above sequence of events implies a cyclic AMP-induced increase in the number of Ca^{2+} binding sites or an increased permeability of superficial membranes to Ca^{2+}. However, the mechanism by which these changes may be achieved remains speculative although it appears probable that phosphorylation of specific proteins by cyclic AMP-activated protein kinases is involved in the overall process (75).

Phosphorylation of sarcolemma preparations from both cardiac and skeletal muscle by a cyclic AMP-dependent protein kinase has been reported (64, 80). The phosphorylated membranes of rabbit skeletal muscle (64) accumulated Ca^{2+} in the presence of oxalate at a greater rate than control membranes, and had a higher Ca^{2+}-stimulated ATPase activity. In the cardiac preparation, phosphorylation resulted in an increased Ca^{2+} binding with the number of binding sites remaining unchanged. If these findings do not represent contamination of the sarcolemma preparations with sarcoplasmic reticulum or actomyosin fragments, the data suggest that phosphorylation of specific sites on the sarcolemma may be functionally linked to a membrane Ca^{2+} transport system, although such a process is expected to be less significant in skeletal than in cardiac muscle. Evidence has also been obtained which supports a role of cyclic AMP in prompting Ca^{2+} sequestration by cardiac sarcoplasmic reticulum, thereby increasing the speed of relaxation. Isolated cardiac sarcoplasmic reticulum has been shown to contain adenyl cyclase and cyclic AMP-dependent protein kinase activity which catalyzed phosphorylation of a protein substrate in the membrane upon addition of cylic AMP (46, 81). Half-maximal stimulation of phosphorylation was achieved with 8.5×10^{-8} M cyclic AMP (81). Other extensive studies with purified preparations of cardiac sarcoplasmic reticulum (32, 33, 66) showed a large specific phosphorylation of serine and threonine residues in membrane protein upon incubation of the vesicles with cyclic AMP-dependent protein kinase and cyclic AMP. Epinephrine-stimulated adenyl cyclase was present in the preparation and enhanced phosphorylation. The initial rates of Ca^{2+} uptake and Ca^{2+}-activated ATPase at micromolar Ca^{2+} concentrations were stimulated 2- to 3-fold by addition of protein kinase, cyclic AMP, Mg^{2+}, and ATP. Solubilization of the phosphorylated vesicles in sodium dodecyl sulfate and fractionation by polyacrylamide gel electrophoresis allowed isolation of a single protein of approximately 22,000 daltons which contained most of the ^{32}P label bound as a phosphoester (65). An increased rate of uptake of Ca^{2+} by sarcoplasmic reticulum will lead to an increased rate of filling of the rapid release pool of Ca^{2+} (Pool II of fig. 1), and therefore will have a feedback stimulatory influence on the rate of contraction induced by epinephrine.

EFFECTS OF EPINEPHRINE IN FACILITATING Ca^{2+} ENTRY INTO MYOCARDIUM

In concluding this review of the literature on the effects of epinephrine and cyclic AMP on cardiac tissue and subcellular components, some new data will be presented which relates to the effect of epinephrine on the mobilization of activator Ca^{2+} for cardiac contractility. In previously reported experiments (76) it was observed that epinephrine addition to the perfused rat heart was able to reverse the inhibition of contractility induced by ruthenium red, which was added to inhibit Ca^{2+} entry across the sarcolemma. These experiments were interpreted as indicating a facilitation of the release of intracellularly bound Ca^{2+} from storage in the sarcoplasmic reticulum by epinephrine without the need for entry of Ca^{2+} across the plasma membrane to act as a trigger for the Ca^{2+} release. However, further studies presented here fail to confirm this interpretation, and suggest instead that epinephrine restores contractility from the hypodynamic state induced by Ca-antagonistic agents by increasing the affinity to Ca^{2+} of sites on the sarcolemma responsible for Ca^{2+} entry.

Figure 2 shows recordings of the left ventricular pressure (LVP) and its derivative (LV dP/dt) from working rat hearts perfused as described previously (76). The initial Ca^{2+} concentration in the perfusion fluid was 1 mM. Figure 2*A* shows that addition of 0.9 mM EGTA to chelate most of the extracellular Ca^{2+} caused a rapid decrease of contractility, which could be reversed by further addition of 1 mM Ca^{2+}. Figure 2*B* shows that a similar inhibition or contractility could be induced by addition of the Ca-antagonistic agent, verapamil (34). Left ventricular pressure development was restored by addition of 0.5 μg/ml of norepinephrine, and a further cycle of inhibition and recovery was achieved by successive additions of EGTA and Ca^{2+}.

The concentration of verapamil required to produce a given inhibition of left ventricular pressure was entirely dependent on the Ca^{2+} concentration of the perfusion fluid, as illustrated by the data in figure 3, which show the log dose-response curves for verapamil at a series of Ca^{2+} concentrations.

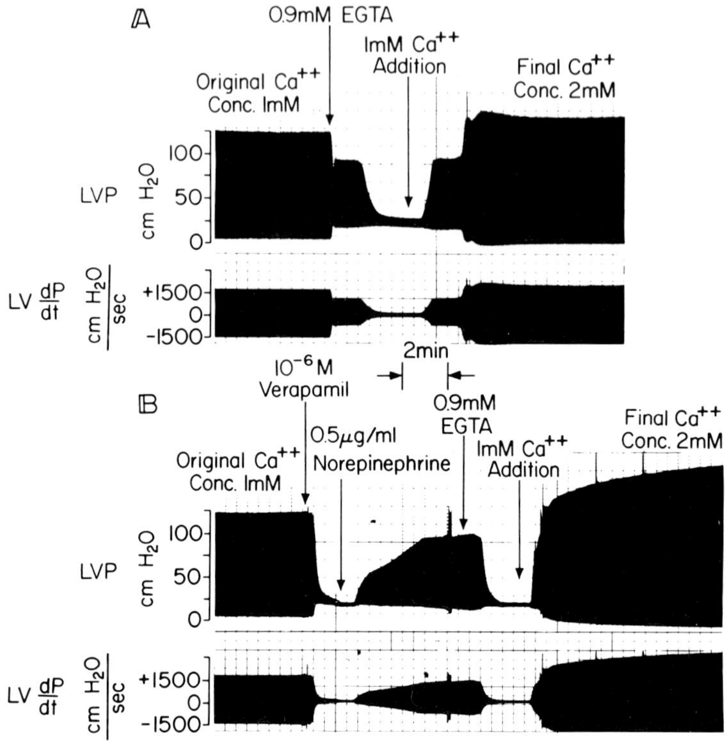

Figure 2 Competition between EGTA and Ca^{2+} (*A*) and verapamil and norepinephrine (*B*) on left ventricular pressure development in perfused rat hearts.

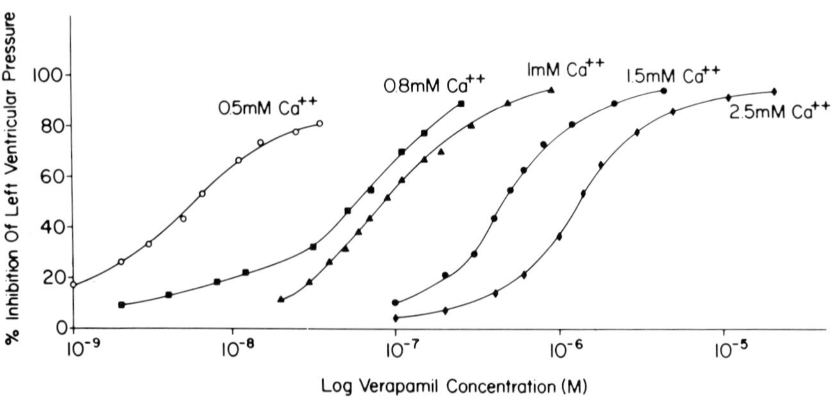

Figure 3 Log dose-response curves for the effects of verapamil in inhibiting left ventricular pressure changes in perfused rat hearts. The calcium concentration in the perfusion fluid was varied from 0.5 to 2.5 mM Ca^{2+}.

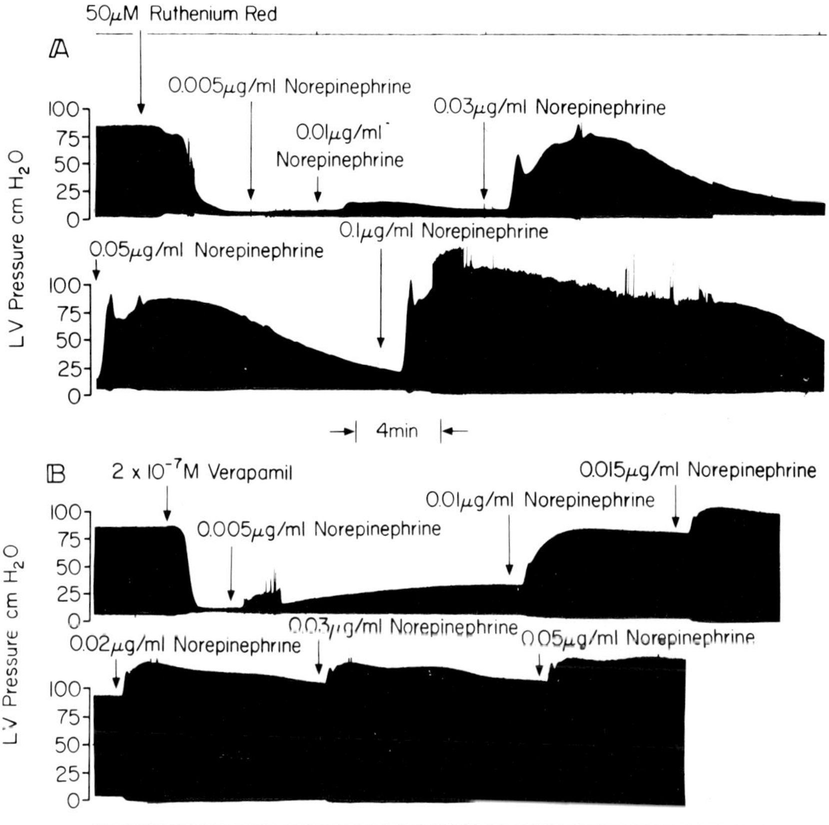

Figure 4 Effects of norepinephrine concentration on the reversal of left ventricular pressure changes induced in perfused rat hearts by addition of 50 μM ruthenium red (*A*), or 2 X 10^{-7} M verapamil (*B*).

The extent of the reversibility of contractile function achieved by norepinephrine after prior inhibition by Ca-antagonistic agents was dependent on the norephinephrine concentration. Figure 4*A* shows a titration of the left ventricular pressure response to norepinephrine in the presence of 50 μM ruthenium red, whereas figure 4*B* shows a similar titration after inhibition by 2 X 10^{-7} M verapamil. The effects of norepinephrine, especially at low concentrations, were transient in the presence of ruthenium red but were sustained in the presence of verapamil, and the hearts responded to lower norepinephrine concentrations than the ruthenium red-treated hearts. These differences probably related to the lower specificity of ruthenium red compared with verapamil in binding to Ca^{2+} sites on the sarcolemma.

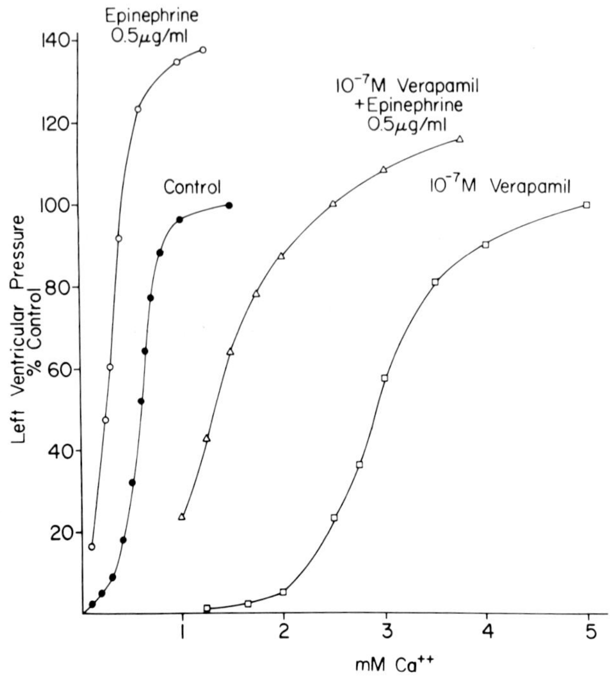

Figure 5 Effects of epinephrine and verapamil on left ventricular pressure changes in perfused rat hearts as a function of Ca^{2+} concentration in the perfusion fluid.

Figure 5 shows Ca^{2+} sensitivity titrations of hearts perfused in the presence of 10^{-7} M verapamil or 0.5 μg/ml of epinephrine compared with a normal control heart. These titrations illustrate clearly the competitive nature of the interaction between verapamil inhibition and epinephrine stimulation of contractility as a function of Ca^{2+} concentration. In these studies, a half-maximal stimulation of left ventricular pressure was obtained with 0.6 mM Ca^{2+} in control hearts. The concentration of Ca^{2+} required for a similar effect was increased to 2.9 mM by verapamil and decreased to 0.3 mM by epinephrine. Addition of epinephrine in the presence of verapamil produced about a 2-fold increased sensitivity to Ca^{2+}. Other experiments have shown that the ionophore, X-537A, which has a cardiotropic action by stimulating the release of norepinephrine from endogenous stores (61), also overcomes the verapamil blockade in a manner similar to exogenously added norepinephrine. Likewise, both catecholamines and X-537A will partially reverse the diminshed sensitivity of perfused hearts to Ca^{2+} observed under acidotic conditions. A decreased binding of Ca^{2+} to low affinity sites (apparent K_m of 0.8 mM) by 10^{-6} M verapamil has been reported previously using an isolated cardiac sarcolemma preparation (77), but attempts to detect either membrane phosphorylation with cyclic AMP and added protein kinase or changes of Ca^{2+} affinity by cyclic AMP have so far been unsuccessful with this preparation (M. Tada, unpublished observations).

Verapamil has been shown to inhibit the slow inward Ca^{2+} current of isolated cat papillary muscle during the plateau phase of the action potential, this effect being reversed by addition of catecholamines (34, 71). These data, together with the present results, stress the importance of Ca^{2+} influx across the sarcolemma as a direct mediator of the increased rate of contraction observed with catecholamines. Therefore, the first effect may be via an augmentation of the Ca^{2+} flux through Pool I (fig. 1), although an increased rate of Ca^{2+} suquestration by the sarcoplasmic reticulum will not only increase the speed of relaxation but may also augment the rate of filling and Ca^{2+} content of Pool II and so indirectly increase the amount of activator Ca^{2+} released by a trigger mechanism consequent to membrane depolarization.

REFERENCES

1. ASHLEY, C. C., and RIDGWAY, E. B. 1968. Stimultaneous recording of membrane potential, calcium transient and tension in single muscle fibres. Nature 219: 1168–1169.
2. ASHLEY, C. C., and IRDGWAY, E. B. 1970. Aequorin-calcium luminescence and its application to muscle physiology, pp. 42–53. *In* A. W. Cuthbert (ed.), A Symposium on Calcium and Cellular Function. St. Martin's Press, New York.
3. BIANCHI, P. 1968. Cell Calcium, Butterwoth, London.
4. BROOKER, G. 1973. Oscillation of cyclic adenosine monophosphate concentration during the myocardial contraction cycle. Science 182: 933–934.
5. BROSTROM, C. O., HUNKELER, F. L., and KREBS, E. G. 1971. The regulation of skeletal muscle phosphorylase kinase by Ca^{++}. J. Biol. Chem. 246: 1961–1967.
6. CARMELIET, E., and VEREECKE, J. 1969. Adrenaline and the plateau phase of the cardiac action potential: Importance of Ca^{++}, Na^{+} and K^{+} conductance. Arch. Europ. J. Physiol. 313: 300–315.
7. CHEUNG, W. Y. 1971. Cyclic 3′,5′-nucleotide phosphodiesterase. Evidence for and properties of a protein activator. J. Biol. Chem. 246: 2859–2869.
8. CHEUNG, W. Y., and WILLIAMSON, J. R. 1965. Kinetics of cyclic adenosine monophosphate changes in rat heart following epinephrine administration. Nature 207: 979–981.
9. DRUMMOND, G. I. 1967. Muscle metabolism. Fortschr. Zool. 18: 360–429.
10. DRUMMOND, G. I., and DUNCAN, L. 1970. Adenyl cyclase in cardiac tissue. J. Biol. Chem. 245: 976–983.
11. DRUMMOND, G. I., DUNCAN, L., and HERTZMAN, E. 1966. Effect of epinephrine on phosphorylase *b* kinase in perfused rat hearts. J. Biol. Chem. 241: 5899–5903.
12. DRUMMOND, G. I., SEVERSON, D. L., and DUNCAN, L. 1971. Adenyl cyclase, kinetic properties and nature of fluoride and hormone stimulation. J. Biol. Chem. 246: 4166–4173.
13. DRUMMOND, G. I., VALLIERES, J., and DRUMMOND, M. 1976. Adenylate cyclase and catecholamine binding in plasma membrane enriched preparations of cardiac and skeletal muscle, pp. 161–182. *In* N. S. Dhalla (ed.), Recent Advances in Studies on Cardiac Structure and Metabolism, Vol. 9. University Park Press, Baltimore.
14. EBASHI, S., and ENDO, M. 1968. Calcium ion and muscle contraction, pp. 123–183.

In J. A. V. BUTLER and D. NOBLE (eds.), Progress in Biophysics and Molecular Biology Pergamon Press, New York.

15. EHRENPREIS, S., FLEISCH, J. H., and MITTAG, T. W. 1969. Approaches to the molecular nature of pharamacological receptors. Pharmacol. Rev. 21: 131–181.
16. ENDO, M., and NAKAJIMA, Y. 1973. Release of calcium induced by "deplorisation" of the sarcoplasmic reticulum membrane. Nature New Biol. 246: 216–218.
17. FAWCETT, D. W., and McNUTT, N. S. 1969. The ultrastructure of the cat myocardium. I. Ventricular papillary muscle. J. Cell Biol. 42: 1–45.
18. FISCHER, E. H., HEILMEYER, L. M. G., JR., and HASCHKE, R. H. 1971. Phosphorylase and the control of glycogen degradation. Curr. Topics Cell. Regul. 4: 211–251.
19. GOREN, E. N., and ROSEN, O. M. 1971. The effect of nucleotides and a nondialyzable factor on the hydrolysis of cyclic AMP by a cyclic nucleotide phosphodiesterase from beef heart. Arch. Biochem. Biophys. 142: 720–723.
20. GOREN, E. N., and ROSEN, O. M. 1972. Purification and properties of a cyclic nucleotide phosphodiesterase from bovine heart. Arch. Biochem. Biophys. 153: 384–397.
21. GREASER, M. L., and GERGELY, J. 1973. Purification and properties of the components from troponin. J. Biol. Chem. 248: 2125–2133.
22. GROSSMAN, A., and FURCHGOTT, R. F. 1964. The effects of various drugs on calcium exchange in the isolated guinea pig left auricle. J. Pharmacol. Exp. Therap. 145: 162–172.
23. HARRIS, P., and OPIE, L. H. (eds.) 1971. Calcium and the Heart. Academic Press, New York.
24. HAUGAARD, N., and HESS, M. E. 1965. Actions of autonomic drugs on phosphorylase activity and function. Pharmacol. Rev. 17: 27–69.
25. HELLAM, D. C., and PODOLSKY, R. J. 1969. Force measurements in skinned muscle fibres. J. Physiol. (London) 200: 807–819.
26. HESS, M., and HAUGAARD, N. 1958. The effect of epinephrine and aminophylline on the phosphorylase activity of perfused contracting heart muscle. J. Pharmacol. Exp. Therap. 122: 169–175.
27. HIMMS-HAGEN, J. 1967. Sympathetic regulation of metabolism. Pharmacol. Rev. 19: 367–461.
28. KAKIUCHI, S., YAMAZAKI, R., AND NAKAJIMA, H. 1970. Properties of a heat-stable phosphodiesterase activating factor isolated from brain extract. Proc. Jap. Acad. 46: 587–592.
29. KASAI, M., and MIYAMOTO, H. 1973. Depolarization induced calcium release from sarcoplasmic reticulum membrane fragments by changing ionic environment. Fed. Eur. Biochem. Soc. Lett. 34: 299–301.
30. KASSEBAUM, D. G., and VAN DYKE, A. R. 1966. Electrophysiological effects of isoproterenol on Purkinje fibers of the heart. Circ. Res. 19: 940–946.
31. KATZ, A. M. 1970. Contractile proteins of the heart. Physiol. Rev. 50: 63–158.
32. KIRCHBERGER, M. A., TADA, M., and KATZ, A. 1974. Adenosine 3′,5′-monophosphate-dependent protein kinase catalysed phosphorylation reaction and its relationship to calcium transport in cardiac sarcoplasmic reticulum. J. Biol. Chem. 249: 6166–6173.
33. KIRCHBERGER, M. A., TADA, M., REPKE, D. I., and KATZ, A. M. 1972. Cyclic adenosine 3′,5′-monophosphate-dependent protein kinase stimulation of calcium uptake by canine cardiac microsomes. J. Molec. Cell. Cardiol. 4: 673–680.
34. KOHLHARDT, M., BAUER, B., KRAUSE, H., and FLECKENSTEIN, A. 1972. Differentiation of the transmembrane Na and Ca channels in mammalian cardiac fibers by the use of specific inhibitors. Pflügers Arch. Ges. Physiol. 335: 309–322.
35. KRAUSE, E. G., HALLE, W., KALLABIS, E., and WOLLENBERGER, A. 1970. Positive chronotropic response of cultured isolated rat heart cells to N^6,2′-O-dibutyryl-3′,5′-adenosine monophosphate. J. Molec. Cell. Cardiol. 1: 1–10.

36. KUKOVETZ, W. R., and PÖCH, G. 1970. Cardiostimulatory effects of cyclic 3′,5′-adenosine monophosphate and its acylated derivatives. Arch Exp. Pharmakol. Pathol. 266: 236–254.
37. LANGER, G. A. 1973. Heart: Excitation-contraction coupling. Ann. Rev. Physiol. 35: 55–86.
38. LEFKOWITZ, R. J., and HABER, E. 1971. A fraction of the ventricular myocardium that has the specificity of the cardiac beta-adrenergic receptor. Proc. Nat. Acad. Sci. U. S. A. 68: 1773–1777.
39. LEVEY, G. S. 1970. Solubilization of myocardial adenyl cyclase. Biochem. Biophys. Res. Commun. 38: 86–92.
40. LEVEY, G. S. 1971. Restoration of norepinephrine responsiveness of solubilized myocardial adenylate cyclase by phosphatidyl-ionsitol. J. Biol. Chem. 246: 7405–7407.
41. LEVEY, G. S. 1972. Restoration of glucagon responsiveness of solubilized myocardial adenyl cyclase by phosphatidylserine. Biochem. Biophys. Res. Commun. 43: 108–113.
42. MARTONOSI, A. 1972. Biochemical and clinical aspects of sarcoplasmic reticulum function, pp. 83–197. *In* F. Bronner and A. Kleinzeller (eds.) Current Topics in Membranes and Transport, Vol. III. Academic Press, New York.
43. MAYER, S. E. 1970. Regulation of cardiac and skeletal muscle glycogen metabolism by biogenic amines, pp. 139–159. *In* Biogenic Amines as Physiological Regulators. Prentice-Hall, Englewood Cliffs, N.J.
44. MORAD, M., and GOLDMAN, Y. 1973. Excitation-contraction coupling in heart muscle: Membrane control of development of tension. Progr. Biophys. Mol. Biol. 27: 259–313.
45. MORAD, M., and ROLETT, E. L. 1972. Relaxing effects of catecholamines on mammalian heart. J. Physiol. (London) 224: 537–558.
46. MORKIN, E., and LA RAIA, P. J. 1974. Biochemical studies in the regulation of myocardial contractility. Seminars in Medicine of the Beth Israel Hospital, Boston. 290: 445–451.
47. NAMM, D. H., MAYER, S. E., and MALTBIE, M. 1968. The role of potassium and calcium ions in the effect of epinephrine on cardiac cyclic adenosine 3′,5′-monophosphate, phosphorylase kinase, and phosphorylase. Molec. Pharmacol. 4: 522–530.
48. NAYLER, W. B. 1970. Ion movements in heart muscle, pp. 75–94. E. E. Bittar (ed.), *In* Membranes and Ion Transport, Vol. II. Wiley Interscience, London.
49. PAGE, E. 1966. Tubular systems in Purkinje cells of the cat heart. J. Ultrastruct. Res. 17: 72–83.
50. PHOL, S. T., KRANS, H. M. J., KOZYREFF, V., BIRNBAUMER, L, and RODBELL, M. 1971. The glucagon-sensitive adenyl cyclase system in plasma membranes of rat liver. VI. Evidence for a role of membrane lipids. J. Biol. Chem. 246: 4447–4454.
51. PORTZEHL, H., CALDWELL, P. C., and RÜEGG, J. C. 1964. The dependence of contraction and relaxation of muscle fibers from the crab *Maia squinado* on the internal concentration of free calcium ions. Biochim. Biophys. Acta 79: 581–591.
52. RASMUSSEN, H. 1970. Cell communication, calcium ion, and cyclic adenosine monophosphate. Science 170: 404–412.
53. RASMUSSEN, H., GOODMAN, D. B. P., and TENENHOUSE, A. 1972. The role of cyclic AMP and calcium in cell activation. C. R. C. Crit Rev. Biochem. 1: 95–148.
54. REUTER, H. 1967. The dependence of slow inward current in Purkinje fibres on the extracellular calcium-concentration. J. Physiol. (London) 192: 479–492.
55. REUTER, H. 1973. Divalent cations as charge carriers in excitable membranes. Progr. Biophys. Mol. Biol. 26: 1–43.
56. REUTER, H., and WOLLERT, U. 1967. Über die Wirkung verschiedener sympathomimetischer Amine auf Kontraktionskraft and ^{45}Ca-Aufrahm isolierter meerschweinchenvorhofs. Arch. Exp. Pharmakol. Pathol. 258: 288–296.
57. ROBISON, G. A., BUTCHER, R. W., ØYE, I., MORGAN, H. E., and SUTHERLAND,

E. W. 1965. The effect of epinephrine on adenosine-3′,5′-phosphate levels in the isolated perfused rat heart. Molec. Pharmacol. 1: 168–177.
58. ROBISON, G. A., DOBBS, J. W., and SUTHERLAND, E. W. 1970. On the nature of receptor sites for biogenic amines, pp. 3–34. *In* J. J. Blum, (ed.), Biogenic Amines as Physiological Regulators. Prentice-Hall, Englewood Cliffs, N. J..
59. ROLETT, E. L. 1974. Adrenergic mechanisms in mammalian myocardium, pp. 219–250. *In* G. A. Langer and A. Brady (eds.), The Mammalian Myocardium. J. Wiley & Sons, Inc., New York.
60. SCARPA, A., and WILLIAMSON, J. R. 1974. Calcium binding and calcium transport by subcellular fractions of heart, pp. 547–585. *In* W. Drabikowski and E. Carafoli (eds.), Calcium Binding Proteins. Elsevier Publishing Company, Amsterdam.
61. SCHAFFER, S. W., SAFER, B., SCARPA, A., and WILLIAMSON, J. R. 1974. Mode of action of the calcium ionophores X-537A and A23187 on cardiac contractility. Biochem. Pharmacol. 23: 1609–1617.
62. SIMPSON, F. O., and RAYNS, D. G. 1968. The relationship between the transverse tubular system and other tubules at the Z disc levels of myocardial cells in the ferret. Amer. J. Anat. 122: 193–208.
63. SKELTON, C. L., LEVEY, G. S., and EPSTEIN, S. E. 1970. Positive inotropic effects of dibutyryl cyclic AMP. Circ. Res. 26: 35–44.
64. SULAKHE, P. V., and DRUMMOND, G. I. 1974. Protein kinase-catalysed phosphorylation of muscle sarcolemma. Arch. Biochem. Biophys. 161: 448–455.
65. TADA, M., KIRCHBERGER, M. A., IORIO, J.-A., and KATZ, A. 1975. Phosphorylation of a 22,000-Dalton component of the cardiac sarcoplasmic reticulum by adenosine 3′:5′-monophosphate-dependent protein kinase. J. Biol. Chem. 250: 2640–2647.
66. TADA, M., KIRCHBERGER, M. A., REPKE, D. I., and KATZ, A. M. 1974. The stimulation of calcium transport in cardiac sarcoplasmic reticulum by adenosine 3′,5′-monophosphate-dependent protein kinase. J. Biol. Chem. 249: 6174–6180.
67. TEO, T. S., and WANG, J. H. 1973. Mechanism of activation of a cyclic adenosine 3′:5′-monophosphate phosphodiesterase from bovine heart by calcium ions: Identification of the protein activator as a Ca^{2+} binding protein. J. Biol. Chem. 248: 5950–5965.
68. TEO, T. S., WANG, T. H., and WANG, J. H. 1973. Purification and properties of the protein activator of bovine heart cyclic adenosine 3′,5′-monophosphate phosphodiesterase. J. Biol. Chem. 248: 588–595.
69. TSIEN, R. W. 1973. Adrenaline-like effects of intracellular iontophoresis of cyclic AMP in cardiac Purkinje fibres. Nature New Biol. . 245: 120–122.
70. TSIEN, R. W., GILES, W., and GREENGARD, P. 1972. Cyclic AMP mediates the effects of adrenaline in cardiac Purkinje fibres. Nature New Biol. 240: 181–183.
71. TRITTHART, H., VOLKMANN, R., WEISS, R., and FLECKENSTEIN, A. 1973. Calcium-mediated action potentials in mammalian myocardium: Alterations of membrane response as induced by changes of Ca_e or by promoters and inhibitors of transmembrane Ca inflow. Naunyn-Schmiedeberg's Arch. Pharmacol. 280: 239–252.
72. VASSORT, G., ROUGIER, O., GARNIER, D., SAUVIAT, M.-P., CORABOEUF, E., and GARGOÏL, Y.-M. 1968. Effets de l'adrénaline sur les courants entrants transmembranaires au cours de l'activité cardiaque. C. R. Acad. Sci. (Paris) 267: 1762–1765.
73. WAUD, D. R. 1968. Pharamacological receptors. Pharmacol. Rev. 20: 49–88.
74. WEBER, A., and MURRAY, J. M. 1973. Molecular control mechanisms in muscle contraction. Physiol. Rev: 53: 612–672.
75. WILLIAMSON, J. R. 1975. Effects of epinephrine on glycogenolysis and myocardial contractility, pp. 605–636. *In* H. Blaschko and A. D. Smith (eds.), Handbook of Physiology, Section VI on Endocrinology, American Physiological Society, The Williams & Wilkins Co., Baltimore.
76. WILLIAMSON, J. R., SCHAFFER, S. W., SCARPA, A., and SAFER, B. 1974. Investigation of the calcium cycle in perfused rat and frog hearts., pp. 375–392. *In* N. S.

Dhalla (ed.), Myocardial Biology, Vol. 4, Recent Advances in Studies on Cardiac Structure and Metabolism. University Park Press, Baltimore.

77. WILLIAMSON, J. R., WOODROW, M. L., and SCARPA, A. 1975. Calcium binding to cardiac sarcolemma, pp. 61–71. *In* A. Fleckenstein and N. S. Dhalla (eds.), Basic Functions of Cations in Myocardia Activity, Vol. 5, Recent Advances in Studies on Cardiac Structure and Metabolism. University Park Press, Baltimore.
78. WINEGRAD, S. 1971. Studies of cardiac muscle with a high permeability to calcium produced by treatment with ethylenediaminetetraacetic acid. J. Gen. Physio. 58: 71–93.
79. WOLLENBERGER, A., BABSKII, E. B., KRAUSE, E.-G., GENZ, S., BLOHM, D., and BOGDANOVA, E. V. 1973. Cyclic changes in levels of cyclic AMP and cyclic GMP in frog myocardium during the cardiac cycle. Biochem. Biophys. Res. Commun. 55: 446–452.
80. WOLLENGERGER, A., KRAUSE, E. G., and WILL, H. 1974. Cyclic AMP-dependent protein kinase activity and calcium binding in a cell membrane-enriched preparation from pig myocardium Abstract 90. *In* Abstracts of the Seventh Annual Meeting of the International Study Group for Research in Cardiac Metabolism, Quebec, 1974.
81. WRAY, H. L., GRAY, R. R., and OLSSON, R. A. 1973. Cyclic adenosine 3′,5′-monophosphate stimulated protein kinase and a substrate associated with cardiac sarcoplasmic reticulum. J. Biol. Chem. 248: 1496–1498.

Mailing address: Dr. J. R. Williamson,
Johnson Research Foundation,
Richards Building G4,
University of Pennsylvania,
Philadelphia, Pennsylvania 19175 (USA).

Recent Advances in Studies on Cardiac Structure and Metabolism
Volume 9
The Sarcolemma
Edited by Paul-Emile Roy and Naranjan S. Dhalla

REGULATION OF CALCIUM TRANSPORT IN CARDIAC SARCOPLASMIC RETICULUM BY CYCLIC AMP-DEPENDENT PROTEIN KINASE[1]

M. TADA,[2] M. A. KIRCHBERGER,[3] and A. M. KATZ[4]

Division of Cardiology,
Department of Medicine and Department of Physiology,
Mount Sinai School of Medicine of the City University of New York,
New York, New York, USA

SUMMARY

A manyfold increase in phosphorylation of cardiac sarcoplasmic reticulum (SR) was seen when SR was incubated in the presence of a bovine cardiac cyclic AMP-dependent protein kinase and cyclic AMP. This phosphoprotein had stability characteristics of a phosphoester in which the phosphate is incorporated largely into serine, and its formation did not require calcium ions, unlike the formation of acyl phosphoprotein intermediate of calcium-transport ATPase which is present within the same membrane. When examined by sodium dodecyl sulfate-polyacrylamide gel electrophoresis, the protein kinase-catalyzed phosphorylation occurred at a 22,000-dalton component of the cardiac sarcoplasmic reticulum. This 22,000-dalton protein has been named "phospholamban" (λαμβανειν = to receive), based on its ability to receive phosphate from ATP.

Phosphorylation of phospholamban by cyclic AMP-dependent protein kinase was associated with the stimulation of calcium transport by the cardiac sarcoplasmic reticulum. This stimulation was accompanied by an increase in the calcium-activated ATPase activity, indicating that the overall rate of calcium transport rather than its efficiency is enhanced by protein kinase. The 22,000-dalton phopholamban was susceptible to trypsin. Brief digestion with trypsin in the presence of 1 M sucrose prevented subsequent phosphorylation of phospholamban, while leaving the calcium pump apparently intact. Incubation of trypsin-treated sarcoplasmic reticulum with cyclic AMP-dependent protein kinase did not result in the stimulation of calcium transport. These results may suggest that phospholamban is a modulator of the calcium pump of the cardiac sarcoplasmic reticulum.

---

[1] This research was supported by the United States Public Health Service, contact NIH-NHLI-72-2973-M; grants HL-13191 and HL-15764 from the National Heart and Lung Institute; grants AA-00316 from the National Institute on Alcoholic Abuse and Alcoholism; a Grant-In-Aid from the New York Heart Association; and gifts from the Jack Martin Fund.

[2] Present address: The First Department of Internal Medicine, Osaka University Medical School, Fukushima-ku, Osaka 553, Japan.

[3] A portion of this work was carried out during the tenure of a New York Heart Association Research Fellowship. Recipient of a Research Career Development Award from the U.S. Public Health Service, HL-00053.

[4] Philip J. and Harriet L. Goodhart Professor of Medicine (Cardiology).

Based on these findings, a possible mechanism by which the cyclic AMP-protein kinase system modulates calcium transport was proposed. This effect of cyclic AMP-dependent protein kinase to catalyze phosphorylation of cardiac sarcoplasmic reticulum may account for the abbreviation of systole that is caused by agents, like epinephrine, which increase cyclic AMP production, and resulting alterations in the distribution of calcium within the myocardial cell may be responsible, at least in part, for the augmentation of myocardial contractility.

Increased cyclic AMP production in response to hormonal stimulation of adenylate cyclase results in activation of cyclic AMP-dependent protein kinase (*e.g.* References 2, 30). Within the myocardium, these reactions are associated with at least two major effects. One is the well documented phosphorylation of phosphorylase *b* kinase and glycogen synthetase and a subsequent cascade of reactions resulting in increased glycogenolysis (2, 16, 19). The other main effect, we have found more recently, is the stimulation of Ca transport by the membranes of the cardiac SR (12, 14, 42) through phosphorylation of a 22,000-dalton protein, which we have named "phospholamban" (13, 39–41). Other investigators have also reported phosphorylation of the cardiac SR membrane by protein kinase which is intrinsic to the membrane (18, 47). These findings indicate that the positive inotropic effects of hormones like catecholamines and glucagon, which increase intracellular cyclic AMP levels, may be mediated, at least in part, by the phosphorylation of phospholamban. In this communication we will review some of our findings on the effects of cyclic AMP-dependent protein kinase on Ca transport of the cardiac SR, and suggest a possible mechanism by which the phosphorylation of phospholamban modulates Ca transport of the cardiac SR.

PHOSPHORYLATION OF CARDIAC SR BY CYCLIC AMP-DEPENDENT PROTEIN KINASE

The membranes of cardiac SR, like those of skeletal SR, contain a Ca-dependent ATPase approximately 100,000 of molecular weight (5), which is at least partly responsible for controlling the intracellular Ca ion concentrations during each contraction-relaxation cycle (3, 21, 36). This Ca-dependent ATPase (E) undergoes phosphorylation to form an acyl phosphoprotein intermediate (E~P), which is labile in hydroxylamine. Protein kinase, in a cyclic AMP-dependent manner, phosphorylates the membrane of the cardiac SR to form another phosphoprotein which is chemically distinct from the ATPase intermediate (fig. 1). This phosphoprotein is stable in hydroxylamine and in hot trichloroacetic acid (TCA), but labile in hot alkali, indicating that the phosphate is present as a phosphoester in which the phosphate is largely bound to serine reidues of the protein (12).

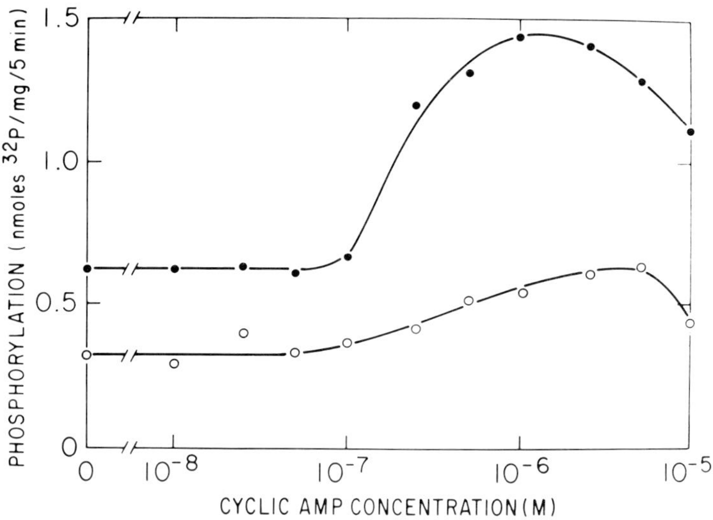

Figure 1 Cyclic AMP concentration dependence of protein kinase-catalyzed phosphorylation of the cardiac SR membranes. The dog cardiac SR (0.5 mg/ml) was incubated for 5 min at 30°C in 40 mM histidine-HCl (pH 6.8), 0.13 M KCl, 5 mM $MgCl_2$, 5 mM [γ-^{32}P]ATP, 2.5 mM Tris-oxalate, 0.7 μM Ca^{2+} (84 μM EGTA and 25 μM $CaCl_2$). Phosphorylation was measured in the presence of increasing concentrations of cyclic AMP in the absence (○) and presence (●) of 0.1 mg of bovine cardiac protein kinase per ml. The reactions were stopped by addition of TCA containing 0.1 mM KH_2PO_4 and bovine serum albumin. The pellet was washed four times, including two mild treatments with 0.5 N NaOH, before transfer to vials for counting ^{32}P. (Reproduced from Kirchberger *et al.* (12) by permission of *Journal of Biological Chemistry*.)

Phosphorylation by both endogenous and exogenous cardiac protein kinase is stimulated by cyclic AMP, half-maximal and maximal stimulation occurring at approximately 10^{-7} M and 10^{-6} M, respectively (fig. 1). These are levels of cyclic AMP that are estimated to occur within the intact myocardium (20, 28). Interestingly, cardiac muscle cytoplasm may contain protein kinase of the same order of magnitude on a molar basis as cyclic AMP (8).

Unlike the formation of Ca-dependent ATPase intermediate which requires Ca ions, the protein kinase-catalyzed phosphorylation does not require Ca^{2+} (fig. 2). In addition to the chemical differences between the phosphoprotein intermediate of ATPase and the phosphoprotein formed by protein kinase, the latter protein is electrophoretically distinct from ATPase protein of 100,000 daltons. Solubilization of phosphorylated SR with SDS and polyacrylamide gel electrophoresis yields a single significant peak of ^{32}P (fig. 3). This peak of radioactivity corresponds to a 22,000-dalton component of the cardiac SR membrane. When SR membranes are phosphorylated in the presence of 25 mM NaF, the amount

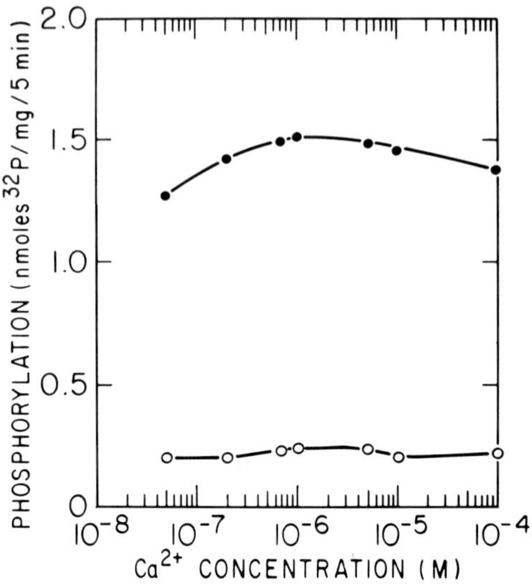

Figure 2 Effect of Ca^{2+} on protein kinase-catalyzed phosphorylation of the cardiac SR. Cardiac SR was incubated under conditions similar to that described in figure 1 in the absence (○) and presence (●) of 1 μM cyclic AMP and 0.1 mg of bovine cardiac protein kinase per ml. Calcium concentrations indicated represent ionized calcium maintained with Ca-EGTA buffers ($CaCl_2$ = 1 mM; EGTA varied between 0.46 and 46 mM). (Reproduced from Kirchberger *et al.* (12) by permission of *Journal of Biological Chemistry.*)

of ^{32}P label in this peak is increased several fold, probably because of inhibition by NaF of intrinsic phosphatase activity which can catalyze dephosphorylation of SR (43). These findings indicate that cyclic AMP-dependent protein kinase catalyzes the phosphorylation of a 22,000-dalton component of the cardiac SR, but not, it appears, of the ATPase protein (100,000 daltons), whose function it controls (see below). These results, however, do not preclude the possibility that this component has become dissociated from ATPase upon incubation with SDS. This 22,000-dalton protein has been named phospholamban, based on its ability to receive phosphate from ATP (13, 39–41). Its phosphorylated form will be designated as phospholamban-P.

CORRELATION BETWEEN PHOSPHORYLATION OF PHOSPHOLAMBAN AND STIMULATION OF Ca-UPTAKE

Phosphorylation of phospholamban by cyclic AMP-dependent protein kinase is associated with an alteration in membrane function: stimulation of Ca transport (12, 14, 42). Protein kinase isolated from both bovine heart and white skeletal muscle increases the rate of Ca uptake by the dog cardiac SR (fig. 4). To

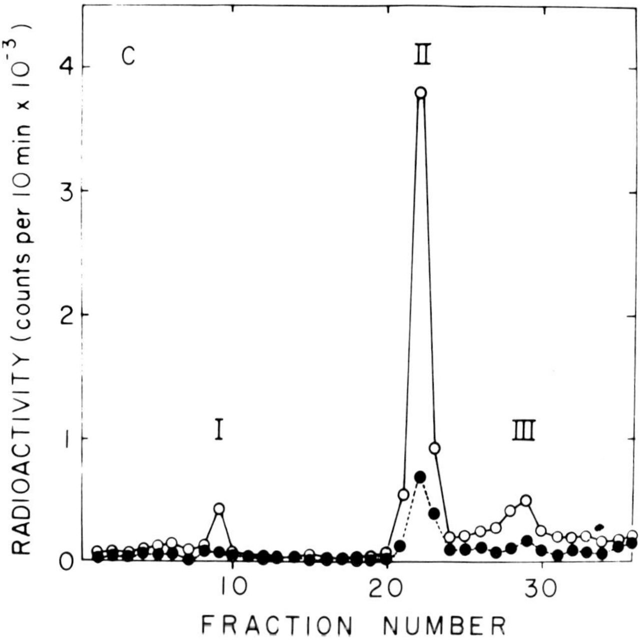

Figure 3 SDS-polyacrylamide gel electrophoresis of the cardiac SR labeled with ^{32}P in the presence of bovine cardiac protein kinase and cyclic AMP. The dog cardiac SR (1.6 mg/ml) was phosphorylated at 25°C for 10 min in 40 mM histidine-HCl (pH 6.8), 0.12 M KCl, 0.5 mM $MgCl_2$, 0.5 mM [γ-^{32}P]ATP, 2.5 mM EGTA, 5 mM NaN_3, 1 μM cyclic AMP, and 0.8 mg/ml of bovine cardiac protein kinase in the presence (○) and absence (●) of 25 mM NaF. After solubilization of the membrane in 2 percent SDS-0.1 mM EGTA-1 percent β-mercaptoethanol at 37°C, the samples, containing approximately 0.11 mg of SR protein, were applied to the 0.1 percent SDS-10 percent polyacrylamide gels. Electrophoresis was performed by the procedures of Weber and Osborn (46). After electrophoresis, the gels were sliced at intervals of 1 mm, and each slice counted for ^{32}P. The top of the gel is at the *left;* proteins migrate toward the anode (*right*). The *right margin* of the figure corresponds to the forefront of the tracking dye. Based on the calibration curve obtained by the simultaneous electrophoresis of marker proteins, the apparent molecular weight of this phosphoprotein component was estimated to be 22,000 ± 1,000 (S.D.). (Reproduced from Tada *et al.* (41) by permission of *Journal of Biological Chemistry.*)

correlate protein kinase-catalyzed phosphorylation of phospholamban with protein kinase-stimulated Ca uptake in the cardiac SR, cardiac SR was incubated in the presence of increasing concentrations of both types of protein kinase (fig. 5). Increasing protein kinase concentrations produce parallel increases in both phosphorylation and Ca uptake although skeletal muscle protein kinase is less active than cardiac protein kinase. A positive correlation is found for both cardiac (r = 0.96, $p < 0.001$) and skeletal muscle protein kinases (r = 0.94, $p <$ 0.001) and the best-fit lines, calculated by the method of least mean squares,

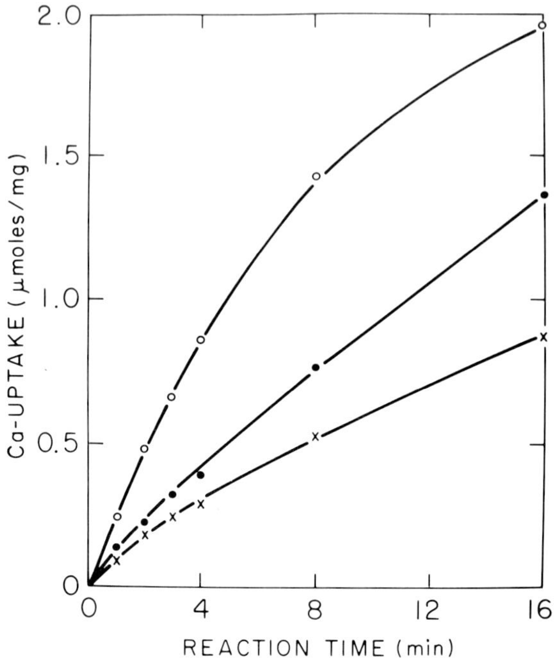

Figure 4 Stimulation of Ca uptake of dog cardiac SR by cyclic AMP-dependent protein kinase. The cardiac SR was preincubated at 25°C for 10 min with bovine cardiac protein kinase and 1 μM cyclic AMP (○) and rabbit skeletal muscle protein kinase and 1 μM cyclic AMP (●) or without protein kinase and cyclic AMP (X) under conditions similar to those described in figure 1, except that Ca-EGTA buffer and Tris-oxalate were omitted and unlabeled ATP was used. At 10 min, an aliquot of the preincubation mixture was added to the Ca uptake assay medium containing 40 mM histidine-HCl (pH 6.8), 0.12 M KCl, 2.5 mM Tris-oxalate, 5 mM ATP, 5 mM $MgCl_2$, and a ^{45}Ca-EGTA buffer that gave a final Ca^{2+} concentration of 0.75 μM ($CaCl_2$ = 125 μM; EGTA = 487 μM). Final SR protein concentration was 0.05 mg/ml. Assays were performed as described previously (11). (Reproduced from Kirchberger *et al.* (12) by permission of *Journal of Biological Chemistry.*)

were virtually superimposable (fig. 6). At saturating concentrations of cardiac protein kinase, approximately 1.4 nmoles of phosphate are incorporated per mg of microsomal proteins. In view of the amount of ATPase intermediate (E~P) formed in the cardiac SR, which is approximately 1.2 nmoles of phosphate/mg (21), a 1:1 stoichiometry appears to exist between phospholamban phosphorylation and E~P formation.

A tight coupling exists between Ca uptake and Ca^{2+}-dependent ATPase activity in skeletal (7, 45, 48) and cardiac (3, 36) muscle SR, so that protein kinase-stimulated Ca uptake could be accompanied by either increased Ca^{2+}-dependent ATPase activity or increased efficiency of Ca transport with no increase in ATPase

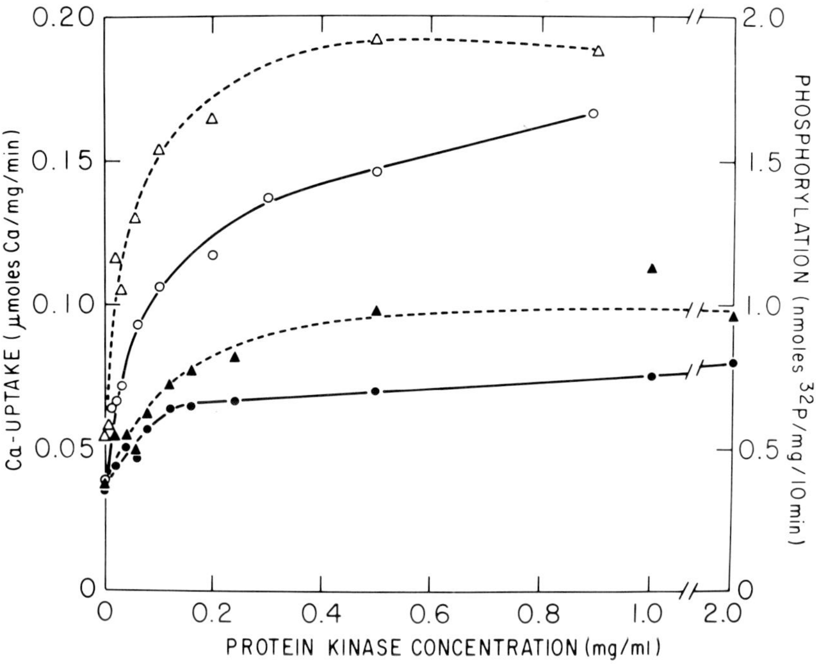

Figure 5 Dependence of phosphorylation (△, ▲) and Ca uptake (○, ●) of dog cardiac SR on the concentration of protein kinase prepared from bovine heart (△, ○) or rabbit skeletal muscle (▲, ●). The rate of Ca uptake was determined by the procedures described in figure 4. For measurement of phosphorylation, the SR membranes from the same preparation were incubated simultaneously with protein kinase under the same conditions as for Ca uptake assay except that [γ-^{32}P]ATP was used, and the formation of phosphoprotein was determined by the procedures described in figure 1. Concentration of protein kinase was varied from 0 to 2 mg/ml, and concentration of cyclic AMP was 1 μM. (Reproduced from Kirchberger *et al.* (12) by permission of *Journal of Biological Chemistry*.)

activity. The enhancement of Ca uptake by protein kinase is found to be accompanied by a concomitant increase in Ca^{2+}-dependent ATPase activity with maintenance of stoichiometric coupling of 2 moles of Ca taken up per mole of ATP hydrolyzed (table I). These findings indicate that the overall rate of Ca transport rather than its efficiency is enhanced by protein kinase.

Marked stimulation of Ca uptake and Ca^{2+}-dependent ATPase activity by protein kinase and cyclic AMP is seen at Ca^{2+} concentrations where both the SR Ca pump and the contractile system of myocardium are fully active (fig. 7). Although the protein kinase-catalyzed phosphorylation of cardiac SR is relatively independent of Ca^{2+} concentration in the physiological range between 0.1 and 10 μM (fig. 2), both Ca^{2+}-dependent ATPase and Ca uptake are significantly

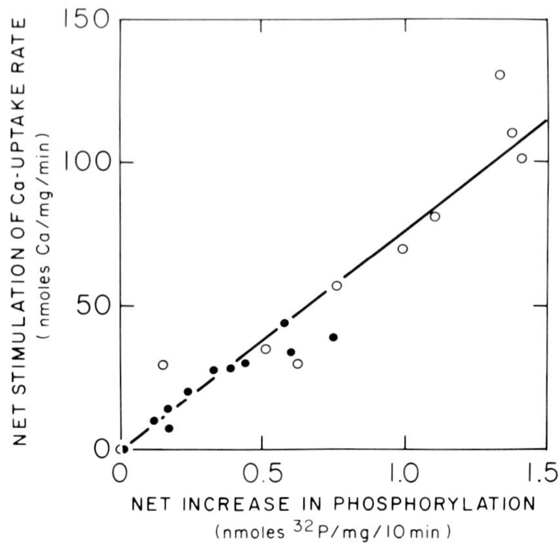

Figure 6 Relationship of net stimulation of the Ca uptake rate by bovine cardiac (○) or rabbit skeletal muscle (●) protein kinase to net increase in protein kinase-catalyzed phosphorylation. Data taken from figure 5. (Reproduced from Kirchberger *et al.* (12) by permission of *Journal of Biological Chemistry.*)

Table I Effects of protein kinase on stoichiometric relationship between Ca uptake and Ca^{2+}-dependent ATPase activity of dog cardiac SR*

Pretreatment	Ca uptake (nmoles Ca/min/mg)	Ca^{2+}-dependent ATPase activity (nmoles Pi/min/mg)	Stoichiometric ratio (Ca/Pi)†
Control	61	33	1.9
Protein kinase	92	52	1.8
Protein kinase + cyclic AMP	143	68	2.1

*Freshly prepared cardiac SR (50 μg/ml) was pretreated for 10 min at 25°C with and without protein kinase (100 μg/ml) and cyclic AMP (1 μM) under conditions similar to those outlined in figure 1 except that Ca-EGTA buffers were omitted. Assays for ATPase and Ca uptake were subsequently started by the addition of Ca-EGTA buffers (Ca^{2+} = 0.75 μM; $CaCl_2$ = 125 μM; EGTA = 487 μM). The ATPase activity was determined by the isobutanol-extraction procedure described by Siegel and Albers (33). (Data reprinted from Tada *et al.* (42) by permission of *Journal of Biological Chemistry.*)

†Calculated by dividing the rate of Ca uptake by the rate of Ca^{2+}-dependent ATPase.

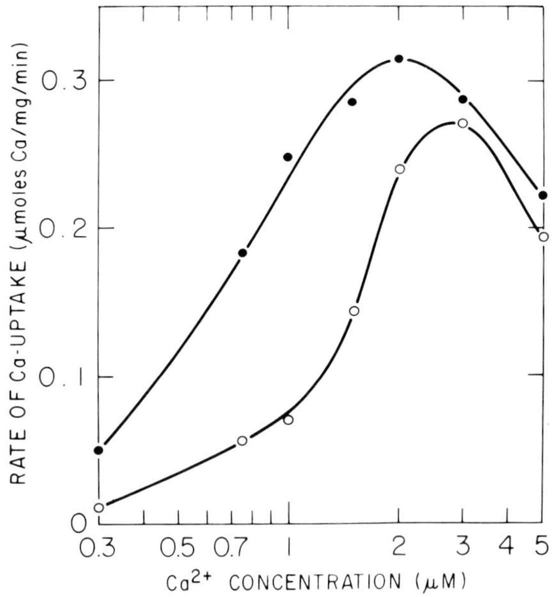

Figure 7 Ca^{2+} dependence of Ca uptake by the cardiac SR pretreated with (•) and without (○) bovine cardiac protein kinase and cyclic AMP. The dog cardiac SR (0.5 mg/ml) was preincubated with bovine cardiac protein kinase (0.2 mg/ml) under conditions described in figure 4. Assay for Ca uptake was started by the addition of aliquots of preincubated SR to assay media (see fig. 4) which contained various concentrations of Ca^{2+}, indicated on the *abscissa*. Ca^{2+} concentrations were maintained by Ca-EGTA buffers containing 125 μM $^{45}CaCl_2$ and various amounts of EGTA. (Reproduced from Tada *et al.* (42) by permission of *Journal of Biological Chemistry.*)

enhanced over the same range of Ca^{2+} concentrations. These results indicate that phosphorylation of phospholamban by cyclic AMP-dependent protein kinase may increase the affinity of the Ca pump for Ca^{2+}.

MECHANISM OF MODULATION OF CARDIAC SR Ca PUMP BY CYCLIC AMP-DEPENDENT PROTEIN KINASE

We have presented strong evidence to indicate that cyclic AMP-dependent protein kinase can regulate Ca transport of the cardiac SR through phosphorylation of a component, phospholamban, of the SR. To examine the nature of this stimulation, the effects of trypsin, which selectively digests phospholamban, were studied. Cardiac SR was incubated with trypsin at a 20:1 weight ratio in the presence of 1 M sucrose, and at various time intervals digestion was stopped by addition of trypsin inhibitor. Inclusion of sucrose prevents effects of trypsin

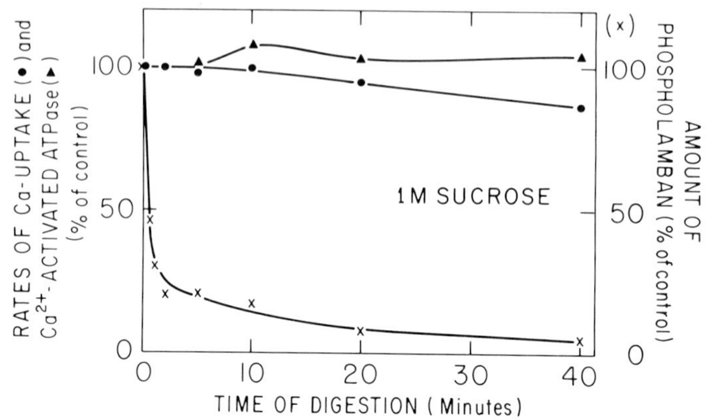

Figure 8 Effect of trypsin digestion in the presence of sucrose on Ca uptake (●), Ca^{2+}-dependent ATPase activity (▲), and phosphorylation of phospholamban (X). The dog cardiac SR (3.7 mg/ml) was incubated at 25°C with trypsin (0.185 mg/ml) in 1.0 M sucrose, 40 mM histidine-HCl (pH 6.8), and 0.12 M KCl. Aliquots taken at various time intervals were added to tubes containing trypsin inhibitor that gave a trypsin to trypsin inhibitor ratio of 1:2 by weight. Assays for Ca uptake and Ca^{2+}-dependent ATPase were carried out subsequently by the procedures described in figure 4 and table I, respectively. For determination of phospholamban phosphorylation, aliquots were incubated with protein kinase (0.5 mg/ml), 1 μM cyclic AMP, and 25 mM NaF under conditions described in figure 3. After 10 min of incubation at 25°C, phosphorylation was terminated by TCA. The amount of phospholamban-P in the solubilized SR membranes was determined by measuring the area of Peak II after SDS-polyacrylamide gel electrophoresis (see fig. 3).

Table II Lack of protein kinase effect on Ca uptake by trypsin-treated cardiac SR*

	Rate of Ca uptake (nmoles Ca/min/mg)	
Treatment	Control	Protein kinase-treated
No trypsin	61	124
Trypsin	62	64

*The dog cardiac SR (1.98 mg/ml) were incubated with and without 0.1 mg/ml of trypsin for 20 min under conditions described in figure 8. Aliquots of control and trypsin-treated SR were incubated with and without protein kinase and cyclic AMP and subsequently assayed for Ca uptake by the procedures described in figure 4.

which would otherwise uncouple the Ca pump activity (9). Although both Ca uptake and Ca^{2+}-dependent ATPase activity are virtually unaffected by incubation with trypsin for 20 min, the ability of phospholamban to be phosphorylated by protein kinase is lost within a few minutes, indicating that trypsin in the presence of sucrose can digest phospholamban while leaving the Ca pump apparently intact (fig. 8). At the same time, Ca uptake by trypsinized microsomes, which have lost 95 percent of their ability to be phosphorylated, was not stimulated by protein kinase and cyclic AMP (table II). Thus, after limited trypsin treatment, basal Ca uptake activity is still retained, suggesting that phospholamban serves as a modulator of the Ca transport rather than as an integral part of the Ca pump.

Figure 9 illustrates a proposed mechanism for modulation of the cardiac SR Ca pump by cyclic AMP-dependent protein kinase. According to this hypothesis, phospholamban serves as a modulator of the Ca pump in the following manner. The ATPase within the SR membrane, which is designated as E and E~P, the latter being its intermediate, is tightly coupled with Ca transport. These reactions are functional in altering intracellular Ca^{2+} concentrations by more than 100-fold with each cardiac contraction-relaxation cycle (*e.g.*, Reference 10). Under the influence of hormones like catecholamines or cyclic AMP, cytoplasmic and/or membrane-bound protein kinase phosphorylates a component of SR, phospholamban, which can stimulate Ca^{2+}-dependent ATPase and Ca uptake by increasing the turnover rate of this process. This hypothesis assumes additional interest in view of the 1:1 stoichiometry existing between phospholamban phosphorylation and E~P formation. Our most recent experiments indicate that phosphoprotein phosphatase isolated from the dog myocardium catalyzes the dephosphorylation of phospholamban-P (43).

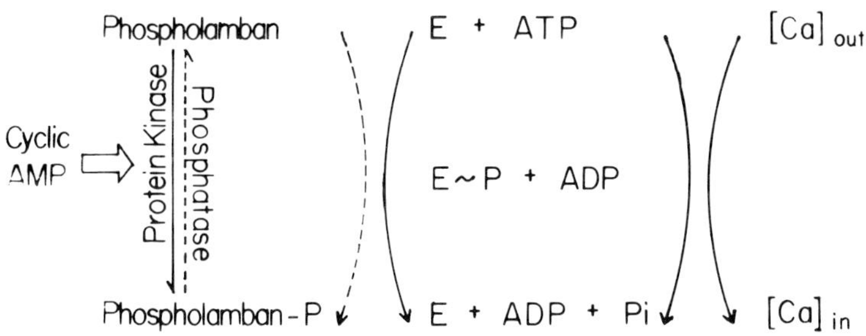

Figure 9 Modulation of cardiac SR calcium transport by cyclic AMP-dependent protein kinase. *E* and *E ~ P* represent cardiac SR Ca^{2+}-dependent ATPase and its phosphorylated intermediate, respectively. $[Ca]_{out}$ and $[Ca]_{in}$ represent calcium located outside and inside, respectively, of the cardiac SR membranes. Therefore, $[Ca]_{out}$ represents cytoplasmic calcium.

Although cyclic AMP is believed to mediate several metabolic actions of catecholamines on heart muscle (*e.g.,* Reference 2), its role in mediating the positive inotropic response to catecholamines, which was first postulated by Sutherland and Rall (37), has not been entirely clarified (6, 15, 17, 31, 34). Considerable evidence, however, exists which suggests such a role for cyclic AMP (4, 38). The evidence summarized in this communication may provide a biochemical basis for the mediation by cyclic AMP of the two principal mechanical effects of catecholamines: abbreviation of systole and augmentation of contractility. The increase in the rate of Ca accumulation by the SR of the intact heart could explain abbreviation of systole because of the increased rate at which Ca would be removed from troponin. Increased Ca uptake by the SR after phosphorylation by protein kinase could also increase the amount of Ca stored in the SR by retaining within the cell some of the Ca which would otherwise be lost during diastole. This increased Ca storage could add to the amount of Ca available for delivery to the contractile proteins in subsequent beats, thus promoting augmentation of myocardial contractility. Such a consideration is supported by the recent findings obtained by Rolett (29), who found that the initial mechanical response of the heart to norepinephrine is a shortening of systole without an increase in tension, the rise in tension occurring later. Augmentation of myocardial contractility due to catecholamines may thus be related, at least in part, to the stimulation of Ca transport by the SR. There is additional evidence, however, that the increased myocardial contractility may also be related to catecholamine-induced enhancement of Ca influx across the plasma membrane (25–27, 32, 44). Additional factors such as phosphorylation of troponin I (1, 24, 35) and light chain of myosin (22, 23) will also have to be considered in explaining the inotropic actions of catecholamines and cyclic AMP.

REFERENCES

1. BAILEY, C., and VILLAR-PALASI, C. 1971. Cyclic AMP-dependent phosphorylation of troponin. Fed. Proc. 30: 1147.
2. BROSTROM, M. A., REIMAN, E. M., WALSH, P. A., and KREBS, E. G. 1970. Cyclic 3′,5′-AMP-stimulated protein kinase from cardiac muscle. Advan. Enzyme Regul. 8: 191–203.
3. ENTMAN, N. L., SNOW, T. R., FREED, D., and SCHWARTZ, A. 1973. Analysis of calcium binding and release by canine cardiac relaxing system (sarcoplasmic reticulum): The use of specific inhibitors to construct a two-component model for calcium binding and transport. J. Biol. Chem. 248: 7762–7772.
4. EPSTEIN, S. E., LEVEY, G. S., and SKELTON, C. L. 1971. Adenyl cyclase and cyclic AMP: Biochemical links in the regulation of myocardial contractility. Circulation 43: 437–450.
5. FANBURG, B. L., and MATSUSHITA, S. 1973. Phosphorylated intermediate of ATPase of isolated cardiac sarcoplasmic reticulum. J. Molec. Cell. Cardiol. 5: 111–115.
6. GARTNER, S. L., and VAHOUNY, G. V. 1972. Effects of epinephrine and cyclic 3′,5′-AMP on perfused rat hearts. Amer. J. Physiol. 222: 1121–1124.

7. HASSELBACH, W., and MAKINOSE, M. 1963. Über den Mechanismus des Calciumtransportes durch die Membranen des sarkoplasmatischen Retikulums. Biochem. Z. 339: 94–111.
8. HOFMANN, F., and KREBS, E. G. 1974. Cyclic AMP-dependent protein kinase: Estimation of the molar ratio of subunits in crude tissue extracts. Fed. Proc. 33: 1324.
9. IKEMOTO, N., SRETER, F. A., and GERGELY, J. 1971. Structural features of the surface of the vesicles of FSR: Lack of functional role in Ca^{2+} uptake and ATPase activity. Arch. Biochem. Biophys. 147: 571–582.
10. INESI, G. 1972. Active transport of calcium ion in sarcoplasmic membranes. Ann. Rev. Biophys. Bioeng. 1: 191–210.
11. KATZ, A. M., REPKE, D. I., UPSHAW, J. E., and POLASCIK, M. A. Biochim. Biophys. Acta 205: 473–490.
12. KIRCHBERGER, M. A., TADA, M., and KATZ, A. M. 1974. Adenosine 3′,5′-monophosphate-dependent protein kinase-catalyzed phosphorylation reaction and its relationship to calcium transport in cardiac sarcoplasmic reticulum. J. Biol. Chem. 249: 6166–6173.
13. KIRCHBERGER, M. A., TADA, M., and KATZ, A. M. 1974. Phospholamban: regulatary protein of the cardiac sarcoplasmic reticulum, pp. 103–115. *In* A. Fleckenstein and N. S. Dhalla (eds.), Recent Advances in Studies on Cardiac Structure and Metabolism, Vol. 5. University Park Press, Baltimore.
14. KIRCHBERGER, M. A., TADA, M., REPKE, D. I., and KATZ, A. M. 1972. Cyclic adenosine 3′,5′-monophosphate-dependent protein kinase stimulation of calcium uptake by canine cardiac microsomes. J. Molec. Cell. Cardiol. 4: 673–680.
15. KJEKSHUS, J. K., HENRY, P. D., and SOBEL, B. E. 1971. Activation of phosphorylase by cyclic AMP without augmentation of contractility in the perfused guinea pig heart. Circ. Res. 29: 468–478.
16. KREBS, E. G. 1972. Protein kinases. Current Top. Cell. Regul. 5: 99–133.
17. LANGSLET, A., and ØYE, I. 1970. The role of cyclic 3′,5′-AMP in the cardiac response to adrenaline. Eur. J. Pharmaco. 12: 137–144.
18. LaRAIA, P. J., and MORKIN, E. 1974. Adenosine 3′,5′-monophosphate-dependent membrane phosphorylation: A possible mechanism for the control of microsomal calcium transport in heart muscle. Circ. Res. 35: 298–306.
19. MAYER, S. E. 1970. Regulation of cardiac and skeletal muscle glycogen metabolism by biogenic amines, pp. 139–159. *In* J. J. Blum (ed.), Biogenic Amines as Physiological Regulators. Prentice-Hall, Englewood Cliffs, N.J.
20. MAYER, S. E., NAMM, D. H., and RICE, L. 1970. Effect of glucagon on cyclic 3′,5′-AMP, phosphorylase activity and contractility of heart muscle of the rat. Circ. Res. 26: 225–233.
21. PANG, D. C., and BRIGGS, F. N. 1973. Reaction mechanism of the cardiac sarcotubule calcium (II) dependent adenosine triphosphatase. Biochemistry 12: 4905–4911.
22. PERRIE, W. T., SMILLIE, L. B., and PERRY, S. V. 1972. A phosphorylated light-chain component of myosin. Biochem. J. 128: 105p–106p.
23. PERRIE, W. T., SMILLIE, L. B., and PERRY, S. V. 1973. A phosphorylated light chain component of myosin from skeletal muscle. Cold Spring Harbor Symp. Quant. Biol. 37: 17–18.
24. REDDY, Y. S., BALLARD, D., GIRI, N. Y., and SCHWARTZ, A. 1973. Phosphorylation of cardiac native tropomyosin and troponin: Inhibitory effect of actomyosin and possible presence of endogenous myofibrillar-located cyclic-AMP-dependent protein kinase. J. Molec. Cell. Cardiol. 5: 461–471.
25. REUTER, H. 1965. Über die Wirkung von Adrenalin auf den cellulären Ca-Umsatz des Meerschweinchenvorhofs. Arch. Exp. Pathol. Pharmakol. 251: 401–412.
26. REUTER, H. 1967. The dependence of slow inward current in Purkinje fibres on the extracellular calcium concentration. J. Physiol. (London) 192: 479–492.
27. REUTER, H. 1974. Localization of *beta* adrenergic receptors, and effects of noradrena-

line and cyclic nucleotides on action potentials, ionic currents and tension in mammalian cardiac muscle. J. Physiol. (London) 242: 429–451.

28. ROBISON, G. A., BUTCHER, R. W., ØYE, I., MORGAN, H. E., and SUTHERLAND, E. W. 1965. The effect of epinephrine on adenosine 3′,5′-phosphate levels in the isolated perfused rat heart. Molec. Pharmacol. 1: 168–177.
29. ROLETT, E. L. 1974. Adrenergic mechanisms in mammalian myocardium, pp. 219–250. *In* G. A. Langer and A. J. Brady (eds.), The Mammalian Myocardium. John Wiley & Sons, Inc., New York.
30. RUBIN, C. S., ERLICHMAN, J., and ROSEN, O. M. 1972. Molecular forms and subunit composition of a cyclic adenosine 3′,5′-monophosphate-dependent protein kinase purified from bovine heart muscle. J. Biol. Chem. 247: 36–44.
31. SHANFIELD, J., FRAZER, A., and HESS, M. E. 1969. The relation of adenosine-3′,5′-phosphate and phosphorylase to the actions of catecholamines and other hormones. J. Pharmacol. Exp. Therap. 169: 315–320.
32. SHIGENOBU, K., and SPERELAKIS, N. 1972. Calcium current induced by catecholamines in chick embryonic hearts whose fast sodium channels are blocked by tetrodotoxin or elevated potassium. Circ. Res. 31: 932–952.
33. SIEGEL, G. J., and ALBERS, R. W. 1967. Sodium-potassium-activated adenosine triphosphatase of *Electrophorus* electric organ. J. Biol. Chem. 242: 4972–4979.
34. SOBEL, B. E., and MAYER, S. E. 1973. Cyclic adenosine monophosphate and cardiac contractility. Circ. Res. 32: 407–414.
35. STULL, J. T., BROSTROM, C. O., and KREBS, E. G. 1972. Phosphorylation of the inhibitor component of troponin by phosphorylase kinase. J. Biol. Chem. 247: 5272–5274.
36. SUKO, J. 1973. The calcium pump of cardiac sarcoplasmic reticulum. Functional alterations at different levels of thyroid state in rabbits. J. Physiol. (London) 228: 563–582.
37. SUTHERLAND, E. W., and RALL, T. W. 1960. The relation of adenosine-3′,5′-phosphate and phosphorylase to the actions of catecholamines and other homones. Pharmacol. Rev. 12: 265–299.
38. SUTHERLAND, E. W., ROBISON, G. A., and BUTCHER, R. W. 1968. Some aspects of the biological role of adenosine 3′,5′-monophosphate (cyclic AMP). Circulation 37: 279–306.
39. TADA, M., and KIRCHBERGER, M. A. Regulation of calcium transport by cyclic AMP: A proposed mechanism for the beta-adrenergic control of myocardial contractility. Acta Cardiologica (in press).
40. TADA, M., KIRCHBERGER, M. A., IORIO, J. M., and KATZ, A. M. 1973. Phosphory-
41. TADA, M., KIRCHBERGER, M. A., and KATZ, A. M. 1975. Phosphorylation of a 22,000-dalton component of the cardiac sarcoplasmic reticulum by adenosine 3′:5′-monophosphate-dependent protein kinase. J. Biol. Chem. 250: 2640–2647.
42. TADA, M., KIRCHBERGER, M. A., REPKE, D. I., and KATZ, A. M. 1974. Stimulation of calcium transport in cardiac sarcoplasmic reticulum by adenosine 3′:5′-monophosphate-dependent protein kinase. J. Biol. Chem. 249: 6174–6180.
43. TADA, M., KIRCHBERGER, M. A., and LI, H.-C. Phosphoprotein phosphatase-catalyzed dephosphorylation of the 22,000 dalton phosphoprotein of cardiac sarcoplasmic reticulum. J. Cyclic Nucleotide Research 1 (in press).
44. WATANABE, A. M., and BESCH, H. R., Jr. 1974. Cyclic adenosine monophosphate modulation of slow calcium influx channels in guinea pig hearts. Circ. Res. 35: 316–324.
45. WEBER, A., HERZ, R., and REISS, I. 1966. Study of the kinetics of calcium transport by isolated fragmented sarcoplasmic reticulum. Biochem. Z. 345: 329–369.
46. WEBER, K., and OSBORN, M. 1969. The reliability of molecular weight determinations by dodecyl sulfate-polyacrylamide gel electrophoresis. J. Biol. Chem. 244: 4406–4412.
47. WRAY, H. L., GRAY, R. R., and OLSSON, R. A. 1973. Cyclic adenosine 3′,5′-mono-

phosphate-stimulated protein kinase and a substrate associated with cardiac sarcoplasmic reticulum. J. Biol. Chem. 248: 1496–1498.
48. YAMADA, S., YAMAMOTO, T., and TONOMURA, Y. 1970. Reaction mechanism of the Ca^{2+}-dependent ATPase of sarcoplasmic reticulum from skeletal muscle. III. Ca^{2+}-uptake and ATP-splitting. J. Biochem. 67: 789–794.

Mailing address: Dr. M. Tada,
The First Department of Internal Medicine,
Osaka University Medical School,
Fukushima-ku, Osaka 553 (Japan).

Recent Advances in Studies on Cardiac Structure and Metabolism
Volume 9
The Sarcolemma
Edited by Paul-Emile Roy and Naranjan S. Dhalla

INCREASED Ca^{2+} BINDING BY A CARDIAC CELL MEMBRANE PREPARATION AFTER CYCLIC AMP-ENHANCED INTRINSIC MEMBRANE PROTEIN PHOSPHORYLATION

H. WILL, B. SCHIRPKE, and A. WOLLENBERGER

Division of Cellular and Molecular Cardiology,
Central Institute of Heart and Circulatory Regulation Research,
Academy of Sciences of the DDR, Berlin-Buch, German Democratic Republic

SUMMARY

Ca^{2+} binding to a sarcolemma-enriched membrane fraction from pig myocardium possessing an intrinsic cyclic AMP-dependent protein kinase occurs at several classes of low affinity binding sites and at two high affinity binding sites with binding constants of about 1.5–1.7 × 10^7 M^{-1} (0.3 nmole of Ca^{2+}/mg protein) and 0.9–2.9 × 10^6 M^{-1} (0.8 nmole of Ca^{2+}/mg protein). Ca^{2+} binding properties are not affected by verapamil and ouabain, whereas ruthenium red depresses Ca^{2+} binding at the low affinity binding sites and La^{3+} ions strongly reduce both low and high affinity Ca^{2+} binding. A profound inhibition of the high affinity Ca^{2+} binding sites was observed in the presence of Na^+ ions, half-maximal inhibition at a free Ca^{2+} concentration of 2 × 10^{-8} M being achieved by 11 mM NaCl. High affinity Ca^{2+} binding is also diminished after pretreatment of the membranes with trypsin and phospholipase A. Phosphorylation of one or two of the membrane proteins by the endogenous cyclic AMP-dependent protein kinase leads at both classes of high affinity Ca^{2+} binding sites to an approximately 4-fold increase in affinity, the number of these sites remaining unchanged. The high affinity Ca^{2+} binding sites may possibly be involved in Ca^{2+} extrusion from the cell and in the relaxation process.

Changes in the concentration of cytoplasmic Ca^{2+} together with that of cyclic AMP coordinate intracellular events in many cells after their activation by specific extracellular signals (15). In accordance with the current concept of the mode of action of cyclic AMP (7), alterations in Ca^{2+} sensitivity of specific proteins after their phosphorylation by cyclic AMP-dependent protein kinases may be one important mechanism underlying the many mutual interactions of the two intracellular messengers (5). With this conception in mind several authors (2, 3, 6, 21, 22) have suggested that the inotropic and relaxation-enhancing effects of the β-adrenergic catecholamines on the heart, which are intimately tied to the role of Ca^{2+} in excitation-contraction coupling and muscular relaxation and usually are associated with an increase in myocardial cyclic AMP level, may result from cyclic AMP-enhanced phosphorylation of protein in subcellular structures concerned with the transport and sequestration of Ca^{2+}. Such phosphorylation was actually demonstrated in preparations of myofibrillar proteins (16) and of membranes (3, 4, 24) from heart muscle.

Because of the important role of the myocardial cell surface membrane in the regulation of the cellular concentration of inorganic ions, in particular Ca^{2+}, the relation between cyclic AMP-enhanced protein phosphorylation in a heart sarcolemma-enriched membrane preparation and the ability of this membrane to bind Ca^{2+} has been investigated (23). The latter property can be assumed to be of significance in the transport of this ion across the myocardial cell membrane, since binding to relevant binding sites is a prerequisite both for active transport and facilitated diffusion (19) and, furthermore, Ca^{2+} binding sites in the basement membrane of myocardial cells are believed to play an important role in the accelerated diffusion of Ca^{2+} through the sarcolemma during the action potential (8). Ca^{2+} binding sites may also be involved in the regulation of other membrane events.

To estimate the amount of Ca^{2+} bound to a cell membrane fraction prepared from pig myocardium according to the method of Natsui and Schwartz (10), the particulate components were sedimented by centrifugation after incubation at 20°C with ^{45}Ca in 120 mM KCl-4 mM HEPES-Tris buffer, pH 6.8, and radioactivity was determined in the pellet and the supernatant, a good balance being obtained in every instance. The concentration of Ca^{2+} was varied over several orders of magnitude, Ca^{2+} concentrations below 10^{-6} M being adjusted by titration with EGTA, using a binding constant for calcium EGTA of 4.4 × 10^5 M (12).

Ca^{2+} binding at free Ca^{2+} concentrations above 1 μM yielded a continuously curved line in the Scatchard plot, indicating an overlap of discrete classes of binding sites differing in number and Ca^{2+} affinity (fig. 1). Even at a free Ca^{2+} concentration of 10^{-3} M the membrane is not saturated with Ca^{2+}, presumably because of the presence of a large number of low affinity binding sites. Ca^{2+} binding is not affected by verapamil or ouabain, but is strongly depressed by La^{3+} ions and, to a lesser extent, by ruthenium red. No attempt was made to resolve Ca^{2+} binding in terms to the various classes of low affinity binding sites involved.

Ca^{2+} binding at a free Ca^{2+} concentration below 1 μM shows a break in the Scatchard plot (fig. 2). This change in slope may be interpreted in different ways. Two distinct Ca^{2+} binding sites may be present in the membrane preparation. Alternately, all high affinity Ca^{2+} binding sites may be the same, but cooperative effects may act in such a fashion that Ca^{2+} is less strongly bound as the membrane becomes more saturated with this ion. If the data are interpreted as indicative of two types of binding sites for Ca^{2+}, then binding constants of about 1.6 × 10^7 M^{-1} and 2.9 × 10^6 M^{-1} can be derived from the plot. (As these

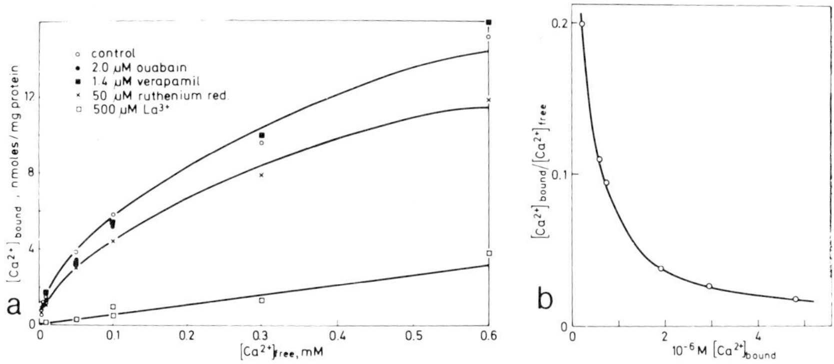

Figure 1 a Binding of Ca^{2+} to a cell membrane-enriched preparation from pig myocardium at free Ca^{2+} concentrations above 10^{-6} M. Influence of ouabain, verapamil, ruthenium red, and La^{3+} ions. The membrane preparation (0.5 mg of protein/ml) was incubated for 10 min at 20°C in the absence and presence of the Ca^{2+} antagonists in 120 mM KCl-4 mM HEPES-Tris buffer, pH 6.8. *b,* Scatchard plot of Ca^{2+} binding based on the control data in *a.*

constants vary somewhat from preparation to preparation, the influence of Ca^{2+} antagonists and phosphorylation on high affinity Ca^{2+} binding described below was always compared with controls measured with the same preparation.) Ruthenium red, ouabain, and verapamil are without effect at the high affinity binding sites, whereas La^{3+} and Na^+ ions are strongly inhibitory.

If 120 mM KCl, usually present in the incubation medium to maintain ionic strength, is replaced by 120 mM NaCl, Ca^{2+} binding to high affinity binding sites is completely suppressed (fig. 3). Na^+ inhibition is reversible and disappears at higher Ca^{2+} concentrations.

Different amounts of KCl were replaced by NaCl to determine the concentration of Na^+ needed to reduce high affinity Ca^{2+} binding to half of the value

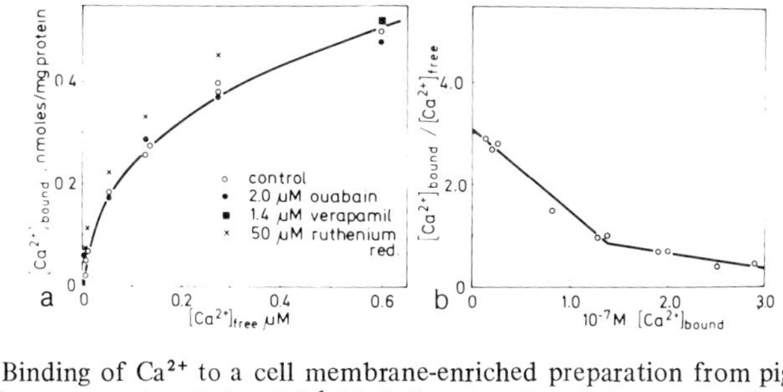

Figure 2 a Binding of Ca^{2+} to a cell membrane-enriched preparation from pig myocardium at free Ca^{2+} concentrations below 10^{-6} M. Influence of ouabain, verapamil, and ruthenium red. Experimental conditions as described in figure *a. b,* Scatchard plot of Ca^{2+} binding based on control data in *a.*

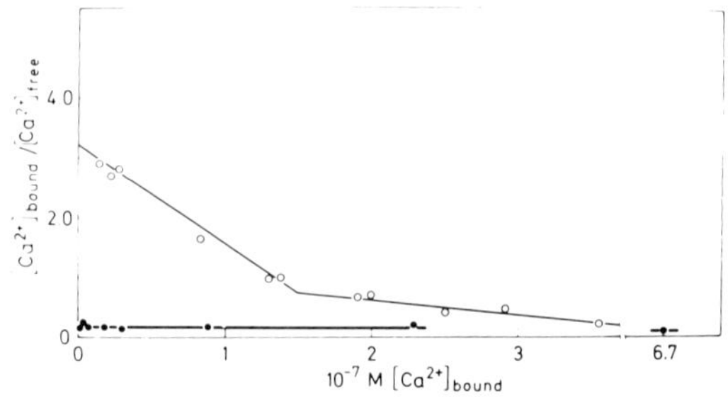

Figure 3 Comparison of Ca^{2+} binding at free Ca^{2+} concentrations below 10^{-6} M in the presence of 120 mM KCl (○) and 120 mM NaCl (●).

measured in the absence of Na^+ ions (fig. 4). This concentration depends on that of the free Ca^{2+} present, becoming higher with increasing Ca^{2+} concentration. At a free Ca^{2+} concentration of 2×10^{-8} M, half-maximal inhibition of Ca^{2+} binding occurs at about 11 mM NaCl.

Ca^{2+} binding to the high affinity binding sites is also diminished after pretreatment of the membranes with trypsin or phospholipase A (table I). Before the binding experiments aliquots of membrane protein were incubated at 37°C for 20 min with 0.1 mg of one of these enzymes per mg of membrane protein and subsequently washed three times with HEPES-Tris buffer. Control samples taken to represent 100 percent of binding activity were treated in the same way but without addition of the enzymes. Since high affinity binding sites are destroyed or releasedd by trypsin, they are probably located on a protein or proteins. However, as Ca^{2+} binding is likewise decreased by pretreatment with phospholipase A, phospholipids seem to be required for the native organization of these binding sites.

To evaluate the influence of membrane phosphorylation by the intrinsic cyclic AMP-dependent protein kinase on Ca^{2+} binding, prior phosphorylation was

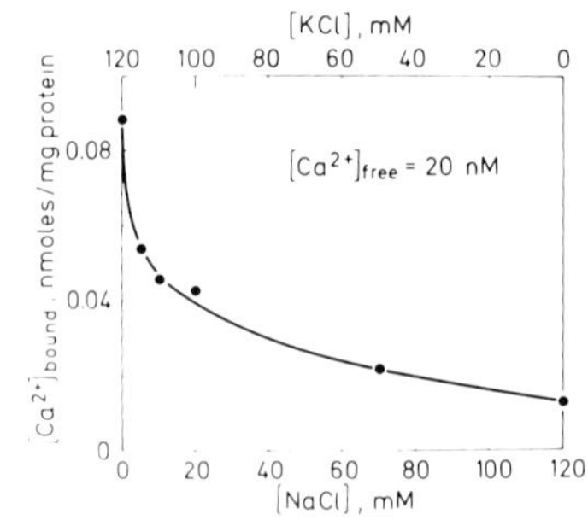

Figure 4 Decrease of Ca^{2+} binding in the presence of different concentrations of NaCl. Various amounts of KCl were replaced by NaCl, holding ionic strength constant at 0.12.

Table I Effect of pretreatment of membrane preparation with trypsin, phospholipase A, and neuraminidase*

	$[Ca^{2+}]$ bound					
	at $[Ca^{2+}]_{free} = 8 \cdot 10^{-8}$ M		at $[Ca^{2+}]_{free} = 2 \cdot 10^{-7}$ M		at $[Ca^{2+}]_{free} = 4 \cdot 10^{-7}$ M	
Pretreatment	M	%	M	%	M	%
Control	$1.78 \cdot 10^{-7}$	100	$2.47 \cdot 10^{-7}$	100	$2.95 \cdot 10^{-7}$	100
Trypsin	$1.45 \cdot 10^{-8}$	8	$3.7 \cdot 10^{-8}$	15	$5.12 \cdot 10^{-8}$	17
Phospholipase A	$1.14 \cdot 10^{-7}$	64	$1.67 \cdot 10^{-7}$	68	$2.75 \cdot 10^{-7}$	93
Neuraminidase	$1.38 \cdot 10^{-7}$	78	$2.26 \cdot 10^{-7}$	92	$3.04 \cdot 10^{-7}$	103

*For experimental conditions see text.

carried out by incubating the samples for 10 min at 37°C in the presence of 16 mM Mg^{2+}, 0.12 mM ATP, and 0.5 μM cyclic AMP. Phosphorylation of the membranes caused an enhancement of Ca^{2+} binding at the low affinity binding sites (fig. 5*a*). In this experiment, approximately 150 pmoles of [^{32}P] phosphate were incorporated per mg of protein.

Ca^{2+} binding at the high affinity binding sites is likewise increased (fig. 5*b*). This effect is due entirely to an increase in the Ca^{2+} affinity of these sites, the maximum number of sites remaining unchanged. It was calculated that phosphorylation of membrane protein enhanced by the endogenous cyclic AMP-dependent protein kinase was associated with an approximately 4-fold increase in the affinity for Ca^{2+} at both high affinity binding sites, total Ca^{2+} binding capacity (0.30 and 0.77 nmole/mg protein) remaining unchanged.

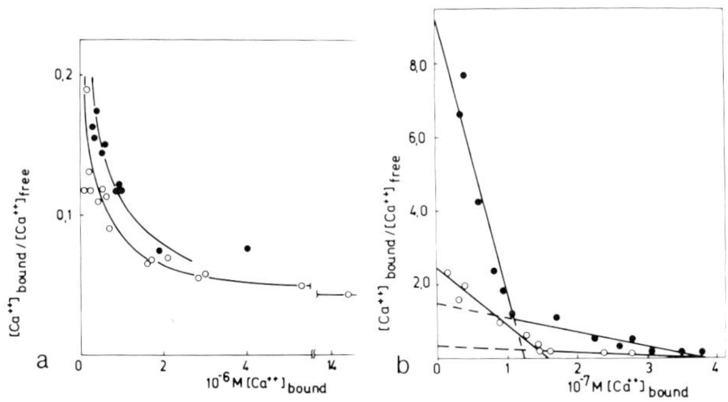

Figure 5 Increase in Ca^{2+} binding to a cell membrane-enriched preparation from pig myocardium after phosphorylation of the membranes for 10 min in the presence of 5 X 10^7 M cyclic AMP (20).

From these experiments a number of conclusions concerning the nature and function of the high affinity binding sites can be drawn.

1. Considering their susceptibility toward Na^+ ions, the high affinity Ca^{2+} binding sites present in the membrane preparation investigated seem to be different from Ca^{2+} binding sites responsible for active transport in sarcoplasmic reticulum and mitochondria. Although Na^+ ions compete with Ca^{2+} ions in these systems (1, 13), the competition is not nearly as strong as that observed for the high affinity Ca^{2+} binding sites in the presently investigated membrane preparation, which on account of its relatively high (Na^+,K^+)-ATPase activity may provisionally be classified as a subcellular fraction rich in pieces of cell surface membrane. The affinity of Na^+ ions towards the high affinity Ca^{2+} binding sites is close to the Na^+ affinity toward (Na^+,K^+)-ATPase (9, 18). However, the high affinity Ca^{2+} binding sites do not seem to be associated with this enzyme, since preliminary experiments revealed that in the course of further purification of the preparation to yield (Na^+,K^+)-ATPase of higher specific activity (14), the high affinity Ca^{2+} binding sites were lost.

2. The values of the dissociation constants for Ca^{2+} (reciprocal values of the above listed binding constants), which equal the ambient Ca^{2+} concentration at which half of the binding sites are occupied by the cation, lie in the concentration range within which the free Ca^{2+} concentration fluctuates in the myoplasm during the contraction-relaxation cycle and are several orders of magnitude below the extracellular Ca^{2+} concentration. This fact, together with the almost complete inhibition of the high affinity Ca^{2+} binding by Na^+ concentrations above 100 mM, suggests (on the assumption that we are dealing indeed with properties of the myocardial cell surface membrane) that these sites are located *in situ* on the internal side of this membrane and may be involved in a Na^+-coupled transport of Ca^{2+} out of the cell and/or the regulation of other specific membrane functions by Ca^{2+}.

3. The increase in the affinity of these binding sites for Ca^{2+} caused by cyclic AMP-stimulated protein phosphorylation can then be expected to be associated with an acceleration of Ca^{2+} efflux and/or with an increase in other membrane activities regulated by Ca^{2+}.

4. It is a potentially intriguing feature of the high affinity of Ca^{2+} binding sites, shown to be in all likelihood located on protein, that they may be controlled in opposite directions by two important classes of cardioactive agents, catecholamines and cardiac glycosides. Catecholamines, by raising intracellular cyclic AMP levels and thus promoting protein phosphorylation, may well enhance Ca^{2+} sensitivity of these sites, whereas cardioactive steroids, although not directly influencing high affinity Ca^{2+} binding sites, can be expected to reduce their interaction with Ca^{2+} ions by promoting, through their inhibitory effect on (Na^+,K^+)-ATPase, the accumulation of Na^+ ions close to the internal side of the

cell surface membrane. Calling to mind a main difference in the mechanical response of heart muscle to these two classes of agents, catecholamines promoting (11, 17) and glycosides opposing (17) relaxation, one is led independently of other arguments to the speculation that the high affinity Ca^{2+} binding sites may be involved in the relaxation process.

REFERENCES

1. DRANSFELD, H., GREEFF, K., HESS, D., and SCHORN, A. 1967. Die Abhängigkeit der Ca^{2+}-Aufnahme isolierter Mitochondrien des Herzmuskels von der Na^+- und K^+-Konzentration als mögliche Ursache der inotropen Digitaliswirkung. Experientia 23: 375–379.
2. KATZ, A. M., and REPKE, D. I. 1973. Calcium-membrane interactions in the myocardium: Effects of ouabain, epinephrine and 3′,5′-cyclic adenosine monophosphate. Amer. J. Cardiol. 31: 193–201.
3. KIRCHBERGER, M. A., TADA, M., REPKE, D. I., and KATZ, A. 1972. Cyclic adenosine 3′,5′-monophosphate dependent protein kinase stimulation of calcium uptake by canine cardiac microsomes. J. Molec. Cell. Cardiol. 4: 673–680.
4. KRAUSE, E.-G., WILL, H., PELOUCH, V., and WOLLENBERGER, A. 1973. Cyclic AMP-dependent protein kinase activity in a cell membrane enriched subcellular fraction of pig myocardium. Acta Biol. Med. Germ. 31: K37–44.
5. KREBS, E. G. 1972. Protein kinases. Current Top. Cell. Regul. 5: 99–133.
6. KUKOVETZ, W. R., and POCH, G. 1972. The positive inotropic effect of cyclic AMP. Advan. Cycl. Nucl. Res. 1: 261–290.
7. KUO, J. P., and GREENGARD, P. 1969. Cyclic nucleotide-dependent protein kinase. IV. Widespread occurrence of adenosine 3′,5′-monophosphate-dependent protein kinase in various tissues and phyla of the animal kingdom. Proc. Nat. Acad. Sci. U. S. A. 64: 1349–1355.
8. LANGER, G. A. 1973. Heart: Excitation-contraction coupling. Ann. Rev. Physiol. 35: 55–86.
9. LINDENMAYER, G. E., SCHWARTZ, A., and THOMPSON , H. K., JR. 1974. A kinetic description for sodium and potassium effects on (Na^+ + K^+)-adenosine triphosphatase: A model for two-nonequivalent site potassium activation and an analysis of multiequivalent site models for sodium activation. J. Physiol. (London) 236: 1–28.
10. MATSUI, H., and SCHWARTZ, A. 1966. Purification and properties of a highly active ouabain-sensitive Na^+,K^+-dependent adenosine-triphosphatase from cardiac tissue. Biochim. Biophys. Acta 128: 380–390.
11. MORAD, M., and ROLLET, E. L. 1972. Relaxing effect of catecholamines on mammalian heart. H. Physiol. (London) 224: 537–558.
12. OGAWA, Y. 1968. The apparent binding constant of glycoletherdiaminetetraacetic acid for calcium at neutral pH. J. Biochem. 64: 25–257.
13. PALMER, R. P., and POSEY, V. A. 1967. Ion effects on Ca^{2+} accumulation by cardiac sarcoplasmic reticulum. J. Gen. Physiol. 50: 2085–2096.
14. PITTS, B. J. R., LANE, L. K., and SCHWARTZ, A. 1973. Nature of the transport ATPase-digitalis complex. VI. Purification of acid ouabain binding to a highly active cardiac Na^+,K^+-ATPase. Biochem. Biophys. Res. Commun. 53: 1060–1066.
15. RASMUSSEN, H. 1970. Cell communication, calcium ions and cyclic adenosine monophosphate. Science 170: 404–412.
16. REDDY, Y. S., BALLARD, D., GIRI, N. Y., and SCHWARTZ, A. 1973. Phosphorylation of cardiac native tropomyosin and troponin: Inhibitory effect of actomyosin

and possible presence of endogenous myofibrillar-located cyclic AMP-dependent protein kinase. J. Molec. Cell. Cardiol. 5: 461–471.
17. REITER, M. 1972. Differences in the inotropic cardiac effects of noradrenaline and dihydor-ouabain. Arch. Pharmacol. 275: 243–250.
18. ROBISON, J. D. 1974. The affinity of the (Na^+ + K^+)-dependent ATPase for Na^+ measured by Na^+ modified enzyme inactivation. Fed. Bur. Biochem. Soc. Lett. 38: 325–328.
19. STEIN, W. D. 1967. The Movement of Molecules Across Cell Membranes. Academic Press, New York.
20. WILL, H., SCHIRPKE, B., and WOLLENBERGER, A. 1973. Binding of calcium to a cell membrane-enriched preparations from pig myocardium: Increase in calcium affinity upon membrane protein phosphorylation enhanced by a membrane-bound cyclic AMP-dependent protein kinase. Acta Biol. Med. Germ. 31: K45–52.
21. WOLLENBERGER, A. 1975. The role of cyclic AMP in the adrenergic control of the heart. *In* W. G. Nayler (ed.), Contraction and Relaxation of the Myocardium. Academic Press, London and New York, in press.
22. WOLLENBERGER, A., and KRAUSE, E.-G. 1973. Cyclic nucleotides and the regulation of the heart beat, pp. 170–191. *In* R. A. Maxwell (ed.), Proceedings of the Fifth International Congress of Pharmacology, Vol. 5. Karger, Basel.
23. WOLLENBERGER, A., WILL, H., and KRAUSE, E.-G. 1975. Adenosine 3′,5′-monophosphate, the myocardial cell membrane, and calcium, pp. 81–93. Recent Advances in Studies on Cardiac Structure and Metabolism, Vol. 5. University Park Press, Baltimore.
24. WRAY, H. L., GRAY, R. R., and OLSSON, R. A. 1973. Cyclic adenosine 3′,5′-monophosphate-stimulated protein kinase and a substrate associated with cardiac sarcoplasmic reticulum. J. Biol. Chem. 248: 1496–1498.

Mailing address: Dr. A. Wollenberger,
Division of Cellular and Molecular Cardiology,
Central Institute of Heart and Circulatory Regulation Research,
Academy of Sciences of the DDR,
Berlin-Buch (German Democratic Republic).

Recent Advances in Studies on Cardiac Structure and Metabolism
Volume 9
The Sarcolemma
Edited by Paul-Emile Roy and Naranjan S. Dhalla

CORRELATION BETWEEN CHANGES IN ENDOGENOUS CATECHOLAMINE RELEASE FROM THE HEART AND VARIOUS PHYSIOLOGICAL RESPONSES IN ANESTHETIZED DOGS[1]

N. YAMAGUCHI, J. de CHAMPLAIN[2], and R. NADEAU[3]

Centre de Recherche en Sciences Neurologiques
Départment de physiologie, Faculté de médecine,
Université de Montréal,
Montréal, Québec, Canada

SUMMARY

Correlations between the concentration of endogenous catecholamines in coronary sinus blood and various physiological responses to adrenergic stimulation were studied in anesthetized dogs. Plasma catecholamine levels both in coronary sinus and aortic blood were measured by means of a modification of the radiometric-enzymatic assay for tissue catecholamines, and found to be 0.769 ± 0.110 (S.E.) and 0.972 ± 0.127 (S.E.) ng/ml, respectively, under basal conditions. Circulating catecholamine levels increased after the intracoronary injection of 6-hydroxydopamine, which has an indirect sympathomimetic action. Both the increment in catecholamine release from the heart and the intensity of physiological response were dose-related. A frequency-dependent increase in coronary sinus blood catecholamine levels occurred during right cardioaccelerator nerve stimulation. Maximum values were observed with a stimulation frequency of 10 Hz, at which maximum physiological responses were obtained. A highly significant correlation was found between endogenous catecholamine levels in coronary sinus blood and the dp/dt of left ventricular pressure (r = 0.750, $p < 0.001$), as well as mean coronary blood flow ($r = 0.706, p < 0.001$). The present preparation may prove to be a reliable means of studying the role of the sympathetic system in regulating cardiac function.

Correlations between the amount of endogenous adrenergic transmitter released and the intensity of physiological responses *in vivo* have been reported by a limited number of investigators (7, 9). This type of investigation has been

[1] This work was supported by grants from the Medical Research Council of Canada and from the Quebec Heart Foundation.
[2] Member of the Medical Research Council Group in Neurological Sciences at the University of Montreal.
[3] Associate of the Medical Research Council of Canada.

prevented mainly by the lack of sensitive and accurate methods for the measurement of circulating catecholamines.

The recent development of radiometric-enzymatic assay techniques has provided new means for the detection of small quantities of catecholamines (3–5). With such techniques it has become possible to design experiments in which changes in endogenous catecholamine released from the heart could be correlated with various parameters of the cardiac response after sympathetic stimulation or after injection of sympathomimetic drugs.

The present report demonstrates a close relationship between endogenous catecholamine levels in coronary sinus blood and the dp/dt of left ventricular pressure, mean coronary blood flow, and heart rate in anesthetized dogs.

MATERIALS AND METHODS

Preparation of dog

Fifteen mongrel dogs weighing 10–15 kg were anesthetized with sodium pentobarbital (35 mg/kg i.v.). Artificial respiration was maintained through a tracheal tube by means of a Harvard pump. Bilateral thoracotomy was performed at the level of the fourth intercostal space. The coronary sinus and left femoral artery were cannulated and connected with polyethylene tubes to the left jugular vein and to the distal right femoral artery, respectively. These extracorporeal bypasses permitted frequent collectons of blood samples through a three-way stopcock attached to each tubing. Sodium heparin (200 U/kg i.v. initially, and 100 U/kg every 2 hr) was administered.

Aortic pressure, left ventricular pressure, and its first derivative (dp/dt), heart rate, mean coronary blood flow in the left anterior descending coronary artery, and a peripheral lead electrocardiogram (aVR) were recorded with a Grass polygraph (model 7). Left ventricular pressure was measured by means of a catheter-tip manometer (Millar Mikro-Tip transducers, model PC-470) and mean coronary artery blood flow with an electromagnetic blood flowmeter (Statham model SP 2202). The first derivative of left ventricular pressure (dp/dt) was obtained by using a differentiator circuit patched on an analog computer (Electronic Associates, Inc., model TR-20). The response of this differentiator was linear (±5 percent) through 60 Hz/sec. The right stellate ganglion was dissected and the cardioaccelerator nerves distal to the stellate ganglion were stimulated with bipolar platinum electrodes at supramaximal intensity (10 V) for a period of 1 min (2 msec duration, 1, 3, 10, and 30 Hz/sec) using a Grass stimulator (model S-4).

For measurement of endogenous catecholamines, blood samples of 4 ml were collected simultaneously from the coronary sinus and aorta in about 30 sec. In the experiments with electrical stimulation, blood samples were collected during each stimulation period, *i.e.*, from 30 sec after the onset of stimulation to the end of stimulation. In the experiments involving 6-hydroxydopamine (6-OH-DA) (Labkemi AB, Göteborg, Stockholm), injections were made into the left anterior descending coronary artery through previously inserted fine polyethylene tubing by means of a hand-driven syringe in about 30 sec. Various concentrations of 6-OH-DA were prepared immediately before the injection in 1 ml of Ringer's solution. Blood samples were taken at 1, 3, 5, 10, and 30 min after the beginning of the injection.

Preparation of blood samples and catecholamine assay

Immediately after sampling, blood was centrifuged at −2°C, 10,000 r.p.m. for 20 min, and plasma proteins were removed by adding 2 N perchloric acid containing 1 percent EGTA and 2.0 μM magnesium chloride in a proportion of 1 ml of the acid to 9 ml of plasma. An aliquot of supernatant fluid (300 μl) obtained from the plasma-acid mixture was used for the catecholamine assay according to the method of Coyle and Henry (3). The principle of this technique is based on the conversion of noradrenaline and adrenaline to tritiated normetanephrine and metanephrine in the presence of catechol-O-methyltransferase and tritiated S-adenosylmethionine as a labeled methyl donnor. Tritiated vanillin formed from the normetanephrine by oxidation with sodium periodide was counted in a Packard liquid scintillation counter (Tri-Carb). The concentration of plasma catecholamines (ng/ml) was obtained by correcting with given amount of internal standard of noradrenaline. The assay procedure for noradrenaline showed linearity with various amounts of noradrenaline ranging from 0.1 to at least 20 ng. For statistical evaluation of the difference between various results, Student's *t* test was used.

RESULTS

Changes in circulating catecholamine levels produced by sympathetic cardiac nerve stimulation

The mean catecholamine levels obtained from 15 dogs under basal conditions were found to be 0.769 ± 0.110 (S.E.) and 0.972 ± 0.127 (S.E.) ng/ml of the plasma in coronary sinus and aortic blood, respectively. The difference between these two values was statistically significant ($p < 0.05$). The control circulating catecholamine levels, both in coronary sinus and aortic blood, were variable during a given experimental period; however, these variations were not statistically significant.

Circulating catecholamine levels in coronary sinus blood significantly increased 3–4-fold during the stimulation (10 Hz, 1 min) of the right cardiac accelerator nerves, whereas those in the aortic blood did not change significantly. This catecholamine overflow from the heart was reproducible upon repetition of the same stimulation three times at 60-min intervals (fig. 1).

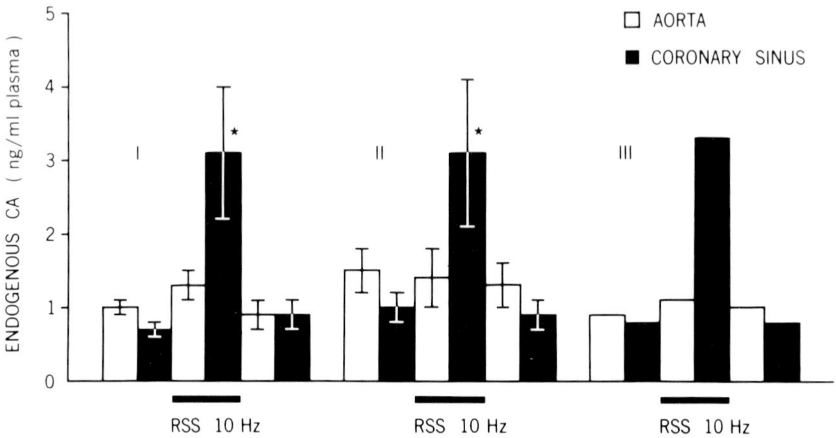

Figure 1 Circulating catecholamine levels in the coronary sinus (*black columns*) and aortic (*white columns*) blood. Blood samples were collected simultaneously from the coronary sinus and aorta. *Solid lines* under each section represent the stimulation of right cardio-accelerator nerves (10 Hz, 1 min, 2 msec, supramaximal voltage). The same stimulation was repeated three times with an interval of 60 min between each stimulation. Controls taken 2 min before and 30 min after the stimulation are shown for each stimulation. In *I* and *II*, each column represents the mean ± S.E.M. of values obtained in five dogs. In *III*, each column represents the mean values obtained in two dogs. The selective increase in endogenous catecholamine levels in the coronary sinus blood during the stimulation periods was significant ($p < 0.05$). *RSS* = represents the right stellate ganglion stimulation.

In a separate series of experiments, the right cardiac nerves were stimulated with different frequencies ranging from 1 to 30 Hz. Frequency-dependent increases of catecholamine levels in coronary sinus blood were obtained up to 10 Hz, but the values obtained at 30 Hz were lower than those observed at 10 Hz (fig. 2). Circulating catecholamine levels in aortic blood did not change significantly in these experiments.

Drug-induced changes in circulating catecholamines

In another group of experiments, changes in circulating catecholamine levels were studied by injecting 6-OH-DA at three different doses of 10, 100, and 1,000 μg directly into the left anterior descending coronary artery. Catecholamine release from the heart was observed after each injection of 6-OH-DA and was dose dependent (fig. 3). The amount of catecholamine released was maximum at 1 min after each injection. Increased catecholamine levels in coronary sinus blood were maintained at a significantly higher level ($p < 0.05$) up to 10 min after injections of 1,000 μg.

In the aortic blood, circulating catecholamine levels were also augmented, showing a maximum increase at 1 min after the injection of 1,000 μg of

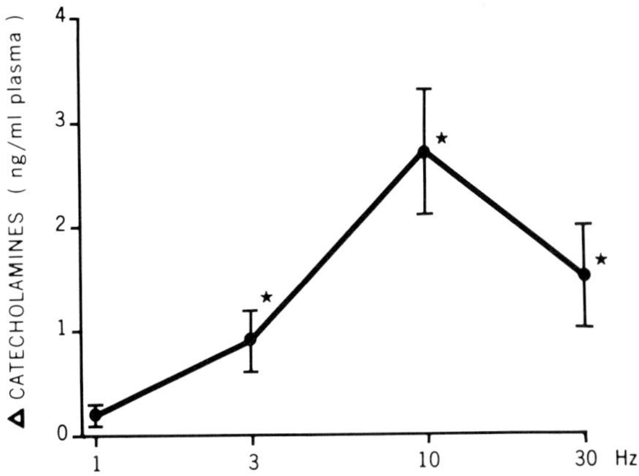

Figure 2 Frequency-response curve of endogenous catecholamine released from the heart into coronary sinus blood during electrical stimulation of the right cardioaccelerator nerves. Each point represents the mean ± S.E.M. of values obtained in five dogs expressed as the increment above control levels (*, $p < 0.05$).

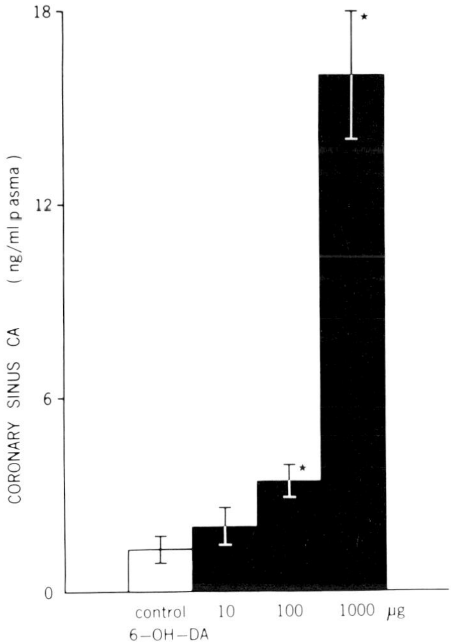

Figure 3 The effect of various doses of 6-OH-DA injected into the left anterior descending coronary artery. Each column represents the mean ± S.E.M. values of circulating catecholamine levels (*CA*) in coronary sinus blood obtained in five dogs. *Black columns* represent catecholamine levels 1 min after the injections (*, $p < 0.01$). Each injection period lasted 30 sec.

6-OH-DA. Significantly increased levels of catecholamines ($p < 0.05$) were sustained until 3 min after the injection.

Correlations between changes in catecholamine release from heart and physiological responses

Significant increases in left ventricular pressure and its dp/dt, mean coronary blood flow, and heart rate were obtained by either stimulations of right cardioaccelerator nerves (10 Hz, 1 min), or injections of 6-OH-DA (100, 1,000 μg). Linear increases in dp/dt of left ventricular pressure and mean coronary blood flow were observed up to a stimulation frequency of 10 Hz, but the responses to 30 Hz were less than those obtained at 10 Hz. The frequency-response curve of changes in heart rate was found to be hyperbolic. Changes in catecholamine levels in coronary sinus blood produced by cardioaccelerator nerve stimulations and injections of 6-OH-DA correlated well with changes in dp/dt of left ventricular pressure and mean coronary blood flow (figs. 4, 5). Correlation between the increment in heart rate and catecholamine levels was also significant ($p \leqslant 0.05$)

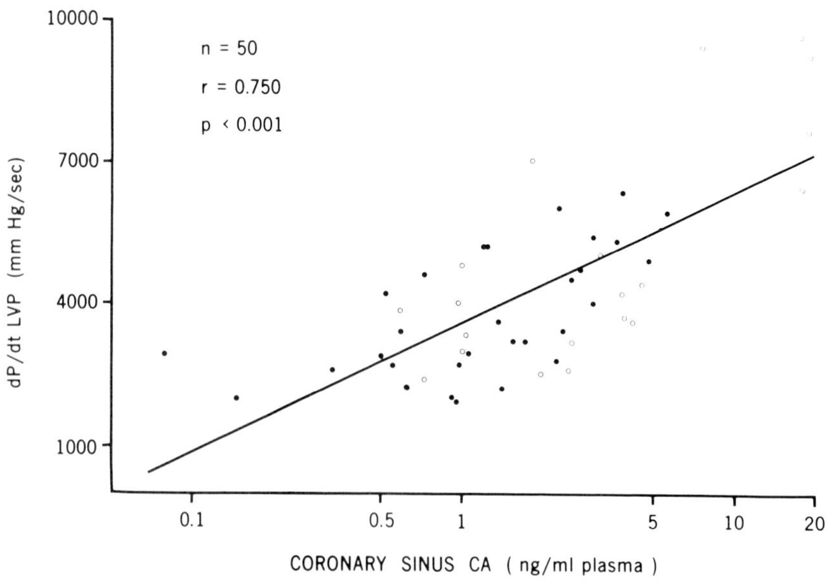

Figure 4 Correlation between endogenous catecholamine levels released from the heart into coronary sinus blood and the corresponding dp/dt of left ventricular pressure. Changes in catecholamine levels in coronary sinus and in dp/dt were produced by electrical stimulations of the right cardioaccelerator nerves (1–30 Hz, 1 min, •) and injections of 6-OH-DA (10, 100, 1,000 μg, ○) into the left anterior descending coronary artery. The points shown in this figure include control values and the peak values obtained in 11 dogs during nervous stimulation and after 6-OH-DA injections. The changes in endogenous catecholamine levels are expressed on a logarithmic scale ($r = 0.750$, $p < 0.001$, $n = 50$).

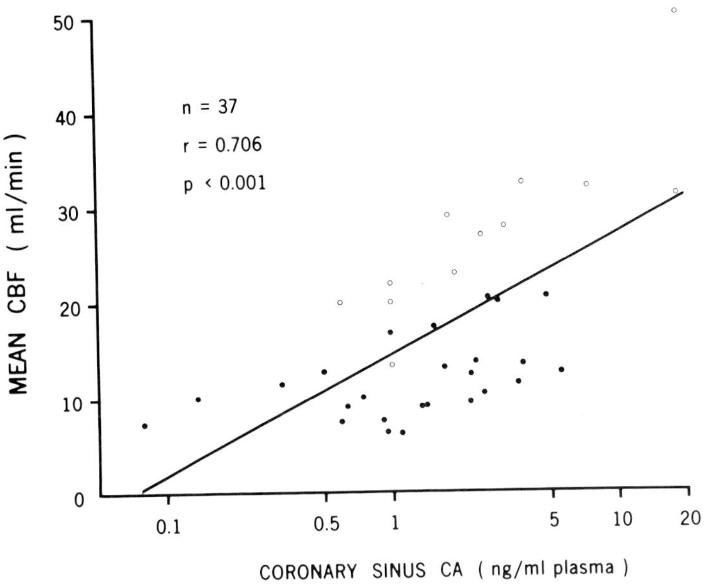

Figure 5 Correlation between endogenous catecholamine levels released from the heart in coronary sinus blood and mean coronary blood flow (*CBF*). The points shown in this figure include control values, the peak values obtained in eight dogs during nervous stimulation (•), and values after injection of 6-OH-DA (○). The changes in endogenous catecholamine levels are expressed on a logarithmic scale ($r = 0.706$, $p < 0.001$, $n = 37$).

but the correlation coefficient was less than those obtained with the other cardiac responses ($r = 0.385$, $n = 20$).

DISCUSSION

The determinations of small amounts of circulating catecholamines were obtained by using an adaptation of the method which Coyle and Henry (3) developed for the measurement of tissue catecholamines. With this technique, noradrenaline and adrenaline cannot be separated since both of them are β-hydroxylated catecholamines resulting in the same final reaction product, [*methyl*-^{3}H] vanillin (3). However, this does not diminish the significance of the present findings since sympathetic nerve endings release noradrenaline when they are activated (10) and since the mean value of circulating adrenaline was reported to be negligible (2, 5).

The ability of the heart to take up and store exogenously administered noradrenaline has been amply demonstrated (1, 11, 12). Moreover, it has been observed that a more densely innervated tissue takes up more noradrenaline per unit of weight (14). In the present experiments, circulating catecholamine levels in the coronary sinus blood were lower than those of aortic blood under basal conditions. This difference is probably related to the ability of the sympathetic fibers within the heart to take up circulating noradrenaline. This explanation is supported by the observation in perfused rabbit or rat hearts that more than 40 percent of infused noradrenaline is removed by the heart (10).

In the present experiments with stimulations at various frequencies ranging from 1 to 30 Hz, responses in dp/dt of left ventricular pressure and mean coronary blood flow increased linearly up to 10 Hz, whereas heart rate increased according to a hyperbolic curve, as shown previously by others (6). The frequency-response curve of catecholamine levels in coronary sinus blood was exponential up to a stimulation frequency of 10 Hz. This difference in the type of frequency-response curve between heart rate and catecholamine levels in coronary sinus blood, *i.e.*, a slight increase in coronary sinus catecholamine levels at 3 Hz along with an important response in heart rate (approximately 90 percent of the maximum response), is probably related to the fact that noradrenaline is released directly at the sinus node, whereas only the overflow is measured in the coronary sinus blood. The correlation coefficient of linear regression obtained with changes between circulating catecholamine levels in coronary sinus blood and heart rate was less than those obtained with the other cardiac parameters. This can be explained by the difference between the two types of frequency-response curves.

Small doses of 6-OH-DA produce short lasting, dose-related indirect sympathomimetic effects resulting from the displacement of noradrenaline from storage sites (8, 13). Since dopamine does not cross-react with noradrenaline to any marked degree in the present method (3), it can be presumed that injection of 6-OH-DA did not interfere with noradrenaline assay in the present experiments.

In conclusion, a close relationship between the changes in endogenous catecholamine release from heart as measured in coronary sinus blood and various physiological responses was obtained in anesthetized dogs. The functional response which correlated most closely with the catecholamine levels was found to be the dp/dt of the left ventricular pressure. This approach appears promising for a better understanding of the role of the sympathetic system in the control of cardiac function.

Acknowledgments The authors wish to acknowledge the skillful technical assistance of Miss Lise Farley, Mrs. Jo-Anne LeGuerrier, and Mrs. Rollande Caron-Benoît.

REFERENCES

1. AXELROD, J., WEIL-MALHERBE, H., and TOMCHICK, R. 1959. The physiological disposition of H^3-epinephrine and its metabolite metanephrine. J. Pharmacol. Exp. Therap. 127: 251–256.

2. CHRISTENSEN, N. J. 1973. Plasma noradrenaline and adrenaline in patients with thyrotoxicosis and myoedema. Clin. Sci. Molec. Med. 45: 163–171.
3. COYLE, J. T., and HENRY, D. 1973. Catecholamines in fetal and newborn rat brain. J. Neurochem. 21: 61–67.
4. ENGELMAN, K., PORTNOY, B., and LOVENBERG, W. 1968. Sensitive and specific double-isotope derivative method for the determination of catecholamines in biological specimens. Amer. J. Med. Sci. 255: 259–268.
5. ENGELMAN, K., and PORTNOY, B. 1970. A sensitive double-isotope derivative assay for norepinephrine and epinephrine. Circ. Res. 26: 53–57.
6. FOLKOW, B., LÖFVING, B., and MELLANDER, S. 1956. Quantitative aspects of the sympathetic neuro-hormonal control of the heart rate. Acta Physiol. Scand. 37: 363–369.
7. GAFFNEY, T. E., CHIDSEY, C. A., and BRAUNWALD, E. 1963. Study of the relationship between the neurotransmitter store and adrenergic nerve block induced by reserpine and guanethidine. Circ. Res. 12: 264–268.
8. HAEUSLER, G. 1971. Early pre- and post- junctional effects of 6-hydroxydopamine. J. Pharmacol. Exp. Therap. 178: 49–62.
9. HARRISON, D. C., CHIDSEY, C. A., GOLDMAN, R., and BRAUNWALD, E. 1963. Relationship between the release and tissue depletion of norepinephrine from the heart by guanethidine and reserpine. Circ. Res. 12: 256–263.
10. IVERSEN, L. L. 1967. The Uptake and Storage of Noradrenaline in Sympathetic Nerves. Cambridge University Press, Cambridge, England.
11. RAAB, W., and GIGEE, A. B. 1955. Specific avidity of the heart muscle to absorb and sotre epinephrine and norepinephrine. Circ. Res. 3:553–558.
12. STRÖMBLAD, B. C. R., and NICKERSON, M. 1961. Accumulation of epinephrine and norepinephrine by some rat tissues. J. Pharmacol. Exp. Therap. 134: 154–159.
13. THOENEN, H., and TRANZER, J. P. 1973. The pharmacology of 6-hydroxydopamine. Ann. Rev. Pharamcol. 13: 169–180.
14. TRENDELENBURG, U. 1969. The pharmacological importance of the uptake mechanism for sympathomimetic amines. Progr. Brain. Res. 31: 73–85.

Mailing address: Dr. R. Nadeau,
Département de physiologie,
Faculté de médecine,
Université de Montréal,
Montréal, Québec (Canada).

Recent Advances in Studies on Cardiac Structure and Metabolism
Volume 9
The Sarcolemma
Edited by Paul-Emile Roy and Naranjan S. Dhalla

METABOLIC ALTERATIONS FOLLOWING CHEMICAL SYMPATHECTOMY WITH 6-HYDROXYDOPAMINE IN THE RAT[1]

A. IMBACH

Département de physiologie, Faculté de médecine,
Université de Montréal,
Montréal, Québec, Canada

SUMMARY

Until recently the role of the sympathetic nervous system on intermediary metabolism has been difficult to study because of the lack of efficient methods of producing generalized sympathectomy. Various metabolic parameters were measured 24 hr, 72 hr, and 1 week after chemical sympathectomy produced by a single intravenous injection of 6-hydroxydopamine (100 mg/kg) in rats. All animals were fasted 16 hr previous to all measurements. A marked increase was found in blood glucose levels in 6-OH-DA-treated animals. Serum free fatty acids (FFA) were 2½ times higher 24 hr after sympathectomy than in control animals, but returned to normal values thereafter. The glycogen content of the liver was 10 times greater in treated animals 24 hr after 6-OH-DA, whereas cardiac glycogen was more than doubled 72 hr after sympathectomy. In contrast, glycogen content of skeletal muscle was only slightly but insignificantly increased. Immunoreactive insulin levels were also increased by 70 percent in sympathectomized animals after 24 hr, but returned to normal values thereafter. These results indicate an important alteration of carbohydrate and lipid metabolism after chemical sympathectomy, thus supporting a role of the sympathetic nervous system in their regulation.

The role of the sympathetic nervous system on intermediary metabolism is very incompletely understood and has been difficult to study until recently because of the lack of efficient methods of producing generalized sympathectomy.

Until now, the role of the sympathetic fibers on intermediary metabolism has been investigated with catecholamine-depleting agents, such as reserpine, which affects other functions and mechanisms unrelated to the sympathetic system (6, 36). Recently 6-hydroxydopamine (6-OH-DA) has been proposed as a new method of producing a sympathectomy because of its unique capacity to induce selective degenerative lesions in terminal adrenergic nerve fibers (33, 34).

This method is used in the present study to see how carbohydrate and lipid metabolism is affected after chemical sympathectomy with 6-OH-DA, clarifying the role of the sympathetic nervous system in their regulation.

[1] This work was supported by a grant from the Medical Research Council of Canada.

METHODS

One hundred ten male Sprague-Dawley rats weighing 200 gm were used in this study. All animals were fed a regular laboratory diet of Purina chow.

Chemical sympathectomy was induced by a single intravenous injection of 100 mg/kg of 6-hydroxydopamine (Labkemi AB, Göteborg, Sweden). These animals were compared with a control group of untreated rats and to adrenalectomized and medullectomized animals. Bilateral adrenalectomy and medullectomy were performed under ether anesthesia. All adrenalectomized animals received 1 percent solution to drink *ad libitum* and half received a daily subcutaneous injection of 100 μg of cortisone acetate suspension (Merck, Sharpe and Dohme, Canada).

Rats were fasted 16 hr before all measurements. Values were obtained 24 hr, 72 hr, and 1 week after the initial preparation. To minimize seasonal changes, each series of experiments included animals of different groups.

Analytical procedures

Blood glucose, serum triglycerides and serum lactate were determined with the use of optical enzymatic tests (Stat-Pack, Calbiochem, LaJolla, Calif.) developed from methods described previously (1, 9, 18). Serum free fatty acids were measured by the method of Duncombe (8). Glycogen content of liver, heart, and skeletal muscle was determined on homogenates by the anthrone method (4). The radioimmunoassay of insulin was based on the double antibody method of Morgan and Lazarow as adapted by Soeldner and Slone (31).

The statistical significance of the difference between various results was estimated by variance analysis.

RESULTS

Effect of chemical sympathectomy on blood glucose and lactate

There is a marked increase in blood glucose after chemical sympathectomy with 6-OH-DA. Twenty-four hours after injection, blood glucose reaches 147 mg/100 ml, compared with 76 mg/100 ml in control animals. The hyperglycemia persists 72 hr after the injection and values are still 70 percent higher than in untreated animals. After 1 week, blood glucose is still slightly elevated but tends to return to normal values.

Medullectomy and adrenalectomy, whether or not compensated by a physiological dose of cortisone, lead to lower levels of blood glucose 24 hr after surgery. Two days later, values are normal (fig. 1).

Blood lactate is slightly but insignificantly increased in all groups after 24 and 72 hr. After 1 week, values are almost doubled in adrenalectomized animals receiving cortisone, but remain unchanged in other groups.

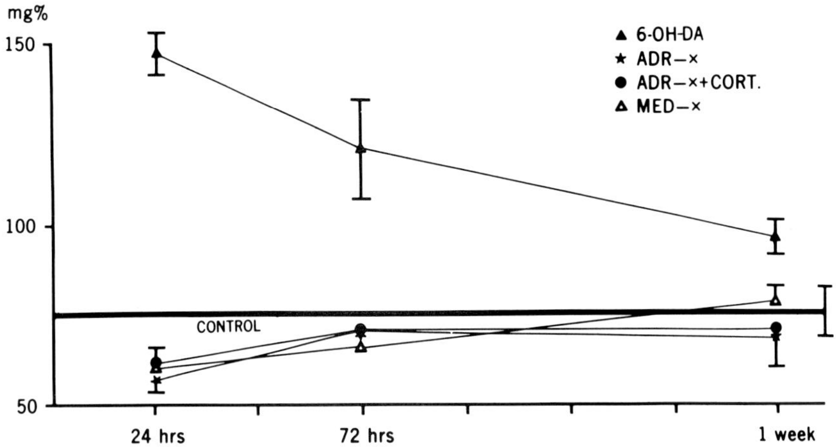

Figure 1 Blood glucose levels of rats 24 hr, 72 hr, and 1 week after sympathectomy with 6-OH-DA (100 mg/kg), adrenalectomy + 1 percent NaCl, adrenalectomy + 1 percent NaCl + 100 μg of cortisone acetate, and medullectomy. Control value is represented by a *solid line.* Each point represents the mean ± S.E. of 8–11 animals.

Effect of 6-OH-DA injection on serum FFA and triglycerides

Serum free fatty acids are 2½ times higher 24 hr after chemical sympathectomy than in control animals, but decrease to normal values thereafter. FFA values double in medullectomized and adrenalectomized animals treated with cortisone; they are 4 times higher 1 week after surgery. In untreated adrenalectomized animals, serum FFA values are normal at 24 hr, rising slowly to reach 3 times normal after 1 week (fig. 2).

Circulating triglycerides are only slightly elevated 24 hr after sympathectomy and become significantly lower than control value after 72 hr. They return to normal values after 1 week. In all other groups, serum triglycerides are twice as high at 24 hr as in the untreated animals. Values are still elevated 72 hr after adrenalectomy and medullectomy, but become normal in the group receiving cortisone.

Changes in glycogen storage after chemical sympathectomy

The liver glycogen level is 10 times higher than control values 24 hr after sympathectomy. This parameter returns to normal 72 hr after the injection of 6-OH-DA. In all other groups, liver glycogen is slightly lower than in untreated animals at all times (fig. 3).

The glycogen content of the heart is almost 50 percent lower at 24 hr in sympathectomized animals than in the control group. However, 2 days later, the

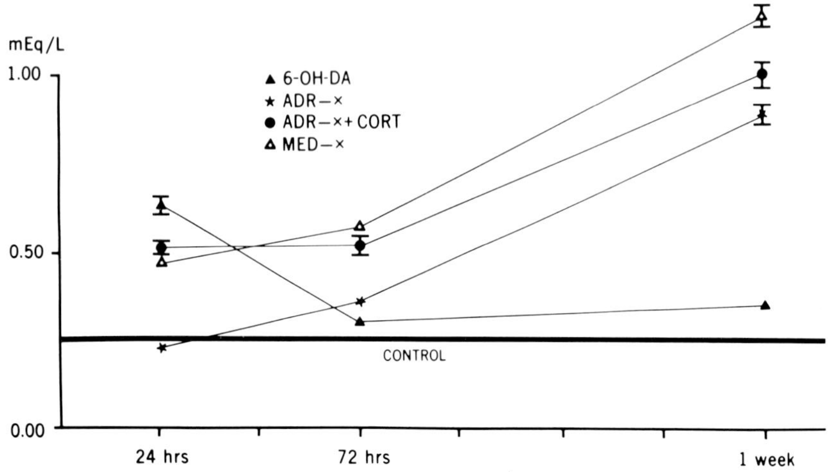

Figure 2 Serum free fatty acids levels after sympathectomy with 6-OH-DA, adrenalectomy + 1 percent NaCl, adrenalectomy + 1 percent NaCl + cortisone acetate, and medullectomy. Control value is represented by a *solid line.* Each point represents the mean ± S.E. of 8–11 animals.

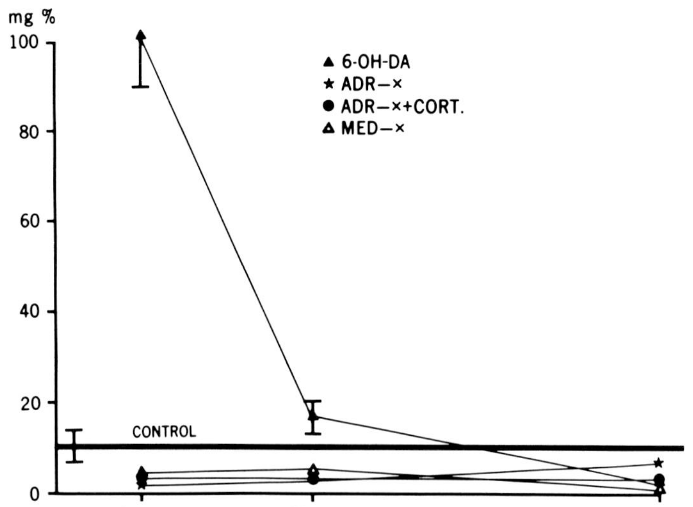

Figure 3 Liver glycogen content of rats 24 hr, 72 hr, and 1 week after sympathectomy with 6-OH-DA, adrenalectomy, and medullectomy. Control value is represent by a *solid line.* Each point represents the mean ± S.E. of 8–11 animals.

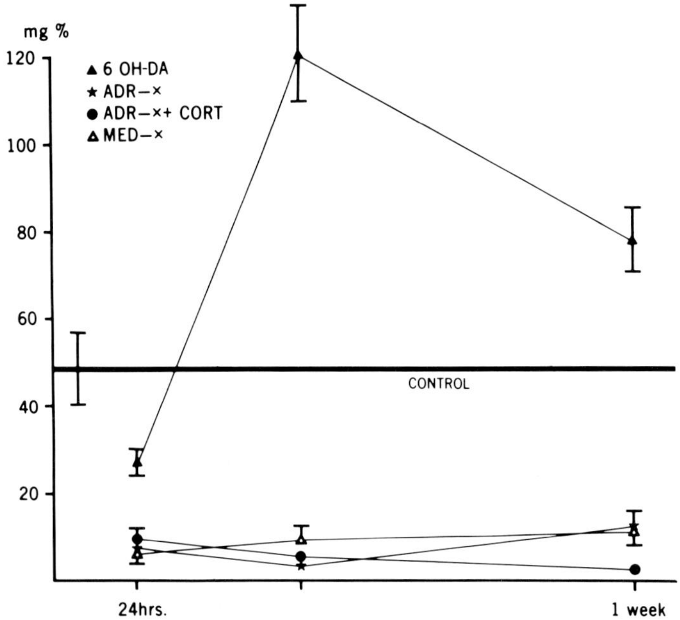

Figure 4 Heart glycogen content of rats 24 hr, 72 hr, and 1 week after sympathectomy with 6-OH-DA, adrenalectomy, and medullectomy. Control value is represented by a *solid line*. Each point represents the mean ± S.E. of 8–11 animals.

heart glycogen is markedly increased to 2½ times the normal value. One week after the injection of 6-OH-DA, the glycogen content of the heart is still 60 percent higher. In adrenalectomized and medullectomized animals, the heart glycogen is depleted severely at all times (fig. 4).

In skeletal muscle, glycogen stores are not altered 24 hr after sympathectomy. Levels are only slightly but insignificantly increased 72 hr after sympathectomy and decreased after 1 week. Adrenalectomy did not alter significantly the glycogen content of skeletal muscle. Values obtained were lower in medullectomized and adrenalectomized animals treated with cortisone.

Serum insulin in sympathectomized animals

Control values of serum insulin in the fasted rat was found to be 7.42 ± 0.92 μU/ml. Twenty-four hours after chemical sympathectomy with 6-OH-DA, serum insulin rises to 12.72 ± 2.12 μU/ml, a 70 percent increase. Values were normal 72 hr and 1 week after injection with 6-OH-DA. Adrenalectomy and medullectomy are followed by a decrease in serum insulin of about 50 percent.

DISCUSSION

Administration of 6-OH-DA was used to produce peripheral sympathectomy since it was established that this compound destroys specifically the adrenergic nerve endings (6, 33, 34) and depletes peripheral tissue stores of norepinephrine (NE) markedly (30). Catecholamine-secreting cells of the adrenal medulla are, however, unaffected by chemical sympathectomy (25), and it was found that the adrenal medulla has the ability to compensate for the role of the sympathetic fibers destroyed by 6-OH-DA (7). This is also in agreement with the reported increase in tyrosine hydroxylase activity in the adrenal medulla of animals treated with 6-OH-DA (25).

This compensatory role of the adrenal medulla can account for the hyperglycemia and the important mobilization of FFA observed in this study 24 hr after the injection of 6-OH-DA. The hyperglycemic effect of epinephrine is well documented (32), as is the lipolytic effect of catecholamines on adipose tissue (2). In fact, there is 3 times more epinephrine excreted in the urine of rats treated with 6-OH-DA than in the urine of control rats, on the day after the injection. One week later there is still 2½ times more epinephrine excreted by the sympathectomized animals (21), thus suggesting an increase in circulating catecholamines after chemical sympathectomy.

The persistence of high blood glucose in the week after sympathectomy could be attributed to the supersensitivity of adrenergic receptors after 6-OH-DA, as reported previously (3, 13, 26).

In spite of the presumed increase in circulating epinephrine, the serum FFA levels return to normal 72 hr after sympathectomy. This lowering could probably be explained by the counteracting effect of hyperglycemia on mobilization of FFA since it has been reported that the stimulation of FFA release by catecholamines is in fact diminished in the presence of glucose (11, 15).

Serum triglycerides were never significantly elevated after 6-OH-DA and we did not observe any significant increase in tissue triglycerides as could be expected at least 24 hr after sympathectomy when serum FFA levels are 2½ times higher than normal.

The high glycogen content of the liver, 24 hr after injection of 6-OH-DA is not consistent with the glycogenolytic effect of epinephrine resulting from a compensation of the adrenal medulla. Far from the expected depletion, liver glycogen was found to be 10 times increased compared with the untreated animals. This increase could be in part attributed to the rise in insulin levels which favor glycogenesis (35). We cannot eliminate the possibility of a compensation by the adrenal cortex liberating glucocorticoids in response to the alerting reaction which could be triggered by the sympathectomy. Glucocorticoids would also favor glycogenesis and gluconeogenesis (35), which could account in part for the observed hyperglycemia. The weight loss of about 10 percent 24 hr after administration of 6-OH-DA could also be the reflection of an important catabolic metabolism taking place during that period.

The increase in immunoreactive insulin levels 24 hr after 6-OH-DA administration is not compatible with the postulated increase in circulating epinephrine. Infused epinephrine has been reported to inhibit the extra insulin secretion otherwise induced by hyperglycemia (19, 28). This inhibition on insulin release appears to be mediated via the α-adrenergic receptors of the pancreas (24, 27), whereas β-adrenergic receptors stimulate insulin release (22, 23). However, the inhibition of epinephrine on insulin release can be partially overcome by a sufficiently strong stimulus (28), which is perhaps the case 24 hr after chemical sympathectomy.

One cannot rule out the possibility that the sympathetic nervous fibers influence directly the release of insulin. Noradrenergic nerve fibers have been visualized by histofluorescence method in the islets of Langerhans, in close relationship to α and β cells (10). It was demonstrated recently that stimulation of the pancreatic nerves increases insulin output by a parasympathetic mechanism since it is blocked by atropine. Stimulation of the same nerve after atropinization shows a sympathetic effect in which glucose-stimulated insulin is inhibited. This suggests both a parasympathetic and sympathetic link between the central nervous system and the endocrine pancreas (20, 29).

The important increase in heart glycogen 3 days after sympathectomy with 6-OH-DA which persists after 1 week seems to be a specific effect of sympathectomy on the heart since similar observations have been made after surgical sympathectomy (5, 17) and reserpine treatment (12). After sympathectomy catecholamines are almost completely depleted in the heart. The norepinephrine content of the heart is reduced by 92 percent 72 hr after 6-OH-DA (16). This depletion is not associated with ultrastructural changes in cardiac muscle cells (14), and there has been no evidence so far of functional changes attributable to the increased glycogen content of the heart.

The mechanism by which cardiac glycogen is increased after sympathectomy is unknown at the present. Previous studies failed to demonstrate changes in transferase and phosphorylase activities after cardiac denervation (5). An increased rate of glycogen synthesis or a diminished glycogenolysis or both were suggested as an explanation (12). However, the I form of glycogen synthetase activity decreases after sympathectomy and under aerobic conditions phosphorylase *a* activity is low (5, 12).

In conclusion, therefore, metabolic alterations after chemical sympathectomy with 6-OH-DA can be attributed partially to the ensuing compensation of the adrenal medulla. A possible compensatory role played by the adrenal cortex is also suggested since the important increase in liver glycogen content is not compatible with the known metabolic effects of medullary hormones. The effect of sympathectomy on the endocrine pancreas deserves further investigation since interesting changes were observed in insulin release after administration of 6-OH-DA. The rise in heart glycogen could be a specific effect of sympathectomy on the heart. Further studies are needed to clarify the mechanism by which cardiac glycogen is increased after sympathectomy and the functional implication of such an increase on cardiac performance.

REFERENCES

1. BERGEMEYER, H.-U. 1963. Methods of Enzymatic Analysis. Academic Press, New York.
2. BUTCHER, R. W. 1966. Cyclic 3′,5′-AMP and the lipolytic effects of hormones in adipose tissue. Pharmacol. Rev. 18: 237–241.
3. BRUS, R., HESS, M. E., and JACOBOWITZ, D. 1970. Effect of 6-hydroxydopamine and thyroxine on chronotropic response to norepinephrine. Europ. J. Pharmacol. 10: 323–327.
4. CARROL, N. V., LONGLEY, R. W., and ROE, J. H. 1956. The determination of glycogen in liver and muscle by use of anthrone reagent. J. Biol. Chem. 220: 583.
5. DAW, J. C., and BERNE, R. M. 1967. Effect of sympathectomy on cardiac UDPG-glycogen transferase activity in the cat. Amer. J. Physiol. 213: 1480–1484.
6. de CHAMPLAIN, J., and NADEAU, R. 1971. 6-Hydroxydopamine, 6-hydroxydopa and degeneration of adrenergic nerves. Fed. Proc. 30: 877–885.
7. de CHAMPLAIN, J., and van AMERINGEN, M. R. 1972. Regulation of blood pressure by sympathetic fibers and adrenal medulla in normotensive and hypertensive rats. Circ. Res. 31: 617–628.
8. DUNCOMBE, W. G. 1964. The colorimetric micro-determination of non-esterified fatty acids in plasma. Clin. Chim. Acta 9: 122.
9. EGGSTEIN, M., and KREUTZ, F. H. 1966. Eine neue Bestimmung der Neutralfette im Blutserum und Gewebe. Klin. Wschr. 44: 262–267.
10. ESTERHUIZEN, A. C., SPRIGGS, T. L. B., and LEVER, J. D. 1968. Nature of islet-cell innervation in the cat pancreas. Diabetes 17: 33–36.
11. FEIGELSON, E. W., PFAFF, W., KARMEN, A., and STEINBERG, D. 1961. The role of plasma free fatty acids in development of fatty liver. J. Clin. Invest. 40: 2171.
12. GERCKEN, G., and DORING, V. 1973. The metabolism of the anoxic heart after surgical denervation or reserpine pretreatment. J. Molec. Cell. Cardiol. 5: 275–286.
13. HAEUSLER, G., HAEFELY, W., and THOENEN, H. 1969. Chemical sympathectomy of the cat with 6-OH-dopamine. J. Pharmacol. Exp. Therap. 170: 50–61.
14. HAGOPIAN, M., GERSHON, M. D., and NUNEZ, E. A. 1973. Ultrastructural changes in cardiac muscle cells after administration of reserpine, parachlorophenylalanine, or 6-hydroxydopamine. Lab. Invest. 29: 562–569.
15. HAVEL, R. J., and GOLDFIEN, A. 1959. The role of the sympathetic nervous system in the metabolism of free fatty acids. J. Lipid Res. 1: 102.
16. JACKS, B. R., de CHAMPLAIN, J., and CORDEAU, J. P. 1972. Effects of 6-hydroxydopamine on putative transmitter substances in the central nervous system. Eur. J. Pharmacol. 18: 353–360.
17. JELLINEK, M., KAYE, M. P., NIGH, C. A., and COOPER, T. 1964. Alterations in chemical composition of canine heart after sympathetic denervation. Amer. J. Physiol. 206: 971–974.
18. KORNBERG, A., and HORECKER, B. L. 1955. Glucose-6-phosphate dehydrogenase. Methods Enzymol. 1: 323–324.
19. KRIS, A. O., MILLER, R. E., WHERRY, F. E., and MASON, J. W. 1966. Inhibition of insulin secretion by infused epinephrine in rhesus monkeys. Endocrinology 78: 87–97.

20. LEBOVITZ, H. E., and FELDMAN, J. M. 1973. Pancreatic biogenic amines and insulin secretion in health and disease. Fed. Proc. 32: 1797–1802.
21. LEDUC, J. 1974. Oxygen consumption and urinary excretion of catecholamines in sympathectomized rats exposed to cold. Proc. Can. Fed. Biol. Soc. 17: 41.
22. LUNDQUIST, I. 1972. Interaction of amines and aminergic blocking agents with blood glucose regulation. I. β-Adrenergic blockade. Eur. J. Pharmacol. 18: 213–224.
23. LUNDQUIST, I. 1972. Interaction of amines and aminergic blocking agents with blood glucose regulation. II. α-Adrenergic blockade. Eur. J. Pharmacol. 18: 225–235.
24. MAYHEW, D. A., WRIGHT, P. H., and ASHMORE, J. 1969. Regulation of insulin secretion. Pharmacol. Rev. 21: 183–212.
25. MUELLER, R. A., THOENEN, H., and AXELROD, J. 1969. Adrenal tyrosine hydroxylase: compensatory increase in activity after chemical sympathectomy. Science 163: 468.
26. NADEAU, R. A., de CHAMPLAIN, J., and TREMBLAY, G. M. 1971. Supersensitivity of the isolated rat heart after chemical sympathectomy with 6-hydroxydopamine. Can. J. Physiol. Pharmacol. 49: 36–44.
27. PORTE, D. 1969. Sympathetic regulation of insulin secretion. Arch. Intern. Med. 123: 252–260.
28. PORTE, D., GRABER, A. L., KUZUYA, T., and WILLIAMS, R. H. 1966. The effect of epinephrine on immunoreactive insulin levels in man. J. Clin. Invest. 45: 228–236.
29. PORTE, D., and ROBERTSON, R. P. 1973. Control of insulin secretion by catecholamines, stress, and the sympathetic nervous system. Fed. Proc. 32: 1792–1796.
30. PORTER, C. C., TOTARO, A., and STONE, C. A. 1963. Effects of 6-OH-DA and some other compounds on the concentration of norepinephrine in the hearts of mice. J. Pharmacol. Exp. Therap. 140: 308–316.
31. SOELDNER, J. S., and SLONE, D. 1965. Critical variables in the radioimmunoassay of serum insulin using the double antibody technic. Diabetes 14: 771.
32. SUTHERLAND, E. W., and ROBISON, G. A. 1966. The role of cyclic-3′,5′-AMP in responses to catecholamines and other hormones. Pharmacol. Rev. 18: 145–161.
33. THOENEN, H., and TRANZER, J. P. 1968. Chemical sympathectomy by selective destruction of adrenergic nerve endings with 6-hydroxydopamine. Arch. Pharmakol. Exp. Pathol. 261: 271.
34. TRANZER, J. P., and THOENEN, H. 1968. An electron microscopic study of selective acute degeneration of sympathetic nerve terminals after administration of 6-hydroxydopamine. Experientia 24: 155.
35. WRIGHT, S. 1973. Physiologie appliquée à la médecine, Ed. 12. Flammarion Médecine-Sciences, Paris.
36. ZAIMIS, E. 1964. Pharmacology of the autonomic nervous system. Ann. Rev. Pharmacol. 4: 365.

Mailing address: Dr. A. Imbach,
Département de physiologie, Faculté de médecine,
Université de Montréal,
Montréal, Québec (Canada).

Recent Advances in Studies on Cardiac Structure and Metabolism
Volume 9
The Sarcolemma
Edited by Paul-Emile Roy and Naranjan S. Dhalla

CAROTID BODY CONTROL OF CORONARY FLOW, MYOCARDIAL OXIDATIVE METABOLISM, AND CARDIAC CATECHOLAMINES IN THE DOG[1]

R. LIMET, E. CHABI,
K. M. A. WELCH, and J. H. KENNEDY

Taub Laboratories for Mechanical Circulatory Support,
Cora and Webb Mading Department of Surgery, and Neurology Department,
Baylor College of Medicine,
Houston, Texas, USA

SUMMARY

In a series of 11 dogs, selective stimulation of the carotid body receptors by hypoxic, hypercapnic, acidotic blood produces bradycardia, an increase of coronary flow, and greater release of norepinephrine from the heart; the coronary resistances are decreased. The same stimulation after vagotomy is no longer accompanied by bradycardia; the decrease of coronary resistance is less marked and the release of norephinephrine is increased.

Since the work of Heymans and Neil (14) on the effects of hypoxic stimulation of the carotid body upon the respiratory centers, the physiological importance of that structure has constantly been reappraised as the centers can be directly excited by the changes causing peripheral chemoreceptor stimulation. It seems well established that, during pure hypoxemia, carotid chemoreceptors activate the respiratory centers before there is time for excitation because of the changes in brain metabolism resulting from hypoxia. Thus, the role of carotid body appears to be a supplementary safety factor responsible for hypertachypnea before the brain becomes hypoxic.

In contrast with the respiratory effects of carotid body function, the study of the cardiovascular responses has produced many conflicting reports. Although an increase in vagal efferent activity is acknowledged by most authors, effects upon vascular resistances have been variable. Authors have found no change (21),

[1] This work was supported in part by Grant HL-13330-04 from the National Heart and Lung Institute and in part by Grant NS-09287 from the National Institute of Neurological Diseases and Stroke and by Grant RR 00350 from the General Clinical Research Centers Branch, Division of Research Resources, National Institutes of Health, Bethesda, Maryland.

increase (8, 11, 17), or decrease (5) of the vascular resistances. If one is concerned with the effects of carotid body stimulation upon the coronary flow, one has to deal not only with the direct effects of carotid body stimulation upon the coronary resistance, but also with the indirect changes resulting from modifications in heart rate, cardiac output, contractility, and afterload, and, hence, in myocardial metabolism. With respect to myocardial inotropism, carotid body stimulation has been diversly reported to have unaltered (23) or decreased (9, 11) contractility; the decrease either persists (11) or changes to an increease (9) after bilateral vagotomy. As far as the direct effects upon coronary tone are concerned, recent studies (1, 12) have reported that pharmacological stimulation of the carotid chemoreceptors resulted in vagal coronary vasodilation. In order to assess the physiological importance of that observation, we investigated the action of carotid body stimulation upon coronary resistance, using the "natural" carotid body stimulus: perfusion by hypoxic, hypercapnic, and acidotic blood.

METHODS

Eleven dogs ranging in weight from 22 to 36 kg were anesthetized by intravenous injection of 12.5 mg/kg of sodium thiopenthal; a light level of anesthesia was then kept by artificial ventilation with a mixture of two-thirds oxygen, one-third nitrous oxide.

The coronary flow was measured by the right heart exclusion-drainage technique; median sternotomy was performed, and cannulas were inserted in both venae cavae. Venous return was drained continously into a cardiotomy reservoir and pumped into the pulmonary artery; therefore, the right ventricle drainage represented only coronary outflow (coronary sinus and right thebesian flows). We have noted previously (unpublished data) that the left thebesian outflow of animals under cardiopulmonary by-pass was less than 5 percent of the total coronary outflow. Both common carotids were isolated and cannulas were inserted and connected to the output of a small roller pump, the input of which was constituted by a Y tube in connection with either the femoral artery or the cardiotomy reservoir bottom. Small branches of external carotid were ligated distal to the receptor structures (internal carotid and occipital artery) and a catheter was introduced into the external carotid and connected with the jugular vein so that an arteriovenous shunt was created that would prevent any change in the carotid body circulation from affecting the brain. Disalvo *et al.* (10) reported that under these conditions the brain flow is kept at a sufficient level by the vertebral circulation.

Pressure in the carotid artery was monitored through a small catheter introduced via the lingual artery. EKG, aortic pressure just above the aortic valves, and left ventricular pressure were recorded continuously. Coronary flows were determined by drainage into a graduated bottle; each measurement was repeated three times and the mean value recorded. Arterial, coronary venous, and carotid blood flows were analyzed as well as hematocrit, pO_2, and pH, at the following periods: Period I, after vagotomy and perfusion of the carotid by hyperoxygenated blood; (pO_2 280 mm Hg ± 32); Period II, after perfusion of the carotid by venous blood from the cardiotomy reservoir (pO_2 30 mm Hg ± 2). Bilateral cervical vagotomy was then performed in seven dogs as follows: Period III, after vagotomy and perfusion of the carotid by hyperoxygenated blood; Period IV, after vagotomy and perfusion of the carotid by venous blood. At the end of the experiement the left ventricle was weighted and the coronary flow values reported to 100 gm of left ventricle (LV); coronary resistance by 100 gm of LV were expressed in dynes·sec·cm^{-5} by dividing mean aortic pressure (dynes per cm^2) above the aortic valves by coronary flow per sec per 100 gm.

In seven animals, measurement of norepinephrine content of arterial and coronary venous blood samples was based on a modification of the fluorometric method of Anton and Sayre (3) after prior extraction with perchloric acid. Release and uptake of norepinephrine by 100 gm of LV per min were computed by multiplying the arteriovenous difference in norepinephrine by the coronary flow per 100 gm of LV per min.

RESULTS

Aortic pressure

Carotid body stimulation resulted in a 4 percent [not significant (N.S.)] decrease of the control values of mean aortic pressure. Vagotomy, by itself was accompanied by a 12 percent (N.S.) decrease. Thereafter carotid body stimulation produced a 5 percent (N.S.) decrease (table 1).

Table I Mean values and SE for periods I–IV (defined in text)*

	Aortic pressure	Heart rate	Coronary flow	Coronary resistance	MVO_2
I	57.8 ± 6.9	130.9 ± 5.8	61.2 ± 3.3	76.8 ± 7.2	4.00 ± 0.21
II	55.7 ± 7.1	122.7 ± 7.2	74.4 ± 6.5 †	60.0 ± 5.4 †	4.94 ± 0.54
III	51.1 ± 4.6	129.0 ± 6.6	67.6 ± 6.8	63.6 ± 6.6	4.24 ± 0.56
IV	48.5 ± 4.7	130.7 ± 7.4	71.4 ± 12.0	61.2 ± 6.0	4.47 ± 0.68

*Pressures are expressed in mm Hg; flows in milliliters per min per 100 gm LV, resistances in dynes per sec per cm^{-5} per 100 cm LV, MVO_2 is milliliters of oxygen per min per 100 gm LV.

†$p < 0.05$.

Heart rate

Carotid body stimulation reduced the heart rate by 6 percent from the mean control value; the change is not significant but disappeared after vagotomy.

Coronary flow

There is a 22 percent increase of the coronary flow per 100 gm of left ventricle during carotid body stimulation ($p < 0.05$); after vagotomy, the same stimulation produced an increase of 2 percent (N.S.) (fig. 1).

Coronary resistances

Carotid body stimulation is accompanied by a 23 percent decrease in coronary resistance ($p < 0.05$) during the control period and a 4 percent decrease (N.S.) after vagotomy (fig. 2).

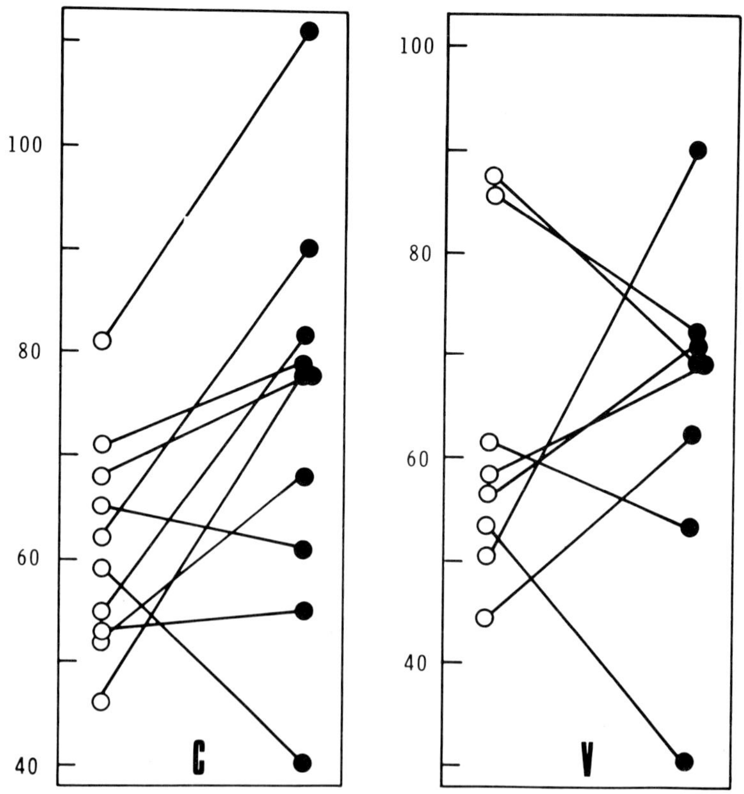

Figure 1 Changes in the coronary flows (milliliters per 100 gm of LV) before (○) and during carotid body stimulation (●). *Left panel* refers to values of the animals with intact vagi, *right panel* refers to values after vagotomy.

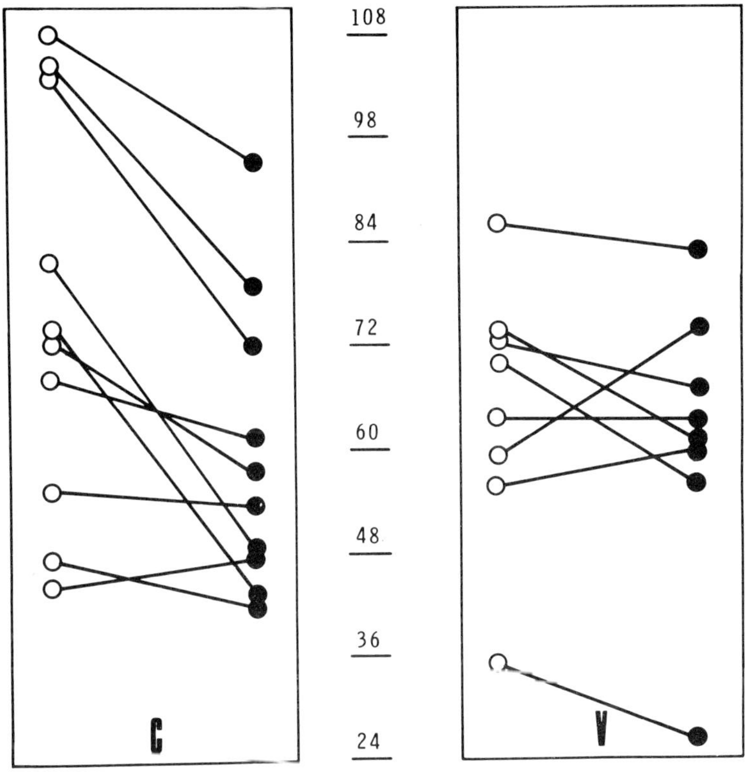

Figure 2 Changes in the coronary resistances (dynes per sec per cm^{-5}) before (○) and during carotid body stimulation (•). *Left panel* refers to values of the animals with intact vagi, *right panel* refers to values after vagotomy.

Myocardial oxygen consumption

Carotid body stimulation is associated with an increase in myocardial oxygen consumption, +24 percent (N.S.) and +6 percent (N.S.), respective, before and after vagotomy.

Catecholamines

During the control conditions there is a release of norepinephrine (NE) by the heart averaging −0.217 γ NE/min/100 gm of LV; carotid body stimulation increases the release to−0.815 γ (N.D.) and −1.702 γ of NE ($p < 0.05$), respectively, before and after vagotomy. Moreover, during the control condition, the carotid body being perfused by hyperoxygenated blood, vagotomy, by itself, converts the release phenomen to an uptake one +0.321 γ of NE ($p < 0.05$) (fig. 3).

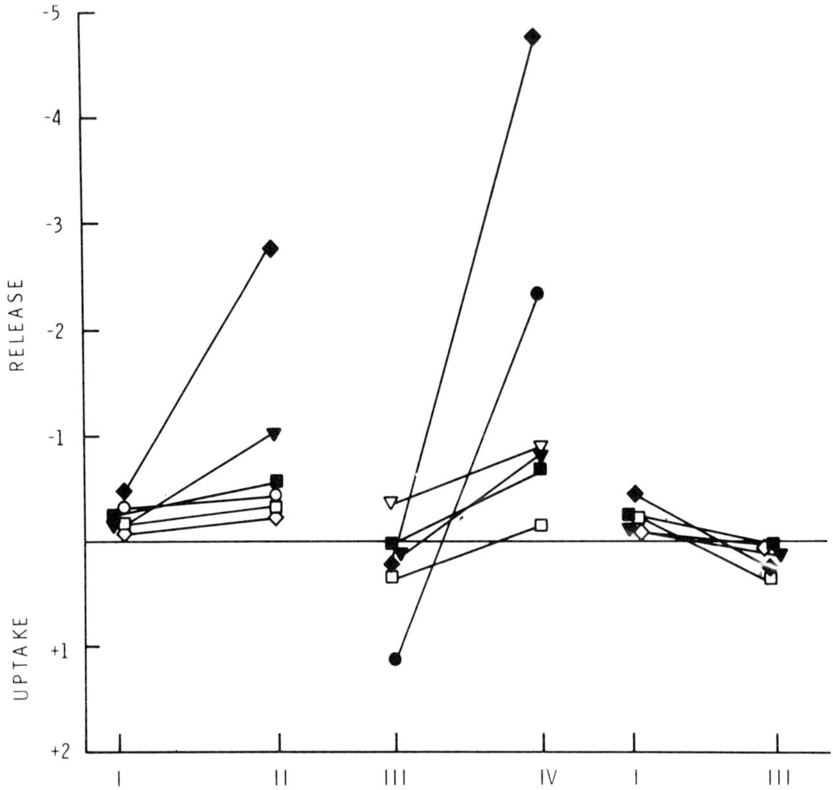

Figure 3 Changes in the heart norepinephrine output (gamma per min per 100 gm of LV) during the four periods defined in the text: from *I* to *II* effects of carotid body stimulation with intact vagi; from *III* to *IV* effects of carotid body stimulation after vagotomy; from *I* to *III* effects of vagotomy by itself.

DISCUSSION

The decrease observed in the aortic pressure after hypoxic stimulation is small but in opposition to many experiments where an increase was reported (8, 11, 17). Daly (8) has pointed out that the constrictory effects of carotid body stimulation could be overwhelmed by simultaneous afferent inhibitory impulses to the vasomotor centers originating from the pulmonary stretch receptors excited by the reflex hyperpnea. Hence, it is necessary to either cut the vagi or to keep the animal under controlled respiration in order to allow the vasoconstrictory response to develop. Since the animals in this experiment were kept constantly under controlled respiration, participation of pulmonary stretch receptors cannot be evoked to explain the lack of increase in aortic pressure (AoP).

The work of Calvelo *et al.* (4) has drawn attention to the fact that the whole pressor response was the sum of various responses in different vascular beds, namely, vasoconstriction in muscle and vasodilation in the skin. Hence, the final pressure response may vary from one experiment to another because of inhomogeneity of the responses of peripheral vascular perfusion beds. Another cause for variations in results is the action of the carotid body stimulation upon adrenal secretion; according to Anichkov *et al.* (2), carotid sinus stimulation in the rat produces an increase of norepinephrine released from the adrenals whereas carotid body stimulation results in an increase of epinephrine. In the dog either carotid sinus hypotension or asphyxia yields a predominant (90 percent) epinephrine secretion from the adrenals (2). Because of the relative predominance of β or α vascular smooth muscle receptors, the varying ratio between norepinephrine and epinephrine is sufficient to make the pressure response vary.

The decrease in heart rate is a finding consonant with all experiments (5, 7, 15, 18) since the report of Heymans *et al.* (13) which attested to a vagal stimulation by the chemoreceptors of the carotid body. The abolishment of the effect after vagotomy is proof that the slowing of the heart rate was caused by an increase in the vagal efferents and not to a decrease in the sympathetic efferents.

If the decrease of coronary resistances is noticable, it cannot be attributed solely to an autonomic effect upon the vascular smooth muscle. Indeed there is a simultaneous increase in myocardial oxygen consumption (MVO_2) and the observed decrease in coronary resistance (CR) could be the result of that increased metabolism. Vagotomy decreased the changes in CR more than the changes in MVO_2, and by plotting the changes in MVO_2 against the changes in CR, it appears that the regression line is displaced upwardly by vagotomy (fig. 4). Hence, it seems that with the vagi intact, the carotid body stimulation decreases the CR by a mechanism different from a metabolical one. Since there is vagal simulation, attested to by bradycardial suppressed by vagotomy, it can be said that carotid body stimulation produces a vagal coronary vasodilation that confirms the view expressed by Hackett *et al.* (12) and Abboud (1) after pharmacological stimulation of the carotid body.

By analogy, with the function of the carotid chemoreceptors in the respiratory regulation, it can be said that the coronary dilation due to the hypoxic stimulation of the carotid body also plays the role of a supplementary safety factor. This reflex increases the coronary flow before myocardial hypoxia winds up in a direct coronary vasodilation by itself, in the same way that reflex hyperpnea prevents direct hypoxic stimulation of the respiratory centers.

Since AoP is slightly reduced and heart rate diminished, the increase in MVO_2 after carotid stimulation can be attributed only to an increase in contractility

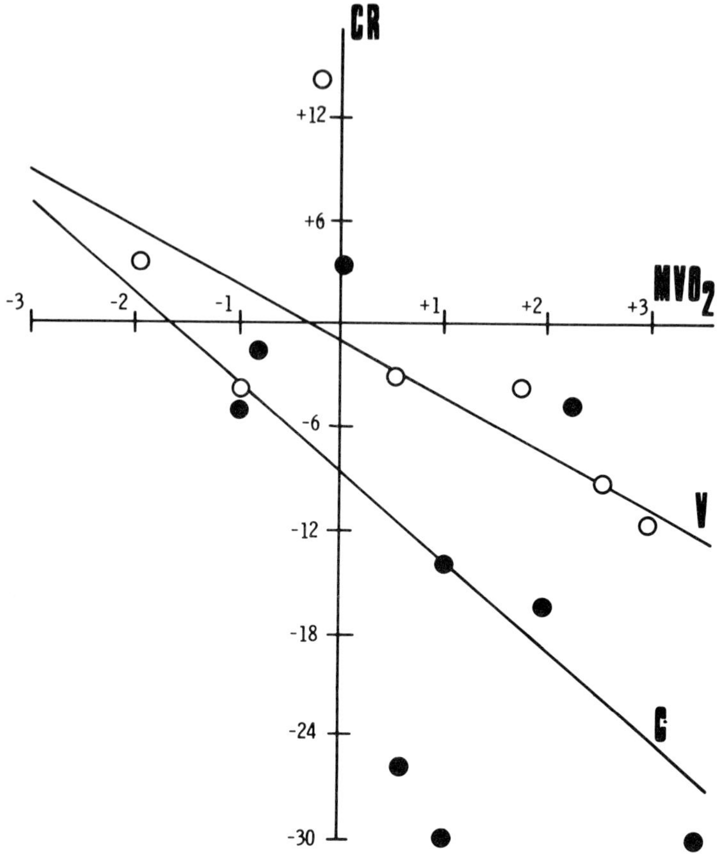

Figure 4 Simultaneous changes from the control values of myocardial oxygen consumption (MVO_2) and CR induced by carotid body stimulation. •, Animals with intact vagi (*C*); ○, vagotomized animals (*V*).

[which occurs despite the negative inotropic effect resulting from vagal stimulation (6, 19)] the increase in contractility and myocardial oxygen consumption should be related to the greater amount of norepinephrine released by the heart. Nevertheless, vagotomy reduces the changes in MVO_2 induced by carotid body stimulation whereas the release of norepinephrine remains increased.

No significant change in ventricular contractility was reported by Stern and Rapaport (23) and Downing *et al.* (11) after carotid body stimulation, although both authors acknowledged an increase of the parasympathetic activity that should have resulted in a negative inotropic effect on the myocardium. De Geest (9) noted that, after vagotomy, carotid body stimulation was followed by an increase of the left ventricular contractility. The variable reported effects of carotid body stimulation upon ventricular performance can be attributed to a simultaneous increase of both parasympathetic and sympathetic activity (as evidenced in our experiment by a greater release of norepinephrine from the heart).

The fact that the release of norepinephrine attributable to carotid body stimulation is amplified after vagotomy can be explained by the disappearance of a muscarinic inhibition of norepinephrine release formerly mediated through the vagi (19). On the other hand, Jellinek *et al.* (16) have reported that bilateral cervical vagotomy resulted in a depletion (−50 to −75 percent of control values) of cardiac norepinephrine stores 3 weeks after the surgical procedure.

What is the cause of the uptake of norepinephrine by the heart during hyperoxic perfusion of the carotid bodies after vagotomy? A rebound effect is conceivable: after the release of norepinephrine because of carotid body stimulation, the sudden vanishing of the chemoreceptor stimulation and of the subsequent sympathetic stimulation could be accompanied by a transitory preponderance of the uptake mechanisms.

From the point of view of the coronary hemodynamics, carotid body stimulation brings an augmentation of coronary flow by a simultaneous increase of the parasympathetic and sympathetic activity of the heart. The direct effect of the parasympathetic neuroeffector is relaxation of the vascular smoooth muscle, whereas norepinephrine, known to have direct coronary constrictory effect (22), yields an increase in myocardial oxygen consumption and, hence, a metabolically induced indirect coronary dilation.

REFERENCES

1. ABBOUD, F. M. 1972. Control of the various components of the peripheral vasculature. Fed. Proc. 31: 1226.
2. ANICHKOV, S. V., MALYGHINA, E. I., POSKALENKO, A. M. and RYZKHENKOV, V. E. 1960. Reflexes from carotid bodies upon the adrenals. Arch. Int. Pharmacol. 129: 156.
3. ANTON, A. H. and SAYRE, D. F. 1962. A study of the factors affecting the aluminum oxide trihydroxyindole procedure for the analysis of catecholamines. J. Pharmacol. Exp. Therap. 136: 360–375.
4. CALVELO, M. G., ABBOUD, F. M., BALLARD, D. R., and ABDEL-SAYED, W. 1970. Reflex vascular responses to stimulation of chemoreceptors with nicotine and cyanide. Circ. Res. 27: 259.
5. COMROE, J. H., and MORTIMER, L. 1964. The respiratory and cardiovascular responses of temporally separated aortic and carotid bodies to cyanide, nicotine, diphenyldiguanide and serotonin. J. Pharmacol. Exp. Therap. 146: 33.
6. DAGGET, W. M., NUGENT, G. C., CARR, P., POWERS, P. C., and HARDA, Y. 1967. Influence of vagal stimulation on ventriculur contractility, oxygen consumption and coronary flow. Amer. J. Physiol. 212: 8.
7. DALY, M. de B., and SCOTT, M. J. 1958. The effects of stimulation of the carotid body chemoreceptors on the heart rate in the dog. J. Physiol. (London) 144: 148.
8. DALY, M. de B., and SCOTT, M. J. 1963. The cardiovascular responses to stimulation of the carotid body chemoreceptors in the dog. J. Physiol. (London) 165: 179.

9. DE GEEST, H., LEVY, M. N., and ZIESKE, H. 1965. Carotid chemoreceptors stimulation and ventricular performance. Amer. J. Physiol. 209: 564.
10. DISALVO, J., PARKER, P. E., SCOTT, J. B., and HADDY, J. B. 1971. Carotid baroreceptors influence on coronary vascular resistance in the anesthetized dog. Amer. J. Physiol. 221: 156.
11. DOWNING, S. E., REMENSNYDER, J. P., and MITCHELL, J. H. 1961. Cardiovascular responses to hypoxic stimulation of the carotid bodies. Circ. Res. 10: 676.
12. HACKETT, J. G., ABBOUD, F. M., MARK, A. L., SCHMID, P. G. and HEISTAD, D. D. 1972. Coronary vascular response to stimulation of chemoreceptors and baroreceptors. Circ. Res. 31: 8.
13. HEYMANS, C., BOUCKAERT, J. J., and DAUTREBANDE, L. 1936. Au sujet du méchanisme de la bradycardie provoquée par la nicotine, la lobéline, le cyanure, le sulfure de sodium les nitrites et la morphine, et de la bradycardie asphyxique. Arch. Int. Pharmacol. 41: 261.
14. HEYMANS, C., and NEIL, E. 1958. Reflexogenic areas of the cardiovascular system. Churchill, London.
15. JACOBS, L., SAMPSON, S. R. and COMROE, J. H. 1971. Carotid sinus versus carotid body origin of nicotine and cyanide bradycardia in the dog. Amer. J. Physiol. 220: 472.
16. JELLINEK, M. KAYE, M. P., KAISER, G. C., and COOPER, T. 1965. Effect of cervical vagosympathectomy on myocardial catecholamines concentration. Amer. Physiol. 209: 951.
17. KORNER, P. I., UTHER, J. B., and WHITE, S. W. 1969. Central nervous integration of the circulatory and respiratory responses to arterial hypoxemia in the rabbit. Circ. Res. 24: 757.
18. LANDGREN, S., and NEIL, E. 1952. The effect of chemical stimulation of the carotid bodies on the heart rate of the cat. Acta Physiol. Scand. 25: 286.
19. LEVY, M. N., and ZIESKE, H. 1969. Comparison of the cardiac effects of vagus nerve stimulation and acetylcholine infusions. Amer. J. Physiol. 216: 890.
20. LOFFELHOLZ, K. and MUSCHOLL, E. 1969. Amuscarinic inhibition of the noradrenaline release evoked by postganglionic sympathetic nerve stimulation. Naunyn-Schmiedeberg's Arch. Pharmakol. Exp. Pathol. 265: 1.
21. MALMEJAC, J. 1964. Activity of the adrenal medulla and its regulation. Physiol. Rev. 44: 186.
22. PITT, B., ELLIOT, E. C., and GREGG, D. E. 1967. Adrenergic receptor activity in the coronary arteries of the anesthetized dog. Circ. Res. 21: 75.
23. STERN, S., and RAPAPORT, E. 1967. Comparison of the reflexes elicited from combined or separate stimulation of the aortic and carotid chemoreceptors on myocardial contractility cardiac output and systemic resistance. Circ. Res. 20: 214.

Mailing address (permanent): Dr. J. H. Kennedy,
Taub Laboratories for Mechanical Circulatory Support,
Cora and Webb Mading Department of Surgery and Neurology Department,
Baylor College of Medicine, Houston, Tex. (USA).

Temporary address:
Department of Macromolecular Science, Olin 518,
Case Western Reserve University, Cleveland, Ohio 44106 (USA).

Sodium-Potassium Pump and Cardiac Glycosides

Recent Advances in Studies on Cardiac Structure and Metabolism
Volume 9
The Sarcolemma
Edited by Paul-Emile Roy and Naranjan S. Dhalla

CHARACTERIZATION OF PARTIALLY PURIFIED HEART SARCOLEMMAL Na^+-K^+-STIMULATED ATPase[1]

P. V. SULAKHE, D. B. McNAMARA, and N. S. DHALLA

Experimental Cardiology Laboratory,
Department of Physiology, Faculty of Medicine,
University of Manitoba,
Winnipeg, Manitoba, Canada

SUMMARY

The K_m value for the dog heart (Na^+-K^+)-ATPase was 0.31 mM (MgATP), whereas the values for the concentrations of K^+ and Na^+ varied from 1.2 to 2.7 mM and 12 to 20 mM for half-maximal activation, respectively. The concentrations of ouabain and calcium for 50 percent inhibition of (Na^+-K^+)-ATPase activity varied from 2.4 to 3.2 μM and 0.5 to 1.2 mM, respectively; the inhibitory effects of these agents were pH dependent. This preparation bound about 50 nmoles of 1-anilino 8-naphthaline sulfonate (ANS)/mg of protein and exhibited fluorescence attributable to the ANS-enzyme complex. Cations such as Na^+,K^+, Ca^{++}, and Mg^{++} increased ANS-enzyme fluorescence intensity and the number of ANS binding sites but decreased the apparent ANS binding constant. The enzyme activity, ANS binding, and ANS-enzyme fluorescence were decreased by phospholipase A, phospholipase C, and trypsin treatments. Although ouabain inhibited enzyme activity and ANS-enzyme fluorescence markedly, it caused only a slight depression in ANS binding. These results extend support for the allosteric nature of the cardiac (Na^+-K^+)-ATPase and provide evidence for conformational changes during its activation by Na^+ and K^+.

Sarcolemmal-bound Na^+-K^+-stimulated, Mg^{++}-dependent ATPase is not only considered to be involved in the active transport of Na^+ and K^+ ions across the cell membrane, but is also believed to be the site of cardiac glycoside action in heart (1, 13, 14). It has been suggested that (Na^+-K^+)-ATPase undergoes conformational changes during its activation by Na^+ and K^+ which subserve the possible mechanism underlying the active transport of ions (10). Although the properties of (Na^+-K^+)-ATPase from different tissues have been studied extensively (13), information with respect to the cardiac muscle enzyme has been derived from preparations with poor specific activity. It was therefore the purpose of this study to examine the kinetic properties of the partially purified

[1] This work was supported by Grant MT-3394 from the Medical Research Council of Canada.

(Na^+-K^+)-ATPase preparation from dog heart. Since ANS has been used extensively for analyzing conformational changes in a wide variety of biomembranes and enzymes, it was considered of interest to study the interaction and binding of ANS with the dog heart (Na^+-K^+)-ATPase preparation under different experimental conditions.

METHODS

Preparation of solubilized (Na^+-K^+)-ATPase enzyme

The hearts were placed in ice-cold medium (0.25 M sucrose, 1 mM EDTA, 10 mM Tris - HCl, pH 7.2–7.4), washed thoroughly to remove blood and homogenized with 15 volumes of the above medium for 40 sec (20 sec × 2) in a Waring Blendor in a cold room (0–4°C). The homogenate was filtered through several layers of gauze, centrifuged at 10,000 × g for 15 min, the residue homogenized (original volume) for 20 sec with the above medium containing 0.1 percent deoxycholate, stirred for 20 min, and spun at 10,000 × *g* for 10 min. The sediment was discarded and the clear supernatant, after filtration, was centrifuged at 70,000 × *g* for 1 hr. The residue was suspended in 0.25 M sucrose, 1 mM EDTA, 10 mM Tris-HCl, pH 7.2–7.4, containing 0.05 percent deoxycholate with the aid of a glass-Teflon homogenizer, stirred at 0–4°C for 15–20 min, and then centrifuged at 10,000 × *g* for 10 min. The residue was discarded and the supernatant, after filtration, was centrifuged at 70,000 × *g* for 1 hr. The residue thus obtained was thoroughly washed and suspended in 1 mM Tris-EDTA solution (pH 7.0) and centrifuged at 70,000 × *g* for 1 hr. This process was repeated twice, the sediment suspended in 1 mM Tris-EDTA. This preparation was treated further as follows: the suspension was stirred slowly in an ice bath with 2 M sodium iodide, 5–10 mM $MgCl_2$, 15 mM EDTA, 100 mM Tris-HCl, pH 8.0, for 45 min, diluted 2.5 times with 1 mM Tris-EDTA, and centrifuged at 38,000 × *g* for 45 min. The final residue was obtained by a repeated cycle of washing, resuspending, and centrifuging in 1 mM Tris-HCl, and used immediately. The principle of this procedure is based on the method described by Matsui and Schwartz (8).

Determination of enzyme activities

About 50–100 μg of the protein preparation was employed for each reaction in this study. Each enzyme reaction was linear with respect to the time and protein concentration. Fresh membrane preparations were used for each experiment.

For routine estimation of the Na^+-K^+-stimulated ATPase activity, the fractions were incubated in a total volume of 1 ml containing 50 mM Tris-HCl, pH 7.4 –7.6, 4 mM $MgCl_2$, 100 mM NaCl, and 10 mM KCl in the presence or absence of ouabain (2 mM). After 5 min of preincubation at 37°C, the reaction was started by addition of ATP (4 mM final concentration) and the fraction was further incubated for 10 min. The reaction was stopped by the addition of 1 ml of 12 percent cold trichloroacetic acid, the fraction was centrifuged, and the Pi in the clear supernatants was determined. The difference of the activities in the absence and presence of ouabain is referred to as Na^+-K^+-stimulated Mg^{++}-dependent, ouabain-sensitive ATPase [(Na^+-K^+)-ATPase] and the values thus obtained agreed very well with the ATP hydrolysis because of the presence of Na^+ and K^+ in the incubation medium. The estimation of Pi was carried out by the method of Taussky and Shorr (16). Usually 0.5 ml or 1.0 ml of the clear supernatant was employed for the estimation of phosphate. Each time a standard curve for Pi was constructed to ensure the reliability of phosphate values. Not more than 15 percent of the added ATP was hydrolyzed under the present experimental conditions. This eliminated the necessity for the use of ATP regenerating system during the assay of the ATPase activity. Protein concentration was estimated by the method of Lowry *et al.* (7).

Determination of ANS binding

ANS binding measurements were carried out by three separate methods. In method A, the protein (0.6–0.8 mg/ml) was incubated in the medium containing 50 mM Tris-HCl, pH 7.4, in the presence of 50 μM ANS in the absence and presence of different cations. At the end of a 10-min incubation period (room temperature), the suspension was filtered through Millipore filters (0.45 μm) and the concentration of ANS in the protein-free filtrate was determined spectrophotometrically at 270 nm with a Zeiss PM QII spectrophotometer (molar extinction coefficient 6.6×10^3). In method B, at the end of the incubation, the suspension was centriguged at $35{,}000 \times g$ for 20–30 min and the concentration of ANS was estimated in the supernatant as in Method A The readings were corrected for residual absorption attributable to protein. In Method C the reaction was terminated by Millipore filtration and the amount of ANS in the filtrate was estimated, after the addition of 95 percent ethanol, by fluorometry with an Aminco-Bowman spectrophotofluorometer (excitation at 375 nm and emission at 480 nm) using the standard calibration curve for ANS. Simultaneously, duplicate incubations were carried out in the cuvettes to analyze the fluorescent intensities of the bound ANS. From these observations the calculations were made for the proportions of bound and free ANS present in the incubation mixture. This approach is similar to that described by Weidekaam *et al.* (19).

In treatments with phospholipases and trypsin enzyme preparations (2 mg/ml) were incubated in 50 mM Tris-maleate, pH 7.6, 2 mM $CaCl_2$ with 200 μg of phospholipase A or C per ml for 15 min at room temperature. At the end of the incubation, EGTA was added to final concentrations of 5 mM and the suspension was centrifuged at $35{,}000 \times g$ for 30 min. The residue was washed by suspending in 1 mM Tris-EDTA, pH 7.0, and centrifuging as above. The sediment was then suspended in Tris-EDTA solution. The control tubes without added enzymes were run simultaneously. For the treatment with trypsin, the enzyme suspension (2 mg/ml) was incubated with 100 μg of trypsin/ml for 15 min. At the end of the incubation, a 4–5-fold excess of trypsin inhibitor was added, followed by centrifugation as above. Washing of the sediment was as described above. The control tubes containing only trypsin inhibitor were treated similarly.

Determination of ANS-enzyme fluorescence

Fluorescence was measured essentially by the method described by Vanderkooi and Martonosi (18) by employing an Aminco-Bowman spectrophotofluorometer at room temperature. Excitation and emission wave lengths were 380 and 490 nm, respectively. To study the effects of phospholipase C and trypsin, these agents were added into the reaction cuvette. The concentration of ANS used in most of the experiments was 50μM and the medium was 80 mM Tris-HCl, pH 7.3–7.4. Fluorescence data were presented in most cases as relative fluorescence intensities.

Each observation was made by employing three to four preparations. All changes in the experimental conditions are mentioned in the text.

RESULTS

In one series of experiments, the activity of heart (Na^+-K^+)-ATPase preparation was studied in the presence of different concentrations of MgATP, K^+, or Na^+ and the results are shown in figs. 1–3. In all three cases the activity curves were sigmoidal and the *n* values obtained by Hill plots were found to be greather than 1. The value for K_m (MgATP) was 0.31 ± 0.7 mM, whereas the concentrations of K^+ and Na^+ for half-maximal activation were 1.2–2.7 mM and 12–20 mM, respectively. Both ouabain and Ca^{++} were observed to markedly inhibit the (Na^+-K^+)-ATPase activity of the dog heart enzyme preparation (fig. 4); however, the inhibitory effect of these agents was found to be pH dependent (fig. 5). The concentrations of ouabain and calcium for 50 percent inhibition were 2.4–3.2 μM and 0.5–1.2 mM, respectively. Various sympathomimetic agents such as epinephrine and isoproterenol, and even cyclic AMP failed to alter the ATPase activity of the dog heart preparation.

Addition of the dog heart enzyme preparation to the cuvette containing ANS markedly increased the fluorescence (fig. 6), which was further increased by Mg^{++} ATP, and Na^+. The increase in fluorescence due to Na^+ was reduced by the addition of K^+. When ouabain was added into the cuvette, it abolished the fluorescence increase attributable to Na^+ and K^+. This inhibitory effect of ouabain was not seen with inactive preparations which were kept at room temperature for 2–3 days in 1 mM Tris-EDTA, pH 7.0, containing 0.002 percent

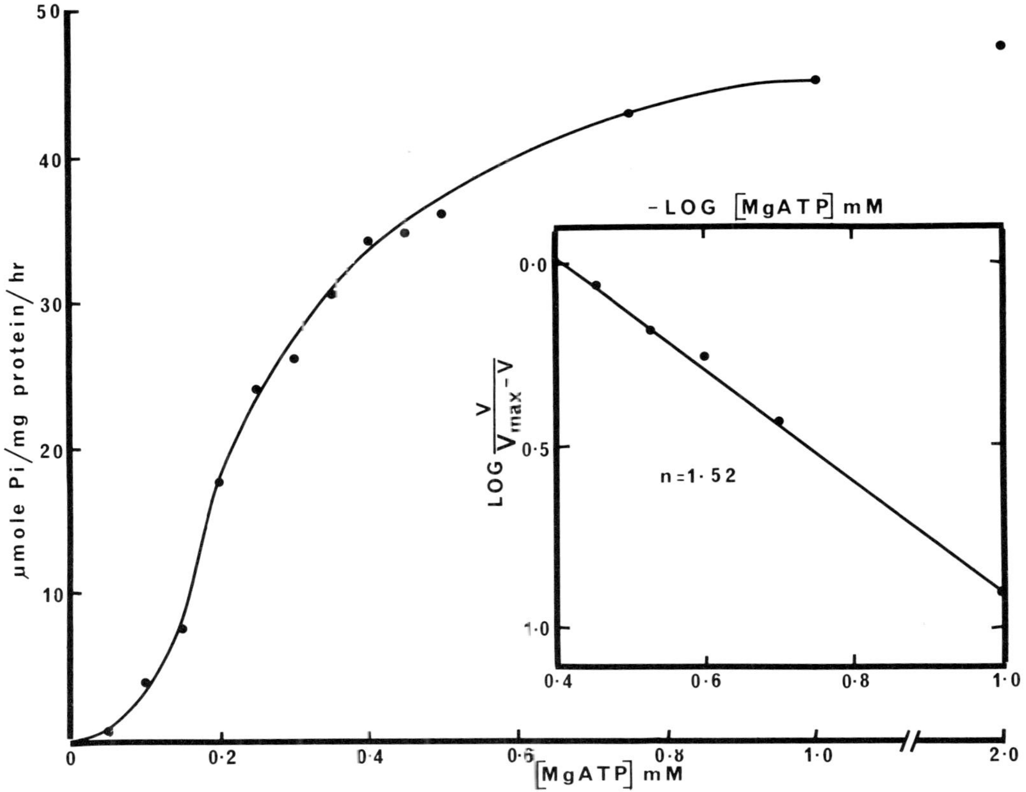

Figure 1 Effect of varying MgATP concentration (Mg^{++}/ATP = 1:1) on the (Na^+-K^+)-ATPase activity of the dog heart solubilized enzyme preparation. *Insert* is a Hill plot of this data. Conditions of incubation were the same as in Methods except for varying MgATP concentrations.

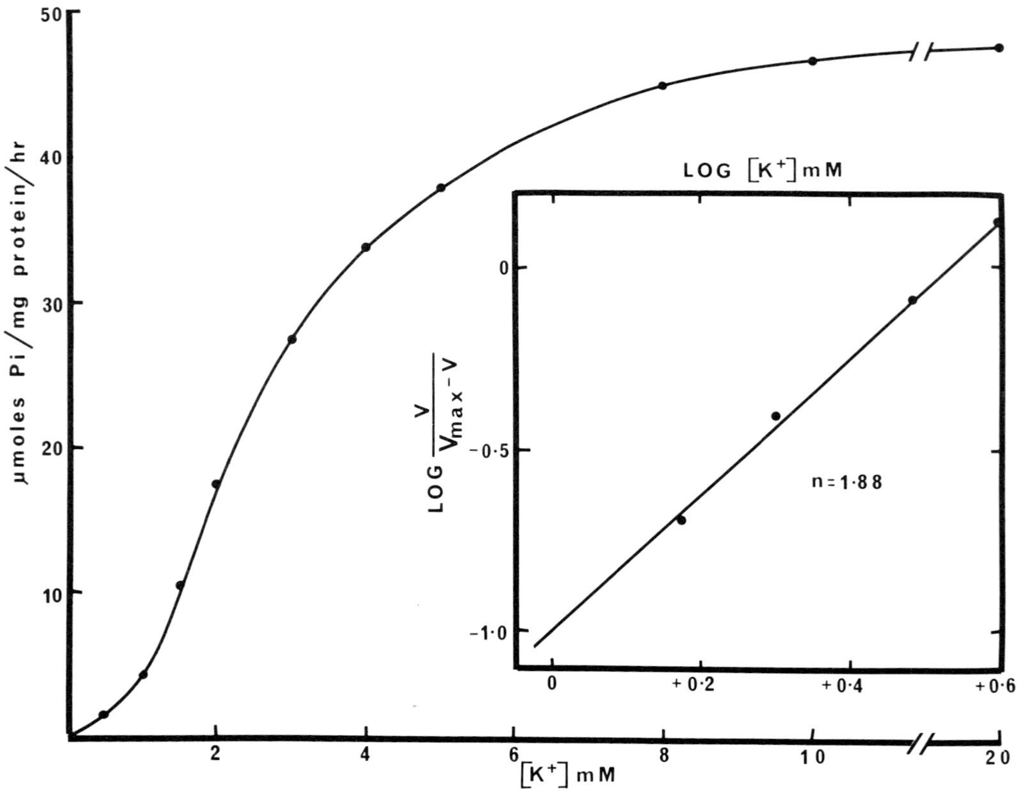

Figure 2 Effect of varying K^+ concentrations on the $(Na^+\text{-}K^+)$-ATPase activity of the dog heart solubilized enzyme preparation. *Insert* is a Hill plot of the data. Conditions of incubation were the same as in Methods except for varying K^+ concentrations.

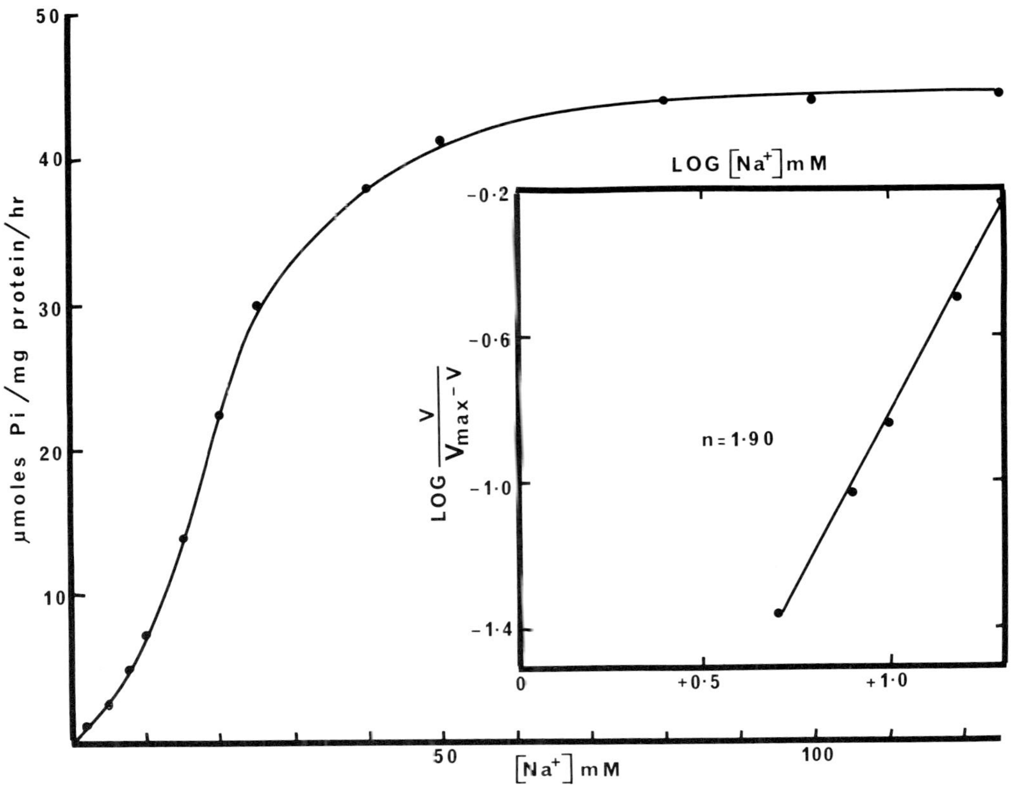

Figure 3 Effect of varying Na$^+$ concentrations on the (Na$^+$-K$^+$)-ATPase activity of the dog heart solubilized enzyme preparation. *Insert* is a Hill plot of the data. Conditions of incubation were the same as in Methods except for varying Na$^+$ concentrations.

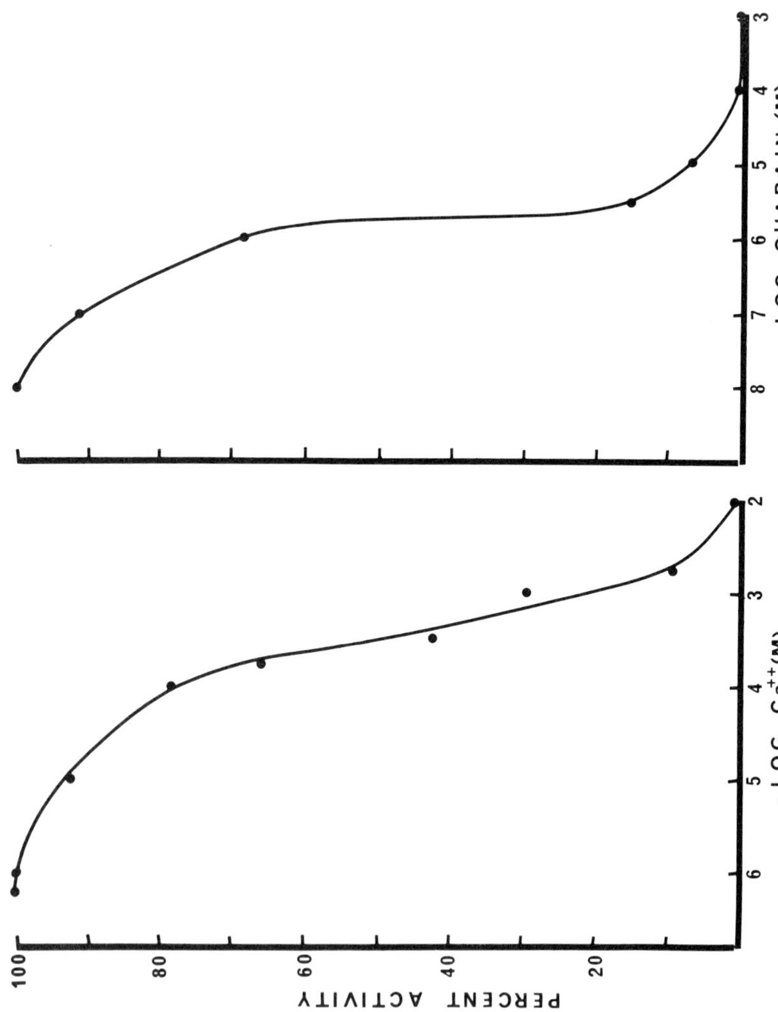

Figure 4 Effect of different concentrations of Ca^{++} or ouabain on the (Na^{+}-K^{+})-ATPase activity of the dog heart solubilized enzyme preparation. Conditions of incubation were the same as in Methods. Each value is an average of five experiments.

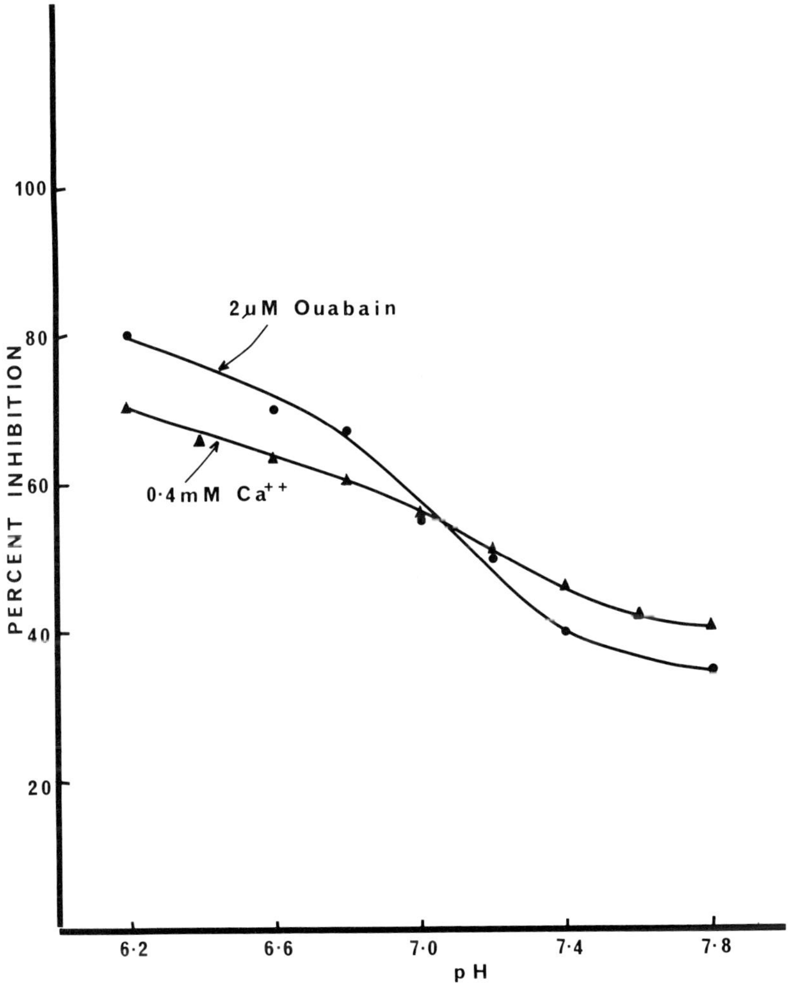

Figure 5 Influence of pH on the Ca^{++}- or ouabain-induced inhibition of the (Na^{+}-K^{+})-ATPase activity of the dog heart solubilized enzyme preparation. Conditions of incubation were the same as in Methods.

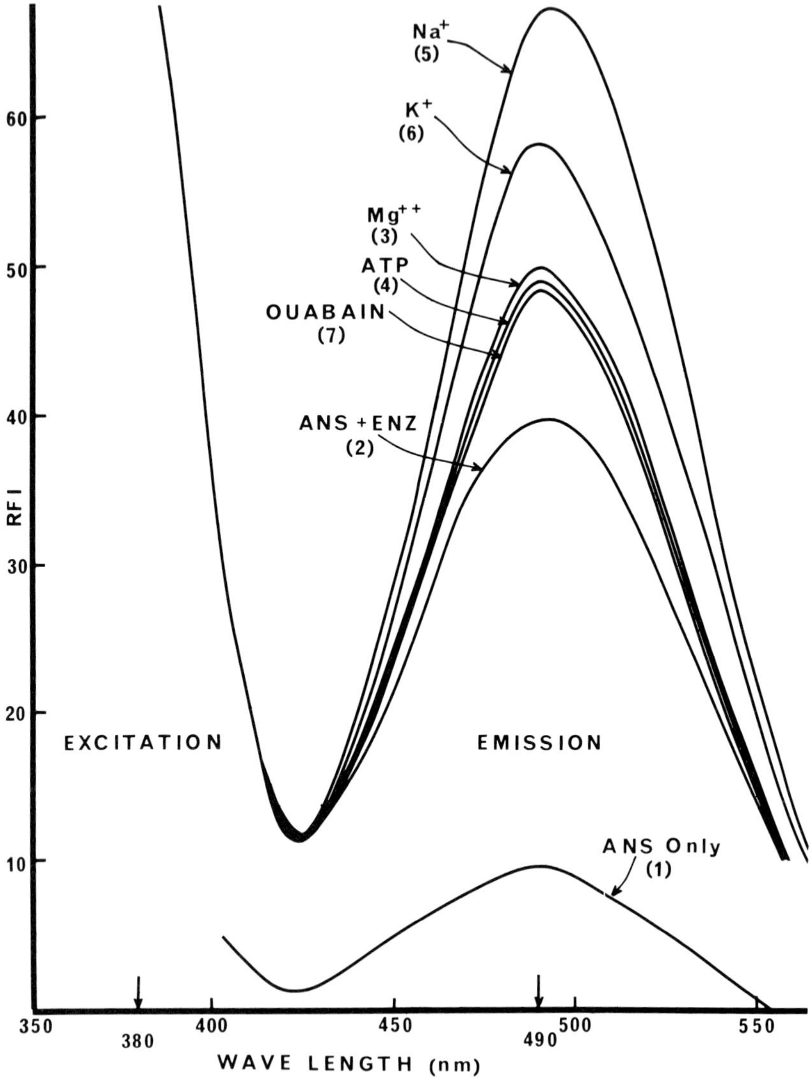

Figure 6 Sequential changes in the ANS-enzyme fluorescence spectrum attributable to the addition of reaction ingredients for (Na^+-K^+)-ATPase activity. The sequence of additions was: *(1)*, 80 mM Tris-HCl, pH 7.3, 50 μM ANS; *(2)*, 80 μg of enzyme preparation; *(3)*, 5 mM $MgCl_2$; *(4)*, 4 mM ATP; *(5)*, 50 mM NaCl; *(6)*, 5 mM KCl; *(7)*, 0.15 mM ouabain. Total volume was 1 ml.

sodium azide to prevent bacterial growth. In one experiment we employed a preparation obtained by deoxycholate treatment before NaI treatment. Although ouabain was found to produce an inhibitory effect on the deoxycholate preparation, addition of K^+ to the cuvette containing buffer, ANS, Mg^{++} Na^+, and ATP further increased the fluorescence, unlike the NaI preparation. The fluorescent intensity of the ANS-enzyme complex was found to decrease during the ATP hydrolysis when this interaction was studied at different intervals after starting the (Na^+-K^+)-ATPase reaction (fig. 7). However, the inhibitory effect of ouabain was still apparent.

The fluorescence attributable to the addition of fresh membranes in cuvette containing buffer and ANS was increased upon addition of Na^+ or K^+ (fig. 8). The Lineweaver-Burk plot of the data by employing different concentrations of preparation protein in the absence or presence of Na^+ essentially produced straight lines (fig. 9). However, increase in Na^+ concentration was found to increase F_{max} and decrease K_{app} for ANS (fig. 10). It should be mentioned that K_{app} for ANS under the experimental conditions employed in this study was 87 μM in the absence of any cation. K^+ was also found to have effects similar to that seen with Na^+. Likewise, both Mg^{++} and Ca^{++} increased ANS-enzyme fluorescence and the double-reciprocal plots of changes in fluorescence verus cation concentration produced straight lines (fig. 11).

Since phospholipids and protein composition play an important role in determining the activities of membrane-bound enzymes, the effects of phospholipase and trypsin treatments were studied in the ANS-enzyme fluorescence. The fluorescence intensity was markedly less in preparations after treatment with phospholipase A (table I). The double-reciprocal plot of these data showed no difference in F_{max} values between control and phospholipase A treated preparations. Addition of phospholipase C in the cuvette decreased the ANS-enzyme fluorescence; however, the presence of ATP in the medium partially prevented this change (fig. 12). Although trypsin also decreased ANS-enzyme fluorescence, the presence of ATP in the medium did not affect this change appreciably (fig. 13).

The enzyme preparation employed in the present investigation was purified by a factor of 15–20. The preparation appeared vesicular under an electron microscope and was found to contain about 20 percent lipid. In a series of experiments, we observed that ANS in concentrations up to 100 μM had no effect on the activity of (Na^+-K^+)-ATPase.

As can be seen from table II, the partially purified (Na^+-K^+)-ATPase bound about 50 nmoles of ANS/mg of protein. The binding of ANS estimated by three separate methods was very similar. Method A, involving Millipore filtration and spectrophotometry, was employed most frequently because of its simplicity. Both monovalent and divalent cations increased the amount of ANS bound to the enzyme preparation. However, when the cations of the same groups were compared over wide ranges of concentrations there were no significant differ-

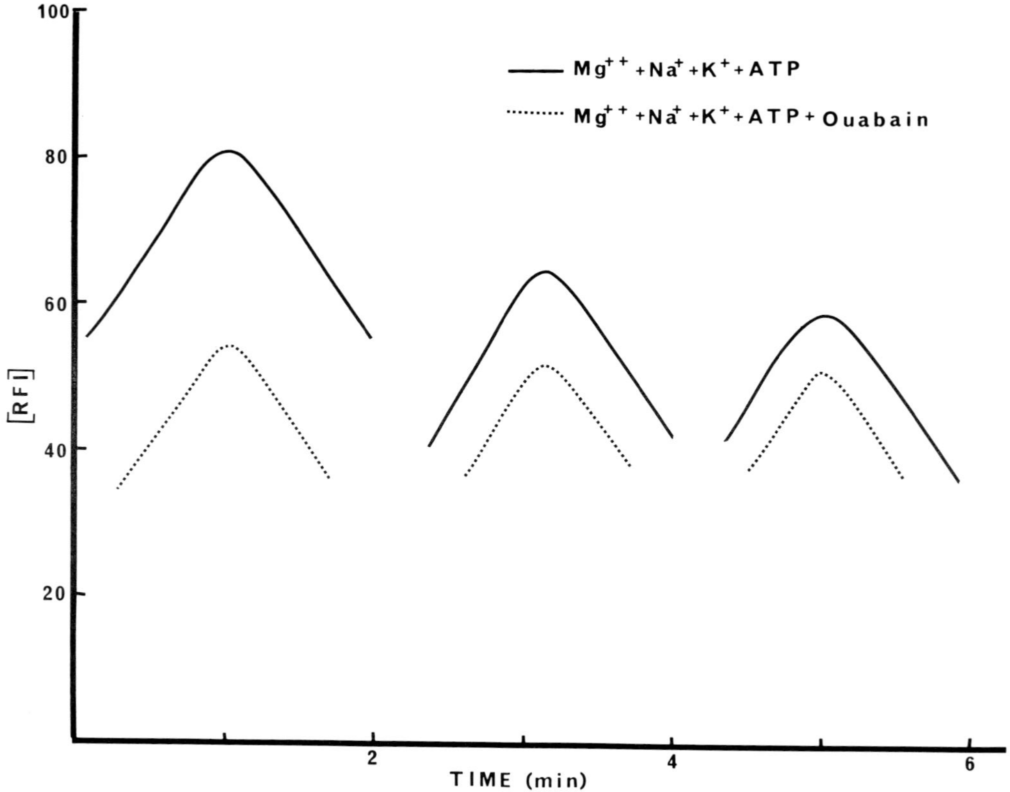

Figure 7 ANS-enzyme fluorescence spectra in the absence and presence of ouabain at different time intervals after starting the reaction for $(Na^{+}$-$K^{+})$-ATPase activity. The fluorescence spectral peaks were taken at 1, 3, and 5 min after starting the reaction and the peaks shown represent only the top portion of the spectra similar to that shown in figure 6. The reaction conditions are also similar to those in figure 6.

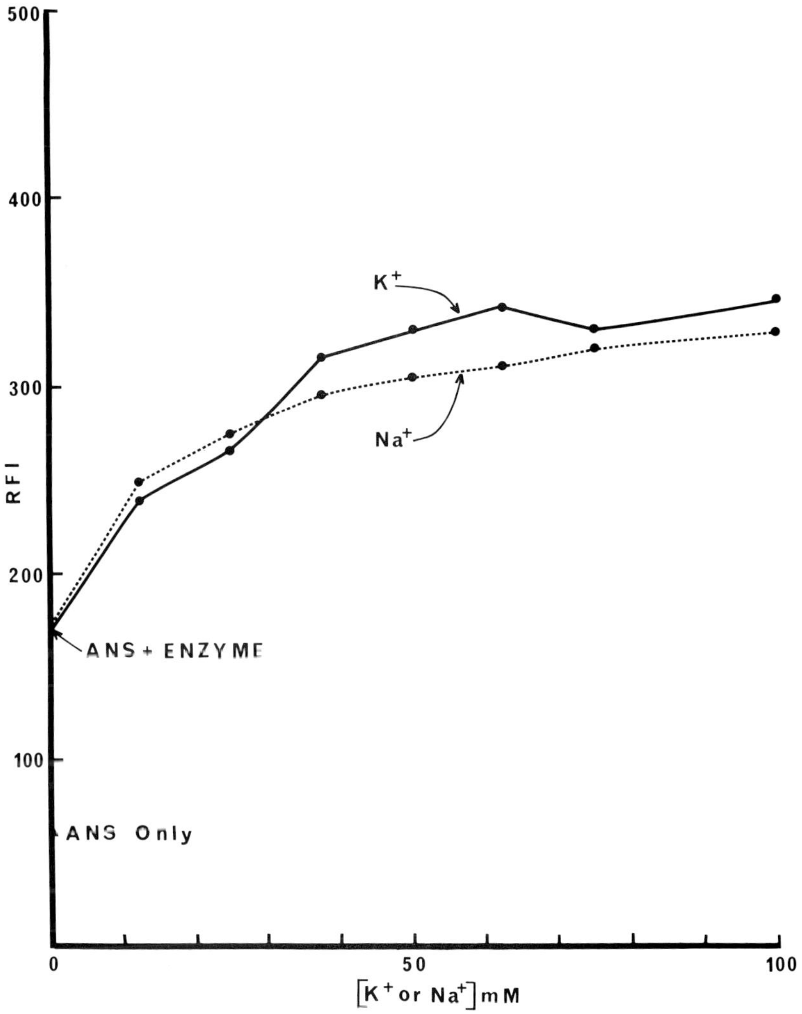

Figure 8 Effect of different concentrations of Na^+ and K^+ on the ANS-enzyme fluorescence.

ences in the abilities of these ions to increase ANS binding. In general, divalent metal ions were effective at lower concentrations (K_{app} 0.3 mM) compared with monovalent ions (K_{app} 15–20 mM).

In some experiments membrane preparation was titrated with increasing concentrations of ANS (5–200 μM) while in other experiments fixed amounts of ANS (50 μM) were titrated with increasing amounts of protein (0.2–1.8 mg/ml). The data were analyzed by Scatchard plots (ratio of bound ANS to free ANS

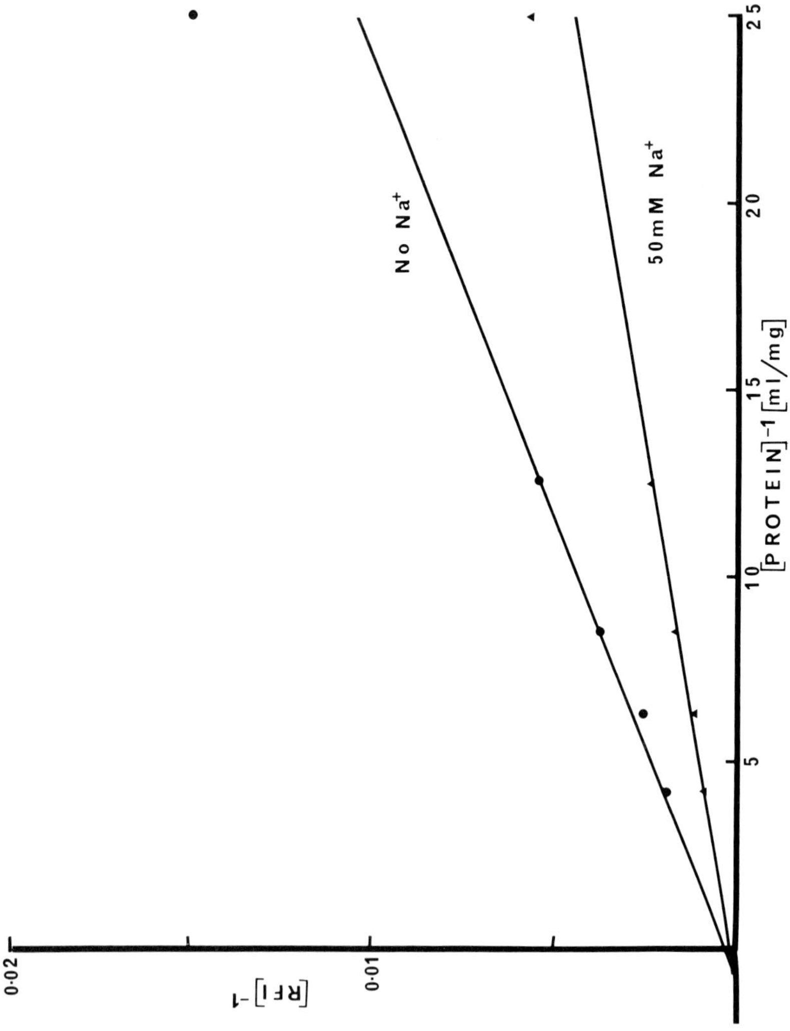

Figure 9 Double reciprocal plot of the ANS-enzyme fluorescence versus protein concentration in the absence or presence of 50 mM Na^+.

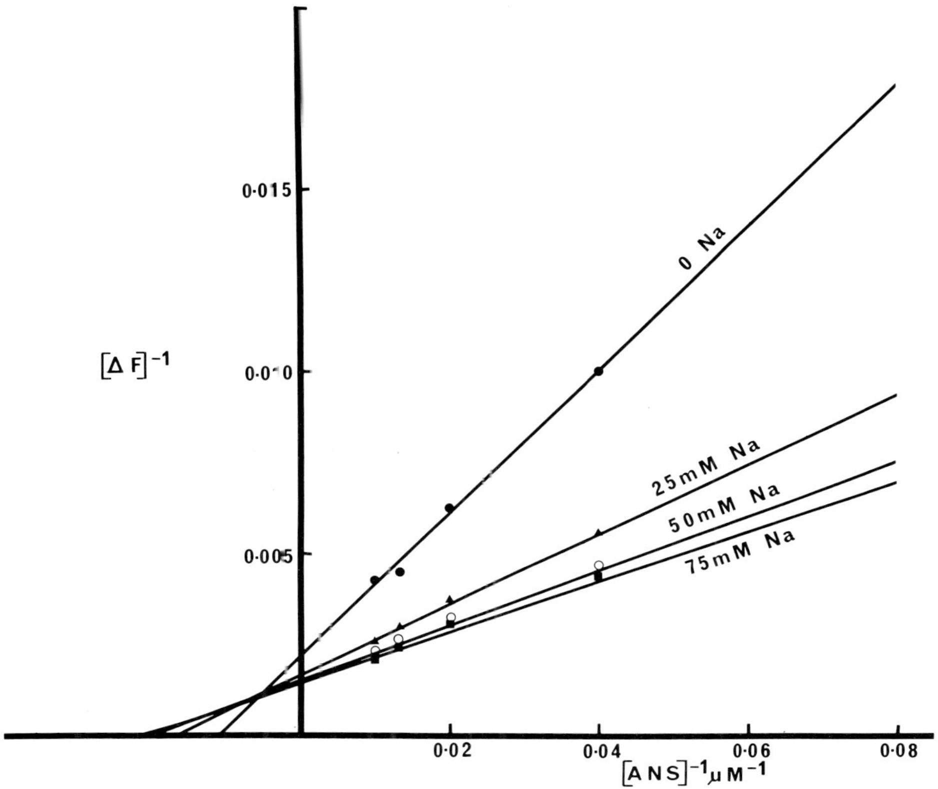

Figure 10 Double reciprocal plot of the changes in ANS-enzyme fluorescence due to various concentrations of Na^+ versus ANS concentration.

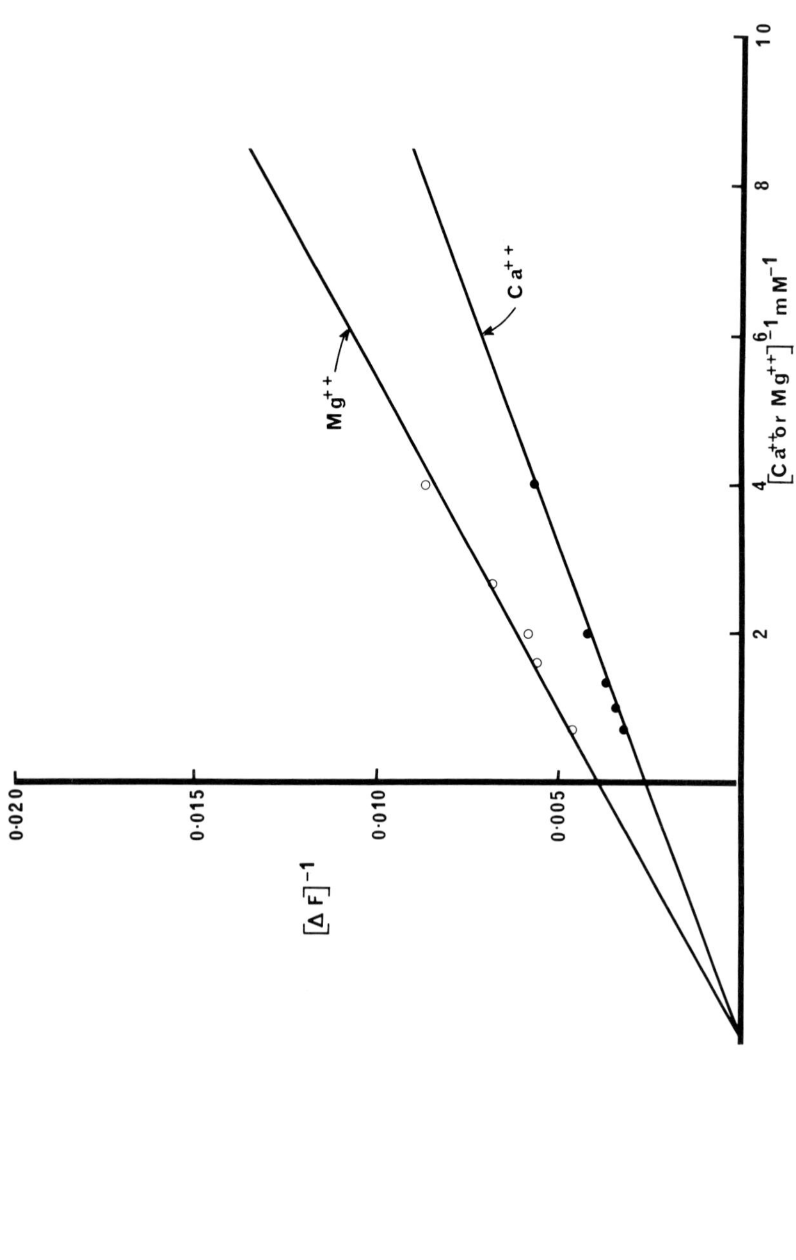

Figure 11 Double reciprocal plot of the changes in ANS-enzyme fluorescence versus Ca^{++} or Mg^{++} concentration.

Table I Relative fluorescence intensity of ANS-enzyme complex with or without phospholipase A treatment

Enzyme concentration (μg protein/ml)	Relative fluorescence intensity			
	10 μM ANS		50 μM ANS	
	Control	Phospholipase A treated	Control	Phospholipase A treated
40	24	12	54	32
80	46	22	104	64
160	78	45	180	116
240	100	64	250	160

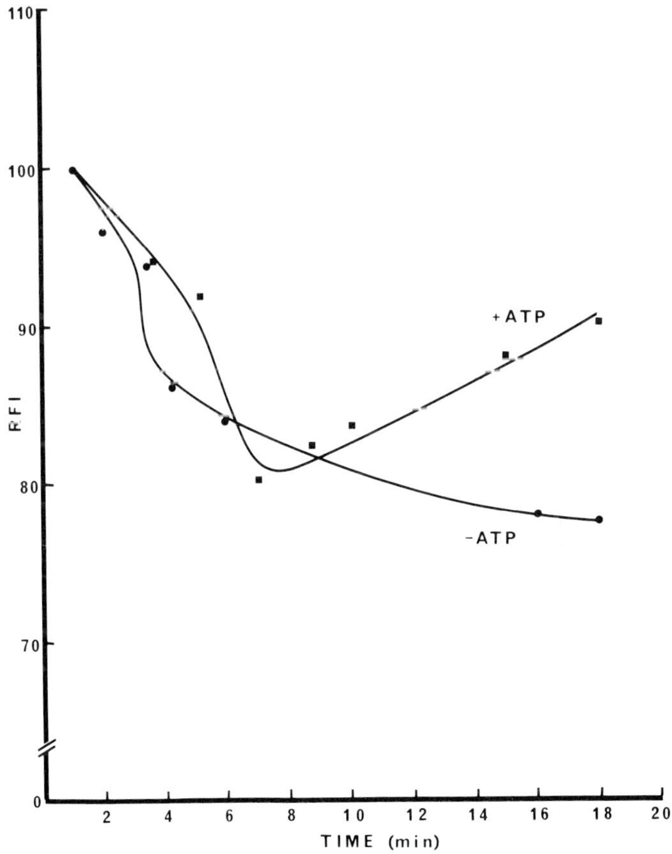

Figure 12 Effect of trypsin (125 μg/ml) on the relative fluorescence intensity (*RFI*) of the ANS-enzyme complex in the absence or presence of ATP (2.5 mM).

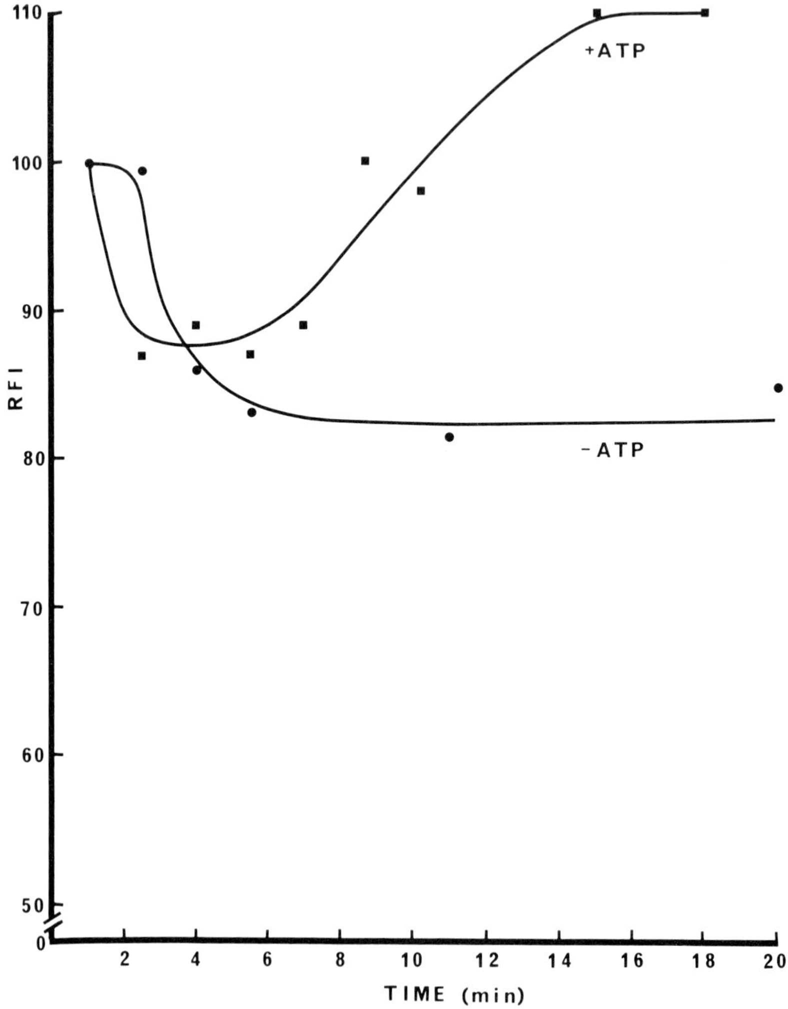

Figure 13 Effect of phospholipase C (75 μg/ml) on the relative fluorescence intensity (*RFI*) of the ANS-enzyme complex in the absence or presence of ATP (2.5 mM). The reaction was carried out in medium containing 80 mM Tris-HCl, pH 7.4, 0.5 mM $CaCl_2$, 125 μg of protein (enzyme), 50 μM ANS. Total volume was 1 ml and the temperature was 23.5°C.

Table II ANS binding by cardiac (Na^+-K^+)-ATPase by different methods*

Additions	ANS binding (moles ANS/mg protein × 10^{-8}) Method A	Method B	Method C
None	4.7	4.5	5.0
5 mM Mg^{++}	7.8	7.8	7.5
5 mM Ca^{++}	7.4	7.5	7.6
40 mM Na^+	6.8	7.0	6.9
40 mM K^+	7.0	6.9	6.8

*The enzyme preparation (0.58–0.78 mg/ml) was incubated at room temperature in 50 mM Tris-HCl, pH 7.4, in the absence or presence of different ions. The reaction was terminated by either centrifugation (Method A) or by Millipore filtration (Methods A and C). The amount of ANS bound to the enzyme was estimated either by spectrophotometry (Methods A and B) or by spectrophotoflurometry (Method C).

versus moles of ANS bound per mg of protein). The plots were linear and the slopes of the lines were used to obtain apparent binding constants (K_{app}). The intersecting point on the abscissa gave the number of binding sites (n) of the preparation. The summary of data from three experiments is presented in table III. Both divalent and monovalent cations increased the number of binding sites while decreasing the binding constants appreciably.

The ANS binding by the enzyme preparation was reduced after phospholipase A or C and trypsin treatments (table IV). In another set of experiments the ANS binding under the ionic conditions of the ATPase reaction, *i.e.*, in the presence of Mg^{++}, Na^+, K^+, and ouabain, was studied in order to compare it with the

Table III ANS binding characteristics of (Na^+-K^+)-ATPase*

Additions	Number of binding sites (n) (moles ANS/mg protein × 10^{-8})	Binding constant, K_{app} (M × 10^{-5})
50 mM Tris-Cl, pH 7.4	6.0	7.5
10 mM $MgCl_2$	9.7	3.9
10 mM $CaCl_2$	10.0	3.7
50 mM NaCl	8.2	4.8
50 mM KCl	8.8	5.1

*Number of binding sites and K_{app} calculated were as described in Methods. Method B was used for ANS binding measurement.

Table IV Influence of phospholipase A, C, and trypsin on binding on ANS and (Na^+-K^+)-ATPase K^+)-ATPase Activity*

	ANS binding (moles ANS/mg protein X 10^{-8})	(Na^+ - K^+)-ATPase activity (μmoles Pi/mg protein/hr)
Control	5.0	48.2
Phospholipase A treatment	2.7	20.0
Phospholipase C treatment	2.9	19.5
Trypsin treatment	3.2	17.3

*For binding assay: 50 mM Tris -Cl, pH 7.4, 60 μM ANS; Method C was employed for ANS binding measurement; protein concentrations varied from 0.46 to 0.54 mg/ml. For ATPase activity, protein concentrations were 50–60 μg/ml, respectively. Temperature was 37°C.

activation of the ATPase system. It was necessary to omit ATP from the ANS binding assay because of its interference with spectrophotometry at 270 nm. Ouabain (0.1 mM) slightly reduced the binding of ANS in the presence of Mg^{++} Na^+, and K^+, although the same concentration of the drug inhibited (Na^+-K^+)-ATPase completely (table V).

Table V Effects of Mg^{++}, Na^+, K^+, and ouabain on ANS binding and ATPase activity*

Medium	ANS binding (moles ANS/mg protein X 10^{-8})	ATPase activity (μmoles Pi/mg protein/hr)
50 mM Tris-Cl, pH 7.4	4.7	
+ 5 mM Mg^{++}	7.5	14
+ 5 mM Mg^{++} + 100 Na	7.9	16
+ 5 mM Mg^{++} + 20 Mm K^+	8.0	16
+ 5 mM Mg^{++} + 100 mM Na^+ + 20 mM K^+	7.8	49
+ 5 mM Mg^{++} + 100 mM Na^+ + 20 mM K^+ + 0.1 mM ouabain	7.3	15
+ 100 mM Na^+	7.1	
+ 20 mM K^+	6.6	
+ 0.1 mM ouabain	4.8	

*Conditions for ANS binding: 50 μM ANS, room temperature. Protein concentrations were 0.38–0.43 mg/ml. Method A was used. Conditions for ATPase assay: 37°C, protein concentrations were 45–60 μg/ml.

DISCUSSION

The specific activities of (Na^+-K^+)-ATPase of the solubilized enzyme preparations obtained after deoxycholate and NaI treatments were higher than that reported for the heart sarcolemmal preparation (9). The low activity of this enzyme in the sarcolemmal fraction may reflect more the "native state" of these membranes because (Na^+-K^+)-ATPase is a latent enzyme which is activated by a variety of structure-disrupting agents (14). At any rate, it was interesting to observe that the (Na^+-K^+)-ATPase of the sarcolemmal fraction showed similarities in many kinetic properties, such as concentrations of Na^+ or K^+ for 50 percent of the maximal activities, and concentrations of ouabain or calcium for 50 percent inhibition, with some minor differences. Although the value of K_m for the solubilized (Na^+-K^+)-ATPase preparation was lower than that for the sarcolemmal membrane fraction (9), this should not be interpreted to mean that the enzymes in these preparations are different from each other. In this regard it should be pointed out that since the preparations employed in these studies are vesicular in shape, the diffusion of substrate may have an important role in the activity of the enzyme. Thus, different shapes of the vesicles may alter the accessibility of substrate to the catalytic sites without changing the properties of the enzyme *per se*. We therefore believe that the enzyme preparation employed in this study is of heart sarcolemmal origin.

From the apparent sigmoidal shape of the curves for substrate velocity reactions using varying amounts of Na^+, K^+, and MgATP in the incubation medium, the heart (Na^+-K^+)-ATPase can be considered to exhibit an allosteric nature of activation kinetics. This is further supported by the fact that n values obtained from Hill plots of the data were greater than 1. These results indicate cooperativity of interaction between the activating ions as well as of the substrate. The interaction of this enzyme with ouabain, a specific inhibitor, under different experimental conditions has also been suggested to provide evidence for an allosteric site (2, 3). The rat brain (Na^+-K^+)-ATPase had also been shown to be allosteric in nature (11, 15). It is perhaps this allosteric transition which accounts for the ability of this enzyme to adapt to varying conditions in its role as a "Na^+-K^+ pump" at the level of the cell membrane.

The inhibition of (Na^+-K^+)-ATPase by ouabain has been suggested to be the molecular mechanism of the positive inotropic action of this agent (13). Although Ca^{++} has been also found to have an inhibitory effect on this enzyme system, the contribution of this effect towards positive inotropism due to calcium can not be stated with certainty. This is because of the fact that calcium is known to possess a direct action on the contractile unit. Although some investigators (4, 6) have shown that (Na^+-K^+)-ATPase is inhibited by catecholamines and tyramine, we have failed to show any effect on this enzyme of various sympathomimetic agents or cyclic AMP which are known to have inotropic effects. This discrepancy in results may be due to differences in the source of (Na^+-K^+)-ATPase since these workers (4, 6) have employed microsomal preparations in their studies. At any rate, we do not consider that the inhibition of sarcolemmal (Na^+-K^+)-ATPase is a universal requirement for the augmentation of myocardial contractility.

The binding of ANS to partially purified cardiac (Na^+-K^+)-ATPase showed many similarities with published reports for a variety of biomembranes (12, 18, 19). The binding of ANS is very sensitive to the conditions of the incubation medium, especially to cations. Both monovalent and divalent metal ions enhanced the ANS binding to the vesicular preparation by increasing the number of binding sites (n) as well as the affinity of the sites towards ANS. Although similar observations were made by Rubalcava *et al.* (12) with red cell membranes, different findings are noted with other biological membranes (5, 18). The reasons for such discrepancies are not yet clear but may reflect the differences in the lipoprotein makeup of these membranes. Phospholipases, as well as trypsin, reduced the ANS binding, ANS-enzyme fluorescence, and (Na^+-K^+)-ATPase activity. This supports the contention that membrane integrity is necessary for the full expression of the catalytic reactivity. The data on ANS binding confirms the hypothesis that both protein and lipid contribute to the binding reaction. At present, there are a number of reports which suggest either predominantly lipid or protein contribution for ANS binding sites on the biomembranes (18, 19). This question is not yet resolved. We would suggest that the lipoprotein portion (hydrophobic parts) of these membranes are important regions for ANS binding sites.

In a series of experiments, ANS binding and ANS-enzyme fluorescence were studied under the ionic conditions of the ATPase assay. Mg^{++} (5 mM) enhanced the ANS binding and ANS-enzyme fluorescence which were further increased by the addition of Na^+ (100 mM), K^+ (20 mM) or both. Addition of ouabain had only slight depressant effect on ANS binding, although the enzymatic activity and ANS fluorescence were markedly inhibited by the same concentration of this glycoside. Furthermore, calcium which inhibited $(Na^+ - K^+)$-ATPase activity was found to increase ANS binding and ANS-enzyme fluorescence. Since binding of ANS was studied at 10 min, whereas fluorescence measurements were recorded in less than 1 min, it is difficult to compare these two observations. Although ATPase activity is linear up to 20 min of the incubation time under the present assay conditions, ANS binding reaches a steady state by 10 min and is essentially complete by that time. However, it is apparent from the data that under certain conditions there is no correlation between amounts of ANS binding, changes in ANS fluorescence, and (Na^+-K^+)-ATPase activity. These data also suggest that nonspecific binding of ANS sites other than $(Na^+-K^+$-ATPase of this preparation respond to ionic environment in a manner analogous to (Na^+-K^+)-ATPase. The high base line fluorescence intensity of the ANS-enzyme complex is also a reflection of nonspecific ANS binding sites not associated with enzyme sites.

The experiments reported in this study reveal that the membrane ATPase activity and ANS-enzyme fluorescence are increased by Na^+ and K^+. Furthermore, these effects of Na^+ and K^+ are markedly inhibited by ouabain. The observed changes in ANS-enzyme fluorescence were not due to artifacts of the method employed for fluorescence measurement since inactive preparations failed to exhibit these alterations. Although it is recognized that only a small amount of the total ANS binding to the heart preparation used in our study may be involved in eliciting specific fluorescence changes due to interaction with (Na^+-K^+)-ATPase, the results reported here can be interpreted to provide evidence for conformational changes due to Na^+ and K^+ action and ouabain-induced enzyme inhibition.

REFERENCES

1. ALBERS, R. W. 1967. Biochemical aspects of active transport. Ann. Rev. Biochem. 36: 727–756.
2. ALLEN, J. C., LINDENMAYER, G. E., and SCHWARTZ, A. 1970. An allosteric explanation for ouabain-induced time dependent inhibition of sodium, potassium-adenosine triphosphatase. Arch. Biochem. Biophys. 141: 322–328.
3. ALLEN, J. C., and SCHWARTZ, A. 1970. Effects of potassium, temperature and time on ouabain interaction with cardiac Na^+, K^+ ATPase: Further evidence supporting an allosteric site. J. Molec. Cell. Cardiol. 1: 39–45.
4. GIBSON, K., and HARRIS, P. 1970. The influence of catecholamines and related substances on hyman myocardial (Na^+, K^+)-ATPase. Res. Commun. Chem. Pathol. Pharmacol. 1: 370–374.
5. GOMPERTS, B., LANTELME, F., and STOCK, R. 1970. Ion association reactions with biological membranes studied with the fluorescent dye 1-anilino-8-naphthalene sulfonate. J. Membrane Biol. 3: 241–266.
6. LEE, K. S., and YU, D. H. 1963. A study of the sodium-and-potassium-activated adenosine-triphosphatase activity of heart microsomal fraction. Biochem. Pharmacol. 12: 1253–1264.
7. LOWRY, D. H., ROSEBROUGH, N. J., FARR, A. L., and RANDALL, R. J. 1951. Protein measurement with the Folin phenol reagent. J. Biol. Chem. 193: 265–275.
8. MATSUI, H., and SCHWARTZ, A. 1966. Purification and properties of a highly active ouabain sensitive Na^+,K^+-dependent adenosinetriphosphatase from cardiac tissue. Biochim. Biophys. Acta 128: 380–390.
9. McNAMARA, D. B., SULAKHE, P. V., SINGH, J. N., and DHALLA, N. S. 1974. Properties of heart sarcolemmal Na^+ - K^+ ATPase. J. Biochem. 75: 795–803.
10. NAGAI, K., LINDENMAYER, G. E., and SCHWARTZ, A. 1970. Direct evidence for conformational nature of the Na^+ K^+-ATPase system: Fluorescence and circular dichroism studies. Arch. Biochem. Biophys. 139: 252–254.
11. ROBINSON, J. D. 1968. Allosteric interactions with the (Na^+ + K^+)-dependent adenosine triphosphatase. Nature 220: 1325–1326.
12. RUBALCAVA, B., DE MUNOZ, D. M., and GITLER, C. 1969. Interaction of fluorescent probes with membranes. I. Effect of ions on erythrocyte membranes. Biochemistry 8: 2742–2747.
13. SCHWARTZ, A., LINDENMAYER, G. E., and ALLEN, J. C. 1972. The Na^+,K^+-ATPase membrane transport system: Importance in cellular function, pp. 1–83. *In* F. Bonner and A. Kleinzeller (eds.), Current Topics in Membranes and Transport, Vol. 3, Academic Press, New York.
14. SKOU, J. C. 1965. Enzymatic basis for active transport of Na^+ and K^+ across cell membranes. Physiol. Rev. 45: 596–617.

15. SQUIRES, R. F. 1965. On the interactions of Na^+, K^+, Mg^{++}, and ATP with the Na^+ plus K^+ activated ATPase from rat brain. Biochem. Biophys. Res. Commun. 19: 27–32.
16. TAUSSKY, H., and SHORR, E. 1953. A microcolorimetric method for the determination of inorganic phosphorus. J. Biol. Chem. 202: 675–685.
17. VANDERKOOI, J., and MARTONOSI, A. 1969. Sarcoplasmic reticulum. VIII. Use of 8-anilino-1-naphthalene sulfonate as conformational probe on biological membranes. Arch. Biochem. Biophys. 133: 153–163.
18. VANDERKOOI, J., and MARTONOSI, A. 1971. Sarcoplasmic reticulum. XII. The interaction of 8-anilino-1-naphthalene sulfonate with skeletal muscle microsomes. Arch. Biochem. Biophys. 144: 87–98.
19. WEIDEKAAM, E., WALLACH, D. F. H., and FISCHER, H. 1971. The effect of phospholipase A upon the interaction of 1-anilinonaphthalene-8-sulfonate with erythrocyte membranes. Biochim. Biophys. Acta 241: 770–778.

Mailing Address: Dr. N. S. Dhalla
Professor of Physiology, Faculty of Medicine,
University of Manitoba.
Winnipeg, Manitoba R3E OW3 (Canada).

Recent Advances in Studies in Cardiac Structure and Metabolism
Volume 9
The Sarcolemma
Edited by Paul-Emile Roy and Naranjan S. Dhalla

STUDIES ON SPECIFIC PROPERTIES OF DOG HEART (Na^+-K^+)-ATPase

M. FEDELEŠOVÁ, A. DŽURBA, and A. ZIEGELHÖFFER

Department of Biochemistry,
Institute of Experimental Surgery,
Slovak Academy of Sciences,
Bratislava, Czechoslovakia

SUMMARY

Three different methods used for partial purification of (Na^+-K^+)-ATPase, Method a, Matsui and Schwartz (5), Method b, Fedelešová *et al.* (3), and Method c, Pitts *et al.* (6), were compared with respect to specific activities, yields, and recovery of the enzyme in preparations from the dog heart. Highest specific activities of $(Na^+ + K^+)$-ATPase, as well as the ratio of $(Na^+, + K^+) + Mg^{2+}$ to Mg^{2+}-ATPase, and the second best recovery and yield have been found in the preparation obtained by Method b. However, using Method c, which includes two deoxycholate treatments, no high $(Na^+ + K^+)$-ATPase activity could be obtained after the second treatment although the yields of protein did not differ from those reported by Pitts *et al.* (6). Therefore, the other specific properties of the dog heart $(Na^+ + K^+)$-ATPase were compared only in enzyme obtained by Lubrol extraction (Method b) and after NaI treatment (Method a). The Na^+ to K^+ concentration ratio necessary for optimal stimulation of $(Na^+ + K^+)$-ATPase activity as well as the pH optimum of the enzyme proved to be similar in NaI- and Lubrol-treated preparations, whereas in the latter the sensitivity to ouabain was 4.55 times higher than in NaI-treated preparation. The K_m values for $MgATP^{2-}$, K_h values for Na^+ and K^+ activation, and S_{50} values for Na^+ activation were similar in preparations obtained by Methods a and b, whereas the values of (n) for K^+ and Na^+ activation and S_{50} for K^+ activation were slightly higher in the NaI-treated fraction.

The activity of $(Na^+ + K^+)$-ATPase in the heart varies in different species, and one of the lowest levels of activity is found in the dog heart (1). Therefore, it would be of particular interest to further purify this enzyme in order to carry out more precise and detailed enzymological and functional studies.

The dog heart preparation obtained by NaI treatment according to the method of Matsui and Schwartz (5) shows relatively low $(Na^+ + K^+)$-ATPase activities. Therefore attempts were made to purify the enzyme with the aid of nonionic detergents. Such a method was worked out by Shirachi (7) and slightly modified for the dog heart enzyme by us (3).

A third method, based on combination of differential centrifugation with deoxycholate, NaI, and glyceral treatment, was described recently by Pitts *et al.*

(6). This method promises the purest enzyme with highest $(Na^+ + K^+)$-ATPase activity, but little information is available to state whether this preparation exhibits specific properties similar to those obtained by other methods.

In the present study the three methods for partial purification of $(Na^+ + K^+)$-ATPase mentioned above have been compared with respect to specific activity, yield, and recovery of the enzyme in preparations from dog heart. In addition, specific properties such as pH optimum, K_m for $MgATP^{2-}$ complex, sensitivity of the enzyme to ouabain, as well as allosteric interactions among the binding sites for Na^+ and K^+ were investigated.

MATERIALS AND METHODS

Preparation of microsomal fractions

Adult mongrel dogs were used. The hearts were removed under pentobarbital anesthesia. Approximately 5–10 gm of dog heart left ventricles were freed from fat and connective tissue and washed thoroughly in an ice-cold medium (0.25 M sucrose, 1 mM EDTA, 10 mM Tris-HCl, pH 7.4) to remove blood. A further procedure for preparation of the sarcolemna-enriched microsomal fraction when it was later used for purification with NaI treatment or with Lubrol treatment was described in some of our earlier studies (2). This was a slight modification of the method reported by Matsui and Schwartz (5). The whole procedure for isolation of the glycerol-treated enzyme was carried out according to the original method described by Pitts *et al.* (6).

The NaI treatment of the microsomal fraction was also carried out essentially according to the method of Matsui and Schwartz (5) with minor modifications described in one of our earlier papers (2).

Lubrol extraction of microsomal fractions

Solution of Lubrol WX within a concentration range of 0.1–1.6 percent (w/v) in 0.32 M sucrose and 1 mM EDTA, pH adjusted to 7.6 with Tris buffer, was added, according to the method of Shirachi (7), to an equal volume of the prepared microsomal fraction. This suspension was homogenized in an Ultra/Turrax homogenizer (Janke and Kunkel, GRF).

The suspension was kept on ice for 15 min and mixed intermittently. It was then centrifuged in a VAC 60 (Janetzky, GDR) at $100{,}000 \times g$ for 1 hr. The supernatant was stored at $-10°C$.

Assays

The conventional medium used in assaying total ATPase contained 50 mM Tris HCl, pH 7.4, 100 mM NaCl, 20 mM KCl, 5 mM $MgCl_2$, 4 mM Tris-ATP, and the enzyme. In determining the initial reaction velocities to establish reaction kinetics, an ATP regeneration system consisting of 2.5 mM phosphoenolpyruvic acid and 0.2 mg pyruvate kinase was added to the conventional assay medium. Any further variations from these assay conditions are indicated in the figures. The medium without ATP was preincubated at 37°C for 1 min and the reaction was allowed to run for 10 min at 37°C.

It was terminated by the addition of 1 ml of 10 percent trichloroacetic acid. The inorganic phosphate, in an aliquot of the upper phase obtained after centrifugation, was determined by the method of Taussky and Shorr (8). Protein concentrations were measured by the method of Lowry *et al.* (4). Appropriate sucrose-Lubrol blanks were run for all experiments.

Materials

All water used was tap-distilled, then passed through a bed of mixed anion and cation exchange resins and finally double distilled in a quartz still. Lubrol WX was obtained from I.C.I. Organics, Inc. (Providence, R. I.). The Tris-ATP (Sigma, Ltd.) solution was adjusted to pH 7.4. Ouabain was obtained from Sigma, Ltd., the phosphoenolpyruvic acid and pyruvate kinase were purchased from Boehringer, Ltd. Other reagents or chemicals used were analytical reagent grade.

RESULTS AND DISCUSSION

For comparative purposes and also for checking efficiency of single preparation procedure, the yields, ATPase activities, and values of recovery of $(Na^+ + K^+)$-ATPase obtained by all procedures employed and/or some of their main steps are presented in table I. Since the microsomal fraction obtained by differential centriguation according to the method of Matsui and Schwartz (5) represented the first main step (Preparation A) for both on the NaI and Lubrol treatment purification techniques (Preparations B and C); no other "intersteps" of the latter methods were taken into consideration. Concerning the method of Pitts *et al.* (6), five main intersteps, namely: the homogenate obtained after first stirring with deoxycholate (DOC) termed, D_1; the 95,000 X *g* fraction frozen overnight (D_2); the NaI-treated enzyme frozen overnight at −20°C (D_3); the 80,000 X *g* DOC-solubilized enzyme fraction (D_4); and the glycerol-percipitated enzyme fraction (D_5) were examined as to their yield. ATPase activity and recovery of (Na^+-K^+)-ATPase. Using the above preparation technique we did not succeed in obtaining high ATPase activities after the NaI treatment step (D_3); on the contrary, we found diminution of $(Na^+ + K^+)$-ATPase and total ATPase activities, in spite of the fact that the recovery of protein in fractions D_4 and D_5 was very similar to that reported in the original paper of Pitts *et al.* (6). However, it also appears from table I that the NaI-treated Matsui-Schwartz preparation (Preparation B), and the NaI-treated preparation frozen overnight (D_3) according to the method of Pitts *et al.* (6) give similar yields of 0.65 and 0.70, whereas the latter exhibits higher total and $(Na^+$

Table I Yields, ATPase activities, and total recovery of different preparations from dog heart*

Experimental preparations	Yield	$(Na^+ + K^+) + Mg^{2+}$-ATPase	Mg^{2+}-ATPase	$\frac{(Na^+ + K^+) + Mg^{2+}}{Mg^{2+}}$	Recovery of $(Na^+ + K^+)$-ATPase (%)
A	1.60 ± 0.11	32.50 ± 2.50	25.00 ± 2.10	1.30	100.00
B	0.65 ± 0.07	24.00 ± 1.80	12.00 ± 0.90	2.00	65.00
C	0.27 ± 0.06	31.20 ± 3.00	6.00 ± 0.50	5.16	56.10
D_1	118.00 ± 41.60	32.20 ± 1.30	28.60 ± 5.80	1.13	100.00
D_2	3.30 ± 1.35	33.60 ± 4.40	22.30 ± 4.80	1.48	8.75
D_3	0.70 ± 0.30	30.60 ± 1.80	11.10 ± 2.90	2.76	3.20
D_4	0.28 ± 0.01	6.00 ± 0.30	3.00 ± 0.09	2.00	0.20
D_5	0.14 ± 0.01				

*The results are means ±S.E. of five to eight experiments. The yields are expressed in milligrams of protein g^{-1} and the specific activities in micromoles of Pi per mg of protein per hr. Microsomal fractions were prepared from dog heart tissue by differential centrifugation, sodium iodide, DOC, glycerol, or Lubrol treatment, and their ATPase activities were determined as indicated under Methods. A, Microsomal fraction obtained by differential centrifugation according to the method of Matsui-Schwartz (5); B, sodium iodide-treated microsomal fraction of type A; C, Lubrol-extracted microsomal fraction of type A; D, preparation according to Pitts *et al.* (6). Single steps are indicated as: D_1, homogenate after first stirring with DOC; D_2, the 95,000 × *g* enzyme frozen overnight; D_3, the NaI-treated enzyme frozen overnight at −20°C; D_4, the 80,000 × *g* DOC-solubilized enzyme; D_5, the glycerol-precipitated enzyme.

$+ K^+)$-ATPase activities. Taking the recovery of $(Na^+ + K^+)$-ATPase and the ratio of total to Mg^{2+}-ATPase as criteria of suitability of the purification procedures employed, the Lubrol treatment seems to yield the best preparation. Moreover, it gives yields of protein as high as those of interstep D_4 in Pitts' preparation with near-to-maximal specific activities of total ATPase, whereas the specific activities of $(Na^+ + K^+)$-ATPase are about 25.2 μmoles of Pi/mg of protein/hr and thus highly exceed the activities of the enzyme in the other preparations.

Other specific properties of the dog heart $(Na^+ + K^+)$-ATPase preparations were compared only with enzymes obtained by Lubrol extraction and NaI treatment (table II). Because of the low specific activities of $(Na^+ + K^+)$-ATPase in preparations treated for a second time with DOC and in glycerol-precipitated preparations (D_4 and D_5), only some properties of the enzyme in the first DOC- and NaI-treated fraction (Preparation D_3) were examined. The sodium to potassium ratio necessary for optimal stimulation of $(Na^+ + K^+)$-ATPase activity proved to be similar, *i.e.*, $[Na^+]:[K^+]$ = 100:20 mM for both the NaI-treated and the Lubrol-extracted preparations (Preparations B, C, and D_3).

pH sensitivity of $(Na^+ + K^+)$-ATPase activities were examined in Tris-malate buffer in a pH range from 6.4 to 8.0 (in preparation D_3 in a pH range from 7.4 to 7.8), the optimum being in all cases 7.6. As we have recently reported (3), Lubrol treatment apparently does not influence the pH optimum; it lowers,

Table II Kenetic parameters of different preparations from dog heart

Microsomal preparations	V_{max}	S_{50}	K_m	K_h	*n*
Sodium iodide treated					
For ATP	10.5		0.66		
For K^+	4.7	0.25		0.07	1.88
For Na^+	11.9	2.23		3.00	1.36
Lubrol extract					
For ATP	33.0		0.73		
For K^+	12.5	0.17		0.07	1.50
For Na^+	22.7	2.45		3.07	1.25

*Results are obtained from Linewiaver-Burk plots and Hill plots in which the straight lines are drawn by the method of least squares; each experimental point represents the average of five or more experiments performed in duplicate. Values from the Hill plots of *n*, S_{50}, and K_h were calculated from the equation for the straight lines. Preparation of various microsomal fractions as well as the determination of the enzyme activities were described in Methods.

however, the sensitivity of the enzyme to the influence of pH values differing from the pH optimum.

Since both NaI-treated preparations (B and D_3) not only gave similar yields and specific activities of $(Na^+ + K^+)$-ATPase (table I) but did not differ either in the $[Na^+]:[K^+]$ ratio necessary for optimal stimulation or in the pH optimum, for further specific properties we examined only the NaI-treated preparation according to the method of Matsui and Schwartz (5) and the Lubrol-extracted enzyme preparations B and C).

The $(Na^+ + K^+)$-ATPase activity of the Lubrol-treated preparation was found to be 4.55 times more sensitive to ouabain, *i.e.*, a 50 percent inhibition occurred at a concentration of 0.88 μmole of ouabain, whereas in the case of NaI-treated microsomes, the same extent of inhibition occurred only at as high a concentration as 4 μM of ouabain (3).

It appears from an overall comparison of the kinetic parameters of $(Na^+ + K^+)$-ATPase that, besides all the V_{max} and n values as well as S_{50} values for K^+ activation, there were but minor differences between properties of the enzyme in the former two preparations, as has already been reported (3).

Sensitivity to NaN_3 of $(Na^+ + K^+)$-ATPase activity in Lubrol extracts and in both NaI-treated preparations (B and D_3 could be demonstrated; it varied in all cases between 1.5 and 0.5 percent inhibition.

Our results suggest that, although the yields, recovery, specific activities of $(Na^+ + K^+)$-ATPase, ratio of specific activities of $(Na^+ + K^+) + Mg^{2+}$ to Mg^{2+}-ATPase, as well as joint changes in kinetics and ouabain sensitivity differed depending upon the procedures employed for preparation, most of the specific properties of $(Na^+ + K^+)$-ATPase remained unaltered or were modified only slightly, at least in the NaI-treated and in Lubrol-extracted preparations.

REFERENCES

1. AKERA, T., LARSEN, F., and BRODY, T. 1969. The effect of ouabain on sodium- and potassium-activated adenosine triphosphatase from the hearts of several mammalian species. J. Pharmacol. Exp. Therap. 170: 17–26.
2. FEDELEŠOVÁ, M., ZIEGELHÖFFER, A., LUKNÁROVÁ, O., DŽURBA, A., and KOSTOLANSKÝ, Š. 1973. Influence of K^+-Mg^{++}-(D,L)-aspartate on various ATPase activities of the dog heart. Arzneim. Forsch. (Drug. Res.) 23 (8): 1048–1053.
3. FEDELEŠOVÁ, M., DŽURBA, A., and ZIEGELHÖFFER, A. 1974. Partial purification of dog heart $(Na^+ + K^+)$-ATPase by Lubrol. Acta Biol. Med. Germ. 33: 169–180.
4. LOWRY, O. H., ROSEBOROUGH, N. J., FARR, A. L., and RANDALL, R. J. 1951. Protein measurement with the Folin phenol reagent. J. Biol. Chem. 193: 265–275.
5. MATSUI, H., and SCHWARTZ, A. 1966. Purification and properties of a highly active ouabain sensitive Na^+ ++ K^+ dependent adenosine triphosphatase from cardiac tissue. Biochim. Biophys. Acta 128: 380–390.
6. PITTS, B. J. R., LANE, L. K., and SCHWARTZ, A. 1970. Nature of the transport ATPase-digitalis complex. VI. Purification of and ouabain binding to a highly active cardiac Na^+, K^+-ATPase. Biochem. Biophys. Res. Commun. 53: 1060–1065.

7. SHIRACHI, D., ALLARD, A., and TREVOR, A. 1970. Partial purification and ouabain sensitivity of Lubrol-extracted sodium-potassium transport adenosine triphosphatases from brain and cardiac tissues. Biochem. Pharmacol. 19: 2893–2906.
8. TAUSSKY, H., and SHORR, E. 1953. A microcolorimetric method for the determination of inorganic phosphorus. J. Biol. Chem. 202: 675–685.

Mailing address: Dr. M. Fedelešová,
Department of Biochemistry,
Institute of Experimental Surgery,
Slovak Academy of Sciences,
Bratislava (Czechoslovakia).

Recent Advances in Studies on Cardiac Structure and Metabolism
Volume 9
The Sarcolemma
Edited by Paul-Emile Roy and Naranjan S. Dhalla

ON THE SARCOLEMMAL SITE OF ACTION OF CARDIAC GLYCOSIDES

H. LÜLLMANN and T. PETERS

Department of Pharmacology,
University of Kiel,
Kiel, Germany

SUMMARY

An electron microscopic study of the guinea pig myocardial cell was conducted. By the use of $LaCl_3$ two morphologically distinct tubular systems could be demonstrated. The larger one represents the transverse tubular system which originates from the plasma membrane and is continous with the extracellular space (ECS). In the heart muscle cell it forms a three-dimensional network and is closely apposed to the myofilaments at the level of the Z and I bands. A comparatively small and sparsely developed tubular system which is not accessible from the ECS represents the sarcoplasmic reticulum (SR).

By means of sucrose-density centrifugation in a discontinuous gradient two main microsomal fractions were obtained (F1 and F3). The pre-existing structure of F1 could be identified as the transverse tubular system, whereas F3 originates from the SR.

The morphological features (*e.g.*, an enclosed basement membrane) and the fact that after perfusion of hearts with [^{3}H]inulin and preparation of microsomes the ECS marker is almost completely retained in F1 indicate that upon fractionation and vesiculation inside-out vesicles are formed which trap the former extracellular fluid. Moreover, the former extracellular surface is hidden and no longer accessible for membrane-impermeable compounds.

After perfusion of hearts with [^{3}H]ouabain and fractionation and centrifugation the radioactive material was found to accumulate in F1, the kinetics being identical with that of binding to the intact tissue and with the time course of the inotropic action of ouabain.

The inside-out vesicles derived from the plasma membrane are characterized by the presence of a (Na^+-K^+)-ATPase activity, an outward Ca pump, a high binding capacity for Ca, and a low Ca permeability.

Since the (Na^+-K^+)-ATPase is known to react with cardiac glycosides (CG) at the outer cell surface it was not surprising to find a complete lack of inhibition of the enzyme by the membrane-impermeable CG ouabain if administered to the intact sarcolemmal microsome. After disruption of the microsomes, however, a considerable inhibition of the (Na^+-K^+)-ATPase became demonstrable.

Similarly, the Ca permeability of the plasma membranes could only be enhanced by CG if interacting with the former outside. Although ouabain again failed to exert any action, the membrane-permeable digoxin increased the Ca permeability of the intact vesicular membrane.

The rate of Ca transport as an indicator of the Ca pump activity remained unaffected by CG provided from either side of the membrane.

The passive Ca binding to the plasma membrane, which results in a concentration up to 5 mM in this structure, is thought to represent the Ca pool essential for excitation-contraction coupling. The CG are thought to alter the Ca binding in this pool, thus rendering depolarization more effective in releasing Ca^{++}.

Present reports which deal with the site and the mode of action of CG remain controversial in spite of the fact that during the past years numerous efforts have been made to gain more detailed information on cellular distribution and action of this class of drugs (for review see References 6 and 12). It is well accepted that CG at low concentrations increase the Ca exchange rate and stimulate (Na^+-K^+)-ATPase. Higher concentrations of CG will lead to an inhibition of the Na^+-K^+ pump, an enhanced K^+ efflux, and an increased passive Ca^{++} permeability (6, 12). Recently an influence of ouabain on a superficially located Ca pool could be observed (2, 7, 9). However, it is not yet decided whether CG display their inotropic and toxic actions exclusively at the plasma membrane (7) or also at intracellular sites (3).

The data to be presented in this paper are thought to add to the understanding of the modes of action of CG and to define the sites at which these actions are exerted on a cellular level. The investigations were conducted in order to 1) identify the cellular structures which are involved in the binding of CG under special consideration of the morphological particularities of the heart muscle cell (10), 2) investigate the physiologically occurring processes located at this site, 3) study the influence of CG on these processes, and, finally, 4) correlate the findings to the possible inotropic and toxic modes of action of CG.

METHODS

Preparation for electron microscopy[1]

Hearts were perfused according to the method of Langendorff with HEPES buffer either with or without $LaCl_3$ in a concentration of 10^{-4} or 10^{-3} M. After different time periods perfusion was continued for 30 min with cold fixative (3 percent glutaraldehyde in 0.1 M cacodylate buffer, pH 7.4, containing 0.2 M sucrose). Small cubes of tissue were prepared, postfixed with unbuffered 2 percent OsO_4 for 2 hr, and further processed for electron microscopy, using araldite as the embedding medium. Thin sections were stained with uranyl acetate and lead citrate. Electron micrographs were taken with a Philips E 300.

Pellets were covered with the same fixative as mentioned above. After fixation for 2 hr, the pellets were removed carefully from the centrifuge tubes and postfixed. For cutting the blocks were oriented so that all the pellets from the top to the bottom were included in the thin sections. Further processing was performed as indicated above.

---

[1] The electron microscopic investigations were carried out in cooperation with Dr. Renate Lüllman-Rauch, Department of Anatomy, University of Kiel.

Preparation of plasma membranes and sarcoplasmic reticulum

Hearts were excised from guinea pigs of either sex weighing 500–700 gm. The hearts were washed free of blood by a 2-min perfusion period according to the method of Langendorff and collected in ice-cold 0.3 M sucrose solution. Myocardial muscle (15 gm) was minced and homogenized in 3 times its volume (w/v) of 0.3 M sucrose solution in a Virtis homogenizer four times for 7 sec at 20-sec intervals. The homogenate was transferred to cellulose-nitrate centrifuge tubes (Beckman) and a sucrose gradient of the following combination was established: 4 ml of 0.4 M, 4 ml of 0.6 M, 4 ml of 0.8 M, and 8 ml of 1.0 M sucrose solution were added by means of a capillary tube. The sucrose solutions were buffered with 10 mM imidazole and HCl to a final pH of 7.4. The sucrose gradients were then centrifuged in an SW 27 rotor in a Beckmann-Spinco centrifuge for 50 min at 12,500 r.p.m. The dense zones which after centrifugation appeared at the interphases of the 0.3/0.4 M, 0.4/0.6 M, and 0.6/0.8 M sucrose solution were carefully removed and centrifuged again for 50 min at 50,000 r.p.m. in a 60 Ti rotor (Beckman, Spinco Division). After dissolution in 0.3 M sucrose solution, the pellets were pottered and either used for 24 hr or frozen quickly in liquid air, stored at −20°C, and used over a period of about 10 days. Figure 1 gives a schematic presentation of the main preparation steps.

Perfusion of Langendorff hearts with [^{3}H]inulin or [^{3}H]sucrose and successive preparation of membrane fractions

Coronary perfusion was performed with Tyrode solution of the usual composition and a pH 7.4. The calcium concentration was 1.8 mM. After an equilibration period of 30 min [^{3}H]inulin or [^{3}H]sucrose was added in a final concentration of 5×10^{-6} M. After a perfusion period of 120 min the hearts were minced and homogenized rapidly in 3 times their volume of 0.3 M sucrose solution. Homogenization and centrifugation were carried out as mentioned above. The

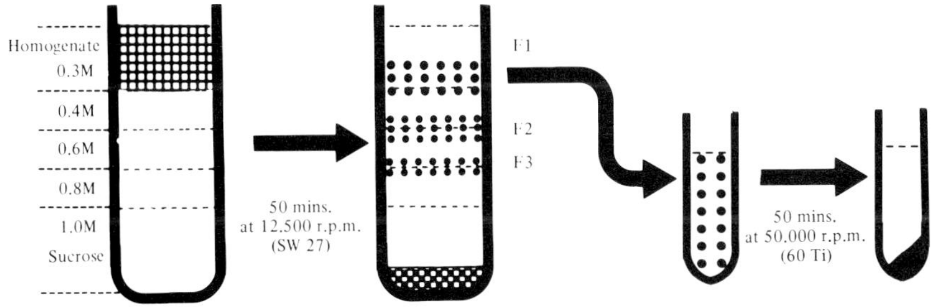

Figure 1 Schematic representation of the preparation procedure of plasma membrane (F1) and sarcoplasmatic (F3) fractions. For details see Methods.

pellets were weighed and dissolved in scintillation vials by adding 2 ml of Soluene (Packard Instruments). Radioactivity was determined as usual.

Perfusion of Langendorff hearts with [^{3}H]ouabain

In these experiments the hearts were perfused for time intervals ranging from 5 to 120 min. Otherwise, experiments proceeded as mentioned above. Ouabain was applied at a concentration of 1×10^{-7} M.

Determination of enzyme activities

Total ATPase activity. In a standard procedure 0.1 mg of membrane protein was incubated with 1 ml of either intracellular (I) or extracellular (II) composition: Medium I, KCl 100 mM, imidazole 10 mM, $MgCl_2$ 3mM, $CaCl_2$ 3×10^{-5} M, NaH_2PO_4 23 mM, ATP 1 mM, pH 7.4; Medium II, NaCl 100 mM, KCl 5 mM, $MgCl_2$ 3 mM, EDTA 1 mM, Tris 20 mM, pH 7.4.

The ATP hydrolysis was measured by spectrophotometric determination of the NADH consumption. In order to eliminate interferences of ions or drugs with the enzymatic ATPase assay for any of the conditions chosen, a parallel test was conducted in the following way: membrane protein was incubated in a volume of 5 ml of the respective medium but without the reactants used for the activity determination, then an aliquot of 1 ml was transfered into a cuvette prepared with the necessary concentrations of reactants containing the ions required for the activation of this test set (Mg^{2+} and K^+). The immediate consumption of NADH was recorded. The activity of a (Na^+-K^+)-ATPase was determined by the addition of 10^{-4} M ouabain or the same concentration of either digitoxin or digoxin, or by omission of Na and K in the 5-ml medium.

(Na^+-K^+)-ATPase activity. The plasma membrane fraction was diluted by a hypotonic buffer solution (1 mM EDTA, 10 mM Tris HCl, pH 7.4), 1:10, and sonicated in the presence of 0.1 percent Na-deoxycholate for 15 sec at 200 W. The resulting suspension was centrifuged at 50,000 r.p.m. for 60 min in a 65 rotor (Beckman, Spinco Division). The supernatant was discarded and the pellet resuspended and washed three times with the above mentioned buffer solution, in which the Na-deoxycholate was omitted in the third wash. The resulting pellet was pottered in a solution of 1 mM EDTA and buffered with Tris HCl to a pH of 7.4. Aliquots of this suspension were subjected to the ATPase assay as described above.

Measurement of Ca transport

Vesicle protein (0.02 mg/ml) was incubated in Medium I which now additionally contained ^{45}Ca for a total period of 60 min. After different time intervals 1 ml was removed and suctioned through ultrafilters (pore size 0.45 μm). The radioactivity of filter and filtrate were determined in the usual way (^{45}Ca; specific activity 27 mCi/mg).

Determination of passive Ca binding to plasma membrane preparation

Vesicle protein in a concentration of 0.5 mg/ml was incubated for 30 min in a medium containing different amounts of calcium (10^{-7}–10^{-2} M), and buffered with imidazole HCl to a pH of 7.4. Mg and ATP were present in a concentration of 1.5 mM and 1 mM, respectively. ATP hydrolysis in these experiments was checked as mentioned under (Na^+-K^+)-ATPase activity.

Statistical analysis

Student's *t*-test was used as a test of significance between mean values. Significance was chosen at the $p < 0.5$ level.

Drugs and chemicals applied

Ouabain, digoxin, digitoxin, and Na_2-ATP were obtained from Merck AG, Darmstadt, Germany; [^{3}H]ouabain (specific activity 10 Ci/mmole), [^{3}H]inulin (specific activity 100–500 mCi/gm), and [^{3}H]sucrose were obtained from NEN, Chicago. Na–pentobarbital was a gift of Knoll AG; lactate, phosphoenolepyruvate, nicotinamide adenine dinucleotide (reduced form), lactate dehydrogenase, and pyruvate kinase were obtained from Boehringer, Mannheim, Germany.

RESULTS AND DISCUSSION

Morphology of guinea pig cardiac muscle

The most striking feature of the cardiac muscle cell is the appearance of an abundant amount of holes and tubules, the diameters of which range from 2,000 to 4,000 Å (figs. 2 and 3). These structures are regularly arranged at the level of the Z and I bands of the sarcomeric units. By means of perfusion of guinea pig hearts with HEPES solution containing lanthanum those broad tubular systems become stained and thus more pronounced.

It can be seen in figure 2 that these transverse tubules are invaginations of the sarcolemma, which penetrate the cell and branch in several directions. In some places this transverse tubular system is accompanied by small tubular structures, the interior of which remain unstained by lanthanum. These tubules represent the sarcoplasmatic reticulum (for details see legend of figure 2). From these electron microscopic studies it can be concluded that the heart muscle cell contains two major tubular systems: 1) the large transverse tubular system which is continuous with the plasma membrane, thus representing part of the ECS, and 2) the small tubular branches of the sarcoplasmatic reticulum with a diameter of about 400 Å, which are not accessible from the ECS.

Fractionation of cardiac tissue and isolation of microsomal and tubular structures

Upon homogenization and fractionation of the crude homogenate in a discontinuous sucrose density gradient (0.4, 0.6, 0.8, 1.0 M sucrose), three distinct

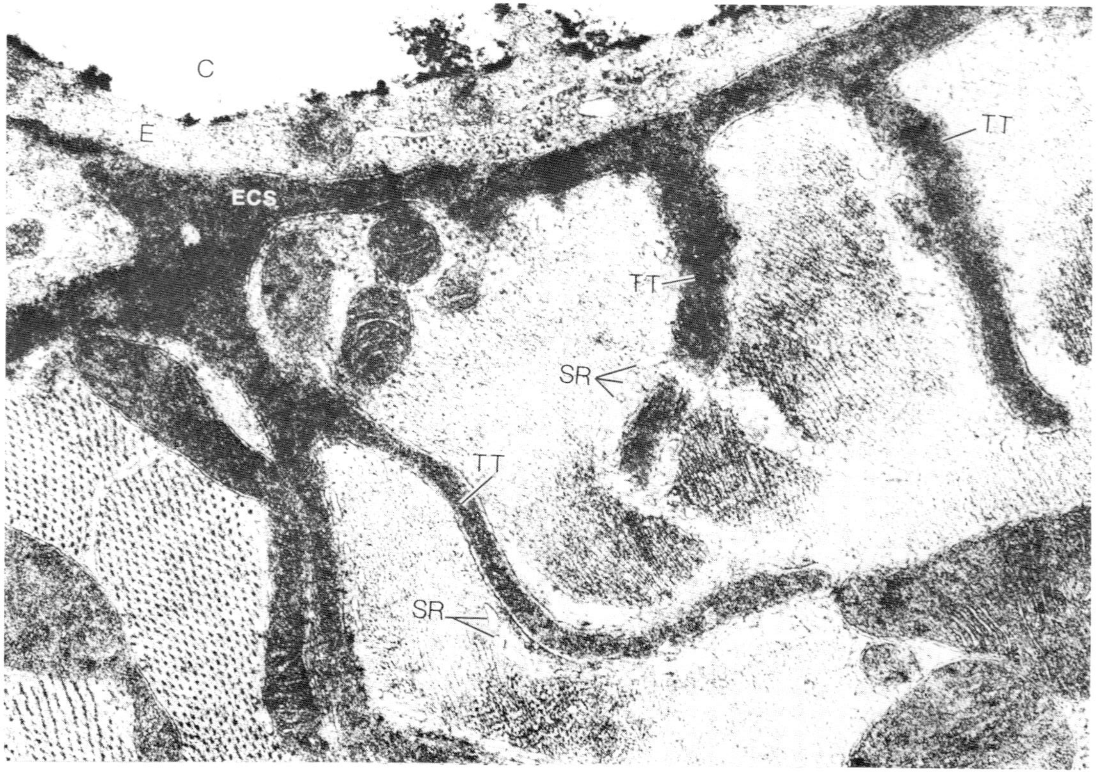

Figure 2 Part of a guinea pig myocardial cell from a Langendorff heart perfused with lanthanum. Capillary lumen (*C*), extracellular space (*ECS*) and transverse tubules (*TT*) are filled with electron dense material, whereas the cisternae of the sarcoplasmic reticulum (*SR*) are empty. On this cross-section the communication between ECS and TT is evident. *E*, Capillary endothelium. Preparation fixed with glutaraldehyde and osmium, section stained with uranyl acetate and lead citrate (× 36,900).

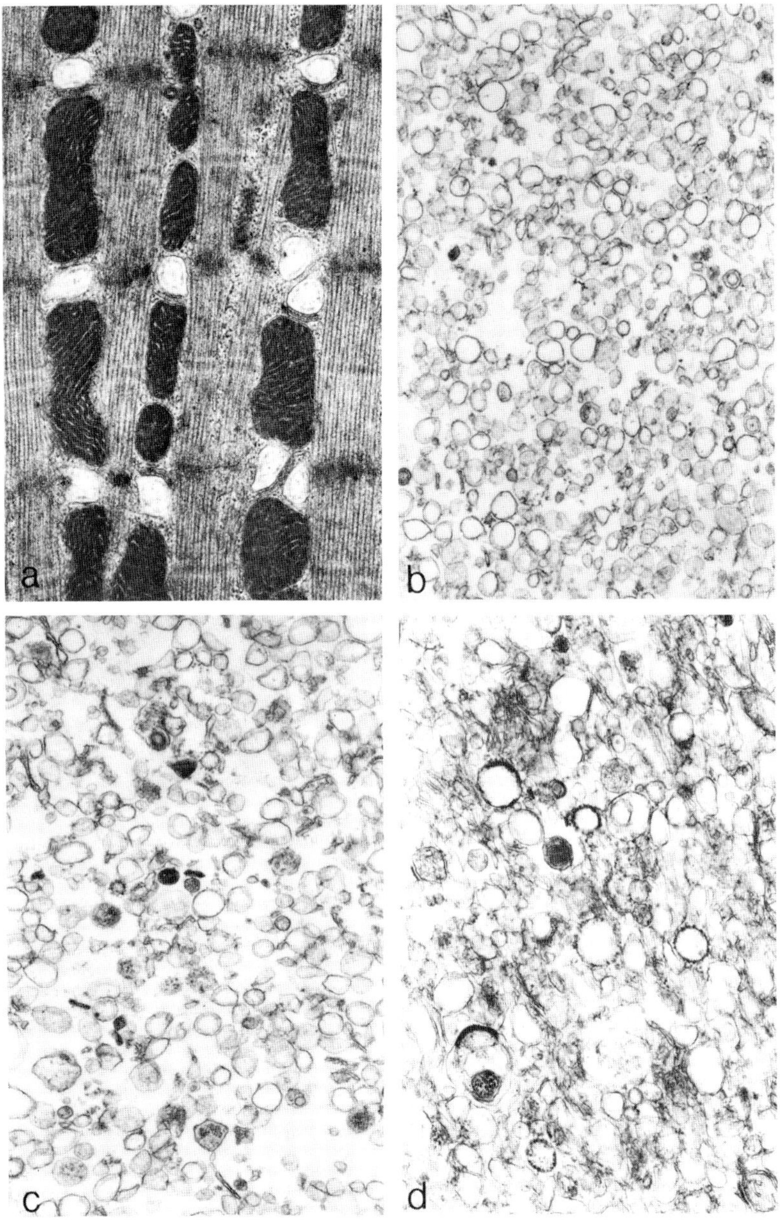

Figure 3 Comparative representation of the ultrastructure of an intact cardiac cell and the fractions obtained from sucrose density gradient centrifugation. *a,* Part of a myocardial cell from a perfused Langendorff heart. Several T tubules in cross-section are visible. They are lined by a basement membrane (X 20 500). *b, c, d,* fractions F1, F2, F3. F1 is the most homogenous fraction (X 20,500). F1 and F2 contain vesicles, most of which are lined at the inner surface by a layer of amorphous material. F3 contains tubular and vesicular structures and some larger vesicles resembling rough endoplasmic reticulum.

bands are obtained in the 0.3/0.4, 0.4/0.6, and 0.6/0.8 M interphases (fig. 1). The ultrastructural appearances of the respective pellets are presented in figure 3*b*, *c*, and *d*. There exist considerable differences concerning shape (vesicular, tubular) as well as size of these structures. The 0.3/0.4 M interphase contains a vesicular fraction (F1) of high homogeneity and purity (fig. 3), the vesicular diameters ranging between 2,000 and 4,000 Å. The inner surface of the vesicles is lined by a basement membrane. In contrast, from the 0.6/0.8 M interphase a pellet is obtained which consists mainly of small vesicles (diameter about 400 Å) and tubules (F3), the inner surface of both structures being devoid of a basement membrane. The 0.4/0.6 M fraction represents a mixture of F1 and F3.

Comparison between tubular structures of intact myocardium and structures obtained by density gradient centrifugation

An ultrastructural comparison of the morphological appearance of intact cells and fractions F1 and F3 (fig. 3) reveals the following structural similarities. Concerning the size (1,000–2,000 Å) and structural properties of the vesicular membrane of F1, a pronounced similarity exists between this fraction and the transverse tubular system of cardiac muscle. The electron microscopic appearance suggests that the sarcolemmal invaginations have to be considered as the *in vivo* pre-existing structures of F1. The size and shape of F3 support the possibility that the pre-existing structure of this fraction is the sarcoplasmatic reticulum.

Perfusion of Langendorff hearts with tritiated inulin, sucrose, and ouabain, successive fractionation of the tissues, and determination of radioactive material in different microsomal fractions

In order to further substantiate the morphological findings the attempt was made to obtain more detailed information by means of radioactively labeled compounds, which mark the ECS but do not penetrate into the cell. It was expected that these compounds after perfusion and fractionation of a Langendorff heart would label microsomal fractions of extracellular origin but not fractions originating from intracellular tubular systems. In a first set of experiments Langendorff hearts were perfused for 120 min with [^{3}H]inulin or [^{3}H]sucrose. After homogenization of the hearts and fractionation in a density gradient the obtained fractions (F1, F2, F3) were resuspended in 0.3 M sucrose and centrifuged to form pellets. The amount of radioactive material in pellets and supernatants was determined in the usual way. The results of these experiments are depicted in figure 4. In the case of [^{3}H]inulin as well as in the case of [^{3}H]sucrose, F1 contained a concentration of radioactivity which was identical with that originally determined in the perfusion fluid. A far lower concentration was detected in F2, and the concentration of radioactivity found in F3 was negligibly small and can be attributed to contamination resulting from the

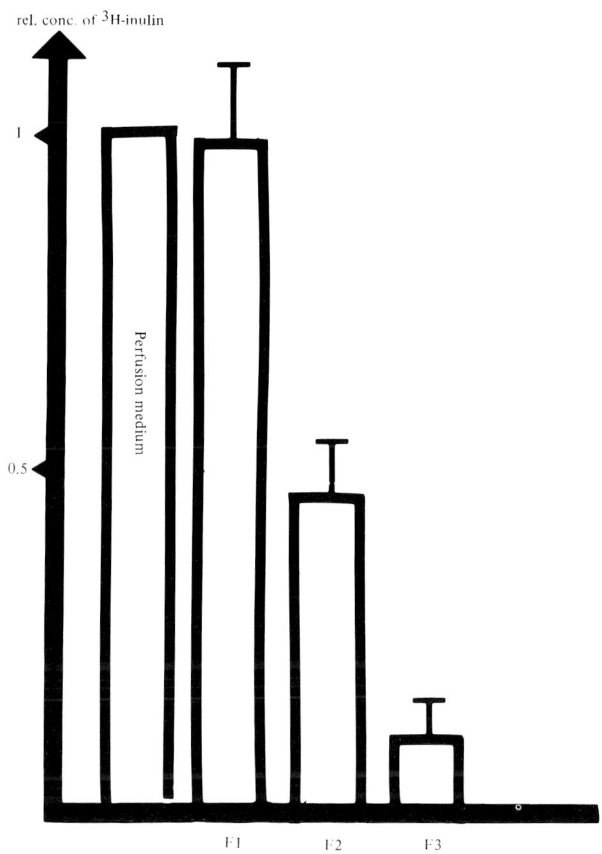

Figure 4 Concentration of [^{3}H]inulin in the perfusion medium in relation to the concentrations in fractions F1, F2, and F3 after Langendorff perfusion. *Ordinate,* the concentration in the perfusion medium is set equal to 1. *Vertical bars on top of each column* indicate ± S.E.M. of n = 6 experiments. For details see text.

process of fractionation. In control experiments hearts were minced and [^{3}H]inulin or [^{3}H]sucrose was added immediately before homogenization. In this case less than 3 percent of the radioactivity, as compared with the perfusion experiment, was bound. The findings indicate that before the process of vesiculation, the structures forming F1 were continuous with the ECS and freely accessible for [^{3}H]inulin and [^{3}H]sucrose. For F3, however, it becomes obvious that a continuity with the ECS has never existed.

A second set of experiments was designed to study the microsomal distribution of [^{3}H]ouabain under the same conditions as mentioned above. Moreover, the perfusion was performed for different time intervals in order to additionally obtain kinetic data on the time course of binding of this drug to the intact

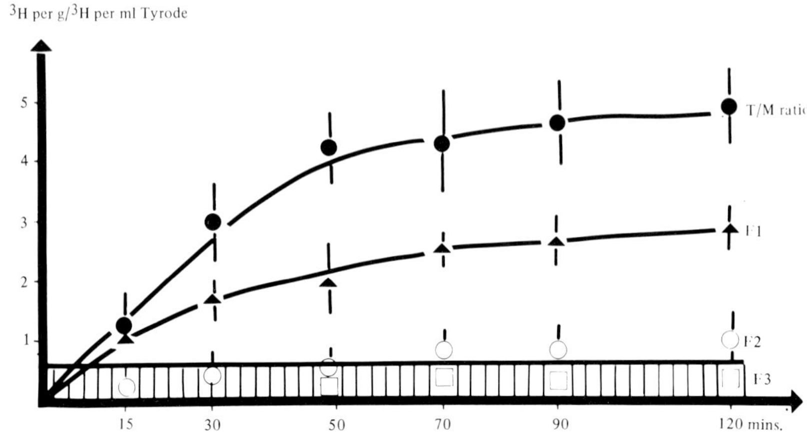

Figure 5 Uptake kinetics of [^{3}H]ouabain in heart muscle and fractions F1, F2, and F3 after Langendorff perfusion for different time intervals. *Ordinate,* ratio of ^{3}H per gm of pellet or tissue to ^{3}H per ml of Tyrode solution. *Abscissa,* time of perfusion in minutes. The points of the curves are the means ± S.E.M. The *hatched area* indicates the size of the ECS.

tissues as well as of accumulation in those parts of the tissue, which later form microsomes. A presentation of the results is given in figure 5. The ratio ^{3}H per g of tissue: ^{3}H per ml Tyrode solution (T/M) of almost 5 demonstrates, as do experiments carried out on isolated papillary muscles (8), that [^{3}H] ouabain at the concentration applied becomes adsorbed to cardiac tissue. Similar to the findings with ECS markers, ouabain accumulates mainly in F1, the time course of binding to this fraction being identical with that of binding to the intact tissue (T/M ratio). The content of radioactivity in F2 and F3 is by far lower than in F1, which may also be because of contamination during homogenization and additionally because F2 and F3 contain minor amounts of F1.

Schematic presentation of formation of different microsomal fractions and of "trapping mechanism"

In figure 6 a scheme is drawn which may explain the formation of microsomes from morphologically different tubular systems. During the process of homogenization the transverse and longitudinal invaginations of the plasma membrane become disrupted. Because of immediate vesiculation their content, namely extracellular fluid, becomes trapped. If this extracellular fluid contains hydrophilic molecules like inulin, sucrose, or ouabain, these compounds will not be able to leave the microsomes because they cannot penetrate their intact membranes. These hydrophilic compounds simply become trapped. In contrast, structures which never had any continuity with the ECS, like the sarcoplasmatic reticulum, accordingly will not contain extracellular material after vesiculation.

The results obtained so far may serve as an explanation for the controversial findings concerning an intra- or extracellular site of action for cardiac glycosides. The fact that [^{3}H]ouabain is detectable in a microsomal fraction cannot be taken as a proof for an intracellular site of action. The reason for this lies in the peculiarity of ventricular myocardial cells to contain an abundance of sarcolemmal invaginations which, upon homogenization, form a microsomal fraction with extracellular contents.

Evidence for side-specific interaction of cardiac glycosides with microsomal plasma membrane fraction

The location of the (Na^+-K^+)-ATPase system in the plasmalemma is well documented. It also could be shown that the site of interaction of this enzyme with CG is located strictly at the outer surface of the sarcolemma. From the foregoing paragraphs it could be noted that in the isolated sarcolemmal microsomes the former extracellular surface of the cells now faces the intravesicular lumen and that the former intracellular surface is now freely accessible from the incubation medium. This indicates that hydrophilic glycosides like ouabain should not be

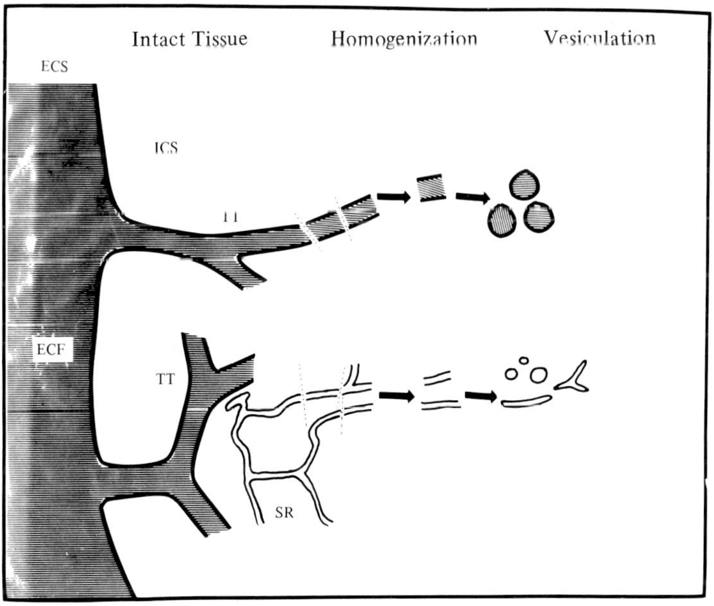

Figure 6 Illustration of the "trapping" mechanism and the formation of inside-out plasma membrane vesicles. *ECS,* extracellular space; *ICS,* intracellular space; *ECF* (*hatched area*), extracellular fluid; *TT,* transverse tubule; *SR,* sarcoplasmic reticulum.

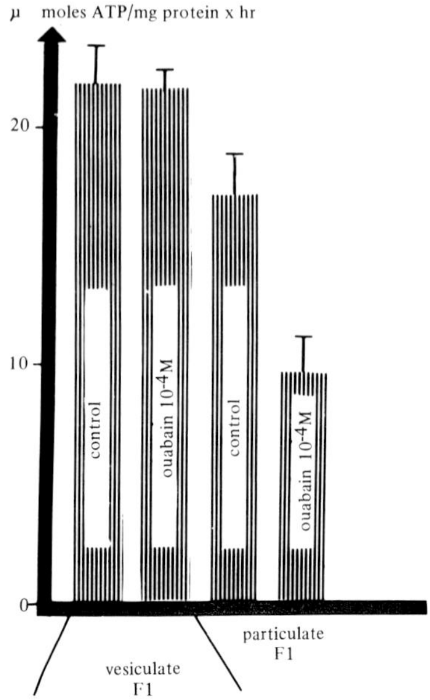

Figure 7 Demonstration of the (Na^+-K^+)-ATPase activity. *Ordinate:* micromoles of ATP split per mg of protein per hr. *Vesiculate F1,* Intact sarcolemmal microsomes; *particulate F1,* Disrupted sarcolemmal microsomes after sonication in the presence of Na deoxycholate and EDTA. *Vertical bars on top of the columns,* +S.E.M.

able to exert their ATPase inhibitory effect if they approach the membrane from the "wrong" side. In fact, the results shown in figure 7 support this view. Upon addition of ouabain up to 1×10^{-4} M to a suspension of intact vesicles, no inhibition of the total ATPase activity can be detected. However, after sonication of the preparation in the presence of EDTA and Na-deoxycholate, which leads to a disruption of the vesicles, ouabain inhibits the enzyme activity by about 40 percent as compared with the respective controls. This finding clearly demonstrates the presence of a high concentration of (Na^+-K^+)-ATPase in the sarcolemmal fraction.

Sarcolemmal microsome as model for studying active outward transport of calcium from cardiac cell

The existence of an active outward transport of calcium from the cardiac cell has been suggested (5). Recently Stam *et al.* (11) were able to trace an active transport ATPase system for Ca in a cardiac membrane fraction enriched with sarcolemma. Their results were restricted to the assay of enzymatic activity during active Ca movement but the membranous fraction did not allow a detection of the amount of Ca transported across the membrane.

In analogy to experiments dealing with the active transport of Ca into the isolated sarcoplasmic reticulum, we investigated the Ca accumulation into sarcolemmal microsomes (1). As can be recognized from figure 8 there exists a considerable Ca pump activity. In the presence of 3×10^{-5} M Ca in the incubation medium about 300 nmoles of Ca are taken up per mg of protein. This process is temperature dependent and it can be influenced by cardiac glycosides. The amount of Ca stored in the microsomes is depressed by 10^{-6} M digoxin. The Ca accumulation remains, however, unaffected by ouabain. This can be considered as an additional information on the side specificity of the action of cardiac glycosides on the sarcolemma, furthermore, since the Ca permeability of the plasma membrane seems to be impaired by cardiac glycosides only if these compounds are allowed to contact the former extracellular surface of the sarcolemmal microsomes. It should be mentioned that control experiments with digoxin and ouabain in the sarcoplasmatic reticular fraction did not reveal any interference with Ca accumulation. Furthermore, in a sarcolemmal preparation obtained from rat hearts, the Ca outward transport remained unaffected by cardiac glycosides which evidences the maintainance of species specificities in sarcolemmal microsomes.

It is possible to calculate from the initial rate of Ca outward transport the amount of Ca which is extruded during one contraction-relaxation cycle. Under

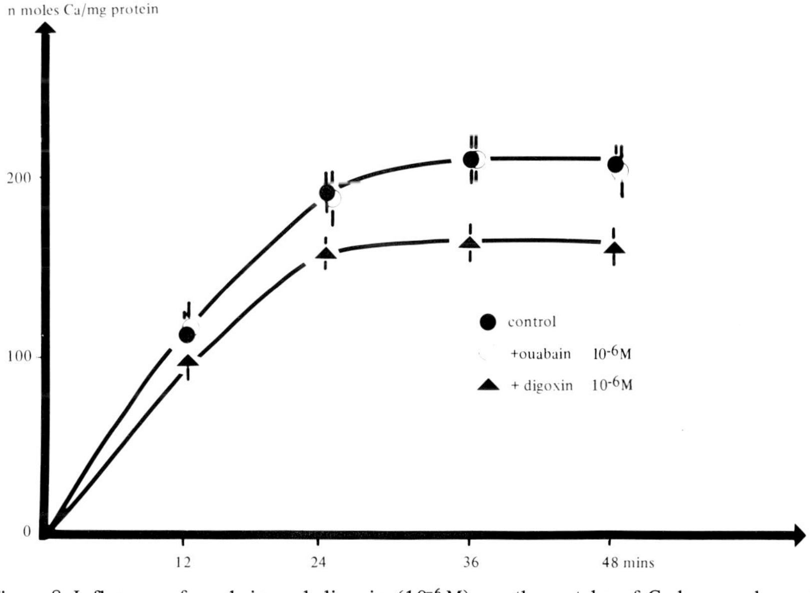

Figure 8 Influence of ouabain and digoxin (10^{-6} M) on the uptake of Ca by sarcolemmal microsomes. *Ordinate,* nanomoles of Ca per mg of protein; *abscissa,* time in minutes. The *points of the curves* represent the means ± S.E.M.

optimum conditions the highest amount was determined to be 5×10^{-6} moles of Ca/kg cells. This pump capacity would not account for the complete sequestration of the Ca provided for the activation of the contractile system (10^{-5}–10^{-4} moles of Ca/kg cells). This makes it probable that the Ca pump is responsible for the maintainance of the end diastolic Ca concentration in the cytosol by extruding leaking Ca and the Ca, which influxes per beat.

Since digoxin in the above mentioned experiments did not slow the rate of Ca outward transport, but only decreased the equilibrium level, the conclusion can be drawn that the Ca pump is not involved in the action of digoxin. An inhibition of the pump would result in a slower rate of Ca transport; however, the equilibrium level of the control experiment should be attained as well. A decreased equilibrium level at an unaffected rate fits better into the concept of a decreased effectivity of the Ca pump. This would, *e.g.*, happen if the permeability of the membrane for Ca increases so that the leak of Ca will be enhanced. This interpretation is in accordance with the well known fact that cardiac glycosides at higher concentrations increase the permeability of the heart cell membrane for Ca (6).

Passive calcium binding to plasma membrane fraction and significance of this process for electromechanical coupling

As can be seen from figure 9, there exists a correlation between the concentration of Ca in the incubation medium and the binding of the cation to the

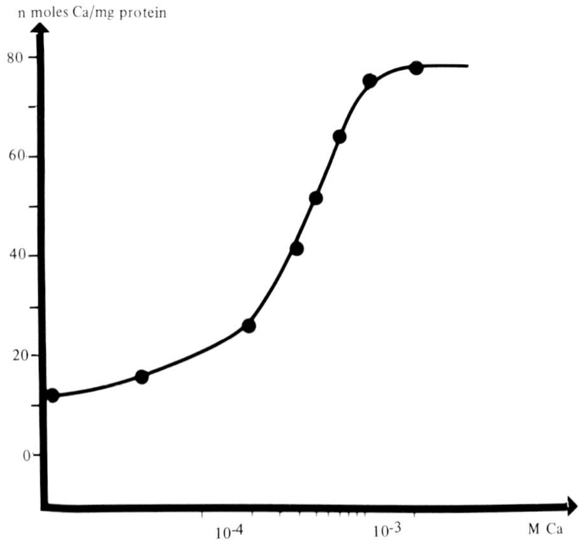

Figure 9 Passive Ca binding to sarcolemmal microsomes. *Ordinate,* nanomoles of Ca per mg of protein, *abscissa,* log of the Ca concentrations in the incubation medium.

membrane fraction. The saturation is attained at 2–5 mM Ca in the medium and amounts to a value of about 100 nM Ca/mg of protein (1). This adsorption of Ca is not energy consuming and ATP is not hydrolyzed; however, its presence considerably enhances the Ca adsorption. Calculating the molarity of Ca in the plasma membrane (based on volume and weight of the preparation and subtracting the intravesicular amount of Ca) yields a value of about 3 mM Ca in the membrane. The greater portion of this Ca participates in the electromechanical coupling (7). Because of the close apposition of the transverse tubular membranes to the contractile elements, the per beat release of this concentration of Ca upon membrane depolarization will provide levels of up to 2×10^{-4} M Ca^{++} in the vicinity of the contractile proteins. It seems most probable that the characteristics of these plasma membranous calcium binding sites can be changed by drugs, *e.g.*, pentobarbital and cardiac glycosides (2, 7, 9). In experiments dealing with rates of decline and recovery in Langendorff hearts obtained upon rapid alterations of the extracellular Ca^{++} concentrations, it could be observed that between the Ca available for the contractile system and the membrane Ca pool on the one hand, and the membrane Ca pool and the extracellular space on the other hand, the equilibration proceeds at extremely high rates. The half-lives of these processes could be determined to be in the range of 3–7 sec (Lenschow, V., unpublished observation).

Conclusional remarks on possible pharmacological and physiological properties of plasma membranous tubular system

It could be demonstrated that the tubular systems are both morphologically and functionally continuous with the ECS. Their membranes are structurally identical with those of the outer plasma membrane. Extracellular space markers like inulin and sucrose are found to be present in the tubules at a concentration almost identical with that in the perfusion medium. The cardiac glycoside, [^{3}H]ouabain also diffuses into these structures and becomes adsorbed to their extracellular surface. Time course of adsorption to the intact tissue and time course of adsorption to the tubular membranes are identical and agree well with the time course of the onset of action of ouabain ($t_{1/2}$ at 1×10^{-7} M ouabain = 21 min).

The tubular membranes are in close apposition to the Z and I bands of the myofilaments (fig. 10). The properties of the membrane so far investigated are (Na^+-K^+)-ATPase activity, Ca outward transport, passive Ca binding, and low Ca permeability. Recent reports (2, 7, 9) gave evidence that drugs can influence a superficially located Ca pool which is responsible for the regulation of contractile force in heart muscle. This Ca pool may be identical with the plasma membrane-bound Ca fraction, since its exchange rate (<1 min) and its rate of adaptation to altered extracellular Ca concentrations (3–7 sec) would not allow a correlation with intracellular Ca storage sites. The authors cited above (2, 7, 9) could demonstrate an influence of ouabain on this particular Ca pool. From the

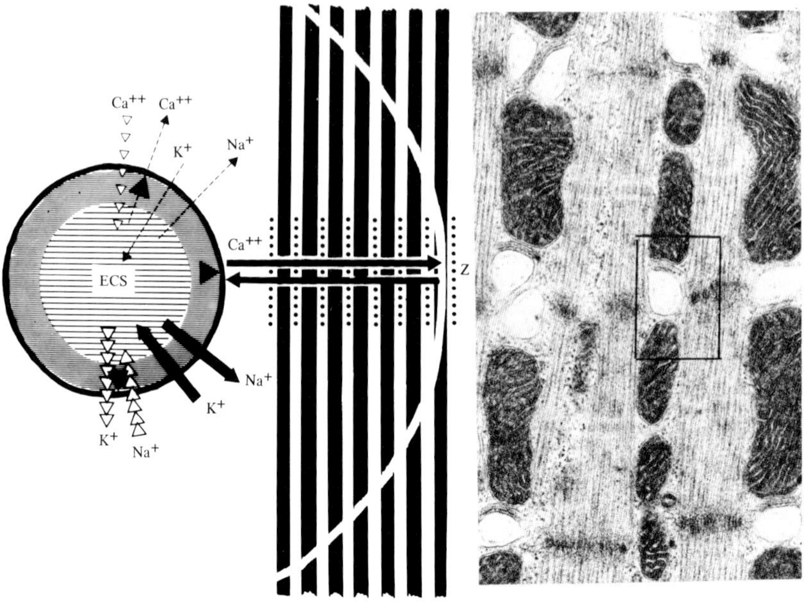

Figure 10 Schematic representation of the properties of the transverse tubular membranes in relation to excitation-contraction coupling and to the action of cardiac glycosides. From the ultrastructure depicted on the *right,* the area of the heart muscle cell can be taken, which is schematically drawn as "one excitation-contraction unit" on the *left. Large solid arrows,* K^+, Na^+, and Ca^{++} leaks. Δ, Ion transport by (Na^+-K^+)-ATPase. ▵, Ca^{++} transport by the outwardly directed Ca^{++} pump. *ECS,* extracellular space; *dotted area,* Z bands of the myofilaments; *white circle,* diffusion of Ca^{++} during 100 msec (distance between the tubular membrane and the circle equals 0.5 μm). ▲, Mechanisms affected by cardiac glycosides. For further details see text.

data presented in this paper the conclusion can be drawn that upon depolarization this pool would provide a fairly high concentration ($>1 \times 10^{-4}$ M) of Ca for the activation of the contractile elements. As shown in figure 10, upon depolarization this Ca will diffuse into the cytosol as long as the membrane remains depolarized. Assuming an apparent diffusion coefficient on the order of 1×10^{-5} cm^2 sec^{-1}, a Ca^{++} concentration sufficiently high to activate the contractile system will be established in a circle with a radius of about 0.5μm within 100 msec (see fig. 10). The particular morphological arrangement of the tubular system adjacent to the myofibrils warrants a more or less complete activation of the actomyosin. Upon repolarization the diffusion process is reversed because of a restitution of the Ca binding sites. Cardiac glycosides possess the ability to labilize the Ca membrane complex which will result in an increased per beat release of Ca^{++} (2) and an enhancement of peak tension. At higher concentrations cardiac glycosides will enhance passive permeabilities for K^+ (4) and for Ca^{++} . These alterations, together with the inhibition of the active Na^+ and K^+ transport, may lead to the toxic effects induced by cardiac glycosides. An influence of cardiac glycosides on the active outward transport of Ca could not be demonstrated in our experiments. Based on the calculated amount of Ca which is transported per beat by this system ($<5 \times 10^{-6}$ M/kg cells) one could assume that this Ca pump will only counteract the passive Ca^{++} influx and the leaking Ca^{++} . The per beat release, however, will exceed the pump capacity by far, which suggests that rebinding probably occurs upon repolarization of the membrane and the reappearance of membrane-located Ca binding sites. In conclusion, CG, in addition to a labilization of the membranous Ca store, increase the passive permeability for K and Ca and inhibit the (Na^+-K^+)-ATPase. They do not affect the active outward transport of Ca. It should be stressed that cardiac glycosides can exert their action only if approaching the membrane from the extracellular surface.

REFERENCES

1. BRADE, H., LÜBBERS, H., and PETERS, T. 1974. Isolation and properties of a microsomal plasmamembrane fraction from guinea-pig ventricular muscle. Naunyn-Schmiedeberg's Arch. Pharmacol. 282: R 11.
2. CARRIER, G. O., LÜLLMANN, H., NEUBAUER, L., and PETERS, T. 1974. The significance of a fast exchanging superficial calcium fraction for the regulation of contractile force in heart muscle. J. Molec. Cell. Cardiol. 6: 333–347.
3. DUTTA, S., GOSWAMI, S., DATTA, D. K., LINDOWER, J. O., and MARKS, B. H. 1968. The uptake and binding of six radiolabelled cardiac glycosides by guinea-pig hearts and by isolated sarcoplasmic reticulum. J. Pharmacol. Exp. Therap. 164: 10–21.
4. KLAUS, W., KUSCHINSKY, G., and LÜLLMANN, H. 1962. Über den Zusammenhang zwischen positiv inotroper Wirkung von Digitoxigenin, Kaliumflux und intrazellulären lonenkonzentrationen im Herzmuskel. Naunyn-Schmiedeberg's Arch. Exp. Pathol. Pharmacol. 242: 480–496.
5. LAHRTZ, H. G., LÜLLMANN, H., and van ZWIETEN, P. A. 1967. Calcium transport in isolated guinea-pig atria during metabolic inhibition. Biochim. Biophys. Acta 135: 701–709.
6. LEE, K. S., and KLAUS, W. 1971. The subcellular basis for the mechanism of inotropic action of cardiac glycosides. Pharmacol. Rev. 23, 193–261.
7. LÜLLMANN, H., and PETERS, T. 1973. Studies on the site of action of cardiac glycosides. *In* O. Storstein (ed.), Symposium on Digitalis, Oslo, 1973. Gyldendal Norsk Forlag, Oslo, Norway.
8. LÜLLMANN, H., PETERS, T., and RAVENS, U. 1975. Studies on the kinetics of ^{3}H-ouabain uptake and exchange in the isolated papillary muscle of the guinea-pig. Brit. Pharm. 53: 99–107.
9. NAYLER, W. 1973. An effect of ouabain on superficially-located stores of calcium in cardiac muscle. J. Molec. Cell. Cardiol. 5: 101–110.
10. SPERELAKIS, N., and RUBIO, R. 1971. An orderly lattice of axial tubules which interconnect adjacent transverse tubules in guinea-pig ventricular myocardium. J. Molec. Cell. Cardiol. 2: 211–220.

11. STAM, A. C. JR., WEGLICKI, W. B., GERTZ, E. W., and SONNENBLICK, E. H. 1973. A calcium-stimulated, ouabain inhibited ATPase in a myocardial fraction enriched with sarcolemma. Biochim. Biophys. Acta 298: 927–931.
12. STORSTEIN, O. 1973. *In* O. Storstein (ed.), Digitalis. Gyldendal Norsk Forlag, Oslo, Norway.

Mailing address: Dr. H. Lüllmann and Dr. T. Peters,
Department of Pharmacology
University of Kiel, 23 Kiel,
Hospitalstrasse 4–6 (Germany).

Recent Advances in Studies on Cardiac Structure and Metabolism
Volume 9
The Sarcolemma
Edited by Paul-Emile Roy and Naranjan S. Dhalla

THE CARDIAC GLYCOSIDE RECEPTOR: ITS PROPERTIES AND ITS CORRELATION TO NUCLEOTIDE BINDING SITES, PHOSPHOINTERMEDIATE, AND ($Na^+ + K^+$)-ATPase ACTIVITY[1]

E. ERDMANN, R. PATZELT, and W. SCHONER

Medizinische Klinik der Universität Göttingen, Götttingen,
and Institut für Biochemie und Endokrinologie der Universität Giessen,
Giessen, Germany

SUMMARY

Binding of ouabain to its receptor is followed by a directly proportional inhibition of (Na^+ + K^+)-ATPase. The stoichiometry of nucleotide binding site, phosphate acceptor site, ouabain binding site in (Na^+ + K^+)-ATPase is 1:1:1. Inactivation of (Na^+ + K^+)-ATPase by S-[2,4-dinitrophenyl]-6-mercaptopurine riboside-5′-triphosphate, which probably forms a thioether derivative of ATP with the enzyme, results in the appearance of a low affinity binding site. It is concluded that the covalently bound ATP fixes the enzyme in a E_1 conformational state which is characterized by a low affinity for ouabain.

The positive inotropic effect of cardiac glycosides on heart muscle cells *in vitro* can be correlated temporally to the binding of ouabain to Na^+ + K^+-activated ATPase and to the dissociation from the enzyme (1). Since it has been shown that cardiac glycosides exert their action as inhibitors of active sodium transport at the outer side of the cell membranes (2), and since ATP, which provides the energy for the transport, is split at the inner side of the cell membrane (7, 17), it seemed worthwhile to study the relationship between the number of glycoside binding sites and nucleotide binding sites and their interactions in (Na^+ + K^+)-ATPase preparations.

METHODS AND MATERIALS

For the preparations of Na^+ + K^+-activated ATPase from human cardiac tissue the method of Matsui and Schwartz (12) was adapted. (Na^+ + K^+)-ATPase from

[1] This work has been supported through DFG Scho 139/7 and SFB 89 (Kardiologie).

bovine brain was purified and quantitated as described earlier (15). One enzyme unit is defined as the amount of enzyme hydrolyzing 1 μmole of ATP/min at 37°C. Protein was measured according to the method of Lowry *et al.* (11).

The determination of the number of [^{3}H]ouabain binding sites and of the dissociation constant of the ouabain receptor complex has been described earlier (3), as well as the method for the determination of the number of nucleotide binding sites (10). The amount of [^{32}P]phosphointermediate formed from γ-[^{32}P]ATP was measured according to the method of Post *et al.* (14) with minor modifications. The ATP analog, S-[2,4-dinitrophenyl]-6-mercaptopurine riboside-5′-triphosphate, was synthesized essentially according to the method of Faust *et al.* (6) with some modifications to be reported elsewhere.[2] [^{3}H]Ouabain (specific activity 13 Ci/mmole) was from NEN, Dreieichenhain, Germany; [^{14}C]ADP (specific activity 535 Ci/mole) and γ-[^{32}P]ATP (specific activity 300–500 Ci/mole) were from Amersham-Buchler, Braunschweig, Germany. All other chemicals were of analytical grade and obtained from Boehringer, Mannheim or E. Merck AG, Darmstadt, Germany.

RESULTS AND DISCUSSION

To ascertain whether the cardiac glycoside receptor in human heart behaves similarly to that of other organs and mammals, the binding of [^{3}H]ouabain to a cell membrane preparation of human heart was studied. There is a single type of a high affinity binding site in human heart muscle cell membranes as is evident from the straight line of the Scatchard plot (fig. 1). The dissociation constant (K_D) of the ouabain receptor complex was calculated to be 0.25×10^{-8} M. This value agrees very well with that determined for the dissociation constant in human red blood cells of 0.22×10^{-8} M. Apparently the same type of receptor seems to exist in all cell membranes of a given species (3). In agreement with previous experiments with bovine brain cell membranes, human heart cell membranes bind maximally 130 pmoles of [^{3}H]ouabain/($Na^+ + K^+$)-ATPase unit also. The extent of binding of ouabain to the cardiac glycoside receptor site correlates very well with the extent of the inhibition of $Na^+ + K^+$-activated ATPase (fig. 2). In conclusion, the findings on human cardiac cell membranes confirm the results obtained for ouabain binding to other cell membrane preparations (5). The data of figure 2 indicate an intimate relationship between ($Na^+ + K^+$)-ATPase and ouabain binding sites. In order to get information on the stoichiometry of the possible subunits of ($Na^+ + K^+$)-ATPase, the ouabain receptor, phosphointermediate, and nucleotide binding site were measured (table I). The number of nucleotide binding sites was determined with [^{14}C] ADP,

---

[2] We thank Professor Fasold, Frankfurt/Main for his kind gift of some of the ATP analog in the early studies.

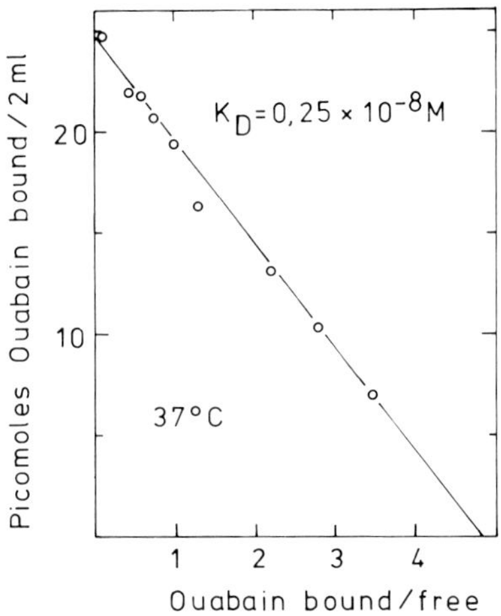

Figure 1 Scatchard plot of ouabain binding to human cardiac cell membranes. Quantities of 2.3 mg of protein [$(Na^+ + K^+)$-ATPase activity = 0.0805 U/mg = 10.5×10^{-12} moles of receptor sites per mg of protein] were incubated in 50 mM imidazole-HCl, pH 7.25, 3 mM $MgCl_2$, 3 mM imidazole-PO_4, and different [^{3}H]ouabain concentrations for 120 min at 37°C. Total volume was 2 ml. The dissociation constant was calculated from the slope of the plot as 0.25×10^{-8} M.

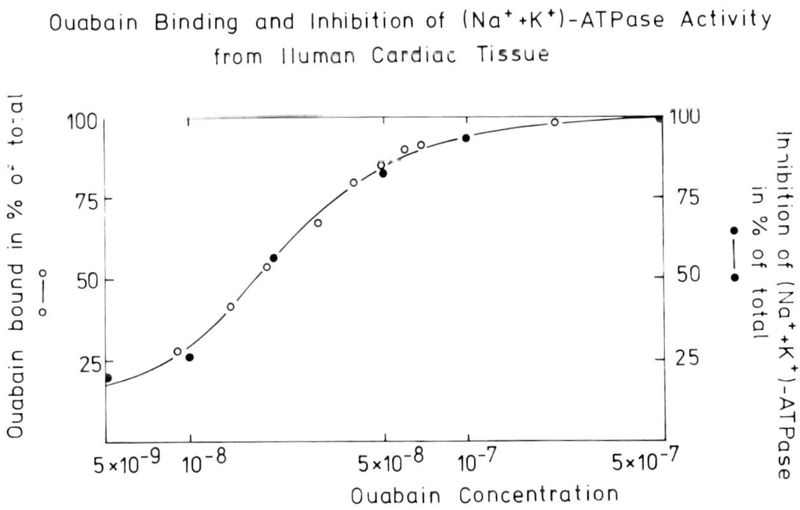

Figure 2 Correlation between the occupation of ouabain receptor sites and the inhibition of $(Na^+ + K^+)$-ATPase in human cardiac cell membranes. Ouabain binding and $(Na^+ + K^+)$-ATPase activity were determined after a 2-hr preincubation period under the conditions of figure 1.

Table I Comparison of capacities of $(Na^+ + K^+)$-ATPase from bovine brain for nucleotide binding, phosphate incorporation, and ouabain binding

	Capacity (pmoles/U)	Dissociation constant (M)
ATP*	143	1.2×10^{-7}
ADP	106	9.0×10^{-7}
E ~ P	103	
Ouabain	130	1.3×10^{-8}

*Data reproduced from Norby and Jensen (13) by permission of *Biochimica and Biophysica Acta.*

since ADP is a cosubstrate of the Na^+-dependent ATP-ADP transphosphorylation reaction, which is a partial reaction of $(Na^+ + K^+)$-ATPase. The results of these experiments indicate that the stoichiometry of the high affinity sites for ADP and ouabain and for the phosphate acceptor groups in $(Na^+ + K^+)$-ATPase from ox brain is 1:1:1.

It is assumed that the inhibition of ATP hydrolysis by ouabain is caused by fixation of the enzyme in an E_2 conformational state (16) which is characterized by a decreased affinity for ATP (8, 16). Since ATP is hydrolyzed at the inner side of the cell membrane (7, 17) and ouabain is bound at the outer site of the cell membrane (2), it appears that the conformational change occurs all over the cell membrane. In order to see whether this interpretation is valid we tried to find out whether ouabain binding at the outer cell surface is hindered by the binding of ATP at the inner cell surface, which occurs in the E_1 conformational state (16).

Binding of ouabain must be studied in the presence of Mg^{2+} which hinders nucleotide binding (10). In order to be sure that ATP is bound at the ATP binding site during the ouabain binding studies, we tried to fixate ATP covalently at the nucleotide binding site. S-[2,4-Dinitrophenyl]-6-mercaptopurine riboside-5′-triphosphate, which has been shown to be an ATP affinity label of actin (6) by forming thioether bonds with the protein (9), proved to be a useful tool for this purpose. The ATP analog inactivated $(Na^+ + K^+)$-ATPase in a pseudo-first-order reaction (fig. 3*A*). Concomitantly the ability to incorporate ^{32}P from γ-[^{32}P]ATP in the presence of Na^+ was lost. (fig. 3*B*). The inactivation was hindered by an excess of ATP (fig. 3 *A*). These findings suggest that the ATP analog is bound covalently to the ATP binding site.

When $(Na^+ + K^+)$-ATPase preparations from bovine brain were partially inactivated by the ATP analog there was a decrease of the number of high affinity

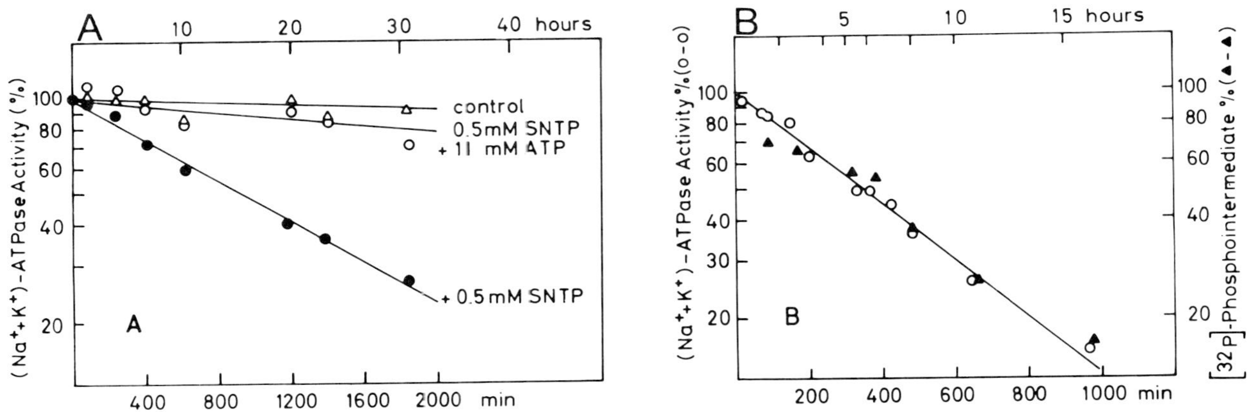

Figure 3 Inactivation of $Na^+ + K^+$-activated ATPase from bovine brain by S-[2,4-dinitrophenyl]-6-mercaptopurine riboside-5′-triphosphate (*SNTP*). A 160-mg quantity of $(Na^+ + K^+)$-ATPase (specific activity 1.2 U/mg) was incubated at 37°C in 40 mM Tris-HCl, pH 8.2, and 0.5 mM SNTP (fig. 3*A*) or 0.79 SNTP (fig. 3*B*). Aliquots of the incubation mixture were assayed for enzymatic activity and phosphorylation capacity at the times indicated.

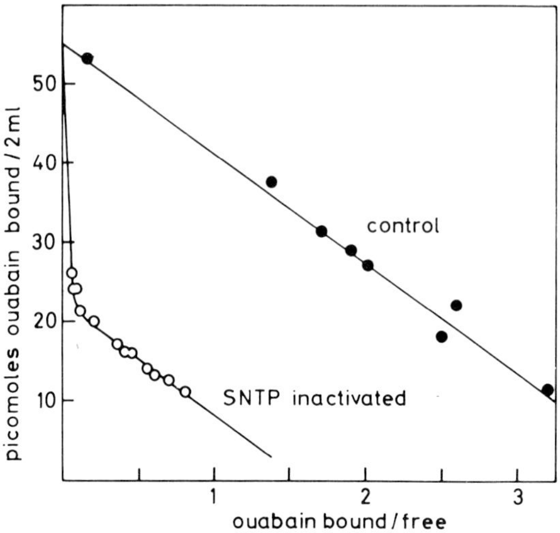

Figure 4 Properties of the ouabain-receptor of ($Na^+ + K^+$)-ATPase from bovine brain after inactivation with S-[2,4-dinitrophenyl]-6-mercaptopurine riboside-5′-triphosphate (SNTP). Scatchard plot.

sites for ouabain together with the decrease of phosphate acceptor sites (fig. 4). In addition, a low affinity site for cardiac glycoside appeared. These findings agree with the concept that upon binding of ATP at the inner surface of the cell membrane there occurs a conformational change. This change results in a decreased affinity of the ouabain receptor site for cardiac glycosides. Similar conclusions can be drawn from studies with ADP-C-P (4).

REFERENCES

1. AKERA, T., BASKIN, S. I., TOBIN, T., and BRODY, T. M. 1973. Ouabain: Temporal relationship between the inotropic effect and the *in vitro* binding to, and the dissociation from ($Na^+ + K^+$)-activated ATPase. Naunyn-Schmiedeberg's Arch. Pharmacol. 277: 151–162.
2. CALDWELL, P. C., and KEYNES, R. D. 1959. The effect of ouabain on the efflux of sodium from a squid giant axon. J. Physiol. (London) 148: 8 P.
3. ERDMANN, E., and SCHONER, W. 1973. Ouabain-receptor interactions in ($Na^+ + K^+$)-ATPase preparations from different tissues and species: Determination of kinetic constatnts and dissociation constants. Biochim. Biophys. Acta 307: 386–398.
4. ERDMANN, E., and SCHONER, W. 1973. Ouabain-receptor interactions in ($Na^+ + K^+$)-ATPase preparations. II. Effect of cations and nucleotides on rate constants and dissociation constants. Biochim. Biophys. Acta 330: 302–315.
5. ERDMANN, E., and SCHONER, W. 1974. Ouabain-receptor interactions in ($Na^+ + K^+$)-ATPase preparations. IV. The molecular structure of different cardiac steroids and other substances and their affinity to the glycoside receptor. Naunyn-Schmiedeberg's Arch. Pharmacol.
6. FAUST, U., FASOLD, H., and ORTANDERL, F. 1974. Synthesis of a protein-reactive

ATP analog and its application for the affinity labeling of rabbit muscle actin. Eur. J. Biochem. 43: 273–279.
7. GLYNN, I. M. 1962. Activation of adenosine triphosphatase activity in cell membrane by external potassium and internal sodium. J. Physiol. (London) 160: 18 P.
8. HEGYVARY, C., and POST, R. L. 1971. Binding of adenosine triphosphate to sodium and potassium ion-stimulated adenosine triphosphatase. J. Biol. Chem. 246: 5234–5240.
9. HULLA, F. W., and FASOLD, H. 1972. Synthesis of an adenosine-5′-monophosphate analog and its use for the affinity labeling of the effector binding site of rabbit skeletal muscle phosphorylase *b*. Biochemistry 11: 1056–1061.
10. KANIIKE, K., ERDMANN, E., and SCHONER, W. 1973. ADP binding to (Na^+ + K^+)-activated ATPase. Biochim. Biophys. Acta 298: 901–905.
11. LOWRY, O. H., ROSEBROUGH, N. J., FARR, A. L., and RANDALL, R. J. 1951. Protein measurement with the Folin phenol reagent. J. Biol. Chem. 193: 265–275.
12. MATSUI, H., and SCHWARTZ, A. 1966. Purification and properties of a highly active ouabain-sensitive Na^+,K^+-dependent adenosine triphosphatase from cardiac tissue. Biochim Biophys. Acta 128: 380–390.
13. NØRBY, J. G., and JENSEN, J. 1971. Binding of ATP to brain microsomal ATPase: Determination of the ATP-binding capacity and the dissociation constant of the enzyme-ATP complex as a function of K^+ concentration. Biochim. Biophys. Acta 233: 104–116.
14. POST, R. L., SEN, A. K., and ROSENTHAL, A. S. 1965. A phosphorylated intermediate in adenosine triphosphate-dependent sodium and potassium transport across kidney membranes. J. Biol. Chem. 240: 1437–1445.
15. SCHONER, W., von ILBERG, C., KRAMER, R., and SEUBERT, W. 1967. On the mechanism of Na^+- and K^+-stimulated hydrolysis of adenosine triphosphate. I. Purification and properties of a Na^+- and K^+-activated ATPase from ox brain. Eur. J. Biochem. 1: 334–343.
16. SEN, A. K., TOBIN, T., and POST, R. L. 1969. A cycle for ouabain inhibition of sodium- and potassium-dependent adenosine triphosphatase. J. Biol. Chem. 244: 6596–6604.
17. WHITTAM, R. 1962. The asymmetrical stimulation of a membrane adenosine triphosphatase in relation to active cation transport. Biochem. J. 84: 110–118.

Mailing address: Dr. E. Erdmann,
Medizinische Klinik der Universität,
D-34 Göttingen, Humboldtallee 1 (Germany).

Recent Advances in Studies on Cardiac Structure and Metabolism
Volume 9
The Sarcolemma
Edited by Paul-Emile Roy and Naranjan S. Dhalla

CARDIAC GLYCOSIDES: RELATIONSHIP AMONG ACTIVE ^{86}Rb UPTAKE, (Na^+,K^+)-ATPase ACTIVITY, AND INOTROPY IN GUINEA PIG HEART[1]

T. AKERA, D. KU, and T. M. BRODY

Department of Pharmacology,
Michigan State University,
East Lansing, Michigan, USA

SUMMARY

Relationships among positive inotropic responses, (Na^+,K^+)-ATPase inhibition, and sodium pump activities were studied using paced Langendorff preparations of guinea pig heart. (Na^+,K^+)-ATPase activity was estimated from the initial velocity of [^{3}H] ouabain binding in ventricular homogenates, and sodium pump activity from ouabain-sensitive ^{86}Rb uptake of ventricular slices. These parameters were measured in control, or in ouabain- or digitoxin-treated hearts either at the time of inotropic response or during the course of drug washout and compared with inotropy in the same heart. Perfusion of ouabain or digitoxin caused an increase in contractile force and a decrease in the initial velocity of ATP-dependent [^{3}H]ouabain binding. The levels of [^{3}H]ouabain binding in homogenates observed after a long incubation period were not different in ouabain-perfused and control homogenates, indicating that the nonlabeled ouabain bound to (Na^+,K^+)-ATPase during Langendorff perfusion was exchangeable with [^{3}H]ouabain. Perfusion of drug-free solution after a 20-min perfusion of the isolated heart with either ouabain or digitoxin resulted in a loss of inotropic response and a recovery of the inhibition of the initial velocity of ATP-dependent [^{3}H]ouabain binding. Inotropic responses to digitoxin during perfusion and subsequent loss during washout were accompanied by a reduction and subsequent recovery of ouabain-sensitive ^{86}Rb uptake. Thus, it would appear that with cardiac glycosides, a relationship exists among cardiac contractile force, the inhibition of cardiac (Na^+,K^+)-ATPase, and the inhibition of the sodium pump activity. The inhibition of (Na^+,K^+)-ATPase and sodium pump because of cardiac glycoside perfusion of the isolated beating heart was reversible.

A close relationship between the inhibition of cardiac (Na^+,K^+)-ATPase activity and the increase in cardiac contractile force induced by cardiac glycosides has been documented by a number of investigators (1–3, 5, 7, 14, 15). Nevertheless, the hypothesis that (Na^+,K^+)-ATPase is the inotropic receptor for cardiac glycosides and that the inhibition of the enzyme is the cause of pharmacological action of these compounds has not been universally accepted (see Reference 10).

[1] This research was supported by United States Public Health Service Grant HL 16052-01 and by the Michigan Heart Association.

Despite a clear demonstration of the inhibition of cardiac (Na^+,K^+)-ATPase activity at the time of the inotropic response to cardiac glycosides (2, 3, 7), subtoxic doses of these drugs failed to produce a consistent accumulation of sodium by cardiac tissue, which may be expected as a consequence of the sodium pump inhibition (see Reference 10). Thus, some investigators speculated that the inhibition of either enzyme activity or the sodium pump *per se* may not be an important factor in the genesis of the inotropic effect of cardiac glycosides (8), or that the magnitude of the enzyme inhibition is not sufficient to cause a significant impairment of the sodium pump activity and hence may have no physiological or pharmacological significance (12). Other investigators concluded that (Na^+,K^+)-ATPase inhibition is related to actions of the cardiac glycosides other than the inotropic effect (16), based upon their observations that the enzyme inhibition persisted several hours after the loss of the inotropic effect of cardiac glycosides in isolated perfused rabbit hearts (13). These latter investigators concluded that ouabain produces reversible or irreversible inhibition of enzyme activity depending upon drug concentration used.

The present study was initiated to determine whether perfusion of the isolated heart with cardiac glycosides results in 1) a reversible or irreversible inhibition of (Na^+,K^+)-ATPase activity and 2) a significant impairment of the sodium pump activity. The data indicate that the development of the inotropic response to ouabain or digitoxin during a Langendorff perfusion of guinea pig heart and the subsequent loss of the response during perfusion with a drug-free solution were accompanied by a development of and recovery from (Na^+,K^+)-ATPase and sodium pump inhibition. The inhibition of both enzyme and cation pump activities was reversible.

METHODS

Hearts were obtained from guinea pigs of either sex weighing 350–500 gm and perfused according to a modified Langendorff method (1) with Krebs-Henseleit (K-H) solution at 30°C at a constant flow rate of 4 ml/min. After the removal of both atria, hearts were electrically driven at 1.5 Hz at a voltage 10 percent above threshold. After a 45-min equilibrium period, either 1.2 μM ouabain or 0.4 μM digitoxin in K-H solution was perfused for 10 or 20 min. For "washout" studies, K-H solution was continued up to 40 min after a 20-min perfusion of the drug-containing K-H solution. Control hearts were perfused with K-H solution for periods comparable with drug-perfused hearts. Inotropic response was estimated from percentage changes in cardiac contractile force monitored with a force-displacement transducer (FT-03, Grass Instruments, Quincy, Mass.).

At a predetermined time, a 0.5-gm portion of the left ventricular tissue was excised, homogenized in 9 volumes of ice-cold 0.32 M sucrose solution (enzyme grade, Schwartz/Mann, Orangeburg, N.Y.), and assayed for ATP-dependent [^{3}H]ouabain binding which is used as an estimate of the (Na$^+$,K$^+$)-ATPase activity (4). The homogenate (4 mg of tissue/ml) was incubated with 0.01 μM [^{3}H]ouabain (specific activity, 13.2 Ci/mmole) in the presence of 5 mM $MgCl_2$, 5 mM Tris-ATP, 200 mM NaCl, and 50 mM Tris-HCl buffer (pH 7.5) at 37°C. The binding reaction was started by the addition of homogenate to a prewarmed incubation mixture and was terminated by filtering aliquots through Millipore filters (pore size, 0.8 μm). The time elapsed between the termination of Langendorff perfusion and the start of the incubation was less than 5 min. Homogenate was kept cold (0–2°C) during this time. These steps were important in order to minimize the dissociation of cardiac glycosides bound to the enzyme during the perfusion of isolated hearts. The radioactivity trapped on the filter (bound ouabain) was assayed using liquid scintillation counting after dissolving the filter with monomethyl ethyleneglycol. Values obtained with the incubation in the absence of ATP (nonspecific binding) were subtracted from those obtained in the presence of ATP to calculate specific, ATP-dependent binding.

Sodium pump activity was estimated from ouabain-sensitive ^{86}Rb uptake according to the method of Bernstein and Israel (6) with slight modifications. Ventricular slices prepared with a Stadie-Riggs tissue slicer (A. H. Thomas Co., Philadelphia, Pa.) were incubated at 0°C for 2 min in K$^+$-free K-H solution for sodium loading and subsequently assayed for ^{86}Rb uptake at 37°C in a K$^+$-free K-H solution containing 2 mM RbCl, to which a tracer amount of ^{86}Rb (specific activity, 0.15 Ci/mmole) was added. The amount of ^{86}Rb accumulated by the tissue slice in the presence of 0.2 mM ouabain (nonspecific uptake) was subtracted from that in the absence of ouabain to calculate the ouabain-sensitive ^{86}Rb uptake. Radioactivity of tissue slices was assayed using a γ scintillation counter.

All data were analyzed statistically using the Student's *t*-test. All *p* values smaller than 0.05 were considered to be statistically significant. Radioactive compounds were purchased from New England Nuclear (Boston, Mass.).

RESULTS

Effects of ouabain on contractile force and initial velocity of [^{3}H] ouabain binding

Perfusion of 1.2 μM ouabain in paced Langendorff preparations of guinea pig hearts produced a gradual increase in contractile force over a period of 20 min (table I). Concomitantly, the initial velocity of ATP-dependent [^{3}H]ouabain binding by the homogenate of ouabain-perfused hearts decreased with time (table I). Since the initial velocity of ATP-dependent [^{3}H]ouabain binding is determined by the concentration of active enzyme and that of [^{3}H]ouabain, and since latter concentration was held constant at 0.01 μM, the decrease in

Table I Effects of ouabain perfusion on contractile force and initial velocity of ATP-dependent [^{3}H] ouabain binding*

	Contractile force (% control)	[^{3}H]Ouabain binding (pmoles/mg tissue/3 min)
Control	1.0 ± 0.7	207.0 ± 6.1
1.2 μM ouabain, 10 min	23 ± 1.8	148.6 ± 7.0
1.2 μM ouabain, 20 min	40 ± 3.7	131.0 ± 7.0
1.2 μM ouabain, 20 min + washout, 8 min	13 ± 1.8	142.4 ± 8.3
1.2 μM ouabain, 20 min + washout, 24 min	10 ± 3.1	155.9 ± 6.1
1.2 μM ouabain, 20 min + washout, 40 min	6.7 ± 4.3	186.0 ± 8.9

*Mean ± S.E. of at least five experiments.

initial velocity of [^{3}H]ouabain binding indicates a decrease in the concentration of active enzyme or a preoccupation of ouabain binding sites by nonlabeled ouabain which took place during the perfusion of isolated hearts (see Reference 4). In control hearts, both contractile force and the initial velocity of [^{3}H] ouabain binding by homogenates remained unchanged during the experimental period. [^{3}H]Ouabain binding by control homogenate was linear during the first 3 min (data not shown).

In the present studies, incubation time (3 min) for the [^{3}H]ouabain binding reaction as well as the time between the termination of Langendorff perfusion and the start of the incubation was minimized in order to reduce the release of nonlabeled ouabain which was bound to the (Na^+,K^+)-ATPase during the perfusion. Such release of bound drug would result in a reactivation of the enzyme activity. Nevertheless, because of the relatively unstable binding of ouabain to guinea pig cardiac (Na^+,K^+)-ATPase (1), it is probable that (Na^+,K^+)-ATPase inhibition is underestimated to some extent in these experiments. In fact, after a 10-min incubation period, the levels of bound [^{3}H]ouabain in ventricular homogenates obtained from ouabain-perfused hearts were not different from those obtained from control hearts. This would indicate that nonlabeled ouabain bound to the (Na^+,K^+)-ATPase during the Langendorff perfusion was reversibly bound and was exchangeable with [^{3}H]ouabain added later to the homogenate during the *in vitro* incubation.

The positive inotropic effect developed during a 20-min perfusion of 1.2 μM ouabain dissipated gradually when the isolated hearts were further perfused with a drug-free K-H solution (table I). The loss of the inotropic response was accompanied by the recovery of the inhibition of ATP-dependent [^{3}H]ouabain binding velocity. Thus, it appears that the binding of nonlabeled ouabain to (Na^+,K^+)-ATPase which occurred in beating isolated heart was reversible.

Effects of digitoxin on contractile force and initial velocity of [^{3}H] ouabain binding

If the binding of cardiac glycosides to (Na^+,K^+)-ATPase and the ensuing inhibition of the enzyme activity is the cause of the inotropic response, then the relationship between (Na^+,K^+)-ATPase inhibition and the developed inotropic response should be similar with various cardiac glycosides. The differences in this relationship with different cardiac glycosides would indicate that these two phenomena are not causally related. Thus, similar experiments as above were repeated with digitoxin, a cardiac glycoside which has been shown to have different time course of inotropic response (9). Table II shows the relationship between cardiac contractile force and the inhibition of [^{3}H]ouabain binding velocity during the development and dissipation of inotropic response induced by a perfusion of 0.4 μM digitoxin and subsequent washout, respectively. Since ouabain and digitoxin bind to the same site on (Na^+,K^+)-ATPase (11), the use of [^{3}H]ouabain to estimate uninhibited (Na^+,K^+)-ATPase can be justified. A gradual development and loss of the inotropic response to digitoxin was accompanied by a gradual inhibition and the recovery of the initial velocity of ATP-dependent [^{3}H]ouabain binding by ventricular homogenate. Thus, there was a good correlation between the (Na^+,K^+)-ATPase inhibition and the inotropic response to cardiac glycosides in isolated guinea pig hearts during the development and dissipation of the effect of glycoside, and the inhibition of [^{3}H]ouabain binding velocity or (Na^+,K^+)-ATPase activity was reversible. The time course of the loss of inotropic response and that of the recovery of the enzyme inhibition during the perfusion with drug-free K-H solution were remarkably similar.

Table II Effects of digitoxin perfusion on contractile force and initial velocity of ATP-dependent [^{3}H] ouabain binding*

	Contractile force (% control)	[^{3}H]Ouabain binding (pmole/mg tissue/3 min)
Control	1.0 ± 0.7	207.0 ± 6.1
0.4 μM digitoxin, 10 min	19.2 ± 1.1	171.3 ± 9.0
0.4 μM digitoxin, 20 min	29.1 ± 2.0	135.6 ± 9.4
0.4 μM digitoxin, 20 min + washout, 8 min	24.3 ± 3.8	167.1 ± 16.9
0.4 μM digitoxin, 20 min + washout, 24 min	16.9 ± 1.6	189.4 ± 22.1
0.4 μM digitoxin, 20 min + washout, 40 min	7.7 ± 2.5	199.2 ± 15.3

*Mean ± S.E. of at least five experiments.

Table III Effects of digitoxin perfusion on contractile force and ouabain-sensitive ^{86}Rb uptake*

	Contractile force (% control)	^{86}Rb uptake (pmole/mg tissue/8 min)
Control	0.0 ± 0.0	336.9 ± 10.5
0.4 μM digitoxin, 10 min	19.2 ± 1.1	147.0 ± 22.2
0.4 μM digitoxin, 20 min	28.2 ± 4.6	141.9 ± 15.9
0.4 μM digitoxin, 20 min + washout, 8 min	25.6 ± 4.2	181.9 ± 27.0
0.4 μM digitoxin, 20 min + washout, 24 min	17.1 ± 1.8	198.5 ± 21.7
0.4 μM digitoxin, 20 min + washout, 40 min	8.5 ± 3.7	241.1 ± 19.2

*Mean ± S.E. of at least five experiments.

Effects of digitoxin on contractile force and sodium pump activity

Table III shows the changes in sodium pump activity estimated from ouabain-sensitive ^{86}Rb uptake by ventricular slices. Isolated guinea pig hearts were perfused with 0.4 μM digitoxin for 20 min and then with digitoxin-free K-H solution. At predetermined times, hearts were removed and ventricular slices were prepared and assayed for ouabain-sensitive ^{86}Rb uptake. Although ouabain-sensitive ^{86}Rb uptake by ventricular slices was linear during the first 16 min under the present experimental conditions, an 8-min incubation period was chosen to minimize the dissociation of digitoxin during the incubation period. The development of the inotropic response to digitoxin was accompanied by a concomitant inhibition of ouabain-sensitive ^{86}Rb uptake, *i.e.*, an inhibition of the sodium pump activity. During perfusion of digitoxin-free K-H solution, the inotropic response and ^{86}Rb uptake returned toward control levels with a similar time course.

DISCUSSION

Allen *et al.* (4) have demonstrated that ATP-dependent [^{3}H]ouabain binding is an accurate estimate of the (Na^+,K^+)-ATPase activity in the presence or absence of cardiac glycosides in canine kidney. These investigators utilized incubation media containing high concentrations of sodium to reduce the dissociation of prebound cardiac glycosides during the incubation for [^{3}H]ouabain binding. Such a procedure, however, would not be successful with guinea pig cardiac enzyme because cardiac glycosides, particularly ouabain, dissociated rapidly from this enzyme even in the presence of high concentrations of sodium. Thus, in the present studies, the initial velocity of the ATP-dependent [^{3}H]ouabain binding was used as an estimate of the (Na^+,K^+)-ATPase activity. With a short incubation period, it was possible to observe inhibited enzyme activity with the guinea pig cardiac homogenate. A 3 min incubation period was chosen because ATP-dependent $[^3H]$ouabain binding with 0.01 μM ouabain was linear during this time period. However, it is probable that the enzyme inhibition was underestimated because of some dissociation of prebound ouabain. With digitoxin perfusion, such underestimation would be less serious since digitoxin dissociates from the enzyme more slowly (17).

Perfusion of the isolated guinea pig heart with either ouabain or digitoxin resulted in a gradual development of a positive inotropic response and a gradual inhibition of (Na^+,K^+)-ATPase. Perfusion of a drug-free K-H solution to wash out the cardiac glycoside, accumulated by the heart during a 20-min perfusion of the drug, resulted in the concomitant loss of the inotropic response and the recovery of enzyme inhibition. These data indicate that perfusion of the isolated guinea pig heart with cardiac glycosides results in a reversible inhibition of (Na^+,K^+)-ATPase and further suggests that such an inhibition and the inotropic response of the heart to cardiac glycosides are closely related. The reversibility of cardiac glycoside binding which took place in Langendorff perfusion of electrically driven heart is also indicated from the exchangeability of the bound glycoside with $[^3H]$ouabain added *in vitro* to the homogenate.

The present data are contradictory to those reported by Okita *et al.* (13) and Ten Eick *et al.* (16) with isolated rabbit heart. The discrepancy is not attributable to the species differences (guinea pig versus rabbits), although there are differences in dissociation rates of ouabain-ATPase complexes obtained from different sources. The time course of the inotropic response to ouabain and the characteristics of interaction of ouabain with cardiac (Na^+,K^+)-ATPase are similar in guinea pig and rabbit (1).

Additionally, the present data indicate clearly that the development of the inotropic response and (Na^+,K^+)-ATPase inhibiton results in a significant inhibition of the sodium pump activity. Thus, it may be concluded that the inotropic response to cardiac glycoside and the inhibition of the sodium pump activity occur simultaneously with a similar time course, and that the sodium pump is inhibited at the time of the inotropic response but before the onset of arrhythmias.

Acknowledgment The authors thank Mr. Carl L. Pew for expert technical assistance.

REFERENCES

1. AKERA, T., BASKIN, S. I., TOBIN, T., and BRODY, T. M. 1973. Ouabain: Temporal relationship between the inotropic effect and the *in vitro* binding to, and dissocia-

tion from ($Na^+ + K^+$)-activated ATPase. Naunyn-Schmiedeberg's Arch. Pharmacol. 277: 151–162.
2. AKERA, T., LARSEN, F.S., and BRODY, T. M. 1969. The effect of ouabain on sodium-and potassium-activated adenosine triphosphatase from the hearts of several mammalian species. J. Pharmacol. Exp. Therap. 170: 17–26.
3. AKERA, T., LARSEN, F. S., and BRODY, T. M. 1970. Correlation of cardiac sodium- and potassium-activated adenosine triphosphatase activity with ouabain-induced inotropic stimulation. J. Pharmacol. Exp. Therap. 173: 145–151.
4. ALLEN, J. C., MARTINES-MALDONADO, M., EKNOYAN, G., SUKI, W. N., and SCHWARTZ, A. 1971. Relation between digitalis binding *in vivo* and inhibition of sodium, potassium-adenosine triphosphatase in canine kidney. Biochem. Pharmacol. 20: 73–80.
5. ALLEN, J. C., and SCHWARTZ, A. 1969. A possible biochemical explanation for the insensitivity of the rat to cardiac glycosides. J. Pharmacol. Exp. Therap. 168: 42–46.
6. BERNSTEIN, J. C., and ISRAEL, Y. 1970. Active transport of ^{86}Rb in human red cells and rat brain slices. J. Pharmacol. Exp. Therap. 174: 323–329.
7. BESCH, H. R., JR., ALLEN, J. C., GLICK, G., and SCHWARTZ, A. 1970. Correlation between the inotropic action of ouabain and its effects on subcellular enzyme systems from canine myocardium. J. Pharmacol. Exp. Therap. 171: 1–12.
8. BESCH, H. R., JR., and SCHWARTZ, A. 1970. On the mechanism of action of digitalis. J. Molec. Cell. Cardiol. 1: 195–199.
9. FRICKE, U., and KLAUS, W. 1971. Über die Wirkung von Strophanthidin-3-bromoacetat am Papillarmuskel des Meerschweinchens. Naunyn-Schmiedelberg's Arch. Pharmakol. 268: 192–199.
10. LEE, K. S., and KLAUS, W. 1971. The subcellular basis for the mechanism of inotropic action of cardiac glycosides. Pharmacol. Rev. 23: 193–261.
11. MATSUI, H., and SCHWARTZ, A. 1968. Mechanism of cardiac glycoside inhibition of the (Na^+-K^+)-dependent ATPase from cardiac tissue. Biochim. Biophys. Acta 151: 655–663.
12. MURTHY, R. V., KIDWAI, A. M., and DANIEL, E. E. 1974. Dissociation of contractile effect and binding and inhibition of Na^+-K^+-adenosine triphosphatase by cardiac glycosides in rabbit myometrium. J. Pharmacol. Exp. Therap. 188: 575–581.
13. OKITA, G. T., RICHARDSON, F., and ROTH-SCHECHTER, B. F. 1973. Dissociation of the positive inotropic action of digitalis from inhibition of sodium- and potassium-activated adenosine triphosphatse. J. Pharmacol. Exp. Therap. 185: 1–11.
14. REPKE, K., EST, M., and PORTIUS, H. J. 1965. Über die Ursache der Speciesunterschiede in der Digitalisempfindlichkeit. Biochem. Pharmacol. 14: 1785–1802.
15. REPKE, K., and PORTIUS, H. J. 1963. Uber den Einfluss verschiedener kardiotonischer Verbindungen auf die Transport ATPase in der Zellmembran des Herzmuskels. Arch. Exp. Pathol. Pharmakol. 245: 59–61.
16. TEN EICK, R. E., BASSETT, A. L., and OKITA, G. T. 1973. Dissociation of electrophysiological and inotropic actions of strophanthidin-3-bromoacetate: Possible role of adenosine triphosphatase in the maintenance of the myocardial transmembrane Na^+ and K^+ gradients. J. Pharmacol. Exp. Therap. 185: 12–23.
17. YODA, A. 1973. Structure-activity relationships of cardiotonic steroids for the inhibition of sodium- and potassium- dependent adenosine triphosphatase. I. Dissociation rate constants of various enzyme-cardiac glycoside complexes formed in the presence of magnesium and phosphate. Molec. Pharmacol. 9: 51–60.

Mailing address: Dr. T. Akera,
Department of Pharmacology,
Michigan State University,
East Lansing, Michigan 48824 (USA).

Recent Advances in Studies on Cardiac Structure and Metabolism
Volume 9
The Sarcolemma
Edited by Paul-Emile Roy and Naranjan S. Dhalla

THE EFFECT OF DIGOXIN ANTIBODY ON THE WASHOUT OF TRITIATED DIGOXIN AND ITS INOTROPIC EFFECT FROM PERFUSED RABBIT HEARTS

R. J. HOESCHEN

Department of Medicine, Section of Cardiology,
University of Manitoba and Health Sciences Centre,
Winnipeg, Manitoba, Canada

SUMMARY

The reversal of digoxin effects by digoxin antibody (Ab) may be mediated by at least two mechanisms. First, the Ab may simply decrease the concentration of free digoxin in the perfusing medium, or second, the Ab may dislodge digoxin from its receptor. To pursue this problem, rabbit hearts were perfused for 20 min with Krebs-Henseleit solution (K-H) followed by a 30-min perfusion with [^{3}H]digoxin in concentrations of 10^{-7}, 5×10^{-7}, and 10^{-6} M. The hearts were then washed out with K-H alone or with K-H containing Ab. During washout the effluent was collected at 30-sec intervals and the concentration of [^{3}H]digoxin measured in each sample. Washout was continued until the positive inotropic effect of digoxin had returned to its previous level. The [^{3}H]digoxin washout curves were analyzed on the basis of statistical criteria and yielded three exponential components. During Ab washout the half-times $t_{1/2}$ of these components were 0.25 ± 0.07, 1.70 ± 0.28, and 19.5 ± 6.9 min, while during K-H washout the values were 0.28 ± 0.02, 1.49 ± 0.22, and 11.8 ± 2.7. The decay of the inotropic effect of digoxin was similar during both washout conditions. The results indicate that the apparent tissue to perfusate concentration gradient for the washout of [^{3}H]digoxin was not increased by Ab. Thus, the Ab was incapable of dislodging digoxin from its receptor or Ab did not come in contact with digoxin bound to its receptor.

It is well documented that the digoxin specific antibody is capable of reversing physiological and toxic effects of digoxin (2, 4–6). The mechanism of antibody reversal of these effects remains uncertain. In most previous studies the reversal of digoxin effects could be explained by a simple reduction in availability of "free" digoxin. However, some observations suggested that the antibody might dislodge digoxin from its receptor site.

Watson and Butler (7), using rat renal cortical slices, indicated that the intracellular to extracellular digoxin ratio could be lowered by elution of labeled digoxin from the slices. In a later report, Gardner *et al.* (3) provided evidence that digoxin specific antibodies removed digoxin from red cell membranes by forming complexes with "free" digoxin which had effluxed from the cells to the incubation medium rather than by dislodging digoxin from its tissue receptor.

To pursue this problem further, the following study was performed. The washout of tritiated digoxin and the inotropic effect from perfused rabbit hearts was measured with and without antibody. If the antibody were capable of dislodging digoxin, then more [^{3}H] digoxin would appear in the perfusate and the inotropic effect would decrease faster during antibody washout when compared with control washout.

METHODS

Rabbits, weighing 1.5–3 kg, were killed by a blow on the head and their hearts removed and placed in chilled K-H solution. The hearts were then mounted on a nonrecirculating perfusion apparatus and perfused retrogradely via the aorta. A cannula was inserted in the right ventricle for collection of the effluent.

A hook was fixed to the apex of the heart and the hook was attached by a monofilament line to a Grass FT03D displacement transducer (Grass Instruments, Quincy, Mass.) connected to a Grass recorder. Sufficient resting tension was applied to the system to produce maximum developed tension at zero time and the hearts were driven at a rate of 180–190 beats/min by a Grass stimulator. Vertical movement of the heart was prevented and the tension recorded was the result of all forces acting along the vertical axis. Contractile force was measured throughout.

The perfusate was K-H solution containing normal rabbit serum (1:1000) except in experimental hearts when a 1:1000 dilution of digoxin rabbit antiserum was used. The 1:1000 dilution of antiserum was known to bind 50 percent of a known amount of digoxin *in vivo*. In some experiments, the antiserum concentration was increased to 1:500. Digoxin antibody had been produced previously by the method of Butler and Chen (1).

Hearts were perfused initially for 20 min to insure stability of the preparation. Then the perfusate was changed to a K-H rabbit serum solution containing tritiated digoxin in concentrations of either 10^{-6}, 5×10^{-7}, or 10^{-7} M digoxin for 30 min. The hearts were then washed out with either control K-H solution or K-H solution containing digoxin antibody.

During washout effluent from the right ventricular cannula was collected at 30-sec intervals until the positive inotropic effect had been washed out, that is until contractile force had returned to the values observed before the addition of digoxin. This was always accomplished within 20 min and no washout period exceeded this duration. The concentration of tritiated digoxin in the samples was determined in a Nuclear Chicago liquid scintillation counter using Aquasol

New England Nuclear (NEN) as the phosphor. The channels ratio method was used for quench correction.

The washout curves were analyzed using the SAAM-23 program and found to demonstrate three exponential components. The ($t_{1/2}$) were calculated for each component.

In some experiments the hearts were kept and homogenized and analyzed for [^{3}H] digoxin. In these experiments a portion of heart was also used for determination of dry to wet weight ratio.

RESULTS

Washout of [^{3}H] digoxin

Figure 1 is a representative semilogarithmic plot of the washout of [^{3}H] digoxin from perfused rabbit hearts. The three exponential components are graphically illustrated. Component *A* probably represents [^{3}H] digoxin washed out of the coronary vasculature. Components *B* and *C* are considered to represent [^{3}H]

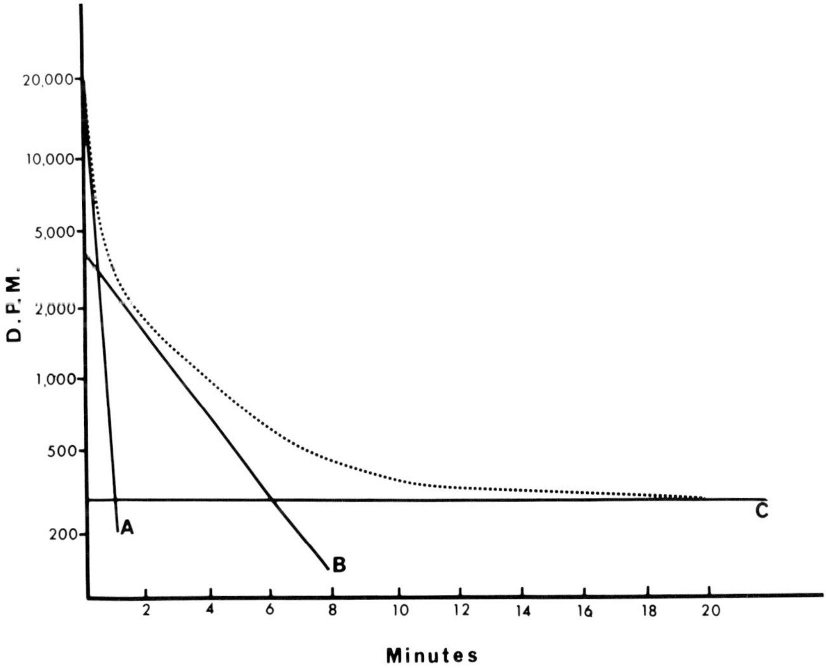

Figure 1 Representative washout curve of [^{3}H] digoxin from perfused rabbit hearts. The interrupted line represents the computer analysis of the raw data. The three exponential components are designated *A*, *B*, and *C*.

digoxin washed out of the hearts from sites with obviously different kinetic characteristics.

The half-times of each component during control washouts at the different digoxin concentrations are shown in figure 2. Perfusing with different digoxin concentrations resulted in no statistical difference in the $t_{1/2}$ of the individual washout components.

Figure 3 illustrates the half-times of the three components obtained during both control and antibody washout periods. The data for all digoxin concentrations (10^{-7} M, 5 X 10^{-7} M, and 10^{-6} M) are included in this figure. The antibody dilution was 1:1000.

The mean half-times in minutes of each component of all control washout experiments were 0.28 ± 0.02 (S.E.), 1.49 ± 0.22, and 11.8 ± 2.7. The values for all antibody washout experiments were 0.25 ± 0.07, 1.70 ± 0.28, 19.5 ± 6.9. None of the apparent differences was significant ($p > 0.1$)

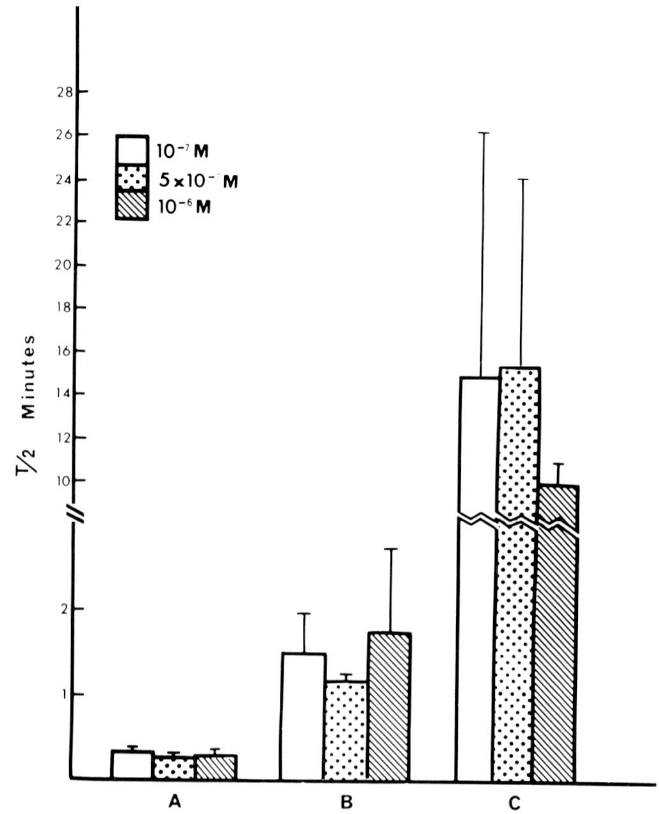

Figure 2 The effect of prior perfusion with different digoxin concentrations on the half-time of the three washout components during control washouts.

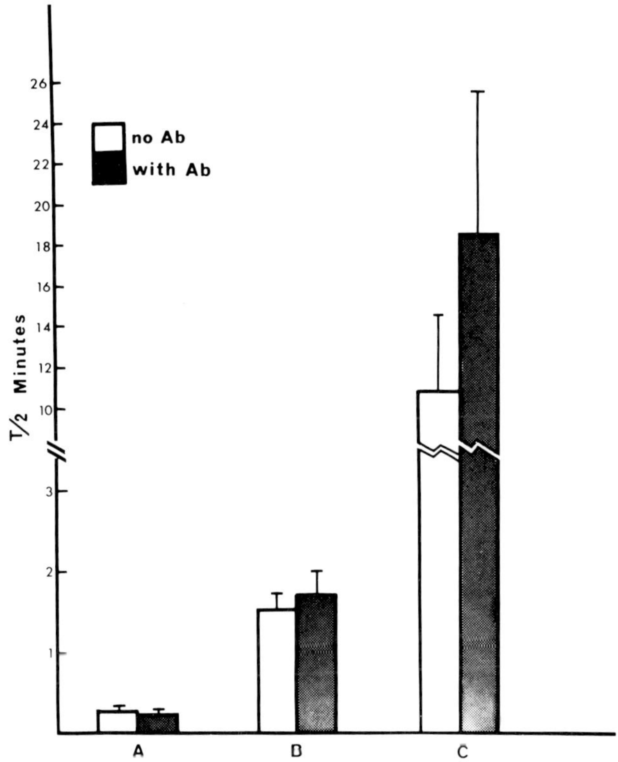

Figure 3 The half-time of the three washout components with and without antibody (1:1000 dilution).

In table I are the results of experiments in which the dilution of antibody was decreased from 1:1000 to 1:500. The digoxin concentration used before washout was 5×10^{-7} M. Again none of the apparent differences were significant ($p > 0.1$).

Table I Half-times of three washout components at different antibody dilutions*

	$t_{1/2}$ (min)		
	A	B	C
Control	0.26 ± 0.02	1.27 ± 0.14	14.8 ± 3.7
Ab (1:1000)	0.15 ± 0.03	1.43 ± 0.10	14.7 ± 3.5
Ab (1:500)	0.21 ± 0.00	1.57 ± 0.35	7.8 ± 0.1

*Values shown are means ± S.E. All hearts had been perfused previously with 5×10^{-7} M digoxin.

Washout of contractility effect

The reversal of the inotropic effect of digoxin during both the control and antibody washout periods is shown in figure 4. In all instances the inotropic effect had been washed out by 5–10 min. In the section labeled *5 × 10⁻⁷ M (Ab × 2)*, figure 4, it may be seen that doubling the amount of antibody did not alter the time required to washout the inotropic effect of digoxin.

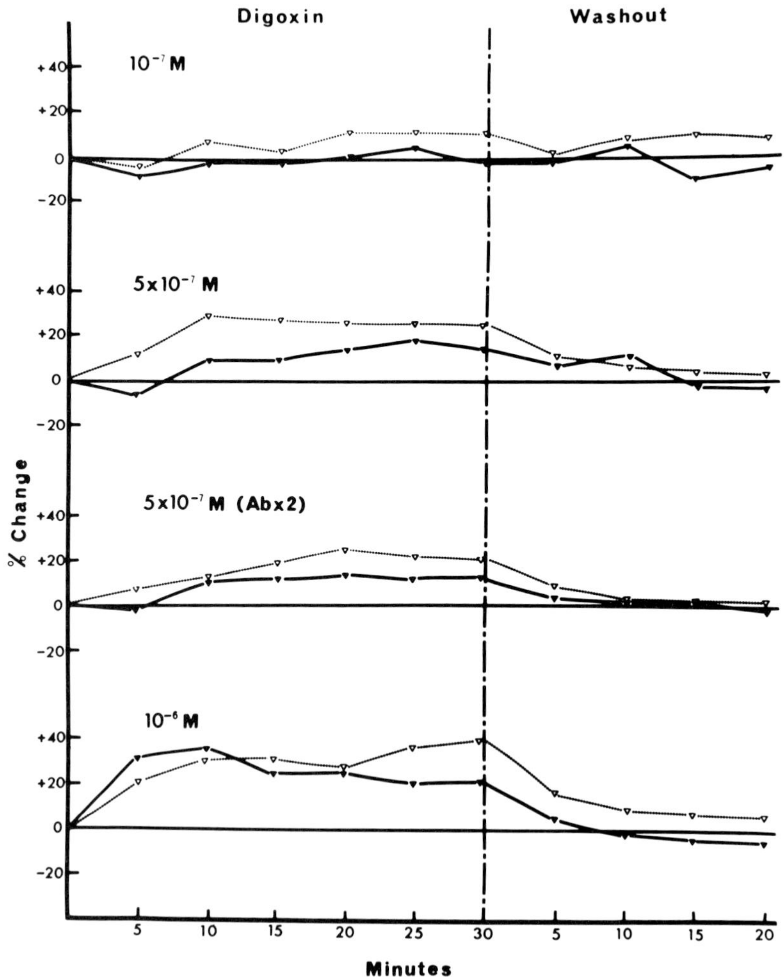

Figure 4 Effect of control and antibody washout on the inotropic effect after perfusion with different concentrations of digoxin. ——, control experiments; – – –, antibody experiments. In this figure the *ordinate* expresses inotropic effect as percentage change from the measurements before perfusion with digoxin.

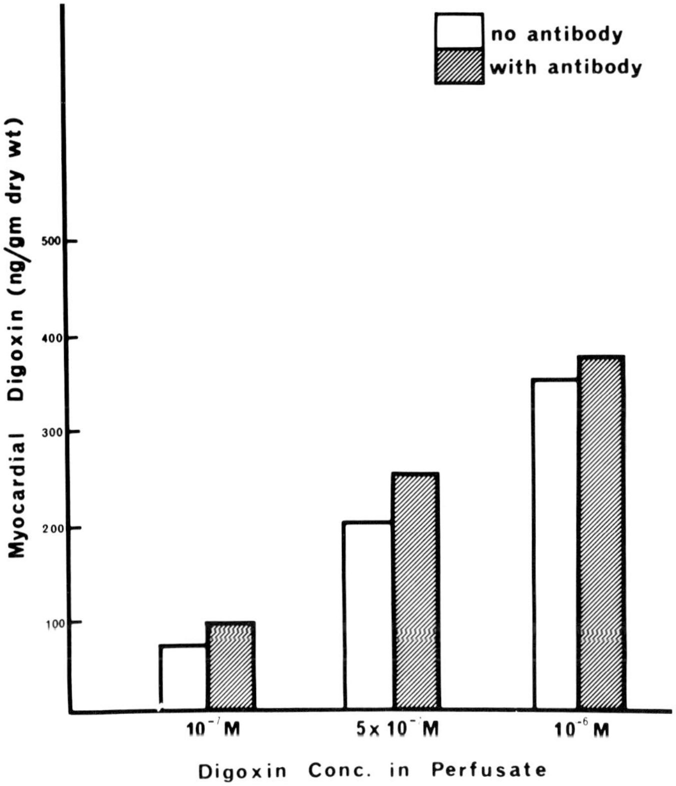

Figure 5 Myocardial [^{3}H]digoxin content after washout with and without antibody and prior perfusion with different concentrations of digoxin.

Myocardial digoxin content after washout

The amount of digoxin remaining in the hearts after 20 min of washout is shown in figure 5. The amount remaining increased with increasing perfusate concentration of digoxin. Since the inotropic effect had been washed out by 10 min, the remaining digoxin was considered to represent nonspecifically bound digoxin. After control washout and with a digoxin concentration of 10^{-7} M, the amount remaining was 72.4 ng/gm dry wt; with 5×10^{-7} M digoxin it was 202.0; with 10^{-6} M it was 353.0.

After washout with antibody the amount of [^{3}H]digoxin remaining in the hearts tended to be higher but the differences were not significant.

DISCUSSION

The present study was designed to determine whether digoxin specific antibody was capable of dislodging digoxin from its receptor site in hearts in which the vasculature was intact. If the antibody were capable of dislodging digoxin, then more [^{3}H] digoxin should have been washed out of the hearts during antibody washout as compared with control washout and the positive inotropic effect should have diminished at a faster rate.

The results indicate that the antibody washout curves did not differ from control washout curves and the amount of [^{3}H] digoxin remaining in the hearts after washout was similar in the two situations. This suggested that digoxin antibody was incapable of dislodging digoxin from its receptor site in a perfused intact heart. The results did not necessarily indicate that the antibody did not possesses the physicochemical ability to dislodge digoxin. It is equally possible that the antibody did not contact the digoxin receptor site via the intact vasculature. This would explain the apparent discrepancy between these data and the data obtained in rat renal cortical slices (7).

The fact that antibody washout did not hasten reversal of the inotropic effect must be reconciled with the observations of others. All previous studies have been done in "closed" systems, that is, incubation baths or intact animals. The present experiments were performed in an "open" nonrecirculating system. In closed systems, such as incubated red cells, the experiments of Gardner and coworkers (3) suggested that digoxin antibody binds "free" digoxin released from the red cells. It is proposed that in the "open" perfused system, most of the [^{3}H] digoxin efflux from the heart was washed out at such a rapid rate during control washout that any potential differences attributable to binding of this "free" digoxin could not be detected during antibody washout.

Thus, this study was not in conflict with previous observations concerning antibody reversal of digoxin effects. However, the results did indicate that digoxin antibody does not dislodge digoxin from its receptor site in a system with an intact vasculature.

REFERENCES

1. BUTLER, V. P., and CHEN J. P. 1967. Digoxin-specific antibodies. Proc. Nat. Acad. Sci. U.S.A. 57: 71–78.
2. CURD, J., SMITH, T. W., JATON, J. C. and HABER, E. 1971. The isolation of digoxin-specific antibody and its use in reversing the effects of digoxin. Proc. Nat. Acad. Sci. U.S.A. 68: 2401–2406.
3. GARDNER, J. D., KIINO, D. R., SWARTZ, T. J., and BUTLER, V. P. 1973. Effects of digoxin-specific antibodies on accumulation and binding of digoxin by human erythrocytes. J. Clin. Invest. 52: 1820–1833.
4. MANDEL, W. J., BIGGER, T. J., and BUTLER, V. P. 1972. The electrophysiologic effects of low and high digoxin concentrations on isolated mammalian cardiac tissue: Reversal by digoxin-specific antibody. J. Clin. Invest. 51: 1378–1387.
5. SCHMIDT, D. T., and BUTLER, V. P. 1971. Immunological protection against digoxin toxicity. J. Clin. Invest. 50: 866–871.
6. SCHMIDT, D. H., and BUTLER, V. P. 1971. Reversal of digoxin toxicity with specific antibodies. J. Clin. Invest. 50: 1738–1744.

7. WATSON, J. F., and BUTLER, V. P. 1972. Biologic activity of digoxin specific antisera. J. Clin. Invest. 51: 638–648.

Mailing address: Dr. R. J. Hoeschen,
Health Sciences Centre, 700 William Avenue,
Winnipeg, Manitoba R3E 0Z3 (Canada).

Recent Advances in Studies on Cardiac Structure and Metabolism
Volume 9
The Sarcolemma
Edited by Paul-Emile Roy and Naranjan S. Dhalla

ON THE MECHANISMS UNDERLYING DIGITALIS TOXICITY IN CARDIAC PURKINJE FIBERS[1]

M. VASSALLE[2] and E. MUSSO

Department of Physiology,
State University of New York,
Downstate Medical Center,
Brooklyn, New York, USA

SUMMARY

The different mechanisms by which digitalis enhances pacemaker automaticity in canine Purkinje fibers has been studied under different conditions. Transmembrane potentials of quiescent, spontaneously active, or electrically driven fibers were recorded in the presence and absence of strophanthidin (10^{-5}–10^{-7} M). In quiescent fibers, strophanthidin caused the membrane potential to depolarize and to develop oscillations of progressively larger magnitude which were followed by spontaneous activity. In spontaneously active preparations strophanthidin enhanced the slope of diastolic depolarization and therefore the rate of discharge. In both cases, there was no evidence of oscillations during diastolic depolarization. A steepening of diastolic depolarizations without superimposed oscillations (and eventually also followed by spontaneous activity) was observed in preparations driven electrically at very slow rates. However, in fibers driven at faster rates (30/min or more) in the presence of ouabain, oscillations developed during diastolic depolarization. These oscillations were frequency dependent in two respects: 1) at slower rates the oscillations remained constant during the drive, the actual magnitude being proportional to the driving rate; 2) at rate of 60/min or more, the oscillations increased in size with each successive beat. Fast rhythms were initiated by a progressive acceleration of the pacemaker potential or abruptly by an extrasystole. When spontaneous activity ceased, the membrane increased slowly toward the control value. We conclude that digitalis increases the frequency of discharge of Purkinje fibers 1) by enhancing diastolic depolarization when the rate of discharge is slow; 2) by developing a new current which modifies diastolic depolarization when diastole is sufficiently short; and 3) by means of the potassium plateau current when the rate of spontaneous discharge is fast. Whether one or more of these mechanisms are operational in affecting Purkinje fiber automaticity depends critically on the rate of discharge of the fibers.

Cardiac glycosides cause many types of arrhythmias both *in vitro* (4, 7, 33) and *in vivo* (24, 32). The mechanisms underlying the altered rhythm are not

[1] This work was supported by grants from the New York Heart Association and from the National Institutes of Health (HL-17451).

[2] Dr. Vassalle is a Career Scientist of the Health Research Council of the City of New York.

completely understood. An increase in slope of diastolic depolarization by cardiac glycosides has been demonstrated repeatedly (4, 7, 16, 33). In Purkinje fiber driven electrically at constant rate such a steepening of diastolic depolarization leads to spontaneous discharge and potential oscillations (33). The oscillatory potentials induced by cardiac glycosides have been studied in detail recently (5, 8, 9, 11, 23). It has been shown that at fast rate of drive, in the presence of digitalis, oscillations appear which grow in size after each successive action potential and may eventually cause spontaneous discharge. At the same time diastolic depolarization was found to be depressed, the apparent increase in slope being in actuality due to superimposed oscillations (9). In another approach, it has been shown that the pacemaker current is depressed by ouabain, and this would be expected to cause an enhancement of diastolic depolarization (2).

Thus, there are uncertainties as to whether the steepening of diastolic depolarizations is caused by an enhancement of the normal pacemaker process or to the development of oscillations. In fact, it is not clear whether the normal pacemaker process is enhanced or depressed by cardiac glycosides. Yet the difference between these mechanisms is substantial and it is possible that the discrepancies in the experimental results reported are due to different conditions imposed on the fiber. Thus, although the mechanisms reported are obviously different, they are not necessarily mutually exclusive or the only ones brought into play. The purpose of this investigation was to test the mechanisms involved in the increased automaticity induced by strophanthidin and to analyze the condition required for the different mechanisms to play a predominant role. A preliminary report has appeared in abstract form (34).

METHODS

Mongrel dogs were anesthetized with pentobarbital sodium (Somnopentyl, Pitman-Moore, 28 mg/kg, i.v.) and the heart quickly excised. Purkinje strands were removed from the ventricles and transferred into a tissue bath perfused with oxygenated (98 percent O_2 and 2 percent CO_2) Tyrode solution.

The composition of the Tyrode solution in micromolar concentration was as follows: NaCl 137; KCl 5.4; $NaHCO_3$ 11.9; NaH_2PO_4 0.5; $MgCl_2$ 1.0; $CaCl_2$ 2.7; dextrose 5.5. In different experiments $[K]_0$ was varied from 2 to 5.4 mM. Warm (37°C) Tyrode solution flowed in the tissue bath at the rate of about 80 drops/min. The preparations were either quiescent, spontaneously active, or electrically stimulated, as will be indicated when appropriate. Electrical stimuli were provided by a Grass S4C stimulator via a Grass S1U5 stimulus isolation unit (Grass Instruments, Quincy, Mass.) and delivered to the preparation by means of steel pins insulated except for the tip. The duration of the stimuli was 1–2 msec and the voltage about 20 percent above threshold. Transmembrane potentials were recorded by means of two microelectrodes of the Ling-Gerard type filled with 3 M KCl. The tip of one microelectrode was immersed in the perfusion fluid and the other was inserted intracellularly. The recording apparatus consisted of a push-pull cathode follower stage, a directly coupled Tektronix type 3A3 dual-trace differential amplifier, and a Tektronix model 565 dual-beam oscilloscope. The traces were photographed with a Kymograph camera (Grass Instruments). The preparations were allowed to equilibrate in Tyrode solution for about 60 min and then control recordings were carried out. Strophanthidin (10^{-5}–10^{-7} M) was dissolved in an aliquot of control Tyrode solution and recordings were repeated during perfusion of the drug. After a variable (3–30 min) period of strophanthidin exposure, the preparation was returned to the Tyrode solution and the recovery studied.

RESULTS

Quiescent fibers

By using quiescent fibers, it should be possible to determine whether digitalis affects pacemaker activity only by the development of transient oscillations or by some other mechanism. If the only mechanism by which digitalis affects pacemaker activity were transient oscillations, the fibers should remain quiescent since transient oscillations only follow an action potential and thus require activity to be already present.

The result of a test of this kind is illustrated in figure 1. The preparation was perfused in 5.4 mM K-Tyrode solution and was therefore quiescent. During the perfusion of 3×10^{-6} M strophanthidin, the preparation began to discharge spontaneously at a slow rate. As shown in figure 1, the initiation of spontaneous activity was brought about by a slow decline in the resting potential upon which oscillations were superimposed. These oscillations became progressively larger and the last one attained the threshold. The ensuing action potential undershot the resting potential and this was followed by the usual diastolic depolarization. No transient oscillations were visible during the diastolic depolarization.

Similar results were obtained in eight trials in six different experiments. In four of the experiments, the preparations were bathed in 2.7 mM K-Tyrode and, once strophanthidin had induced spontaneous activity, the rate of discharge was higher than during the same procedure in 5.4 mM Tyrode. During the recovery period, the fibers became quiescent once more.

Spontaneously active fibers

By using spontaneously active (rather than electrically driven) preparations, it should be possible to determine whether automaticity is enhanced or depressed by strophanthidin. The result of one experiment of this kind is shown in figure

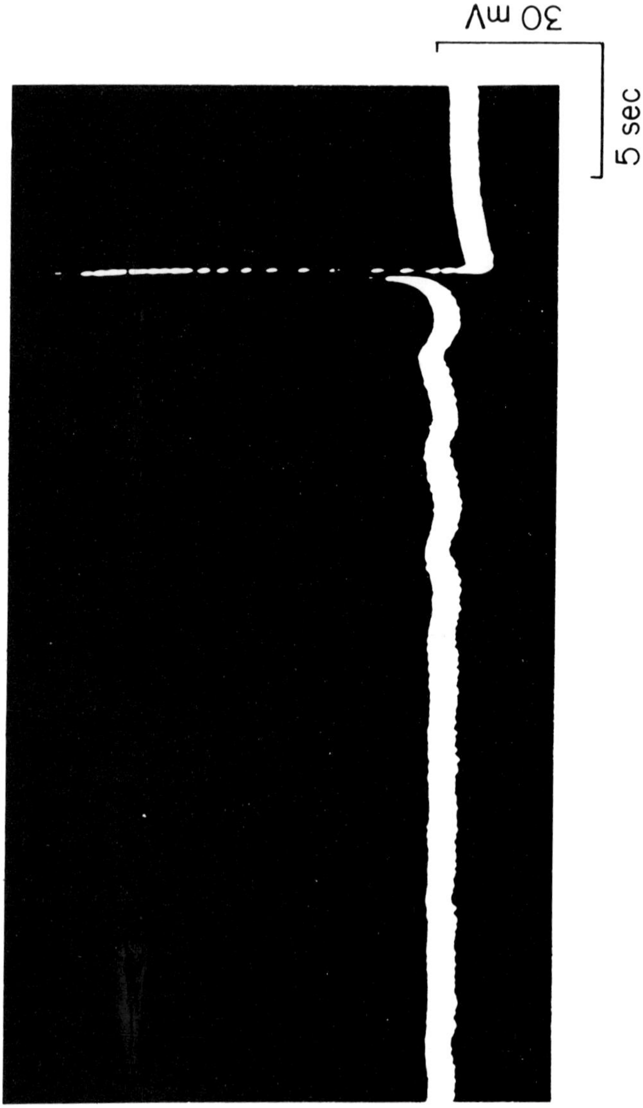

Figure 1 Onset of spontaneous activity induced by strophanthidin; $[K]_o = 5.4$ mM.

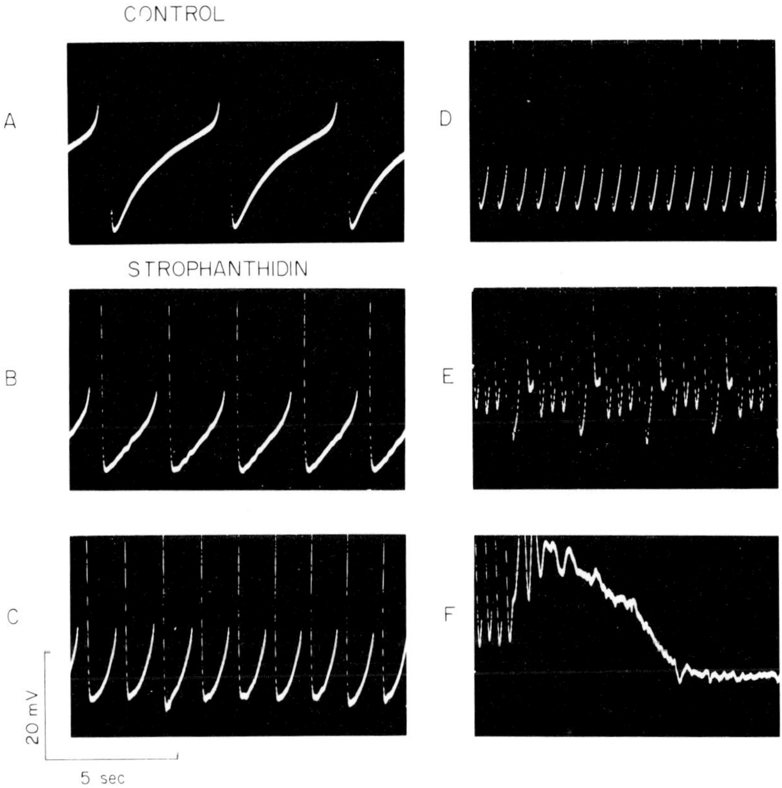

Figure 2 Effect of strophanthidin on the rate of discharge of a spontaneously firing fiber. *Panel A,* only the lower part of the action potential is shown in this and in the other panels. *Panels B–E,* effect of strophanthidin on the pacemaker potential. *Panel F,* recorded during recovery.

2. Only the lower portion of the action potential is shown. It is apparent (fig. 2, *panel A,* control) that diastolic depolarization exhibits the smooth transition into the upstroke of action potential typical of a dominant pacemaker (35). The other panels (*B–E*) were recorded at various times during a 20-min exposure to 0.5×10^{-6} M strophanthidin. It is clear that the rate of depolarization during diastole increased progressively (*panels B–D*) and abnormal rhythms eventually appeared (*panel E*). In *panel E,* an extrasystole appeared during the repolarization phase of every fifth beat and was characterized by a more negative maximum diastolic potential (E_{max}). The beat after the extrasystole, on the contrary, shows the least negative E_{max}. *Panel F* was recorded 42 sec after the termination of the strophanthidin perfusion and shows the cessation of the fast rhythm and the slow increase of the potential to a steady value.

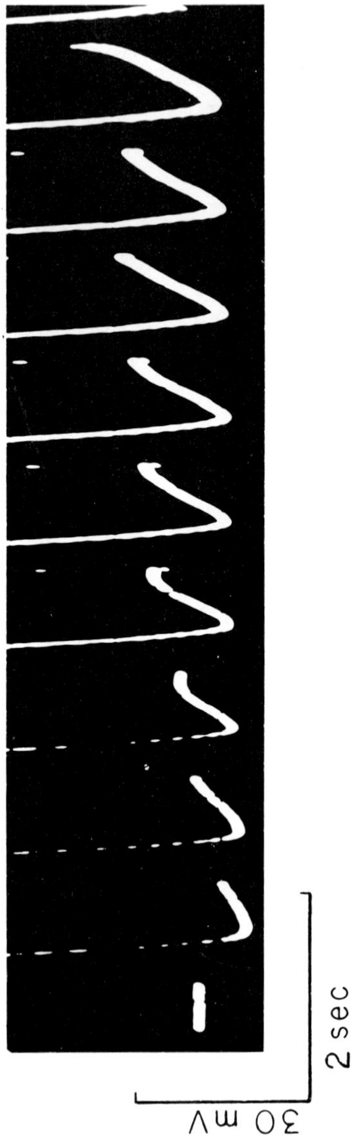

Figure 3 Effect of electrical drive at 60/min on the pacemaker potential of a Purkinje fiber exposed to strophanthidin.

Typically, the strophanthidin poisoning of a spontaneous preparation led to the following progressive changes: steepening of diastolic depolarization; progressive fall in E_{max}; decrease in amplitude of the action potential as the rate became faster; eventual cessation of spontaneous activity at a potential intermediate between the plateau and maximum diastolic potential. On recovery, the potential increased slowly as seen in *panel F* of figure 2. These types of results were obtained in 14 trials in nine experiments.

Driven fibers

In this series of experiments the preparations were not spontaneously active and were stimulated at a given rate for different periods of time. This was done to study the effect of activity and rest on diastolic depolarization. In figure 3, the preparation had been driven in strophanthidin (0.5×10^{-6} M) solution for 16 min and then had been left unstimulated for 1 min. The horizontal trace at the beginning of the figure is the resting potential. As electrical stimulation at 60/min was resumed, the first action potential undershot the resting potential to attain E_{max}. With successive action potentials, the following changes are apparent: 1) a progressive decrease in E_{max}; 2) a progressive steepening of diastolic depolarization; 3) a late diminution of the rate of diastolic depolarization most evident during the third, fourth, fifth, and sixth diastolic intervals; 4) the eventual attainment of the threshold (the last action potential is spontaneous); and 5) an eventual decrease of E_{max} almost to the same value as the resting potential at the beginning of the figure.

The changes in conformation of the action potential induced by strophanthidin administration in a continuously driven Purkinje fiber are illustrated in figure 4. The preparation was driven at all times at 120/min and action potentials recorded during the control period and strophanthidin exposure were superimposed. The figure shows that at the time when relatively little change was apparent at the plateau, E_{max} was decreased and diastolic depolarization steepened.

In the example shown in figure 5, the fiber was continuously perfused in the presence of strophanthidin and was driven at 60/min for several minutes and the last action potential recorded. This action potential was followed by a diastolic depolarization which was initially fast but was not maintained in that the membrane potential increased again. The preparation was allowed to remain quiescent for several seconds and then one action potential was superimposed on the last driven action potential recorded previously. The figure shows that, although the differences at the plateau were minimal (and expected) there was a profound difference during diastole. After the period of quiescence, the potential during diastole decreased continuously as a function of time. Thus, in the presence of strophanthidin, repeated activity leads to a change in diastolic

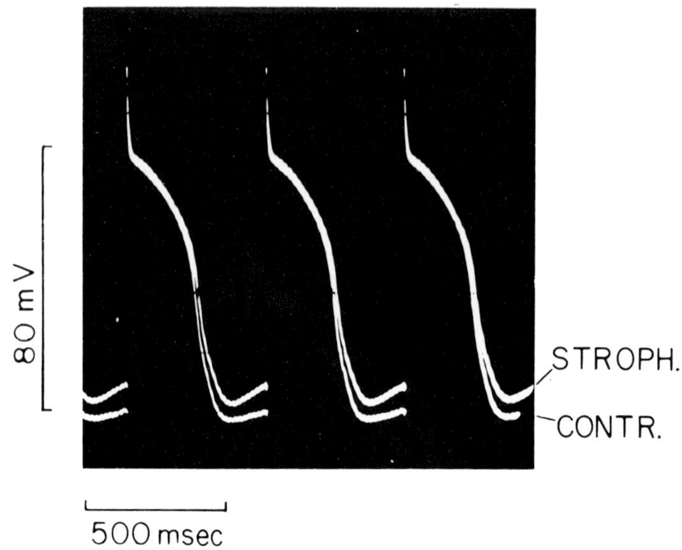

Figure 4 Action potentials of a fiber driven at 120/min in the absence (*CONTR.*) and in the presence of strophanthidin (*STROPH.*).

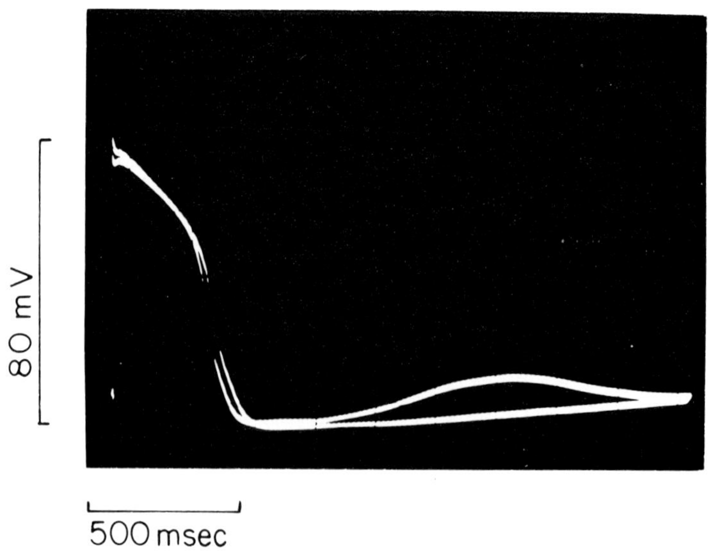

Figure 5 Action potential recorded in the presence of strophanthidin. The last driven action potential is followed by an oscillation which is no longer present after an action potential (*superimposed*) recorded after a pause of several seconds.

depolarization which is dissipated within several seconds after the cessation of the activity.

Fibers driven at a progressively faster rate

Oscillations become more prominent as the rate of drive increases. At a very slow rate they should disappear. Changes in diastolic depolarization as a function of the driving rate in the presence of 0.5 × 10^{-6} M strophanthidin are shown in figure 6. The preparation was driven for 2 min at each of the rates indicated in the figure. The *vertical portion of the traces* is the final part of phase 3 repolarization of the last action potential driven at the rate indicated (60, 36, and 12/min). As the driving rate was decreased from 60 to 12, the slope of the initial part of diastolic depolarization decreased and at 12/min there is no apparent oscillation during diastolic depolarization. Furthermore, E_{max} increased and the potential attained at the end of the diastolic interval was higher as the driving rate was decreased.

This type of change was seen in fibers driven at different rates in eight experiments.

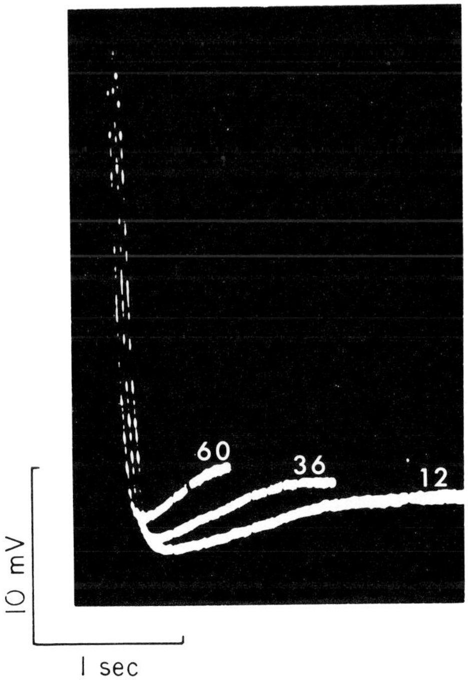

Figure 6 The steepening of the pacemaker potential and the fall in E_{max} as the rate of drive is increased from 12 to 36 and 60/min in the presence of strophanthidin.

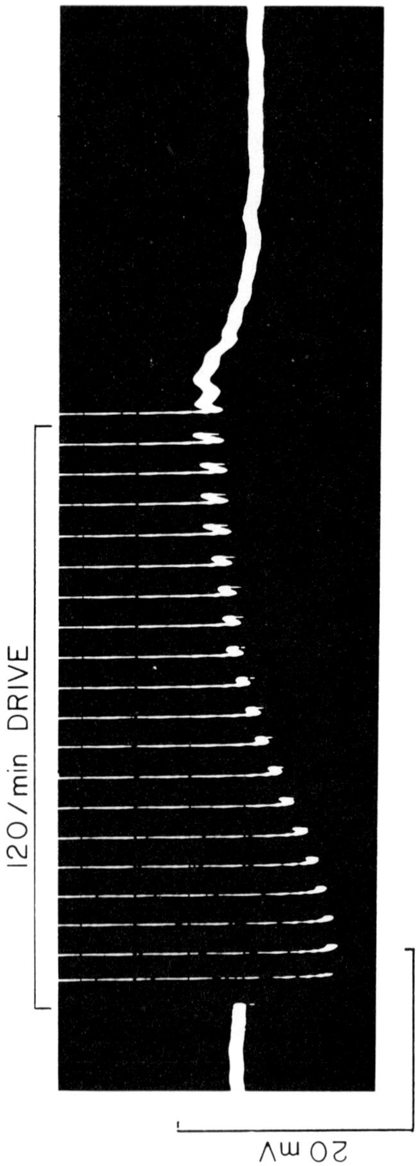

Figure 7 Changes in membrane potential during and after a period of fast drive.

Events following a period of fast drive

The progressive fall of E_{max} with faster driving rates could indicate that the pacemaker current is progressively inactivated when the fiber is repeatedly discharged at a constant rate. In figure 7, a quiescent fiber was driven at 120/min for 10 sec. The first action potential undershot the resting potential as usual but E_{max} progressively decreased (by 10 mV) to a level below the resting potential. At the same time the slope of depolarization during diastole increased. At the end of the drive, one sees irregular oscillations and a slow repolarization to the resting potential. Similar results were obtained when the fibers were driven at the same rate for 15 or 30 sec (11 instances in four experiments). The events illustrated in figure 8 support the possibility that the pacemaker current is inactivated during the drive and reactivated after the drive. The preparation was spontaneously active at 15.4/beats/min. During the electrical drive at 90/min, E_{max} fell and the slope of the depolarization during diastole increased. After the drive, there was an abrupt reversal of the last "diastolic depolarization." However, after that (in contrast to the events shown in figure 7), the membrane potential in this spontaneously active fiber fell slowly toward threshold. The subsequent spontaneous beat shows a maximum diastolic potential far greater than during overdrive and similar to the predrive control. This suggests the

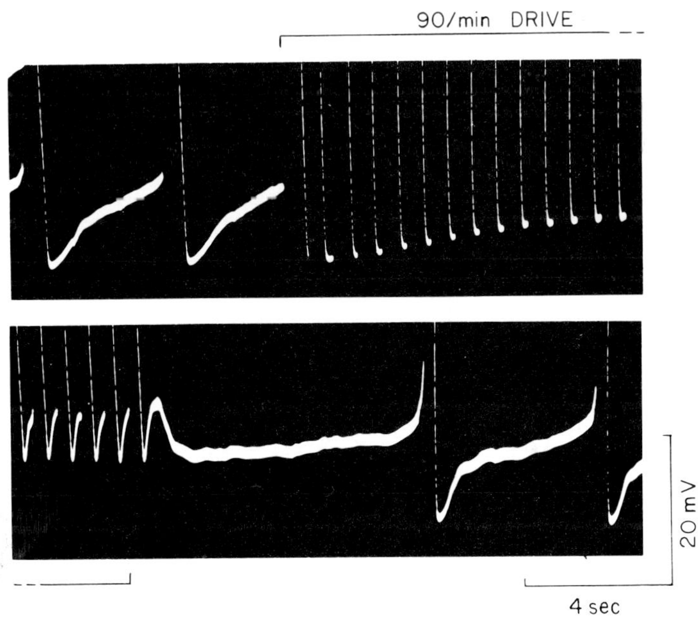

Figure 8 Fast drive in a fiber exposed to strophanthidin. During the drive E_{max} fell to a new steady state value.

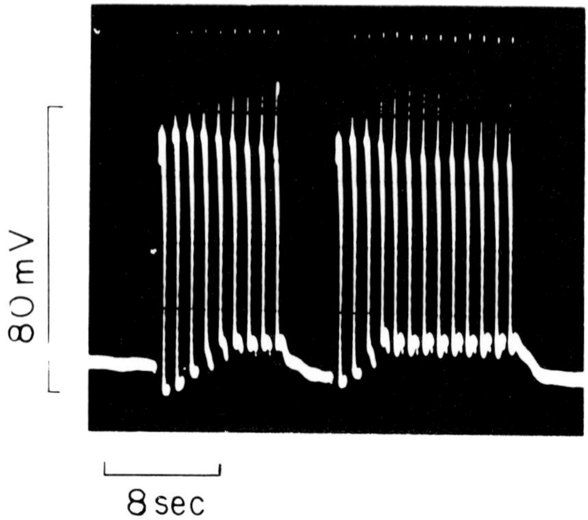

Figure 9 Changes in E_{max} as a function of the duration of the preceding quiescence.

following interpretation. After the end of the drive, the pacemaker current was small as suggested by the small E_{max}. However, in this spontaneous fiber it fell even more during the pause which followed, bringing the potential to threshold. During the subsequent action potential, however, the pacemaker current was fully reactivated as evidenced by the normal E_{max}. The following diastolic depolarization shows the usual time course. Alternative interpretations are presented under Discussion.

If the pacemaker current is reactivated after a drive, the shorter the pause between drives, the smaller E_{max} should be. This is demonstrated clearly in figure 9. The first driven action potential after a period of quiescence of 7 sec shows a maximum diastolic potential larger than the first action potential induced after a second pause at 4 sec. Figure 9 is interesting in another respect. As the preparation was driven at 60/min the maximum diastolic potential fell as usual with each successive beat. However, after the fourth and fifth beats in the first run and after the third beat in the second run, there was an abrupt slowing of phase 3 repolarization. Eventually, E_{max} stabilized to a new level below the original resting potential. On stoppage of the stimulation the potential slowly increased again to the original resting value. Two other findings are worth noticing: 1) the plateau began at progressively more negative potentials during the drive and stabilized to a new value when E_{max} stabilized; and 2) the overshoot remained unaltered in spite of the different "take off" potentials.

Different driving rates and onset of fast rhythms

In Figure 10, it is shown that fast rhythm can be initiated by a mechanism different from a steepening of diastolic depolarization. In the top strip, the fiber was driven at 60/min for 4 beats. When the stimulation was turned off, a spontaneous rhythm began, characterized, as usual, by a fall in E_{max} and a steepening of the pacemaker potential. However, the occurrence of one extrasystole was followed by the onset of a fast thythm which was characterized by small action potentials which oscillate roughly between −50 and 0 mV. The fast rhythm increased progressively until an abrupt slowing occurred. This in turn was followed by a rapid repolarization to the original resting value. In the middle strip the preparation was driven at 30 and 42/min. By selecting this slower rate it was possible to observe the behavior of the pacemaker potential during diastole. The picture makes clear that (in contrast with faster driving rates) at 30/min the oscillatory potentials do not grow in amplitude with successive beats, and also that E_{max} and the plateau level remained unchanged. When the rate was increased to 42/min, immediately the maximum diastolic potential decreased,

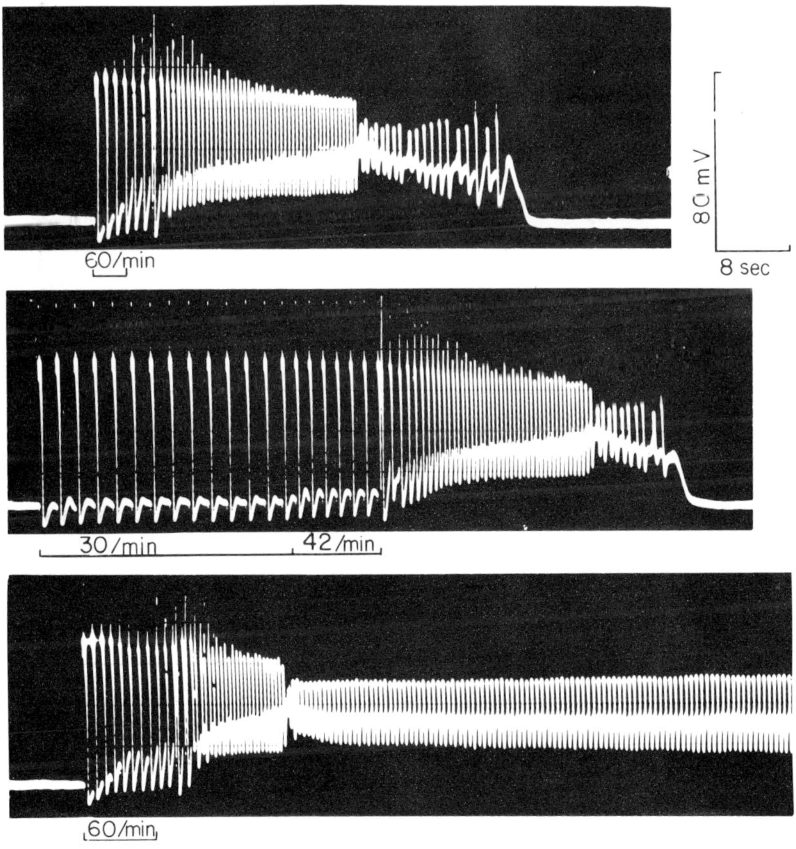

Figure 10 Effect of strophanthidin on the pacemaker potential as a function of the rate and onset of fast rhythms.

the initial slope of the pacemaker potential increased, and the amplitude of the oscillatory potentials was larger. Simultaneously, the plateau level was shifted to more negative values. In spite of the fact that the pacemaker potential did not increase with each successive beat, the occurrence of an extrasystole was sufficient to initiate again a fast rhythm. In the bottom strip the fiber was again driven at 60/min and after the extrasystoles, a regular, sustained fast rhythm was initiated.

Fibers driven at slow rates

Several of the results illustrated above show that in the presence of strophanthidin oscillatory potentials became smaller as the driving rate was decreased. This suggested that in fibers perfused with strophanthidin but driven at very slow rate oscillations might not appear at all. If so, it should be possible to establish how strophanthidin affects diastolic depolarization apart from oscillations. In figure 11, the preparation was driven at 7/min and the control action potentials are shown in *panel A*. In *panel B*, recorded 3 min after the initiation of strophanthi-

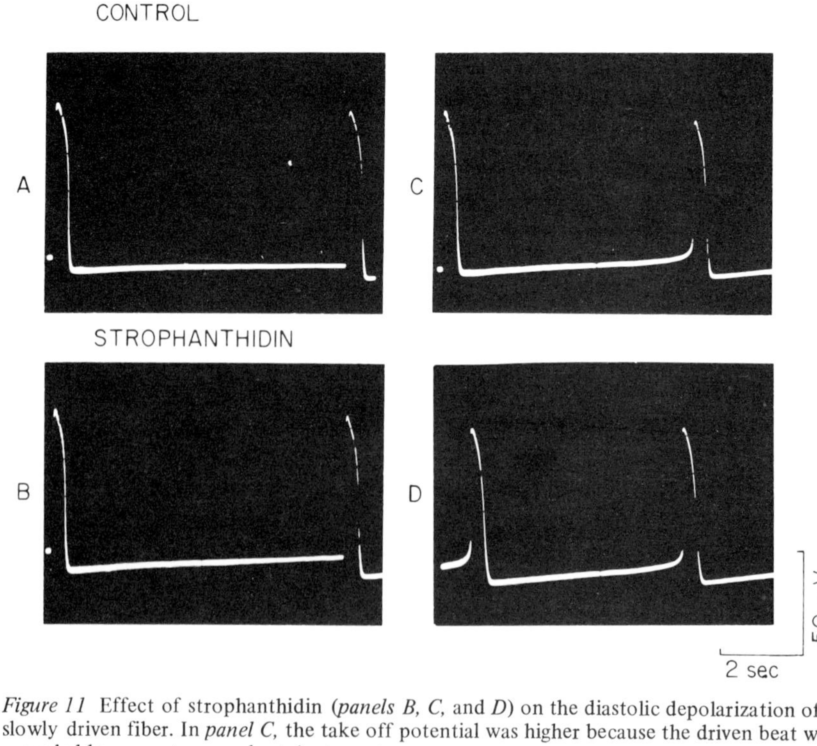

Figure 11 Effect of strophanthidin (*panels B, C,* and *D*) on the diastolic depolarization of a slowly driven fiber. In *panel C,* the take off potential was higher because the driven beat was preceded by a spontaneous beat (not seen).

din perfusion, the rate of diastolic depolarization was obviously enhanced as shown by the difference between the take-off potential and E_{max}. In *panel C*, the first action potential was driven and the second was spontaneous as a result of a further steepening of diastolic depolarization. In *panel D*, the stimulation had been interrupted and both action potentials were spontaneous. No obvious oscillations are seen. After an exposure of 6 min, strophanthidin administration was interrupted and spontaneous activity slowly subsided. In other experiments, the acceleration progressed into fast rhythms during the strophanthidin exposure.

Antiarrhythmic action of high K

It is well known that high potassium counteracts digitalis arrhythmias (25, 31). The events which lead to such an effect are illustrated in figure 12. The preparation had been perfused with strophanthidin in 2.7 mM K-Tyrode and had developed a spontaneous fast rhythm. Quite shortly after removing the strophanthidin and doubling $[K]_0$ to 5.4 mM, the maximal diastolic potential decreased, an effect presumably due to a fall in E_K. As the perfusion with 5.4 mM K-Tyrode continued, the slope of diastolic depolarization decreased progressively; eventually the fiber became quiescent once more. These findings were typically observed in three experiments.

DISCUSSION

The present results show that in Purkinje fibers strophathidin 1) induces spontaneous activity in quiescent fibers; 2) increases the rate of discharge of spontaneously active fibers by steepening diastolic depolarization; 3) induces spontaneous activity in fibers driven at a very slow rate by enhancing diastolic depolarization; 4) causes the development of oscillations in fibers driven at a relatively fast rate; 5) induces spontaneous rhythms characterized by small action potentials in both spontaneous driven preparations provided that the fibers discharge at a relatively fast rate. These findings suggest that strophanthidin affects Purkinje fiber automaticity by different mechanisms depending on the conditions imposed on the fiber. In particular, the rate of discharge of the fiber determines critically the mechanism by which strophanthidin affects Purkinje fiber discharge.

An increase in the rate of spontaneous discharge of pacemakers is usually taken to indicate an increase in automaticity. If this concept is accepted, there is no question that digitalis causes an increase in automaticity (3, 5, 7, 9, 10, 11, 14, 23, 33). As to whether the enhanced automaticity results from changes in the normal process of diastolic depolarization, or, instead, from different processes, the present results provide evidence for the involvement of more than one mechanism.

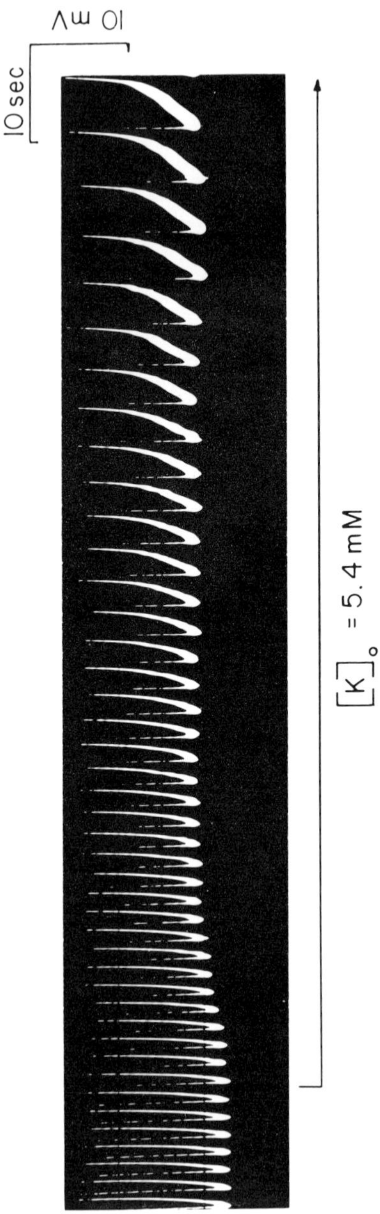

Figure 12 Effect of higher K (5.4 mM) on a fiber intoxicated by strophanthidin in 2.7 mM K-Tyrode.

The results obtained in quiescent fibers demonstrate clearly that strophanthidin affects pacemaker activity of Purkinje fibers independently of the transient oscillations. The onset of activity as shown in figure 1 was due to a slow decline of membrane potential upon which oscillations of increasing magnitude were superimposed until the threshold was attained. The oscillations such as those in figure 1 are different from the oscillations (8, 33) which follow the action potential in the presence of digitalis. Thus, oscillations leading to the onset of automaticity precede (and do not follow, fig. 1) the action potential, whereas transient depolarizations follow (and do not precede, fig. 3) the action potential. Oscillations of increasing magnitude as those in figure 1 constitute a link between rest and activity, whereas the transient oscillations constitute a link between driven and spontaneous action potentials. Epinephrine (21) and low $[K]_0$ (28) also induce spontaneous activity in a manner not unlike that seen in figure 1. Although the mechanisms of these two interventions are different (12, 28), a common consequence is a fall of potassium outward current at resting potential. This may account for the similarity in the oscillations leading to the onset of spontaneous discharge as the net inward leakage current becomes more important. The mechanism by which digitalis induces spontaneous activity is not determined. It could involve either an increase in an inward current or a decrease in potassium outward current. In view of the known relationship between digitalis and calcium, an increase in calcium inward current might be involved in the onset of activity. However, this does not appear probable for several reasons. Thus, digitalis does not increase calcium influx and does not induce contracture in quiescent atria (15). In agreement with the above finding is the fact that ventricular muscle fibers are also particularly resistant to the toxic actions of digitalis (including contracture) as long as they are kept quiescent (33). In addition, digitalis has been found to increase membrane resistance of Purkinje fiber at the time when it enhances spontaneous activity (7). Finally, an increase in outside calcium concentration does not induce spontaneous activity in quiescent Purkinje fibers (27).

Digitalis could, instead, initiate activity by decreasing a potassium outward current at the resting potential. In Purkinje fiber diastolic depolarization results from a time-dependent inactivation of slow potassium current (6, 29) which has been labeled i_{K_2} (18). This pacemaker current is rapidly activated during the action potential (Reference 18, and see fig. 7 in Reference 29) and is responsible for the attainment of the maximum diastolic potential. In a quiescent fiber, this current is partially activated at resting potential (Reference 18; see fig. 4 in Reference 29) and its decrease by strophanthidin could readily account for the onset of spontaneous activity. Alternatively, the initiation of spontaneous activity could be brought about by a fall in the time-independent K current. It also remains to be seen whether digitalis decreases K efflux at resting potential indirectly. This could happen if digitalis were to affect calcium distribution at the cell membrane and thus affect K fluxes. Experimentally, it has been shown that ouabain decreases K efflux both at rest and during activity (22).

Another demonstration of the ability of strophanthidin to enhance pacemaker activity independently of transient oscillations is provided by the results in slowly driven fibers (fig. 11). At a very slow rate oscillations are not seen; yet strophanthidin steepened diastolic depolarization. The maximum diastolic potential also decreased and this would be expected if the K outward current decreases. That diastolic depolarization is indeed enhanced (and not depressed) by strophanthidin is shown by the effect on spontaneously discharging fiber (fig. 2). A depression of diastolic depolarization should have led to an arrest of the spontaneous discharge; instead there was an acceleration. Since the rate was initially rather slow it is unlikely that the acceleration was due to the onset of transient oscillations, at least as long as the rate was not too fast. The administration of ouabain in animals with complete atrial-ventricular block induces an increase in the rate of discharge after an initial decline (10, 36). On occasion, however, digitalis has been shown to cause a marked depression of idioventricular automaticity (24, 32). It must be stressed here that there are no reasons why a spontaneously active fiber (once accelerated by strophanthidin to a critical rate through a steepening of diastolic depolarization) should not sustain a fast rhythm through development of other mechanism (see below). On the contrary, this would be expected since it would matter little whether the rate at which new mechanisms came into play is attained through an imposed drive or a steepening of diastolic depolarization.

As far as the oscillatory behavior during diastole is concerned it is becoming clear that there is a crucial relationship between digitalis, calcium, and transient depolarizations (5, 8, 9). Thus, when Purkinje fibers are driven at constant rate "diastolic depolarization" steepens after each subsequent action potential and on stoppage of the drive oscillations are apparent (9, 33). Increasing the driving rate resulted in more pronounced effect. The oscillations after the drive require a certain number of beats to attain a constant amplitude (9). In the presence of high $[Ca]_0$ alone, Temte and Davis (27) found that diastolic depolarization of a driven fiber becomes steeper as the rate of discharge is increased and the interruption of the drive is also followed by oscillations. After a period of quiescence, driving the fiber at a constant rate resulted in a steepening of diastolic depolarization with each successive beat (27). If increasing the rate in the presence of digitalis leads to an increase in the slope and magnitude of transient depolarizations, decreasing the rate to rather low value should cause a diminution or disappearance of the oscillations. This was borne by the present experiments, inasmuch as, at a very slow rate (7/min), no oscillations were apparent. The present experiments also show another interesting point. Driving at a slow rate oscillations were present and remained constant in slope and amplitude during the drive (fig. 10, *middle strip*, 30/min). Increasing the rate of 42/min (fig. 10, *middle strip*) increased the slope and the magnitude of the oscillations, but again they remained constant during the drive. It was only when the rate was increased to 60/min or higher that transient oscillations increased progressively after each action potential. These results indicate that at lower rate of discharge, diastole was apparently long enough to prevent a frequency-dependent growth in the pacemaker potential after each subsequent action potential. The implications of this finding is that an extrasystole, by shortening the preceding diastole, may be enough to initiate a fast rhythm. This is actually shown in fig. 10. The findings reported in the literature and the present results indicate that both high calcium and digitalis create the condition for an enhancement of the pacemaker potential. However, the duration of diastole is all important in determining whether the enhancement grows progressively to the threshold. The mechanisms of this "diastolic depolarization staircase" are not apparent but there are several reasons to believe that calcium is somehow responsible for this phenomenon, as has been shown (20) for mechanical staircase. Crucial evidence has been provided by Ferrier and Moe (8) who showed that digitalis-induced oscillations are enhanced by higher calcium and disappear in the absence of calcium or in the presence of mangenese. The role of frequency in this connection may be related to movement and/or distribution of calcium as affected by the action potential. As the rate is increased diastole becomes short enough to allow a summation of effects which are reflected in the electrical manifestations. Thus, a fast rate in the presence of strophanthidin not only increases the slope and the magnitude of the pacemaker potential but also simultaneously decreases E_{max} and shifts the plateau to more negative values.

Whether the normal process of diastolic depolarization is depressed or not when the pacemaker potential is steepened by fast drive is not clear. If it is depressed, the depression is frequency related. Thus, if the fall in E_{max} during fast drive is attributable to the inactivation of pacemaker current, the restoration of the resting potential and the maximum diastolic potential within a few seconds from the end of the drive shows that the pacemaker current is promptly reactivated once the rapid activity ceases.

The fall in E_{max} during the drive could be because of a poisoning (26) of the sodium-potassium pump by strophanthidin. In the absence of digitalis, E_{max} decreases initially during a fast drive but (as the drive is continued) increases above the control values. The initial fall has been attributed to an increases in $[K]_0$ and the subsequent increase to the activation of an electrogenic sodium pump (30). In the presence of strophanthidin the fall in E_{max} during the drive is more pronounced and the subsequent increase does not occur (References 3, 14, see also figs. 7 and 8). A poisoning of the pump by digitalis would lead to an accumulation of potassium outside the cell membrane and this would decreases E_K and, therefore, E_{max}. However, the progressive increase in the slope of the pacemaker potential during the drive and the relatively fast return of the potential towards the original resting value argue against an increase in $[K]_0$ as the only cause of the fall in E_{max}. The restoration of the resting potential (fig. 7) and of E_{max} (fig. 8) after a fast drive is too fast to be due to the action of a poisoned pump. These findings and the fact that a fall in E_{max} occurs when the action potential is still little altered otherwise (fig. 4) suggest instead that a frequency-dependent progressive increase of an inward current or a decrease of a K outward current (or both) is responsible for the fall in E_{max}. These changes reach a new steady state during the drive and are dissipated within a few seconds after the drive, unless a fast spontaneous rhythm is precipitated.

The fast rhythms illustrated in figure 10 are quite probably the results of yet another mechanism. Thus, the action potentials occur within a potential range in which a pacemaker current should not vary with time and therefore play no role in rhythmic discharge. The picture resembles the repetitive activity seen in partially depolarized Purkinje fibers (13). This repetitive activity has been shown to be due to a combination of increased leakage current and voltage- and time-dependent changes in the plateau current i_{x_1} (13). In fact, it has been shown that in Purkinje fibers exposed to ouabain application of depolarizing pulses resulted in the appearance of small action potential discharging at fast rate (1). The fast activity on depolarization was attributed to the plateau current i_{x_1}. Our experiments suggest that i_{x_1} might play a role even at higher membrane potentials. Figure 9 is rather interesting in this respect. Experiments by Noble and Tsien (19) have shown that i_{x_1} has a reversal potential of about −70 mV. Under normal circumstances, the activation of the pacemaker current brings the potential to the maximum diastolic value. This happened also in figure 9 with the first action potential after quiescence. However, the reversal of "diastolic depolarization" after the fourth and the fifth beat could be explained if the pacemaker current was inactivated as a function of the rate and i_{x_1} became more important. A new "maximum diastolic potential" is attained with subsequent beats which stabilizes at a potential close to the equilibrium potential for i_{x_1}. During diastole, i_{x_1} is inactivated and since the time constant of its inactivation at these negative potentials is faster than that of i_{K_2}, the "new" diastolic depolarization is faster. At the end of the first run apparently the pacemaker current is reactivated as the potential tends to return to its original value. However, this recovery is incomplete because the first action potential after the pause has a smaller undershoot and it takes only three beats for the reversal of diastolic depolarization.

In conclusion it appears that strophanthidin: 1) may induce a fall in K outward current in a quiescent fiber as a function of the length of exposure; 2) inactivates the pacemaker current as a function of the frequency of discharge; 3) induces a net inward current during diastole which summates to steepen the pacemaker potential; and 4) allows i_{x_1} alone to determine diastolic depolarization during the fast discharge of small action potentials which result from a slow (Na, Ca) current. These mechanisms are brought progressively into play as the rate is increased in a driven preparation and are responsible for the progressive increase in rate in a spontaneous preparation.

The crucial role played by the rate of discharge in the onset of digitalis arrhythmias explains some of the findings reported *in vivo*. For example, the abolition of ouabain-induced coupled extrasystoles with graded vagal stimulation (32) may be explained on the basis that slowing of the heart rate would prevent the growth of the pacemaker potential to the threshold (see fig. 10). Reciprocally, a shortening of diastole by extrastimuli would be expected to increase the number of extrasystoles (17, 32). Also, it would explain why (see fig. 10) coupled extrasystoles may initiate runs of self-sustaining fast rhythms (32).

REFERENCES

1. ARONSON, R. S., and CRANEFIELD, P. F. 1974. The effect of resting potential on the electrical activity of canine cardiac Purkinje fibers exposed to Na-Free solution or to ouabain. Pflügers Arch. ges. Physiol. 347: 101–116.
2. ARONSON, R. S., GELLES, J. M., and HOFFMAN, B. F. 1973. Effect of ouabain on the current underlying spontaneous diastolic depolarization in cardiac Purkinje fibers. Nature New Biol. 245: 118–120.
3. CARPENTIER, R., and VASSALLE, M. 1971. Enhancement and inhibition of a frequency-activated electrogenic sodium pump in cardiac Purkinje fibers, pp. 81–98. *In* F. F. Kao, K. Koizumi, and M. Vassalle (eds.), Research in Physiology *a Liber Memorialis* in Honor of Professor C. McC. Brooks. A. Gaggi Publisher, Bologna.
4. CORABOEUF, E., de LOZÉ, C., and BOISTEL, J. 1953. Action de la digitale sur les potentiels de membrane et d'action du tissu conducteur du coeur de chien étudiée à l'aide de microélectrodes intracellulaires. C. R. Soc. Biol. (Paris) 147: 1169–1172.
5. DAVIS, L. D. 1973. Effect of changes in cycle length on diastolic depolarization produced by ouabain in canine Purkinje fibers. Circ. Res. 32: 206–214.
6. DECK, K. A., and TRAUTWEIN, W. 1964. Ionic currents in cardiac excitation. Pflugers Arch. ges. Physiol. 280: 63–80.
7. DUDEL, J., and TRAUTWEIN, W. 1958. Elektrophysiologische Messungen zur Strophanthinwirkung am Herzmuskel. Naunyn-Schmiedeberg's Arch. Exp. Pathol. Pharmakol. 232: 393–407.
8. FERRIER, G. R., and MOE, G. K. 1973. Effect of calcium on acetylstrophanthidin-induced transient depolarizations in canine Purkinje tissue. Circ. Res. 33: 508–515.
9. FERRIER, G. R., SAUNDERS, J. H., and MENDEZ, C. 1973. A cellular mechanism for the generation of ventricular arrhythmias by acetylstrophanthidin. Circ. Res. 32: 600–609.
10. GREENSPAN, K., VASSALLE, M., and HOFFMAN, B. F. 1962. Ouabain toxicity in complete heart block [Abstract]. Fed. Proc. 21: 134.
11. HASHIMOTO, K., and MOE, G. K. 1973. Transient depolarizations induced by acetylstrophanthidin in specialized tissue of dog atrium and ventricle. Circ. Res. 32: 618–624.
12. HAUSWIRTH, O., NOBLE, D., and TSIEN, R. W. 1968. Adrenaline: Mechanism of action on the pacemaker potential in cardiac Purkinje fibers. Science 162: 916–917.
13. HAUSWIRTH, O., NOBLE, D., and TSIEN, R. W. 1969. The mechanism of oscillatory activity at low membrane potentials in cardiac Purkinje fibres. J. Physiol. (London) 200: 255–265.
14. HOGAN, P. M., WITTENBERG, S. M., and KLOCKE, F. J. 1973. Relationship of stimulation frequency to automaticity in the canine Purkinje fiber during ouabain administration. Circ. Res. 32: 377–384.
15. HOLLAND, W. C., and SEKUL, A. 1961. Influence of K^+ and Ca^{++} on the effect of

ouabain on Ca^{45} entry and contracture in rabbit atria. J. Pharmacol. Exp. Therap. 133: 288–294.
16. KASSEBAUM, D. G. 1963. Electrophysiological effects of strophanthin in the heart. J. Pharmacol. Exp. Therap. 140: 329–338.
17. LOWN, B. 1968. Electrical stimulation to estimate the degree of digitalization. II. Experimental studies. Amer. J. Cardiol. 22: 251–259.
18. NOBLE, D., and TSIEN, R. W. 1968. The kinetics and rectifier properties of the slow potassium current in cardiac Purkinje fibres. J. Physiol. (London) 195: 185–214.
19. NOBLE, D., and TSIEN, R. W. 1969. Outward membrane currents activated in the plateau range of potentials in cardiac Purkinje fibres. J. Physiol. (London) 200: 205–231.
20. OCHI, R., and TRAUTWEIN, W. 1971. The dependence of cardiac contraction on deplorization and slow inward current. Pflüger's Arch. ges. Physiol. 323: 187–203.
21. OTSUKA, M., 1958. Die Wirkung von Adrenalin auf Purkinje-Fasern von Säugetierherzen. Pflügers Arch. ges. Physiol. 266: 512–517.
22. POLIMENI, P. I., and VASSALLE, M. 1971. On the mechanism of ouabain toxicity in Purkinje and ventricular muscle fibers at rest and during activity. Amer. J. Cardiol. 27: 622–629.
23. ROSEN, M. R., GELBAND, H., MERKER, C., and HOFFMAN, B. F. 1973. Mechanisms of digitalis toxicity. Effects of ouabain on phase four of canine Purkinje fiber transmembrane potentials. Circulation 47: 681–689.
24. ROTHBERGER, C. J., and WINTERBERG, H. 1913. Uber den Einfluss von Strophantin auf die Reizbildungsfähigkeit der automatischen Zentren des Herzens. Pflügers Arch. ges. Physiol. 150: 217–261.
25. SAMPSON, J. J., and ANDERSON, E. M. 1932. The treatment of certain cardiac arrhythmias with potassium salts. J. A. M. A. 99: 2257–2261.
26. SCHATZMANN, H. J. 1953. Herzglykoside als Hemmstoffe für den aktiven Kalium- und Natriumtransport durch die Erythrocytenmembran. Helv. Physiol. Pharmacol. Acta 11: 346–354.
27. TEMTE, J. V., and DAVIS, L. D. 1967. Effect of calcium concentration on the transmembrane potentials of Purkinje fibers. Circ. Res. 20: 32–44.
28. VASSALLE, M. 1965. Cardiac pacemaker potentials at different extra- and intracellular K concentrations. Amer. J. Physiol. 208: 770–775.
29. VASSALLE, M. 1966. Analysis of cardiac pacemaker potential using a "voltage clamp" technique. Amer. J. Physiol. 210: 1335–1341.
30. VASSALLE, M. 1970. Electrogenic suppression of automaticity in sheep and dog Purkinje fibers. Circ. Res. 27: 361–377.
31. VASSALLE, M., and GREENSPAN, K. 1963. Effects of potassium on ouabain-induced arrhythmias. Amer. J. Cardiol. 12: 692–701.
32. VASSALLE, M., GREENSPAN, K., and HOFFMAN, B. F. 1963. An analysis of arrhythmias induced by ouabain in intact dogs. Circ. Res. 13: 132–148.
33. VASSALLE, M., KARIS, J., and HOFFMAN, B. F. 1962. Toxic effects of ouabain on Purkinje fibers and ventricular muscle fibers. Amer. J. Physiol. 203: 433–439.
34. VASSALLE, M., and MUSSO, E. 1974. Cardiac Purkinje fibers, automaticity and digitalis [Abstract]. Fed. Proc. 33: 432.
35. WEIDMANN, S. 1951. Effect of current flow on the membrane potential of cardiac muscle. J. Physiol. (London) 115: 227–236.
36. WITTENBERG, S. M., GANDEL, P., HOGAN, P. M., KREUZER, W., and KLOCKE, F. J. 1972. Relationship of heart rate to ventricular automaticity in dogs during ouabain administration. Circ. Res. 30: 167–176.

Mailing address: Dr. M. Vassalle,
Department of Physiology, Box 31,
Downstate Medical Center,
450 Clarkson Avenue,
Brooklyn, N.Y. 11203 (USA).

Pathophysiological Aspects

Recent Advances in Studies on Cardiac Structure and Metabolism
Volume 9
The Sarcolemma
Edited by Paul-Emile Roy and Naranjan S. Dhalla

ROLE OF SARCOLEMMAL CHANGES IN CARDIAC PATHOPHYSIOLOGY[1]

N. S. DHALLA, C. W. TOMLINSON,
J. N. SINGH, S. L. LEE,
D. B. McNAMARA, J. A. C. HARROW, and J. C. YATES

Experimental Cardiology Laboratory,
Department of Physiology,
Faculty of Medicine, University of Manitoba,
Winnipeg, Manitoba, Canada

SUMMARY

Sarcolemmal Ca^{++}-ATPase, Mg^{++}-ATPase, and (Na^+ - K^+)-ATPase activities were increased in late stages of heart failure in myopathic hamsters (BIO 14.6) without any changes in the adenylate cyclase activity. On the other hand, these hamsters at early and moderate stages of heart failure showed depressions in mitochondrial calcium binding and uptake and microsomal calcium binding. Sarcolemmal (Na^+-K^+)-ATPase was decreased in failing hearts because of substrate lack, oxygen lack, and perfusion with Ca^{++}-free, Na^+-free, or K^+-free medium. Both Mg^{++}-ATPase and Ca^{++}-ATPase activities of sarcolemma did not change on perfusing the hearts with substrate-free, hypoxic, Na^+-free, or K^+-free medium. Adenylate cyclase activity decreased on substrate-free or Ca^{++}-free perfusion. Intracellular calcium overload produced by perfusing the hearts with medium containing calcium after Ca^{++}-free perfusion was associated with decrease in all the sarcolemmal-bound enzyme activities. All types of failing hearts employed in this study showed a dramatic shift in the electrolyte composition. Failure of the cardiac muscle to generate contractile force on treatment with trypsin was associated with defects in the functions of sarcolemma, mitochondria, and sarcoplasmic reticulum, whereas such an effect on treatment with phospholipase C was limited to alterations in the activities of sarcolemma. The data suggest that abnormality at the level of sarcolemma plays an important role in the pathogenesis of heart dysfunction; however, the degree and direction of alterations in the sarcolemmal functions seem to be dependent upon the type of heart failure.

Although sarcolemma maintains intracellular concentrations of electrolytes within certain limits as well as prevents leakage of enzymes and other essential constituents of the myocardium, very little work concerning its role in the pathogenesis of heart dysfunction has appeared in the literature. Nonetheless, some of the observations made by various investigators can be interpreted to

[1] This study was supported by the Manitoba Heart Foundation and is Part XVII in a series of papers having the general title "Excitation-Contraction Coupling in Heart."

suggest an abnormality of sarcolemma in heart failure. For example, adenylate cyclase and (Na^+ -K^+)-ATPase, which are associated with cell membrane, were markedly altered in congestive heart failure induced by aortic constriction (18, 24) and in genetically myopathic (BIO 14.6) hamsters (20, 21). Recently, we have isolated heart sarcolemma containing high specific activities of (Na^+ -K^+)-ATPase, Mg^{++}-ATPase, Ca^{++}-ATPase, and adenylate cyclase (14), and have successfully employed this preparation for the first time in identifying sarcolemmal defect in myopathic (UM X 7.1) hamster hearts (17). If sarcolemma is defective in heart failure then this abnormality should be occurring in different types of failing hearts. In the present study, therefore, we have employed hearts which were failing because of genetic defect, oxygen, or substrate lack, intracellular calcium deficiency or overload, as well as electrolyte shifts. In addition, it is the purpose of this investigation to demonstrate whether the failure of the cardiac muscle to generate contractile force and changes in sarcolemma is apparent on treating the cardiac muscle with agents such as phospholipase C and trypsin, which are known to produce dramatic alterations in the phospholipid and protein contents of the membrane systems.

METHODS

Genetically myopathic hamsters (BIO 14.6) at various stages of disease (3, 6) were employed in this study. The failure of hearts in these hamsters has been considered to be due to some genetic defect. Normal, healthy Syrian hamsters served as controls. Normal albino or hooded rats were also used. These animals were decapitated and their hearts removed. For obtaining papillary muscles, healthy dogs and cats were anesthetized with pentobarbital sodium (50 mg/kg, i.p.) and their hearts were quickly removed. The electrical and mechanical events of the papillary muscles were studied according to the method described by Prasad and Callaghan (16). The methods for studying electrical and mechanical properties of isolated perfused rat hearts were the same as described previously (2). The control perfusion medium for rat heart experiments contained 8 mM glucose and was oxygenated with 95 percent O_2 and 5 percent CO_2 gas mixture, whereas hypoxic medium was gassed with 95 percent N_2 and 5 percent CO_2 mixture. Glucose was omitted in the case of substrate-free medium. When Na^+, K^+, or Ca^{++} was omitted from the perfusion medium, the osmolarity was maintained by adding an appropriate amount of sucrose.

The methods for preparing sarcolemma and studying adenylate cyclase, Ca^{++}-ATPase, Mg^{++}-ATPase, and (Na^+ -K^+)-ATPase activities were exactly the same as described earlier (15). The rat hearts, which were employed for determining electrolyte contents, were quickly perfused with 10 ml of ice-cold homogenization solution (0.32 sucrose, 5 mM histidine, pH 7.0). These hearts were then homogenized and the calcium, magnesium, potassium, and sodium contents were measured with a Zeiss atomic absorption spectrophotometer after extraction with 0.5 N HCl. The papillary muscles were also examined with a Zeiss electron microscope (3). The methods for isolating microsomes (fragments of sarcoplasmic reticulum) and mitochondria and determining calcium binding and uptake were similar to those described elsewhere (19).

RESULTS

The criteria of dividing myopathic hamsters (BIO 14.6) in different stages of disease on the basis of their age were based upon the degree of pulmonary edema, liver congestion, cardiac hypertrophy, and elevation in the level of serum creatine phosphokinase. The data in table I reveal that the sarcolemmal Ca^{++}-ATPase, Mg^{++}-ATPase, and (Na^{+} -K^{+})-ATPase activities of the myopathic hearts at late stages of failure were higher ($p < 0.05$) than the control. Unlike Ca^{++}-ATPase and Mg^{++}-ATPase, the activity of sarcolemmal (Na^{+} -K^{+})-ATPase in hearts from the myopathic animals at moderate stages of failure was also elevated. On the other hand, the adenylate cyclase activity of the sarcolemma from myopathic hearts was not different ($p > 0.05$) from the control. Mitochondrial calcium binding and uptake and microsomal calcium binding were progressively decreased at early, moderate, and late stages of failure, whereas microsomal calcium uptake decreased significantly ($p < 0.05$) only in late stages of the disease (table II).

In order to test whether sarcolemmal changes can be demonstrated in other types of failing hearts, we have used isolated rat hearts perfused with substrate-free or hypoxic medium. Both substrate lack and oxygen lack have been shown to produce failure of the heart to generate contractile force (3, 7). It can be seen from figure 1 that both contractile force and rate of change of contraction (dF/dt) decreased markedly over a period of 2 hr on perfusing the heart with substrate-free medium and over a period of 30 min on perfusion with hypoxic medium. The magnitude of the surface electrical activity of the heart started increasing and then decreasing after 45 and 90 min of perfusion with substrate-free medium, respectively. On the other hand, the magnitude of the surface electrical activity started decreasing within 1 min of perfusion with hypoxic medium. The results shown in table III indicate that ^{45}Ca uptake by the hypoxic heart, unlike the substrate-depleted heart, was also decreased significantly ($p < 0.05$). In another series of experiments, calcium binding by sarcolemma obtained from the substrate-depleted hearts and hypoxic hearts was studied and the results are shown in table IV. A marked reduction in calcium binding ability of sarcolemma was noted on perfusing the hearts with substrate-free medium for 2 hr or with hypoxic medium for 30 min.

The sarcolemmal enzyme activities and electrolyte contents of the hearts made to fail on perfusion with substrate-free medium for 2 hr or hypoxic medium for

Table I Sarcolemmal enzyme activities of hearts from control and BIO 14.6 myopathic hamsters at different stages of heart failure*

Enzymes	Control (100–272 days)	Myopathic Early stage (100–150 days)	Myopathic Moderate stage (180–210 days)	Myopathic Late stage (260–275 days)
Adenylate cyclase	240 ± 17	251 ± 21	225 ± 18	221 ± 15
Ca^{++}-ATPase	39.5 ± 1.6	38.1 ± 1.7	42.9 ± 1.2	47.4 ± 2.2
Mg^{++}-ATPase	31.4 ± 1.3	31.7 ± 1.1	35.0 ± 1.6	42.5 ± 2.7
(Na^+ - K^+)-ATPase	7.3 ± 0.9	7.8 ± 0.6	11.4 ± 0.7	14.5 ± 1.4

*Each value is a mean ± S.E. of six to eight experiments. Results for Mg^{++}-ATPase, Ca^{++}-ATPase, and (Na^+ - K^+)-ATPase are shown as micromoles of Pi per mg of protein per hr, whereas those for adenylate cyclase are expressed as picomoles of cyclic AMP formed per mg of protein per min.

Table II Calcium accumulation by subcellular fractions of hearts at varying degrees of congestive heart failure in genetically myopathic hamsters (BIO 14.6)*

Stages of failure	Calcium accumulation (% of control)			
	Mitochondria		Microsomes	
	Binding	Uptake	Binding	Uptake
Early stage (150–170 days old)	83 ± 4.9	80 ± 5.2	74 ± 4.2	97 ± 3.1
Moderate stage (200–230 days old)	50 ± 6.1	56 ± 4.7	50 ± 2.9	93 ± 4.2
Late stage (260–280 days old)	33 ± 3.4	30 ± 2.2	47 ± 1.8	67 ± 3.9

*Each value is a mean ± S.E. of six to eight experiments. Normal healthy hamsters of the respective age group served as controls.

30 min are given in table V. The adenylate cyclase activity was decreased ($p < 0.05$) on perfusion with substrate-free medium but did not change on perfusion with hypoxic medium. Although both Ca^{++} ATPase and Mg^{++}-ATPase activities were slightly elevated in substrate-free and hypoxic heart sarcolemma, these changes were not significant ($p > 0.05$). On the other hand, the sarcolemmal (Na^+ -K^+)-ATPase activity was markedly decreased ($p < 0.05$) in hearts made to fail because of perfusion with substrate-free or hypoxic medium. The potassium and magnesium contents of the substrate-depleted or hypoxic hearts were lower ($p < 0.05$), whereas sodium levels were higher ($p < 0.05$) than the control. In contrast to the substrate-depleted hearts, the hypoxic hearts were found to contain less ($p < 0.05$) calcium than the control.

The isolated rat hearts were unable to generate contractile force within 1 and 12 min on perfusion with Na^+-free medium and K^+-free medium, respectively. We believe that the failure of hearts on perfusion with Na^+-free or K^+-free medium is due to dramatic shifts in sodium and potassium contents. The activity of sarcolemmal (Na^+ -K^+)-ATPase, unlike adenylate cyclase, Ca^{++}-ATPase, or Mg^{++}-ATPase, was decreased ($p < 0.05$) on perfusing the hearts for 20 min with Na^+-free or K^+-free medium (table VI). Both potassium and sodium contents were decreased ($p < 0.05$) and calcium levels were increased in hearts perfused with Na^+-free medium for 20 min (table VI). On the other hand, perfusion with K^+-free medium for 20 min increased sodium and calcium contents and decreased magnesium and potassium levels significantly (table VI).

In another series of experiments changes in the sarcolemmal enzymes and electrolyte contents were studied in hearts perfused with Ca^{++}-free medium as

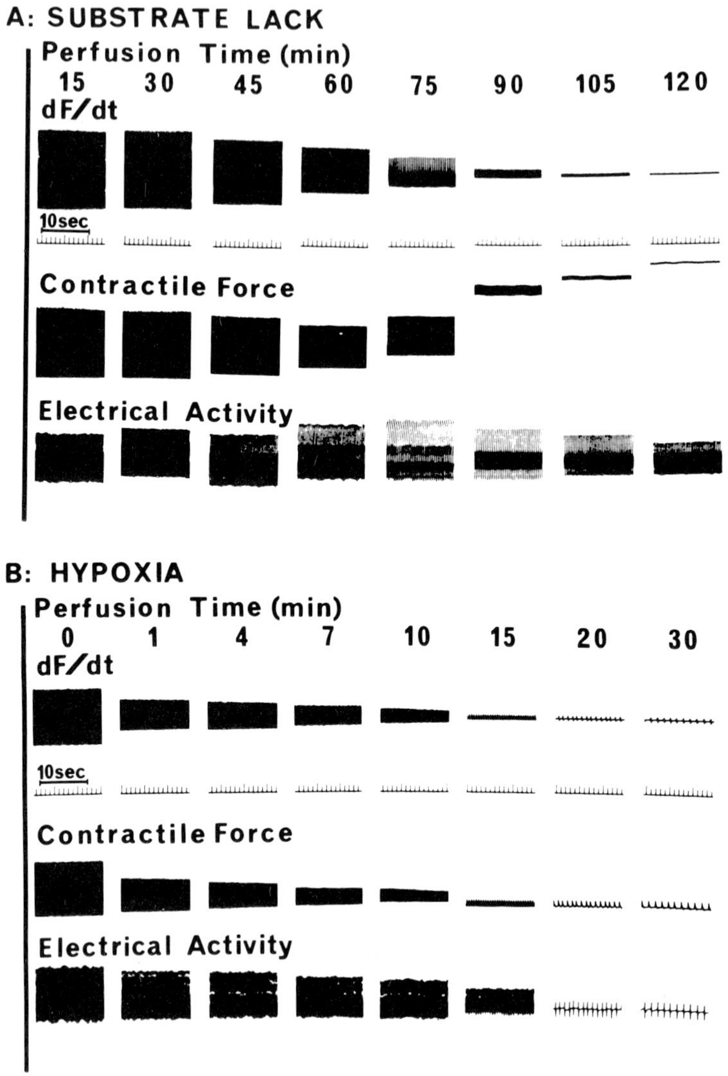

Figure 1 Effect of substrate lack and hypoxia on dF/dt, contractile force and surface electrical activity of the isolated perfused rat heart. *A,* Upper three tracings show the effects of substrate-free medium at different intervals of perfusion; *B*, lower three tracings show the effects of hypoxic medium at different intervals of perfusion.

Table III ^{45}Ca uptake by isolated control, substrate-depleted, and hypoxic rat hearts*

Time of perfusion with ^{45}Ca (min)	^{45}Ca uptake (% of total counts)		
	Control	Substrate-depleted (2-hr perfusion)	Hypoxic (30-min perfusion)
5	5.8 ± 0.4	5.9 ± 0.5	2.9 ± 0.2
10	7.6 ± 0.3	7.4 ± 0.6	3.4 ± 0.4
15	8.2 ± 0.2	7.9 ± 0.4	3.8 ± 0.3

*After perfusion with substrate-free or hypoxic medium, the hearts were switched to a closed circulation apparatus with 50 ml of similar medium containing 5 μCi of ^{45}Ca. The method is described by Dhalla, McNamara, and Sulakhe (4). Each value is a mean ± S.E. of six experiments.

well as in hearts perfused with medium containing 1.25 mM Ca^{++} after perfusion with Ca^{++}-free medium and the results are shown in table VII. The activities of sarcolemmal adenylate cyclase, Mg^{++}-ATPase, and (Na^{+} -K^{+})-ATPase, but not of Ca^{++}-ATPase, were decreased ($p < 0.05$) on perfusing the hearts with Ca^{++}-free medium for 20 min. Perfusing the hearts with medium containing calcium after Ca^{++}-free perfusion for 10 min produced dramatic reductions in the activities of all the above sarcolemmal enzymes. The sodium, potassium, magnesium, and calcium contents were decreased ($p < 0.05$) in hearts perfused with Ca^{++}-free medium. On the other hand, the levels of sodium and calcium were increased (p

Table IV Calcium binding in absence or presence of 4 mM MgATP by sarcolemma obtained from control, substrate-depleted, and hypoxic rat hearts*

Preparation	Calcium binding (nmoles/mg protein/5 min)	
	In absence of MgATP	In presence of MgATP
Control	80 ± 4.2	13 ± 1.4
Substrate-depleted (2-hr perfusion)	36 ± 5.1	7 ± 1.1
Hypoxic (30-min perfusion)	51 ± 7.3	8 ± 0.6

*Each value is a mean ± S.E. of four experiments.

Table V Changes in sarcolemmal enzyme activities and electrolyte contents of isolated rat hearts on perfusion with substrate-free or hypoxic medium*

	Control	Substrate-free (2-hr perfusion)	Hypoxic (30-min perfusion)
Enzyme activities			
Adenylate cyclase	350 ± 21	264 ± 15	369 ± 17
Ca^{++}-ATPase	29.6 ± 1.9	33.7 ± 1.7	32.3 ± 1.4
Mg^{++}-ATPase	24.0 ± 1.7	28.8 ± 1.9	28.3 ± 2.2
(Na^+ - K^+)-ATPase	12.8 ± 1.1	5.9 ± 0.8	6.2 ± 0.4
Electrolyte contents			
Sodium	26.9 ± 1.2	67.3 ± 2.1	58.3 ± 1.7
Potassium	53.1 ± 1.2	31.7 ± 1.5	43.5 ± 1.3
Magnesium	13.9 ± 0.1	12.4 ± 0.2	12.4 ± 0.2
Calcium	1.9 ± 0.1	1.8 ± 0.1	1.3 ± 0.2

*Each value is a mean ± S.E. of six to eight experiments. Result for Mg^{++}-ATPase, Ca^{++}-ATPase, and (Na^+ - K^+)-ATPase are shown as micromoles of Pi per mg of protein per hr, whereas those for adenylate cyclase are expressed as picomoles of cyclic AMP formed per mg of protein per min. Results for electrolyte contents are expressed as micromoles per gm wet wt.

Table VI Changes in sarcolemmal enzyme activities and electrolyte contents of isolated rat hearts on perfusion with Na^+-free or K^+-free medium for 20 min*

	Control	Na^+-free	K^+-free
Enzyme activities			
Adenylate cyclase	370 ± 25	339 ± 21	334 ± 16
Ca^{++}-ATPase	31.3 ± 2.1	35.2 ± 2.4	28.7 ± 0.9
Mg^{++}-ATPase	25.8 ± 2.1	30.1 ± 1.8	23.5 ± 2.0
(Na^+ - K^+)-ATPase	11.2 ± 1.0	6.3 ± 0.6	7.7 ± 0.3
Electrolyte contents			
Sodium	24.9 ± 1.4	12.1 ± 0.3	47.8 ± 1.6
Potassium	53.5 ± 1.5	41.9 ± 3.0	31.9 ± 1.3
Magnesium	14.5 ± 0.1	13.9 ± 0.2	13.3 ± 0.1
Calcium	1.9 ± 0.1	3.2 ± 0.1	3.8 ± 0.2

*Each value is a mean ± S.E. of six or eight experiments. Results for Mg^{++}-ATPase, Ca^{++}-ATPase, and (Na^+ - K^+)-ATPase are shown as micromoles of Pi per mg of protein per hr, whereas those for adenylate cyclase are expressed as picomoles of cyclic AMP formed per mg of protein per min. Results for electrolyte contents are expressed as micromoles per gm wet wt.

Table VII Changes in sarcolemmal enzyme activities and electrolyte contents of isolated rat hearts on perfusion with Ca^{++}-free medium or on reperfusion with medium containing Ca^{++} after Ca^{++}-free perfusion*

	Control	Ca^{++}-free perfusion for 20 min	Reperfusion with Ca^{++} medium for 10 min after Ca^{++}-free perfusion for 10 min
Enzyme activities			
Adenylate cyclase	348 ± 14	271 ± 18	117 ± 9
Ca^{++}-ATPase	30.2 ± 1.7	27.5 ± 1.0	24.2 ± 0.8
Mg^{++}-ATPase	25.0 ± 1.8	19.1 ± 1.2	14.1 ± 2.0
(Na^+ - K^+)-ATPase	11.7 ± 0.5	7.5 ± 0.8	3.4 ± 0.5
Electrolyte contents			
Sodium	22.6 ± 1.7	11.2 ± 0.7	37.8 ± 3
Potassium	54.7 ± 1.3	30.5 ± 1.5	22.0 ± 2.0
Magnesium	13.7 ± 0.1	10.7 ± 0.2	5.8 ± 0.1
Calcium	1.9 ± 0.1	0.8 ± 0.1	3.1 ± 0.1

*Each value is a mean ± S.E. of six to eight experiments. Results for Mg^{++}-ATPase, Ca^{++}-ATPase, and (Na^+ - K^+)-ATPase are shown as micromoles of Pi per mg of protein per hr, whereas those for adenylate cyclase are expressed as picomoles of cyclic AMP formed per mg of protein per min. Results for electrolyte contents are expressed as micromoles per gm wet wt.

< 0.05) and those of potassium and magnesium were decreased in hearts perfused with medium containing calcium after perfusion with Ca^{++}-free medium.

It should be pointed out that the hearts failed to generate contractility within 1 min on perfusion with Ca^{++}-free medium and did not recover upon reperfusion after 10 min of perfusion with Ca^{++}-free medium. Perfusion of the heart with Ca^{++}-free medium has been considered to produce intracellular calcium deficiency, whereas reperfusion of these hearts has been associated with intracellular calcium overload (23). It should also be noted that perfusing the hearts for 20 min with Mg^{++}-free medium did not decrease the contractility or produce any alteration in the activities of sarcolemmal adenylate cyclase, Ca^{++}-ATPase, Mg^{++}-ATPase, and (Na^+ -K^+)-ATPase. Significant reduction in the sodium and magnesium contents was observed in hearts perfused with Mg^{++}-free medium, whereas the level of potassium and calcium did not change appreciably.

In order to investigate whether the sarcolemmal changes are associated with alteration in the membrane composition with respect to phospholipids and proteins, the effects of phospholipase C and trypsin on the cat and dog heart papillary muscles were examined. Treatment with phospholipase C was found to produce dramatic changes in the contractile force and action potentials of the

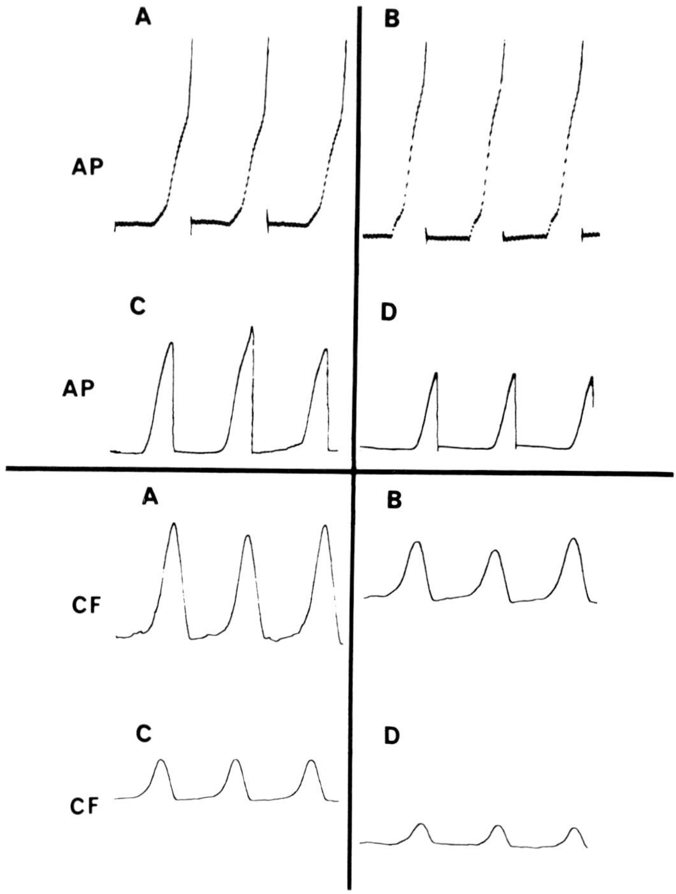

Figure 2 Effect of phospholipase C (0.25 mg/ml) on the action potential (*AP*) and contractile force (*CF*) of the isolated cat papillary muscle. *Upper two tracings,* action potentials: *A*, control; *B, C,* and *D*, after 10, 30, and 60 min of phospholipase treatment, respectively. *Lower two tracings,* contractile force: *A*, control, *B, C,* and *D*, after 10, 30, and 60 min of phospholipase treatment, respectively.

papillary muscles (fig. 2). On the other hand, trypsin markedly decreased contractile force without appreciably changing the action potentials (fig. 3). The data in table VIII indicate that the calcium binding, adenylate cyclase, and (Na^+-K^+)-ATPase activities of sarcolemma obtained from the phospholipase C- or trypsin-treated muscles were decreased ($p < 0.05$). The calcium binding and uptake abilities of both microsomes (fig. 4) and mitochondria (fig. 5) obtained from the trypsin-treated papillary muscle, unlike those from the phospholipase

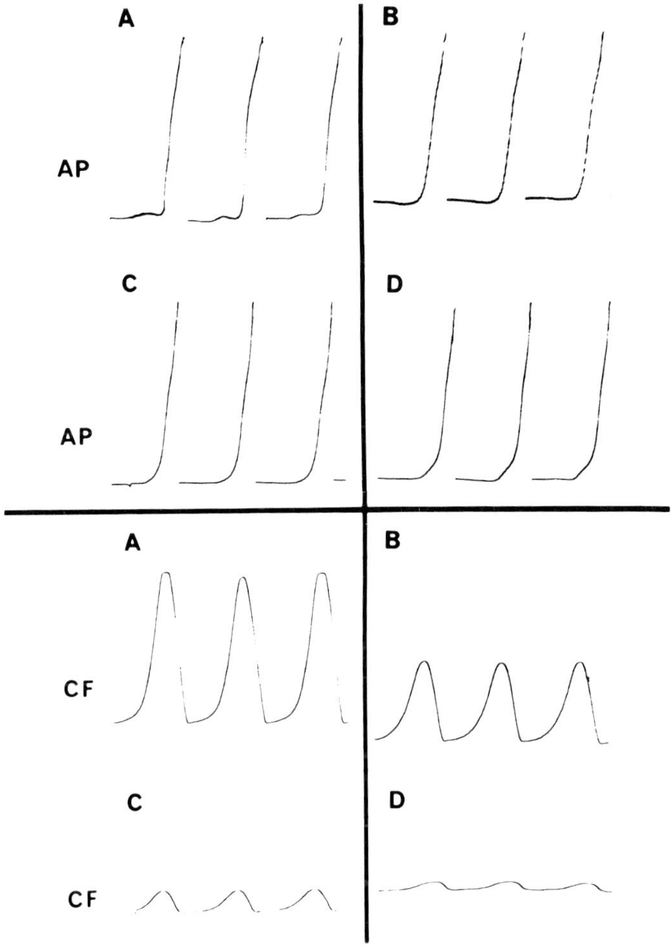

Figure 3 Effect of trypsin (0.1 mg/ml) on the action potential (*AP*) and contractile force (*CF*) of the isolated cat papillary muscle. *Upper two tracings,* action potentials: *A,* control; *B, C,* and *D,* after 10, 30, and 60 min of trypsin treatment, respectively. *Lower two tracings,* contractile force: *A,* control; *B, C,* and *D,* after 10, 30, and 60 min trypsin treatment, respectively.

C-treated muscles, were also significantly decreased ($p < 0.05$). The electron microscopic examination of sections from the phospholipase C-treated papillary muscles did not reveal any change in the ultrastructure, whereas marked damage to mitochondria, myofibrils, and sarcotubular system was seen in sections obtained from the trypsin-treated muscles (fig. 6).

Table VIII Calcium binding, adenylate cyclase, and $(Na^+$-$K^+)$-ATPase activities of sarcolemma obtained from dog papillary muscle with and without phospholipase C or trypsin treatments for 60 min*

Activities	Control	Phospholipase C (0.25 mg/ml)	Trypsin (0.1 mg/ml)
Calcium binding (nmoles/mg/5 min)	88 ± 5.0	21 ± 4.3	32 ± 4.9
Adenylate cyclase (pmoles cyclic AMP/mg/min	290 ± 25	46 ± 7	119 ± 13
$(Na^+ - K^+)$-ATPase (μmoles Pi/mg/hr)	11.2 ± 0.7	2.0 ± 0.4	6.5 ± 0.9

*Each value is a mean ± S.E. of six experiments.

DISCUSSION

A number of investigators have reported dramatic changes in the activities of adenylate cyclase in the absence and presence of NaF or hormones and $(Na^+$-$K^+)$-ATPase in different experimental models of heart failure (1, 5, 8–12, 15, 18, 21, 22). Thus, the observed alterations in the activities of the sarcolemmal enzymes such as adenylate cyclase, $(Na^+$-$K^+)$-ATPase, Ca^{++}-ATPase, and Mg^{++}-ATPase, in hearts failing because of genetic defect, oxygen, or substrate lack, intracellular calcium deficiency or overload, and shifts in sodium and potassium contents are not surprising. However, the degree and the direction of the changes in the sarcolemmal enzymes were found to be dependent upon the type of heart failure. For example, in genetically myopathic hamsters (BIO 14.6) at moderate or late stages of failure, the activities of $(Na^+$-$K^+)$-ATPase, Ca^{++}-ATPase, and Mg^{++}-ATPase were increased. On the other hand, $(Na^+$-$K^+)$-ATPase activity was decreased in all other types of failing hearts employed in this study. Furthermore, Mg^{++}-ATPase activity decreased in hearts failing because of intracellular calcium deficiency produced by perfusion with Ca^{++}-free medium and the activities of both Ca^{++}-ATPase and Mg^{++}-ATPase were reduced in hearts failing because of intracellular calcium overload produced by perfusion with medium containing calcium after perfusion with Ca^{++}-free medium. Likewise, the activity of the sarcolemmal adenylate cyclase was decreased only in hearts made to fail because of substrate lack, intracellular calcium deficiency, and intracellular calcium overload. Although calcium binding by sarcolemma obtained from hypoxic and substrate-depleted hearts was decreased, other membrane events such as surface electrical activity and calcium uptake by whole heart were decreased only under hypoxic conditions.

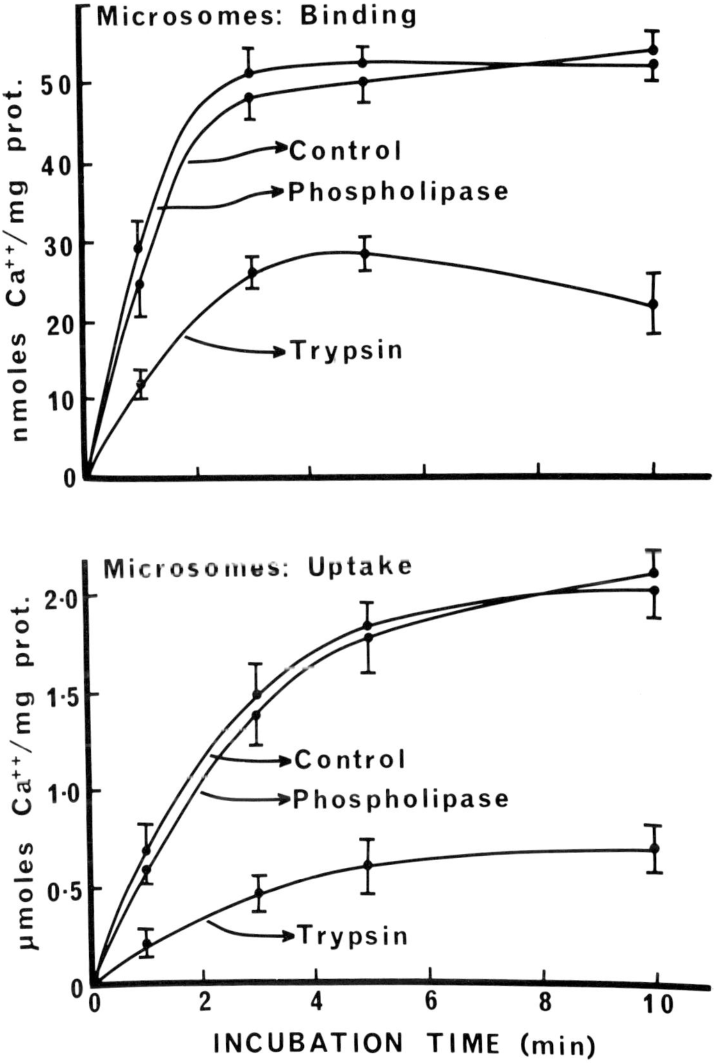

Figure 4 Calcium binding and uptake at different times of incubation by microsomes of dog papillary muscles treated with phospholipase C (0.25 mg/ml) and trypsin (0.1 mg/ml) for 60 min. Each value is a mean ± S.E. of six experiments.

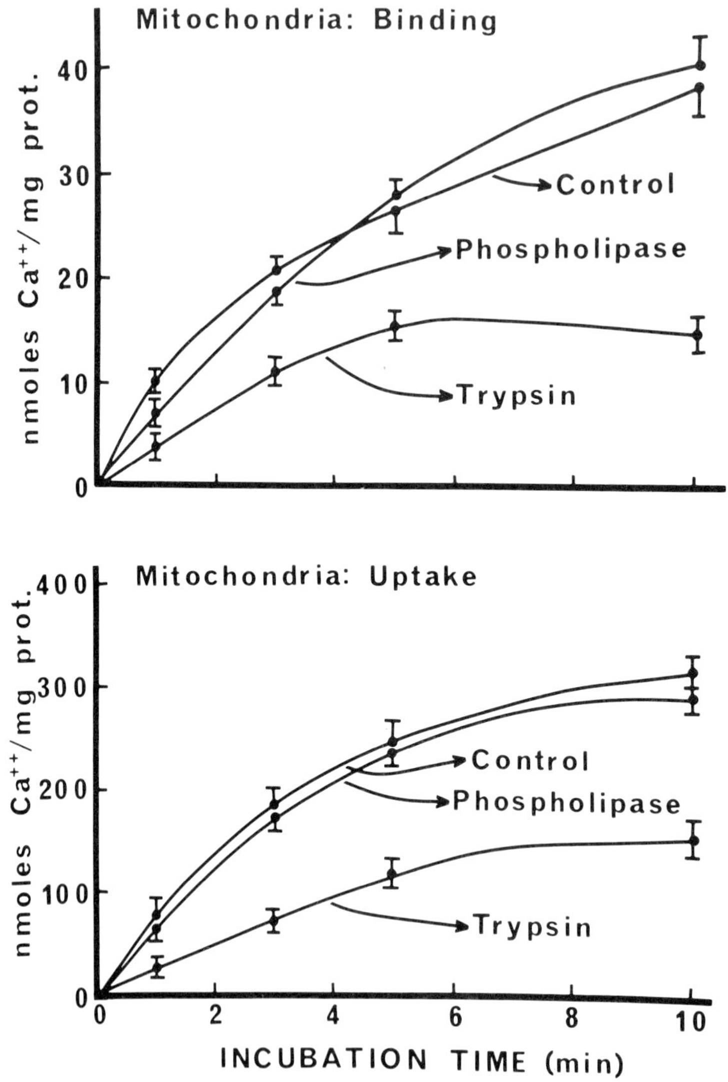

Figure 5 Calcium binding and uptake at different times of incubation by mitochondria of dog papillary muscles treated with phospholipase C (0.25 mg/ml) and trypsin (0.1 mg/ml) for 60 min. Each value is a mean ± S.E. of six experiments.

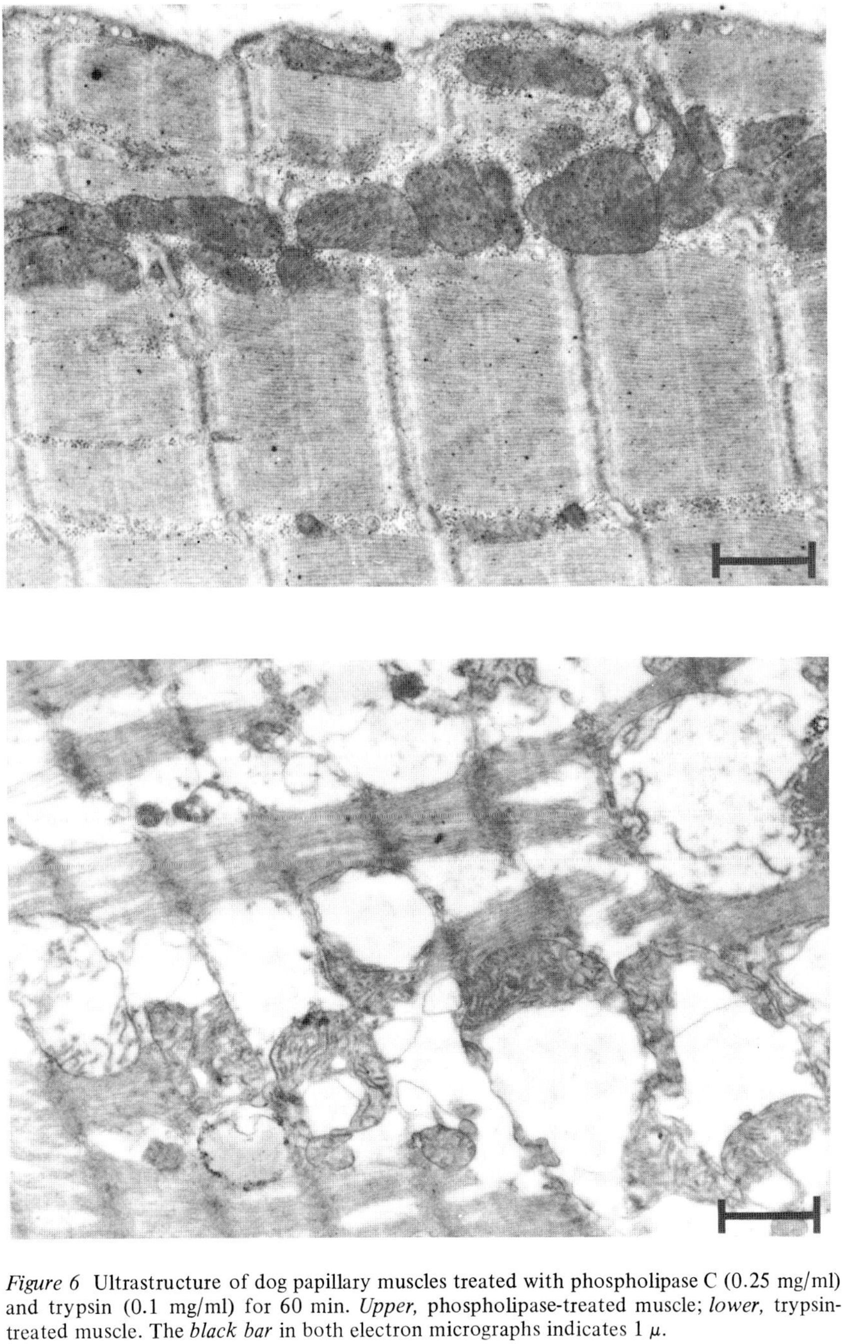

Figure 6 Ultrastructure of dog papillary muscles treated with phospholipase C (0.25 mg/ml) and trypsin (0.1 mg/ml) for 60 min. *Upper,* phospholipase-treated muscle; *lower,* trypsin-treated muscle. The *black bar* in both electron micrographs indicates 1 μ.

The observed changes in the sarcolemmal enzymes in different types of failing hearts could be due to changes in the phospholipids and protein contents of the membrane systems. This contention is supported from our data showing a marked reduction in the sarcolemmal adenylate cyclase, (Na^+ -K^+)-ATPase, and calcium binding activities of cardiac muscles which were treated with trypsin or phospholipase C. It was interesting to note that phospholipase C decreased contractility of cardiac muscle without apparent changes in calcium-transporting abilities of mitochondria and microsomes as well as the ultrastructure of the myocardium. Alterations in the configuration of the action potential by phospholipase C can be interpreted to reflect damage at the level of sarcolemma. On the other hand, trypsin treatment was found to produce damage to the structure of the intracellular organelles as well as to decrease calcium-transporting abilities of mitochondria and microsomes. We therefore consider that the inability of the cardiac muscle to contract on treatment with trypsin is due to its effects on sarcolemma, mitochondria, and sarcotubular system, whereas failure of the heart to generate contractile force on treatment with phospholipase C is limited to its action on the sarcolemma.

The importance of different cations such as calcium, sodium, potassium, and magnesium in myocardial function has now been well recognized. Sarcolemmal damage can be conceived to result in altering the electrolyte contents of the myocardium. In the present study, we have shown dramatic changes in the electrolyte composition of failing rat hearts which showed abnormal sarcolemmal enzyme activities. Electrolyte changes in the failing myopathic hamster hearts (BIO 14.6) have also been reported recently (13). Of particular interest is an increase in sodium and a decrease in potassium content of the failing hearts. These changes can result from a decrease in (Na^+ -K^+)-ATPase activity of the failing heart sarcolemma; however, in the case of myopathic hamsters (BIO 14.6), an increase in (Na^+-K^+)-ATPase activity may represent an adaptive change. In this regard it was of particular interest to find marked changes in mitochondrial and microsomal calcium binding at early stages of the disease when sarcolemmal changes were not observed. Alterations in sarcolemma can also be considered to result in decreasing or increasing calcium influx and to thereby produce intracellular calcium deficiency or overload, respectively. Intracellular calcium deficiency will be associated with decrease in calcium content necessary for the mobilization of the energy stores for maintaining the integrity of the intracellular structures. On the other hand, intracellular calcium overload will be associated with increased splitting of the high energy phosphate compounds which will result in decreased energy state of myocardium and subsequently disruption of the intracellular structures. Abnormality at the level of sarcolemma can also lead to the leakage of various enzymes which are necessary for the operation of myocardial metabolism. Furthermore, changes in the sarcolemmal adenylate cyclase and (Na^+ -K^+)-ATPase, which have been implicated in diverse energy producing and utilizing mechanisms, can be considered to result in abnormal myocardial metabolism. Thus, we suggest that defects at the level of sarcolemmal play a crucial role in the pathogenesis of heart failure and help to explain the alterations in myocardial function and metabolism in this condition.

REFERENCES

1. BALASUBRAMANIAN, V., McNAMARA, D. B., SINGH, J. N., and DHALLA, N. S. 1973. Biochemical basis of heart function. X. Reduction in the Na^+ -K^+ stimulated ATPase activity in failing rat heart due to hypoxia. Can. J. Physiol. Pharmacol. 51: 504–510.
2. DHALLA, N. S., MATOUSHEK, R. F., SUN, C. N., and OLSON, R. E. 1973. Metabolic, ultrastructural and mechanical changes in the isolated rat heart perfused with aerobic medium in the absence or presence of glucose. Can. J. Physiol. Pharmacol. 51: 590–603.
3. DHALLA, N. S., McNAMARA, D. B., BALASUBRAMANIAN, V., GREENLAW, R., and TUCKER, F. R. 1973. Alterations of adenosine triphosphate activities in dystrophic sarcolemma. Res. Commun. Chem. Pathol. Pharmacol. 6: 643–650.
4. DHALLA, N. S., McNAMARA, D. B., and SULAKHE, P. V. 1970. Excitation-contraction coupling in heart. V. Contribution of mitochondria and sarcoplasmic reticulum in the regulation of calcium concentration in the heart. Cardiology 55: 178–191.
5. DHALLA, N. S., SINGH, J. N., FEDELESOVA, M., BALASUBRAMANIAN, V., and McNAMARA, D. B. 1974. Biochemical basis of heart function. XII. Sodium-potassium stimulated adenosine triphosphatase activity in the perfused rat heart made to fail by substrate-lack. Cardiovasc. Res. 8: 227–237.
6. DHALLA, N. S., SULAKHE, P. V., FEDELESOVA, M., and YATES, J. C. 1974. Molecular abnormalities in cardiomyopathy. Advan. Cardiol. 13:
7. DHALLA, N. S., YATES, J. C., WALZ, D. A., McDONALD, V. A., and OLSON, R. E. 1971. Correlation between changes in the endogenous energy stores and myocardial function due to hypoxia in the isolated perfused rat heart. Can. J. Physiol. Pharmacol. 50: 333–345
8. DRANSFELD, H., LIPINSKI, J., and BORSCH-GALETKE, E. 1971. Die Na^+-K^+-activierte transport-ATPase bei experimenteller Herzinsuffizienzy duich Kobaltchloride. Naunyn-Schmeideberg's Arch. Pharmakol. 270: 335–342.
9. FEDELESOVA, M., SULAKHE, P. V., YATES, J. C., and DHALLA, N. S. 1971. Biochemical basis of heart function. IV. Energy metabolism and calcium transport in hearts of vitamin E deficient rats. Can. J. Physiol. Pharmacol. 49: 909–918.
10. GOLD, H. K., PRINDLE, K. H., LEVEY, G. S., and EPSTEIN, S. E. 1970. Effects of experimental heart failure on the capacity of glucagon to augment myocardial contractility and activate adenyl cyclase. J. Clin. Invest. 49: 999–1006.
11. LEVEY, G. S., PRINDLE, K. H., JR., and EPSTEIN, S. E. 1970. Effects of glucagon on adenyl cyclase activity in the left and right ventricles and liver in experimentally-produced isolated right ventricular failure. J. Molec. Cell. Cardiol. 1: 403–419.
12. LINDENMAYER, G. E., SORDAHL, L. A., HARIGAYA, S., ALLEN, J. C., BESCH, H. R., and SCHWARTZ, A. 1971. Some biochemical studies on subcellular systems isolated from fresh recipient human cardiac tissue obtained during transplantation. Amer. J. Cardiol. 27: 277–283.
13. LOSSNITZER, K., and BAJUSZ, E. 1974. Water and electrolyte alterations during the life course of the BIO 14.6 Syrian golden hamsters. A disease model of a hereditary cardiomyopathy. J. Molec. Cell. Cardiol. 6: 163–177.
14. McNAMARA, D. B., SULAKHE, P. V., SINGH, J. N., and DHALLA, N. S. 1973. Adenyl cyclase activity in the hypoxic heart, pp. 303–309. *In* N. S. Dhalla (ed.),

Myocardial Metabolism: Recent Advances in Studies on Cardiac Structure and Metabolism, Vol. 3. University Park Press, Baltimore.

15. McNAMARA, D. B., SULAKHE, P. V., SINGH, J. N., and DHALLA, N. S. 1974. Properties of heart sarcolemmal Na^+-K^+ ATPase. J. Biochem. 75: 795–803.
16. PRASAD, K., and CALLAGHAN, J. C. 1969. Effect of replacement of potassium by rubidium on the transmembrane action potential and contractility of human papillary muscle. Circ. Res. 24: 157–166.
17. SINGH, J. N., DHALLA, N. S., McNAMARA, D. B., BAJUSZ, E. and JASMIN, G. 1975. Membrane alteration in failing hearts of cardiomyopathic hamsters, pp. 259–268. *In* A. Fleckenstein and G. Rona (eds.), Pathophysiology and Morphology of Myocardial Cell Alterations: Recent Advances in Studies on Cardiac Structure and Metabolism, Vol. 6. University Park Press, Baltimore.
18. SOBEL, B. E., HENRY, P. D., ROBISON, A., BLOOR, C., and ROSS, J., JR. 1969. Depressed adenyl cyclase activity in the failing guinea pig heart. Circ. Res. 24: 507–512.
19. SULAKHE, P. V., and DHALLA, N. S. 1971. Excitation-contraction coupling in heart. VII. Calcium accumulation in subcellular particles in congestive heart failure. J. Clin. Invest. 50: 1019–1027.
20. SULAKHE, P. V., and DHALLA, N. S. 1972. Adenyl cyclase activity in failing hearts of genetically myopathic hamsters. Biochem. Med. 6: 471–482.
21. SULAKHE, P. V., and DHALLA, N. S. 1973. Alterations in the activity of cardiac Na^+-K^+-stimulated ATPase in congestive heart failure. Expl. Molec. Pathol. 18: 100–111.
22. YAMAZAKI, M., and OGAWA, K. 1971. Cardiac catecholamine metabolism in heart failure. Jap. Circ. J. 35: 965–971.
23. YATES, J. C., and DHALLA, N. S. 1975. Structural and functional changes associated with failure and recovery of hearts after perfusion with Ca^{2+}-free medium. J. Molec. Cell. Cardiol. 7: 91–103.
24. YAZAKI, Y., and FUJII, J. 1972. Depressed Na^+-K^+-ATPase activity in the failing rabbit heart. Jap. Heart J. 13: 73–83.

Mailing address: Dr. N. S. Dhalla
Professor of Physiology, Faculty of Medicine.
University of Manitoba,
Winnipeg, R3E OW3 (Canada).

Recent Advances in Studies on Cardiac Structure and Metabolism
Volume 9
The Sarcolemma
Edited by Paul-Emile Roy and Naranjan S. Dhalla

SARCOLEMMAL ALTERATIONS IN CARDIAC HYPERTROPHY AND DEGENERATION

V. J. FERRANS, L. M. BUJA, and B. J. MARON

The Section of Pathology and the Cardiology Branch,
National Heart and Lung Institute,
National Institutes of Health,
Bethesda, Maryland, USA

SUMMARY

Sarcolemmal alterations were studied in hypertrophied and/or degenerated ventricular muscle cells from: 1) patients with hypertrophic cardiomyopathy, congestive cardiomyopathy of various causes, aortic valvular disease, or congenital heart diseases associated with obstruction to right ventricular outflow; 2) patients with various neoplasms treated with anthracycline drugs; 3) rabbits given large doses of anthracyclines, and 4) dogs subjected to anoxic cardiac arrest during total cardiopulmonary bypass. Sarcolemmal changes in hypertrophied, nondegenerated cardiac muscle cells consisted of: 1) increase in the area and in the degree of irregularity of cell surfaces, including intercalated discs and T tubules; 2) dilation of T tubules, and 3) formation of large numbers of multiple intercalated discs. Moderately degenerated cardiac muscle cells, with or without associated hypertrophy, showed sarcolemmal changes characterized by: 1) dilation and decreased numbers of T tubules; 2) marked surface irregularity; and 3) dissociation of intercalated discs, associated with development of vesicles in widened disc interspaces, and of variably complex junctional structures formed between two parts of the plasma membrane of the same muscle cell (intracytoplasmic junctions). Sarcolemmal alterations in cells with end-stage degeneration were associated with loss of myofibrils and consisted of: 1) disappearance of T tubules; 2) loss of junctional contacts with adjacent cells; 3) marked surface irregularity; 4) formation of intracytoplasmic junctions, and 5) thickening of basement membranes. These observations show that remodeling of the cell surfaces is an important feature of the cellular responses to the stimuli of hypertrophy or degeneration.

The surfaces of cardiac muscle cells may be regarded as consisting of three components: 1) the sarcolemma proper, which comprises the nonspecialized regions of the surfaces; 2) the T tubules or transverse tubules, which are cylindrical sarcolemmal invaginations that are located at the levels of Z bands and penetrate into the cell interior, and 3) the intercellular junctions (intercalated discs and lateral junctions), which are composed of nexuses, desmosomes, myofibrillar insertion sites, and undifferentiated regions (7). Of the various morphological alterations which the surfaces of cardiac muscle cells undergo in disease states, only those involving intercellular junctions have been reported in detail (8, 15). This communication summarizes our observations on ultrastructural changes involving T tubules and nonspecialized sarcolemmal regions, as well as intercellular junctions, of hypertrophied and/or degenerated cardiac muscle cells in a variety of pathological conditions.

MATERIALS AND METHODS

The observations in this communication were made on myocardium obtained from both humans and experimental animals. Tissues from humans included: 1) crista supraventricularis muscle resected at operation from 59 patients with various congenital cardiac diseases associated with infundibular stenosis (14); 2) ventricular myocardium (obtained at operation from the septum and the left and right ventricular free walls) from 28 patients with hypertrophic cardiomyopathy (9, 18); 3) left ventricular apical myocardium obtained at operation from 16 patients with aortic valvular disease (6 patients with predominant aortic valvular stenosis, 5 patients with pure aortic regurgitation, and 5 patients with combined aortic stenosis and regurgitation) (19, 20); 4) ventricular myocardial biopsies from 31 patients with congestive cardiomyopathies of various types (8); and 5) left ventricular myocardium obtained at necropsy from 13 patients who had various neoplasms and had been treated with chemotherapeutic agents (1, 4). Cardiac tissues from experimental animals consisted of left ventricular myocardium from: 1) dogs which had been subjected to 30 or 45 min of normothermic, anoxic cardiac arrest induced by aortic cross-clamping under complete cardiopulmonary bypass (2, 5), and 2) rabbits which had received large total doses (100–450 mg/m^2 of body surface) of adriamycin, an anthracycline antibiotic with documented antineoplastic and cardiotoxic effects (3).

All tissues were fixed with cold, 3 percent glutaraldehyde in 0.1 M phosphate buffer, pH 7.2, washed with several changes of buffer, postfixed with cold, 1 percent osmium tetroxide in Millonig's phosphate buffer, dehydrated with ethanol and propylene oxide, and embedded in Maraglas (12). Ultrathin sections were stained with lead citrate and uranyl acetate and examined with the electron microscope.

RESULTS

Observations on myocardium from patients with various types of heart diseases

The cardiac muscle cells in virtually all patients studied, except in those with neoplasms, showed changes of hypertrophy, characterized by the presence of muscle cells with transverse diameters exceeding 15 μm. Changes considered indicative of degeneration of cardiac muscle cells also were present in some of these patients. These degenerative changes involved mainly the myofibrils and the sarcoplasmic reticulum. Their prevalence and detailed morphology in these groups of patients are described in detail elsewhere (19, 20).

Alterations in cell surfaces were associated with changes of hypertrophy and of degeneration. The nature of these alterations was similar in the various groups of patients, but the severity varied considerably from one patient to another and among different areas of a given biopsy. In general, however, the extent of alterations in the surfaces of cardiac muscle cells correlated with the degrees of interstitial fibrosis and of cellular hypertrophy and degeneration.

Sarcolemmal changes in hypertrophied but nondegenerated cardiac muscle cells consisted of: 1) irregularities of contour of cell surfaces, including areas of T tubules and intercalated discs; 2) dilation of T tubules, and 3) formation of large numbers of multiple intercalated discs (figs. 1–7).

Dilation (diameters exceeding 5,000 Å) and increased tortuosity of T tubules occurred sometimes to a remarkable degree. These changes, however, were not seen in all hypertrophied hearts. The lumina of dilated T tubules frequently contained connective tissue microfibrils, averaging 150 Å in diameter, in addition to their external laminae or basement membranes. Collagen fibrils, however, were very uncommonly found within T tubules, even in areas of severe fibrosis. A layer of microfibrils usually separated the external laminae of the muscle cells from the sourrounding collagen fibrils in areas of fibrosis. The external laminae of T tubules were morphologically similar to those of unspecialized regions of the sarcolemma. In some hypertrophied muscle cells the external laminae were thickened, occasionally filling completely the lumina of dilated T tubules. Such thickening occasionally occurred in hypertrophied, nondegenerated cells but was more typical of moderately or severely degenerated cells in areas of fibrosis. Small, focal accumulations of electron dense material, similar to that of the Z bands, were closely apposed to the cytoplasmic aspect of the plasma membranes in T tubules and undifferentiated sarcolemmal regions of many hypertrophied, nondegenerated cardiac muscle cells. Present in close relationship to these accumulations were cytoskeletal, 100-Å thick filaments (10).

Irregular surface invaginations or folds, which were inconsistently related to the Z bands of the myofibrils, were present in many hypertrophied cells (fig. 2–5). When viewed in longitudinal sections, such surface folds simulated elongated or elliptically shaped T tubules with their major axes oriented parallel to the longitudinal axis of the cell (figs. 2 and 4). Study of transverse sections showed, however, that these folds extended deep into the cells and that they were not tubular in shape (figs. 3 and 5). Such folds imparted a multilobulated appearance to the cells in which they occurred. In certain cells these folds were present to such an extent that they divided the cells into several large, irregularly shaped areas connected by narrow cytoplasmic bridges. Some of these channels clearly extended from one side of the cell to the other.

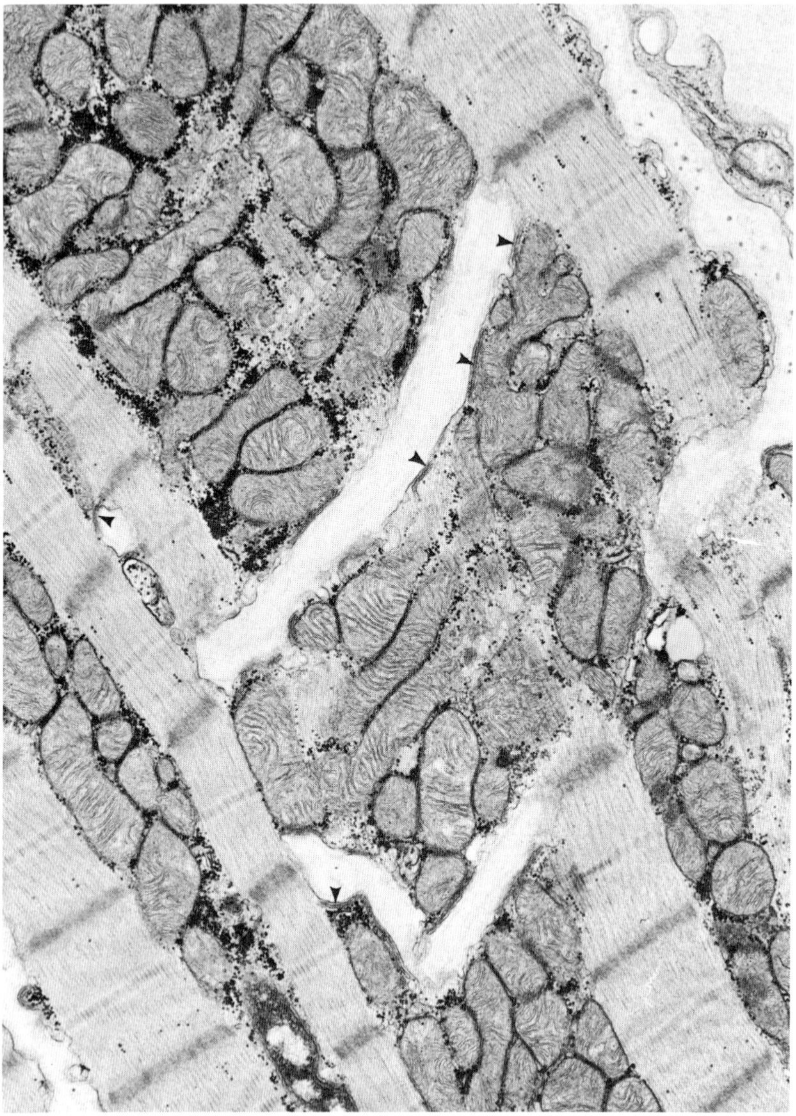

Figure 1 View of normal T tubules in normal-sized (9 μm in width) cardiac muscle cell from crista supraventricularis of 7-year-old boy with ventricular septal defect and aortic regurgitation but no obstruction to right ventricular outflow. These T tubules course somewhat obliquely, show branching, measure about 4,000 Å in diameter, appear circular in transverse section, and are associated with terminal cisterns (*arrowheads*) of sarcoplasmic reticulum (× 13,300).

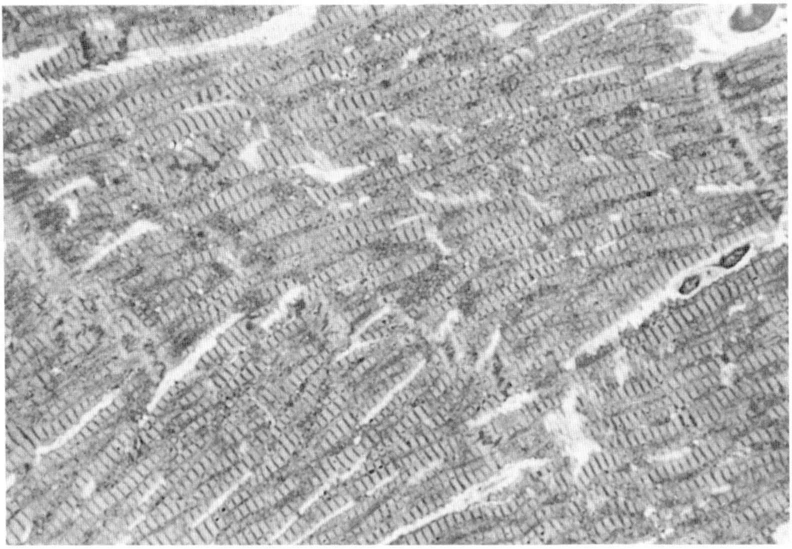

Figure 2 Elongated, longitudinally oriented sarcolemmal invaginations in cardiac muscle cells from crista supraventricularis of 36-year-old man with atrial septal defect and combined valvular and infundibular pulmonic stenosis. Light micrograph of semithin section of plastic-embedded tissue; alkaline toluidine blue stain (X 285).

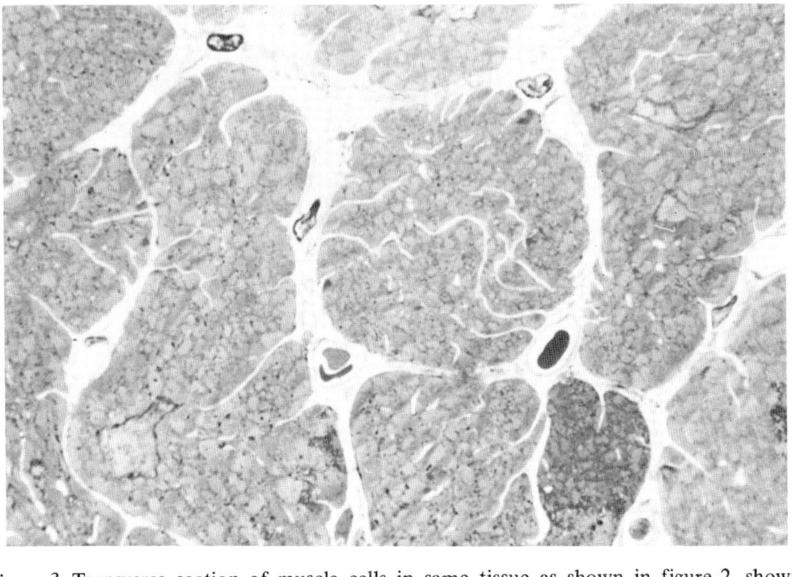

Figure 3 Transverse section of muscle cells in same tissue as shown in figure 2, showing tortuosity and branching of the sarcolemmal invaginations. Light micrograph of semithin section of plastic-embedded tissue; alkaline toluidine blue stain (X 285).

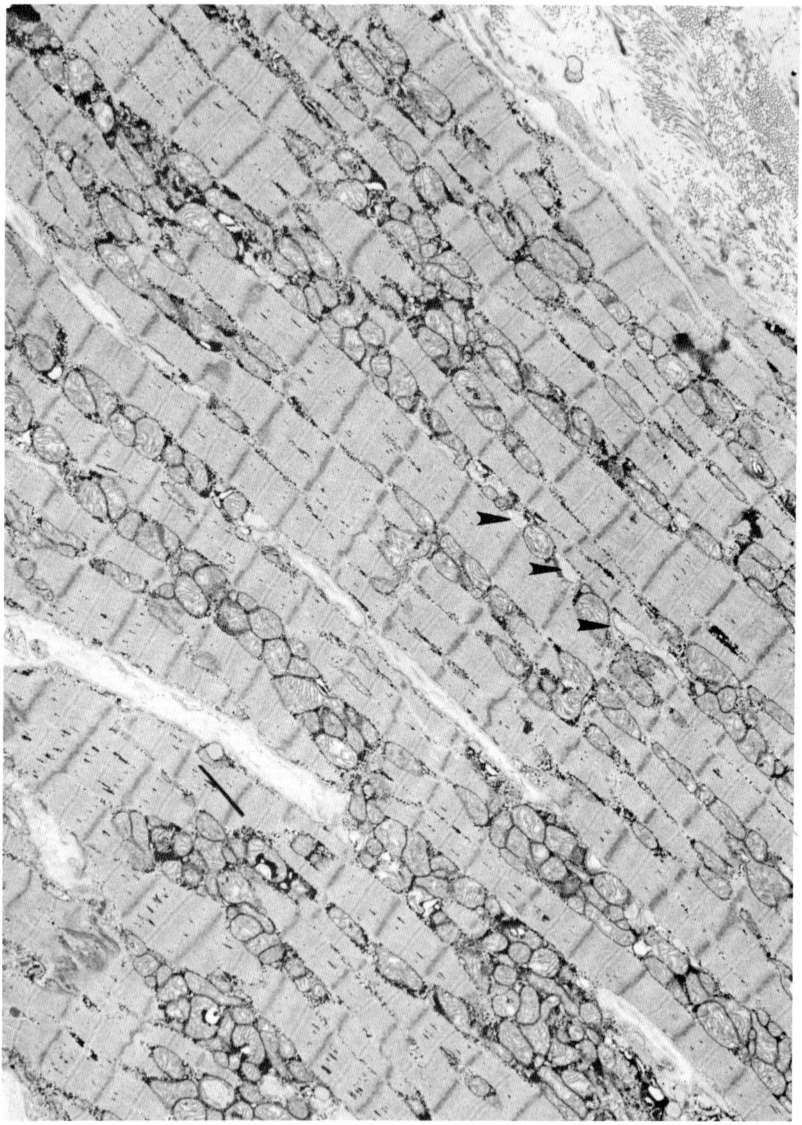

Figure 4 Longitudinally oriented, elongated sarcolemmal invaginations in hypertrophied, nondegenerated cardiac muscle cell. Note focal widening of Z bands and small, tubular sarcolemmal invaginations which are true T tubules (*arrowheads*). Compare with figures 2, 3, and 5, all of which are of the same tissue (× 7,850).

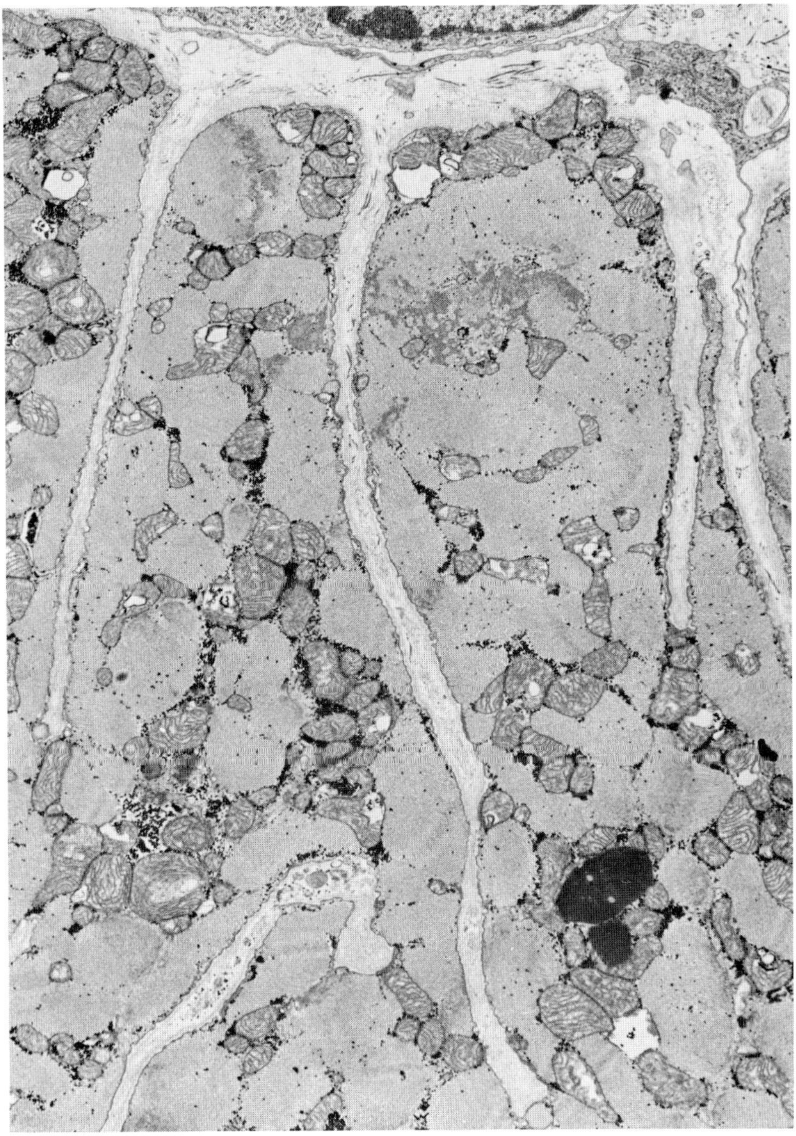

Figure 5 Transverse section through muscle cell in same tissue as shown in figures 2, 3, and 4. Deep sarcolemmal invaginations are present which contain basement membrane material, microfibrils, and a few collagen fibers: compare with figures 3 (× 8,550).

Figure 6 Light micrograph of irregularly shaped, hypertrophied cardiac muscle cell which has a small lateral process (*arrowheads*) with multiple intercalated discs. Ventricular septal myocardium from 18-year-old man with hypertrophic obstructive cardiomyopathy. Semithin section of plastic-embedded tissue; alkaline toluidine blue stain (X 810).

It became evident during the course of this study that many hypertrophied muscle cells were not cylindrical in shape and that they possessed irregularities of contour both at their ends and along their lateral sufraces. Many of these irregularities were obvious in electron micrographs; others, however, were best demonstrated in semifine sections stained with toluidine blue (fig. 6). Changes in the intercellular junctions of hypertrophied cardiac muscle cells consisted of an increased extent of folding of the apposed plasma membranes in regions of intercalated discs. Hypertrophied cells frequently had lateral cytoplasmic processes which interdigitated with those of adjacent cells. Areas of intercellular junctions in these lateral processes formed multiple intercalated discs (fig. 7). These structures, which have been defined as consisting of two parallel segments of intercalated discs separated by 10 or fewer sarcomeres (16), appeared most numerous and irregularly shaped in patients with hypertrophic cardiomyopathy. The shapes of the lateral cytoplasmic processes with which multiple intercalated discs were associated varied according to the plane of sectioning, so that some of these processes appeared as isolated areas of cytoplasm whereas others were obviously connected to the main portion of the cell from which they originated. Details of these relationships have been described elsewhere (17). Areas of sarcoplasm adjacent to multiple intercalated discs often contained many clusters of ribosomes and accumulations of Z band material.

Degenerated cardiac muscle cells showed sarcolemmal changes characterized by: 1) dilation and decrease in number of T tubules; 2) marked surface irregularities, and 3) dissociation of intercalated discs. General morphological changes in moderately or severely degenerated cells consisted of: 1) reduction in number, and even total disappearance, of myofibrils; 2) proliferation of sarcoplasmic reticulum and/or mitochondria; 3) accumulation of concentric lamellar bodies (myelin figures) or of glycogen, and 4) loss of intercellular contacts. As the architecture of degenerating cells became increasingly disrupted, the irregularities of the contours of their surfaces appeared more and more accentuated (figs. 8–11). Many areas of these cells were devoid of T tubules; others contained only dilated, shallow, saccular invaginations which did not bear a spatial relationship to Z bands (fig. 8). The loss of T tubules correlated with that of the myofibrils: T tubules were not found in myofibril-free areas of degenerated cells.

Advanced cellular degenerative changes in severely hypertrophied hearts were accompanied by interstitial fibrosis and by separation or dissociation of adjacent cells from each other (figs. 9–11). The ends of some dissociated cells were characterized by smooth tapering into the interstitium (fig. 9). The ends of

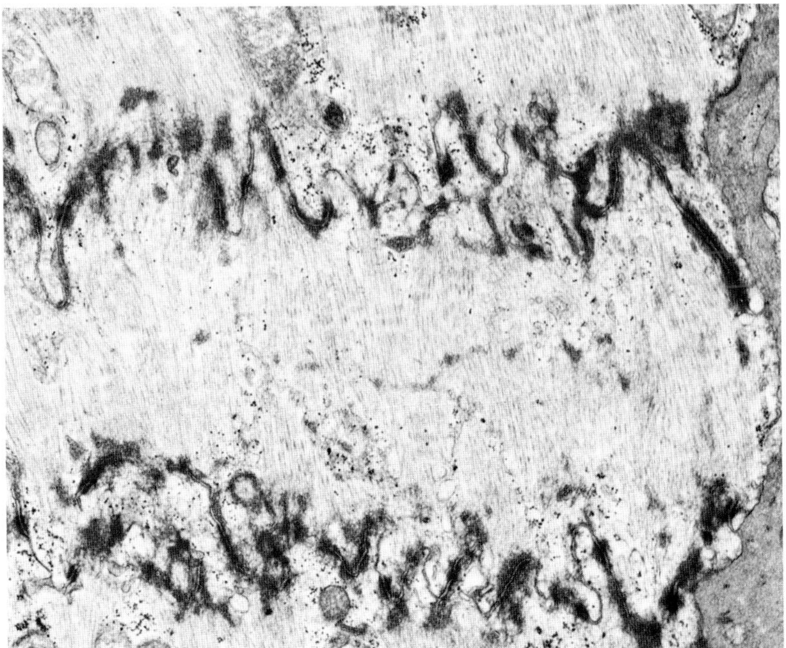

Figure 7 Part of hypertrophied cardiac muscle cell with lateral process similar to that shown in figure 6. Two segments of intercalated disc course parallel to each other in a transverse direction and are separated by only two sarcomeres, both of which appear to be incompletely developed. Same tissue as in figure 6 (× 15,900).

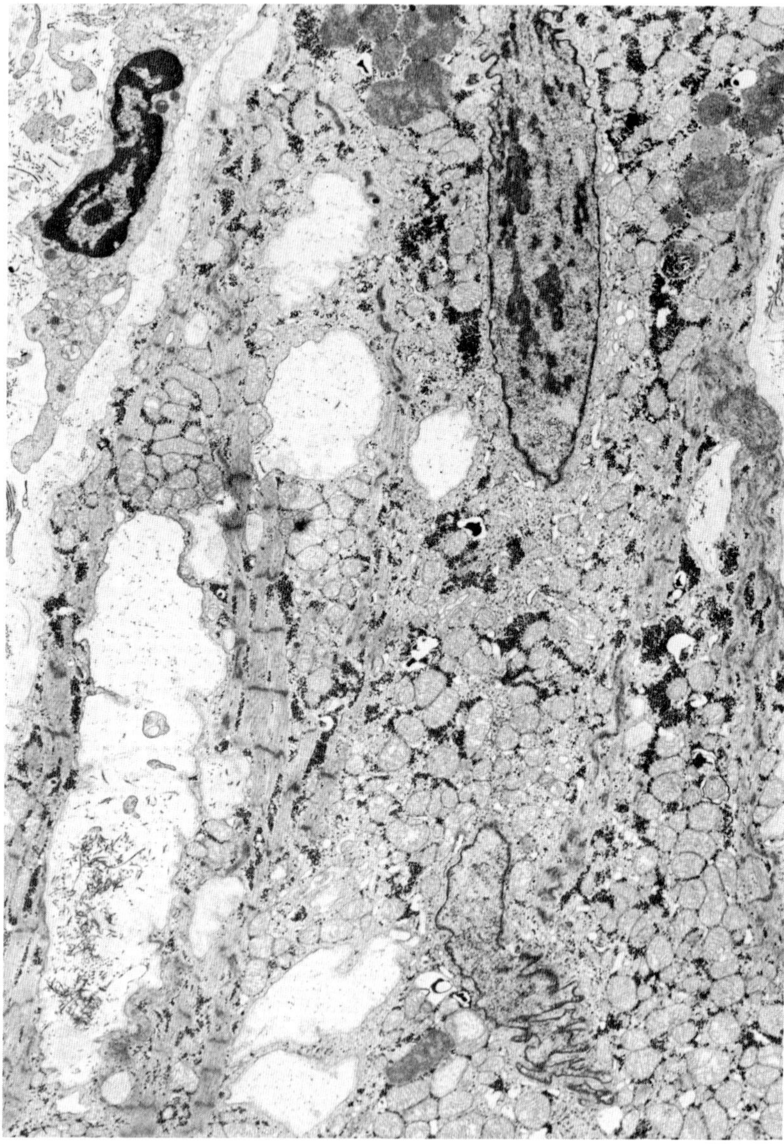

Figure 8 Degenerated cardiac muscle cell from left ventricular apex of 46-year-old man with aortic valvular stenosis and regurgitation shows large, irregularly shaped sarcolemmal invaginations (one of which contains collagen fibrils), proliferation of sarcoplasmic reticulum and mitochondria, and loss of myofibrils. No evidence of an organized transverse tubular system is present in this cell. Note the two nuclei and the lipofuscin granules (× 7,125).

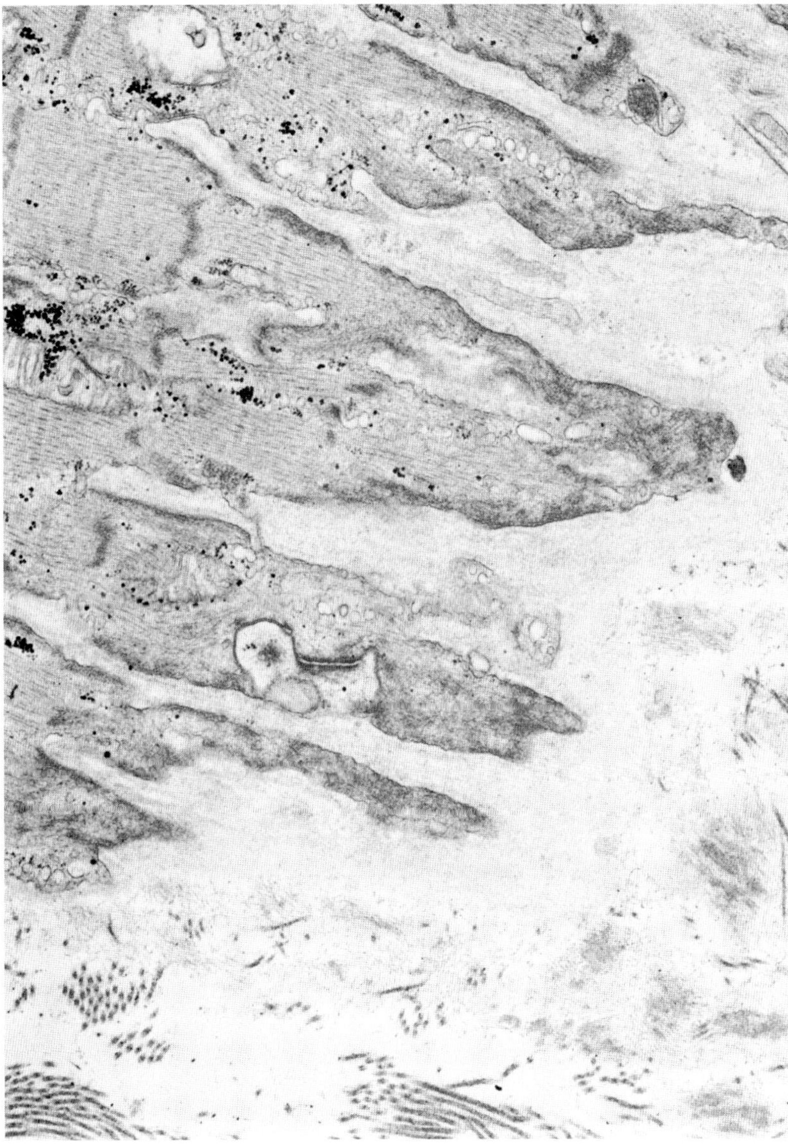

Figure 9 Free end of cardiac muscle cell in area of fibrosis has finger-like projections of cytoplasm, thickened basement membrane, and subsarcolemmal accumulations of Z band material. Only a small area of intercellular junction connects this end of the cell with another cell. Ventricular septal myocardium from 10-year-old boy with combined muscular and fibrous ring type of subaortic stenosis (× 16,500).

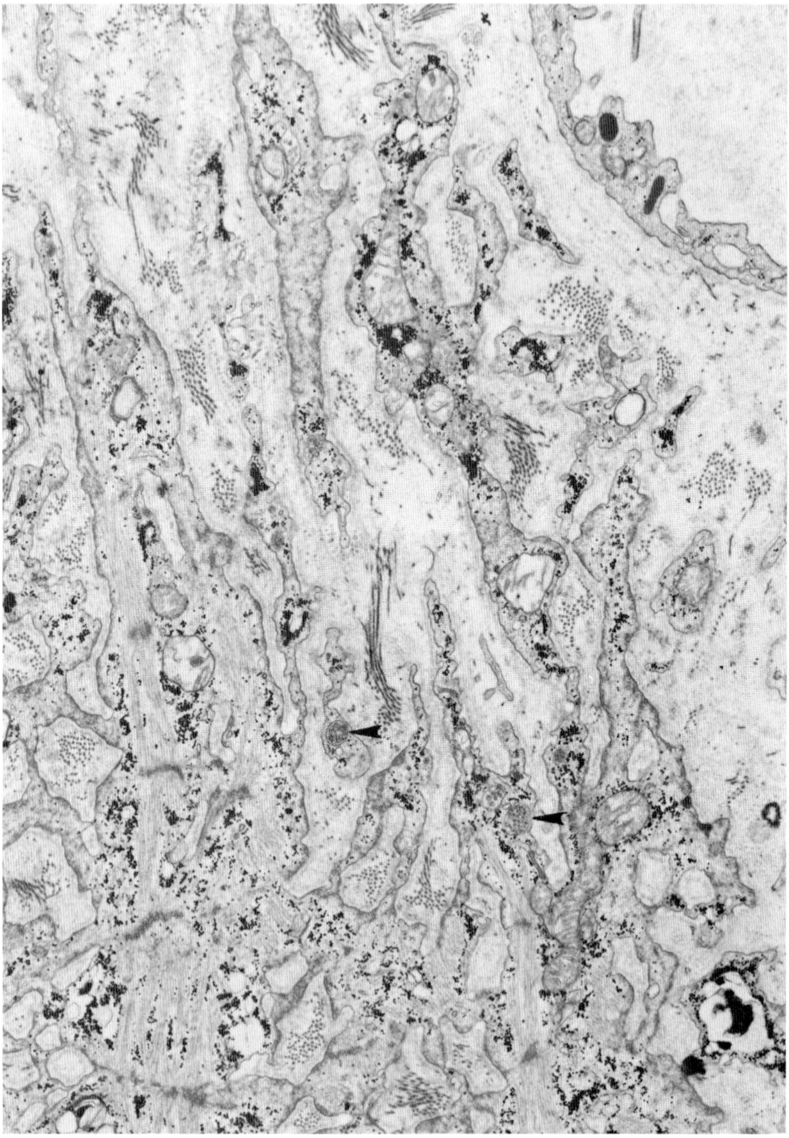

Figure 10 Highly irregular surface at end of cardiac muscle cell which terminates near a capillary in area of fibrosis. Note two plasma membrane processes (*arrowheads*) which represent the only points of specialized contact between this cell and another cell. Small groups of collagen fibrils are present within some sarcolemmal invaginations, and many small accumulations of Z band material are adjacent to the inner aspect of the sarcolemma. Left ventricular apical myocardium from 43-year-old man with aortic valvular stenosis and regurgitation (× 12,350).

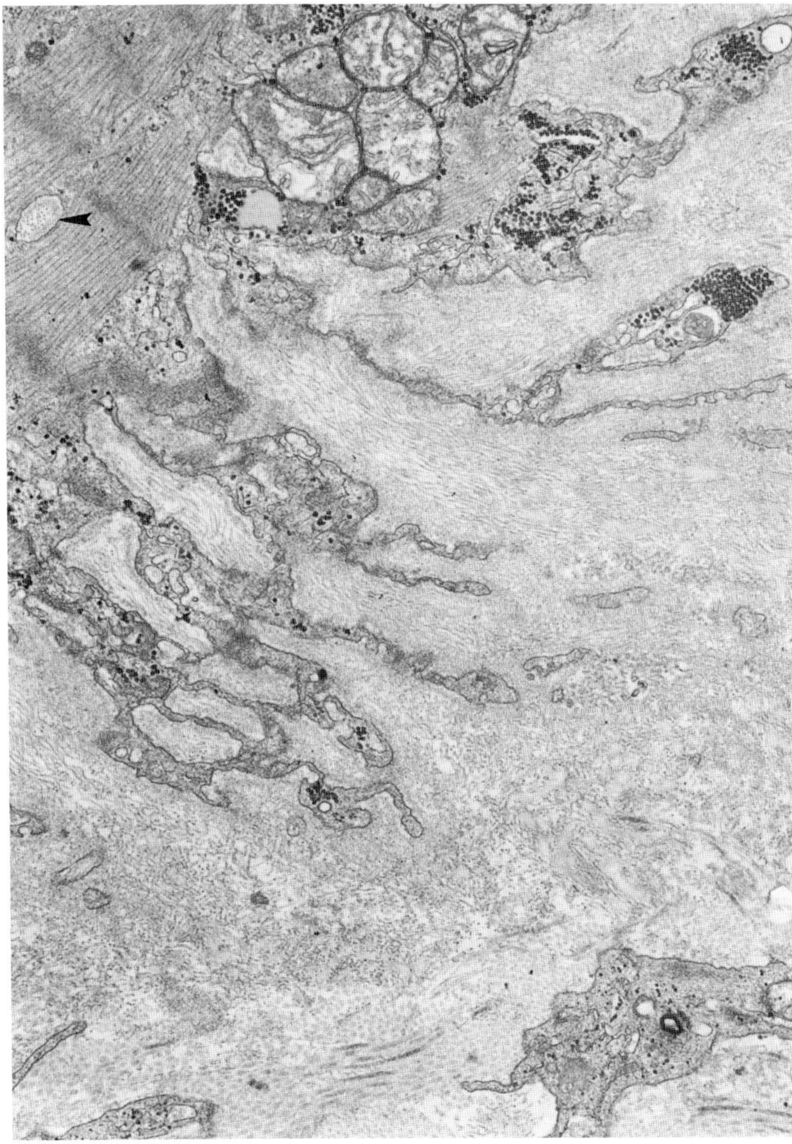

Figure 11 Lateral surface of cardiac muscle cell in area of fibrosis has extensive, irregular sarcolemmal projections which are embedded in a dense stroma composed of microfibrils, small elastic fibrils and collagen fibers. Note the presence of microfibrils within the lumen of a T tubule (*arrowhead*) and the subsarcolemmal accumulations of Z band material. Same tissue as in figure 8 (× 19,950).

other cells (fig. 10) and even the sides (fig. 11), however, exhibited complex elongated cytoplasmic processes and large, redundant sarcolemmal folds. A spectrum of changes involved in the dissociation of these cells was observed. The spaces (normally about 200–300 Å wide) between the apposed membranes of intercalated discs were focally widened (fig. 12). Some of these areas of widening were lined by external laminae and resembled undifferentiated regions of sarcolemma. In other instances, however, these areas contained small portions of nexuses, recognized as such by their characteristic pentalaminar structures, and numerous vesicles which were limited by single, trilaminar membranes and measured from 500 to 1,000 Å in diameter (fig. 12). Vesicles similar to those in the widened disc interspaces also were found adjacent to the surfaces of some cells in areas of fibrosis (fig. 13). These vesicles appeared to be derived from the plasma membranes and formed large clusters that measured up to 5 μm in length. These clusters of vesicles were sometimes associated with pocket- or flap-like infoldings of the plasma membranes of the muscle cells (fig. 13). Such infoldings resembled closely those which are normally present in histiocytes (26). In many areas the extent to which adjacent cells were joined by intercalated discs was markedly reduced, so that only small areas of contact remained between two cells (fig. 9). Many of these residual areas of intercellular contact consisted of small intercalated discs, with prominent nexuses, between the tips of two otherwise isolated cells. In other such areas the only evidence of intercellular contact (fig. 10) consisted of the presence of plasma membrane processes of the type known to extend across intercalated discs (11, 24). These structures (fig. 14) were finger-like processes which, because of the plane of sectioning, most often appeared to be circular in outline. They measured from 2,000–5,000 Å in diameter, were limited by single, trilaminar membranes, and contained clusters of vesicles, 500–900 Å in diameter. Some of these vesicles within the processes were paired and were attached to each other by pentalaminar junctions, which indicated that they contained small areas of nexal contacts (fig. 14).

Another abnormality which was observed frequently in areas of cardiac fibrosis in patients with ventricular hypertrophy and in areas of cardiac edema in patients with neoplasms was the presence of intracytoplasmic junctions (3). These structures have been defined as junctions formed by the apposition of two parts of the plasma membrane of the same cell. Thus, such junctions are intracellular or intracytoplasmic (in contrast to intercellular junctions, which are formed by the apposition of the plasma membranes of two different cells). The complexity of intracytoplasmic junctions varied considerably. Many of these structures had structural features of desmosomes (fig. 15). Others were more complex (fig. 16) and resembled parts of intercalated discs in that they consisted not only of desmosomes, but also of nexuses and myofibrillar insertion sites. Intracytoplasmic junctions composed exclusively of nexuses were not encoun-

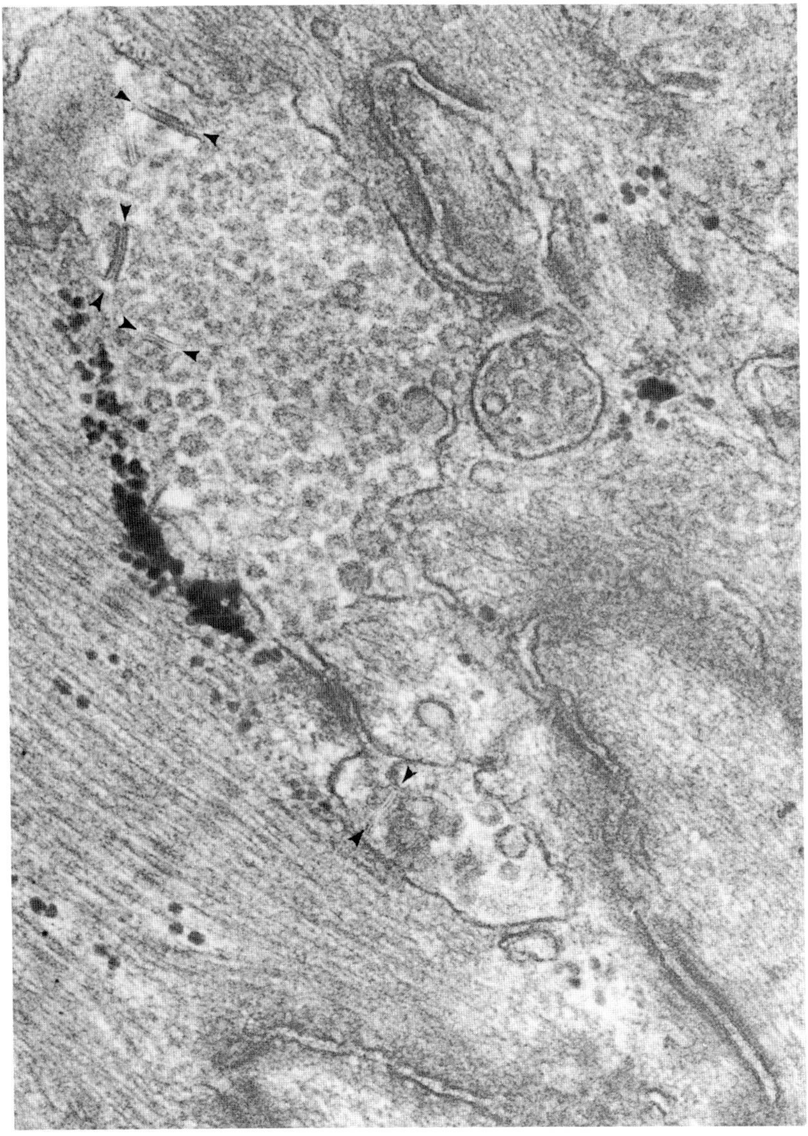

Figure 12 Intercalated disc joining two ventricular muscle cells shows several areas of widening of the space between the apposed membranes. Present in this space are many vesicles and small nexuses (*arrowheads*): same tissue as in figure 10 (× 71,250).

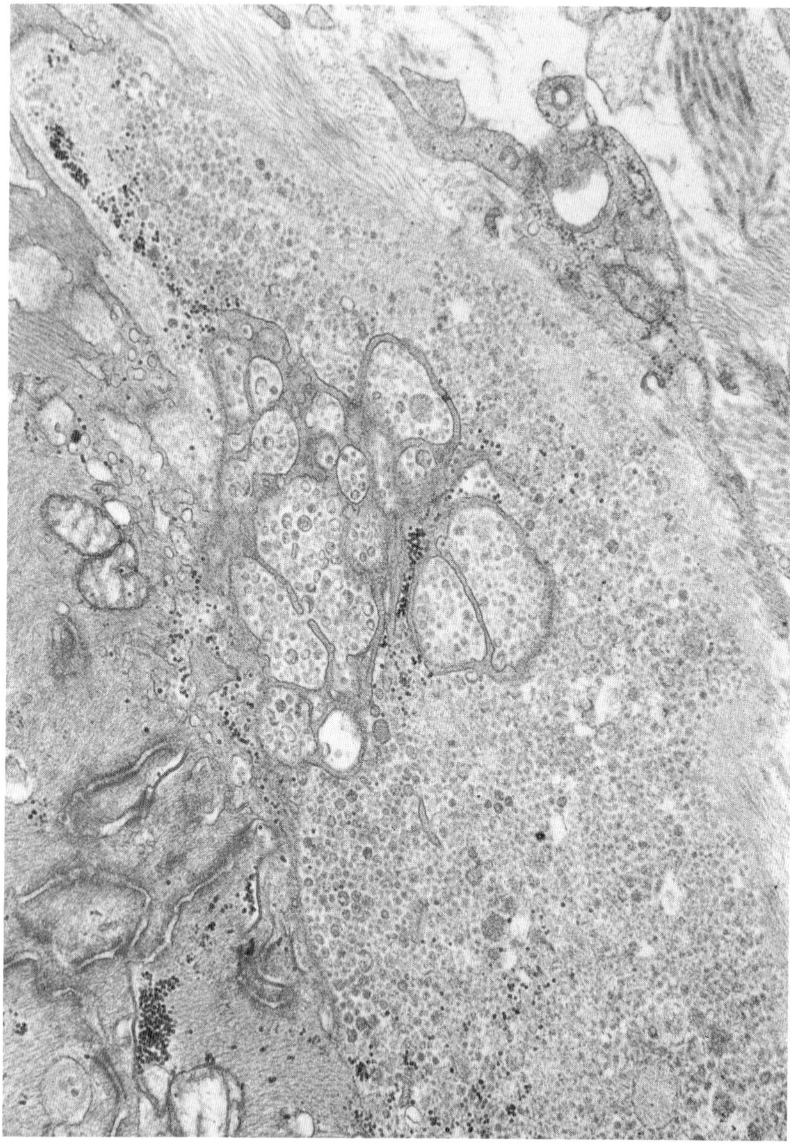

Figure 13 Area of junction between two cardiac muscle cells, showing marked irregularities of the surface of the cell at upper left. This cell has pocket-like folds of plasma membrane which are closely associated with numerous small vesicles. The basement membrane of this cell has been displaced to a position external to that of the vesicles and is immediately adjacent to a layer of connective tissue microfibrils: same tissue as shown in figures 8 and 11 (× 26,100).

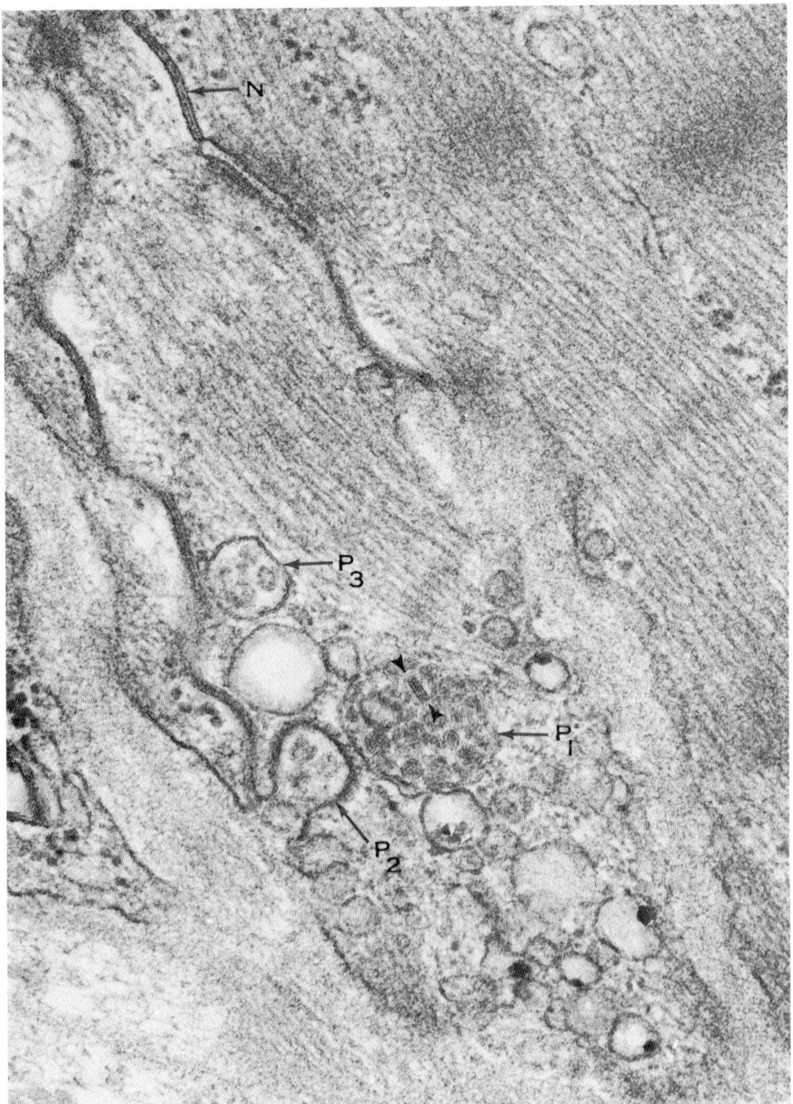

Figure 14 Area of junction between two cardiac muscle cells shows several nexuses (*N*) and three plasma membrane processes (P_1, P_2, *and* P_3) which project from one cell into the other at the level of the disc. The first of these processes (P_1) appears as a circular structure which is limited by a single, trilaminar membrane and contains several vesicles, including one that has a very small nexus (*arrows*). The second process (P_2) is clearly shown to be composed of a cluster of small vesicles protruding into a plasma membrane invagination with a narrow neck. The plasma membrane of this invagination has an accumulation of electron-dense material on its inner leaflet. The third process (P_3) is similar to the second, but its connection with the main portion of the cell is not evident in this section: same tissue as in figure 10 (× 76,500).

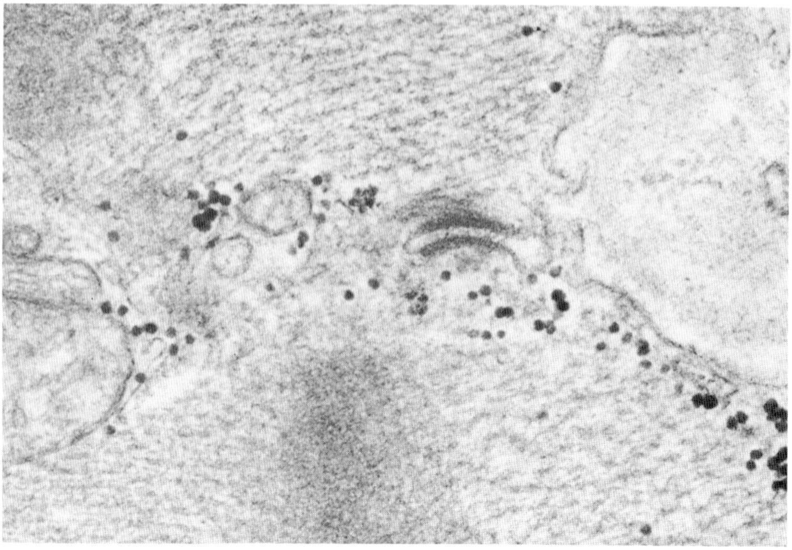

Figure 15 Small, desmosome-like intracytoplasmic junction formed by the apposition of two different areas of the plasma membrane of a single cardiac muscle cell. Note the direct continuity of the cytoplasm on both sides of this desmosome. Ventricular septal myocardium from 47-year-old man with hypertrophic obstructive cardiomyopathy (× 76,950).

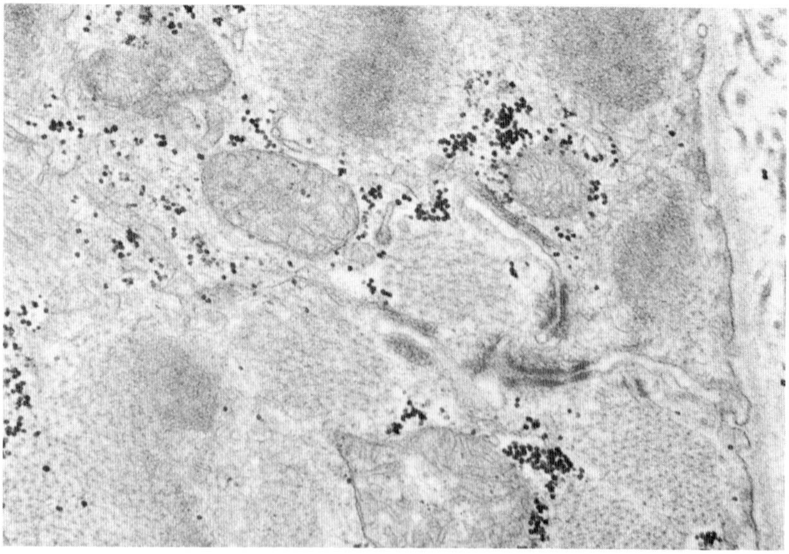

Figure 16 Intracytoplasmic junction in same tissue as in figure 15 is composed of several discontinuous segments of intercalated disc (× 33,750).

tered. Some intracytoplasmic junctions, particularly those which resembled desmosomes, were located along the sides of the cells; others were present at the ends of cells, especially in those cells which terminated in the interstititum in the form of elongated, finger-like processes.

Observations on experimental animals

Previous reports from this laboratory (2, 5) have described cardiac ultrastructural changes in dogs subjected to normothermic, anoxic cardiac arrest by aortic cross-clamping under complete cardiopulmonary bypass. Pertinent to the observations made on human myocardium as described above was the finding of intracytoplasmic junctions in cardiac muscle cells located at the margins of areas of replacement fibrosis which developed 2 or more weeks after the episode of anoxic arrest. Intracytoplasmic junctions in the hearts of these dogs were indistinguishable from those in the human hearts described above.

Rabbits treated with anthracyclines developed areas of focal myocardial damage, with severe interstitial edema, myofibrillar lysis, swelling of sarcoplasmic reticulum, and great dilation of T tubules (fig. 17). Intracytoplasmic junctions were encountered in areas of interstitial edema.

DISCUSSION

The observations reported herein show that the surfaces of cardiac muscle cells undergo important modifications during hypertrophy and degeneration. These modifications involve nonspecialized regions of the sarcolemma as well as regions of T tubules and intercellular junctions. In hypertrophied cells these surface alterations consisted of: 1) increased irregularity of contour of nonspecialized regions of sarcolemma and of sarcolemmal invaginations involved in the formation of T tubules; 2) dilation of T tubules, and 3) increased convolutions of the plasma membranes in areas of intercellular junctions. In degenerated cardiac muscle cells, the alterations present in hypertrophied, nondegenerated cells progressed to: 1) disappearance of T tubules; 2) dissociation of intercellular junctions, and 3) other changes, including the development of cytoplasmic processes, clusters of vesicles, and intracytoplasmic junctions, which resulted from remodeling of areas of plasma membranes previously participating in the formation of intercalated discs.

The extent to which alterations in cell surface morphology were present in hypertrophied, nondegenerated cardiac muscle cells varied considerably from one patient to another and from one area of a given biopsy to another. It was clear from the observations presented in this report that some hypertrophied cardiac muscle cells become highly asymmetrical in shape, whereas other cells retain their cylindrical shape and symmetry. It is also well known that large variations in transverse diameter of adjacent cardiac muscle cells often occur in

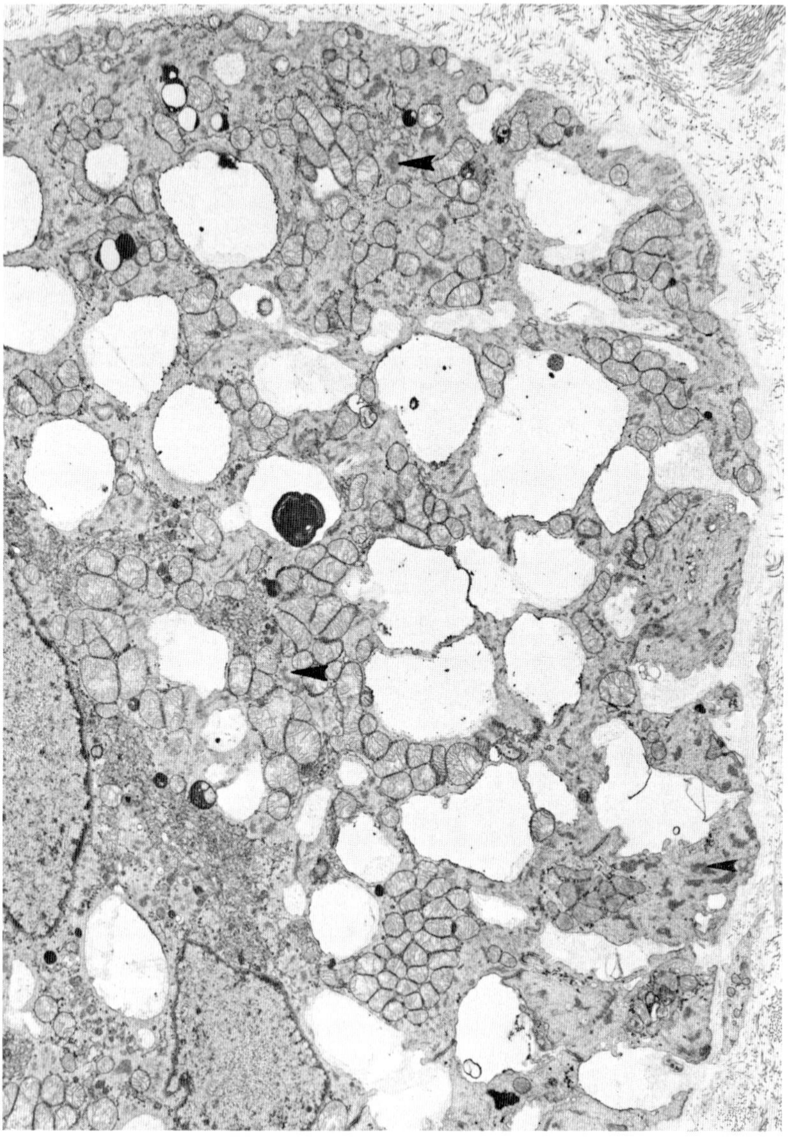

Figure 17 Group of cardiac muscle cells in ventricular septum of rabbit which had received a total dose of 400 mg of adriamycin/m² of body surface. These cells show extensive loss of myofibrils, of which only small patches of Z band material (*arrowheads*) remain, and marked dilation of T tubules. Several areas of communication between dilated T tubules and the interstitium are evident (× 5,700).

hypertrophied hearts. Thus, it is evident that caridac muscle cells can respond unequally (in terms of cell size attained) and asymmetrically (in terms of changes in cell shape) to the stimulus of hypertrophy. It is obvious that the shape of a hypertrophied cardiac muscle cell reflects the manner in which the contractile elements that are newly synthesized during hypertrophy become added to those present before the cell became hypertrophied. It has been shown that normal cardiac muscle cells are irregular in shape (22). Nevertheless, these irregularities are small compared with those which occur in cardiac hypertrophy. Thus, the severity of irregularities of the surfaces of hypertrophied cardiac muscle cells provides an indication of the degree of asymmetry of the growth which the cells undergo during the course of hypertrophy.

One of the most striking alterations in contour of the surfaces of hypertrophied, nondegenerated cells consisted of deep, pocket-like, elongated sarcolemmal infoldings. These infoldings resulted in marked increases in the surface to volume ratio of the cells, and differed from true T tubules in that they were not cylindrical in shape. The shapes of these infoldings suggest that they originate from shallow grooves separating two parallel, longitudinally oriented subsarcolemmal zones of formation of new myofibrils. The overall arrangement of the sarcolemma is stabilized by points of attachment between the plasma membranes and the Z bands of the myofibrils (10). Because of this, the sarcolemma immediately overlying these two areas bulges out as the development of new contractile elements progresses, and the groove separating these two areas becomes deeper. Thus, it would appear that the formation of myofibrils in cells having these deep infoldings proceeds along certain sharply circumscribed locations.

Multiple intercalated discs constituted another morphologically distinct manifestion of irregularity of cell surfaces. We believe that multiple intercalated discs result from asymmetrical but complementary growth of two muscle cells which are jonied to each other by side-to-side intercellular junctions (17). A similar phenomenon of asymmetrical growth of new sarcomeres along the two sides of an end-to-end junction appears to be responsible for the increased convolutions of intercalated discs in hypertrophied myocardium. Changes in a delicate balance between various mechanical forces to which a muscle cell is subject may be "sensed" at a very fine level by different areas of the cell, and such changes may play an important role in determining the ultimate localization of new contractile elements synthesized by growing or hypertrophying muscle cells. Our observations on sarcolemmal changes in hypertrophied human myocardium indicate that some cells add new sarcomeres in a symmetrical manner as they grow, whereas others have sites of preferential growth. We suggest that myofibril formation in hypertrophied cardiac muscle cells having deep sarcolemmal infoldings has taken place in sharply localized areas, and that hypertrophied cells having smoother, more regular surfaces have grown by adding new contractile elements in a more evenly distributed manner.

Sarcolemmal alterations in degenerated muscle cells may be attributed to three factors: loss of myofibrils, disruption of intercellular contacts, and interstitial fibrosis. The disorganization and eventual disappearance of the transverse tubular system was the sarcolemmal alteration most clearly related to loss of myofibrils. Dilation of T tubules occurred, for reasons which are not clear, in certain hypertrophied, nondegenerated cells. In degenerated cells, the T tubules became wide, shallow sarcoplasmic invaginations, particularly in areas of myofibrillar lysis. We believe that this change is related to the disruption of points of anchoring of the T tubules to the Z bands of the myofibrils, so that when the myofibrils are lysed the T tubules retract towards the cell surfaces. The inner aspect of the sarcolemma in degenerated cells often was associated with accumulations of Z band material, which we consider to represent either residues from lysed myofibrils, foci of abortive regeneration of myofibrils, or a mixture of both of these phenomena.

The observations on myocardium of adriamycin-treated rabbits provided a unique demonstration of the dependence of the structural integrity of the transverse tubular system on that of the myofibrils. Adriamycin, an antibiotic of the anthracycline group, binds to DNA and is known to have cardiotoxic effects, including congestive cardiac failure and myocardial degeneration and multifical damage (1). The damaged cardiac muscle cells show lysis of the myofibrils and an acute form of cellular atrophy. As the cell volume decreases, the sarcolemma becomes more and more redundant, and large vacuoles develop within the cells. These vacuoles are continuous with the interstitial space and therefore are sarcolemmal invaginations. They represent redundant membranes of T tubules which have lost their cylindrical shape and their association with myofibrils.

Significant sarcolemmal alterations accompanied the development of interstitial fibrosis and the disruption of intercellular contacts which occurred in late stages of hypertrophy and degeneration. It is uncertain whether the dissociation of cells is caused directly by fibrosis (mechanical effects), by alterations in electrolyte composition (calcium is essential for maintenance of the integrity of intercellular junctions), or by proteolytic enzymes which may be released in the interstitium during inflammation or injury. Regardless of its ultimate cause, the dissociation of cells led to two basic types of sarcolemmal alterations: 1) the conversion of specialized (junctional) to nonspecialized areas of the cell surface, and 2) the rearrangement of some of the specialized junctional structures to form aberrant junctions between two parts of the same cell (intracytoplasmic junctions).

The dissociation of intercellular junctions often appeared to be mediated by the formation of vesicles which filled the widened spaces between the separating disc membranes. These vesicles were similar to those present in plasma membrane processes which are known to extend from one cardiac muscle cell into another at the levels of intercalated discs (11, 23). Some of these vesicles contained small nexuses, which suggests that they were derived from the dissociation of large nexuses. Vesicles of this type (but without nexuses) also were observed in large numbers along the sides of cells in areas of fibrosis, and they also appeared to be derived from the plasma membranes. It is not certain, however, that all of these vesicles originate from the dissociation of side-to-side junctions. Page *et al.* (23) observed similar vesicles (without nexuses) in Purkinje fibers of normal cats; these vesicles formed structures which resembled stalked knobs and projected into the interstitium. We have observed these stalked knobs in human myocardium as part of the morphological spectrum of changes associated with the formation of vesicles in areas of fibrosis, but not in normal ventricular muscle cells. Vesicles of the type just mentioned also have been described by Burkholder *et al.* (6) as "spherical microparticles" in subepithelial, intramembranous, or subendothelial clusters in renal glomeruli from patients with a variety of renal diseases. Burkholder *et al.* considered the possibility that these vesicles were viral particles, but concluded that they were probably derived from plasma membranes in areas of injury. Our observations on myocardium support this interpretation.

The vesicles mentioned above differ clearly from the spherical structures which are evident on study of nexuses in cardiac muscle by the freeze-fracture technique (21). The latter structures are much smaller than the vesicles, and are not visible by transmission electron microscopy of tissue sections without the use of special preparative techniques. It has been recently demonstrated, using freeze-fracture techniques (13), that these spherical structures persist after dissociation of gap junctions in hepatocytes by hypertonic sucrose (even though the sites of previous gap junctions could not be distinguished from nonspecialized areas of plasma membrane in ordinary tissue sections). These observations emphasize the need for the application of newer techniques of morphology to the study of cardiac pathology.

After dissociation of cardiac muscle cells occurs, the areas of plasma membranes previously involved in the formation of intercellular junctions undergo remodeling. Some of these areas form long, finger-like projections which extend in the interstitium, whereas others develop elaborate convolutions and invaginations in which collagen fibrils are entrapped. The close association between the basement membranes of the muscle cells and connective tissue microfibrils appears to be responsible for maintaining the overall shapes of degenerated cells in areas of fibrosis. The cause of thickening of the basement membranes of muscle cells in areas of fibrosis is not clear. It has been shown, however, that the basement membrane material synthesized by epithelial cells has certain resemblances to collagen and that the synthesis of increased amounts of this material is stimulated by cellular injury (25). Thickening of the basement membranes of cardiac muscle cells may represent a similar type of response. Intracytoplasmic junctions appeared to form as a consequence of remodeling of sarcolemmal areas which previously formed parts of intercellular junctions. This was shown by our observations on myocardium of dogs which developed such changes in association with the healing of focal necroses produced by normothermic, anoxic cardiac arrest. These observations were complemented by studies of human myocardial tissue, in which intracytoplasmic junctions also were evident (1, 3). It is not known whether intracytoplasmic junctions are formed either by rearrangement of parts of intercellular junctions (*i.e.*, the apposition of hemidesmosomes) by *de novo* synthesis, or by both of these mechanisms. The origin of the long, complex intracytoplasmic junctions which we have described also is uncertain. We favor the possibility that such junctions, which resemble large, discontinous parts of intercalated discs, actually are parts of discs which are breaking down as a result of cell fusion.

REFERENCES

1. BUJA, L. M., and FERRANS, V. J. 1975. Myocardial injury produced by antineoplastic drugs, pp. 487–497. *In* A. Fleckenstein and G. Rona (eds.), Pathophysiology and Morphology of Myocardial Cell Alterations; Recent Advances in Studies on Cardiac Structure and Metabolism, Vol. 6. University Park Press, Baltimore.
2. BUJA, L. M., FERRANS, V. J., and LEVITSKY, S. 1972. Occurrence of intramitochondrial glycogen in canine myocardium after prolonged anoxic cardiac arrest. J. Molec. Cell. Cardiol. 4: 237–254.
3. BUJA, L. M., FERRANS, V. J., and MARON, B. J. 1974. Intracytoplasmic junctions in cardiac muscle cells. Amer. J. Pathol. 74: 613–648.
4. BUJA, L. M., FERRANS, V. J., MAYER, R. J., ROBERTS, W. C., and HENDERSON, E. S. 1973. Cardiac ultrastructural changes induced by daunorubicin therapy. Cancer 32: 771–788.
5. BUJA, L. M., LEVITSKY, S., FERRANS, V. J., SOUTHER, S. G., ROBERTS, W. C., and MORROW, A. G. 1971. Acute and chronic effects of normothermic anoxia on canine hearts. Light and electron microscopic evaluation. Circulation 43(suppl. 1): 44–50.
6. BURKHOLDER, P. M., HYMAN, L. R., and BARBER, T. A. 1973. Extracellular clusters of spherical microparticles in glomeruli in human renal glomerular diseases. Lab. Invest. 28: 415–425.
7. FAWCETT, D. W., and McNUTT, N. S. 1969. The ultrastructure of the cat myocardium. I. Ventricular papillary muscle. J. Cell Biol. 42: 1–45.
8. FERRANS, V. J., MASSUMI, R. A., SHUGOLL, G. I., ALI, N., and ROBERTS, W. C. 1973. Ultrastructural studies of myocardial biopsies in 45 patients with obstructive or congestive cardiomyopathy, pp. 231–272. *In* E. Bajusz, G. Rona, A. J. Brink, and A. Lochner (eds.), Cardiomyopathies: Recent Advances in Studies on Cardiac Structure and Metabolism, Vol. 2. University Park Press, Baltimore.
9. FERRANS, V. J., MORROW, A. G., and ROBERTS, W. C. 1972. Myocardial ultrastructure in idiopathic hypertrophic subaortic stenosis: A study of operatively excised left ventricular outflow tract muscle in 14 patients. Circulation 45: 769–792.
10. FERRANS, V. J., and ROBERTS, W. C. 1973. Intermyofibrillar and nuclear-myofibrillar connections in human and canine myocardium: an ultrastructural study. J. Molec. Cell. Cardiol. 5: 247–257.
11. FERRANS, V. J., ROBERTS, W. C., SHUGOLL, G. I., MASSUMI, R. A., and ALI, N. 1973. Plasma membrane extensions in intercalated discs of human myocardium and

their relationship to partial dissociations of the discs. J. Molec. Cell. Cardiol. 5: 161–169.
12. FREEMAN, J., and SPURLOCK, B. 1962. A new epoxy embedment for electron microscopy. J. Cell Biol. 13: 437–443.
13. GOODENOUGH, D. A., and GILULA, N. B. 1974. The splitting of hepatocyte gap junctions and zonulae occludentes with hypertonic disaccharides. J. Cell Biol. 61: 575–590.
14. JONES, M., FERRANS, V. J., MORROW, A. G., and ROBERTS, W. C. 1975. Ultrastructure of crista supraventricularis muscle in patients with congenital heart diseases associated with right ventricular outflow tract obstruction. Circulation 51: 39–68.
15. KAWAMURA, K., and JAMES, T. N. 1971. Comparative ultrastructure of cellular junctions in working myocardium and the conduction system under normal and pathologic conditions. J. Molec. Cell. Cardiol. 3: 31–60.
16. LAKS, M. M., MORADY, F., ADOMIAN, G. E., and SWAN, H. J. C. 1970. Presence of widened and multiple intercalated discs in the hypertrophied canine heart. Circ. Res. 27: 391–402.
17. MARON, B. J., and FERRANS, V. J. 1973. Significance of multiple intercalated discs in hypertrophied human myocardium. Amer. J. Pathol. 73: 81–88.
18. MARON, B. J. FERRANS, V. J., HENRY, W. L., CLARK, C. E., REDWOOD, D. R., ROBERTS, W. C., MORROW, A. G., and EPSTEIN, S. E. 1974. Differences in distribution of myocardial abnormalities in patients with obstructive and nonobstructive asymmetric septal hypertrophy (ASH): Light and electron microscopic findings. Circulation 50: 436–446.
19. MARON, B. J., FERRANS, V. J., and ROBERTS, W. C. 1975. Myocardial ultrastructure in patients with chronic aortic valve disease. Amer. J. Cardiol. 35: 725–739.
20. MARON, B. J., FERRANS, V. J., and JONES, M. 1975. The spectrum of degenerative changes in hypertrophied human cardiac muscle cells: An ultrastructural study. *In* P. E. Roy (ed.), Recent Advances in Cardiac Structure and Metabolism. University Park Press, Baltimore. (In press)
21. McNUTT, N. S. 1970. Ultrastructure of intercellular junctions in adult and developing cardiac muscle. Amer. J. Cardiol. 25: 169–183.
22. MUIR, A. R. 1965. Further observations on the cellular structure of cardiac muscle. J. Anat (London) 99: 27–46.
23. PAGE, E. 1967. The occurrence of inclusions within membrane-limited structures that run longitudinally in the cells of mammalian heart muscle. J. Ultrastruct. Res. 17: 63–71.
24. PAGE, E., POWER, B., FOZZARD, H. A., and MEDDOFF, D. A. 1969. Sarcolemmal evaginations with knob-like or stalked projections in Purkinje fibers of the sheep's heart. J. Ultrastruct. Res. 28: 288–300.
25. PIERCE, G. B., and NAKANE, P. K. 1972. Basement membrane: synthesis and role in disease, pp. 125–126. *In* T. Takeuchi, K. Ogawa, and S. Fujita (eds.), Histochemistry and Cytochemistry. Nakanishi Printing Co., Ltd., Kyoto, Japan.
26. TANAKA, Y., and GOODMAN, J. R. 1972. Electron Microscopy of Human Blood Cells, p. 370. Harper and Row, New York.

Mailing address: Dr. V. J. Ferrans,
Section of Pathology,
National Heart and Lung Institute,
National Institutes of Health,
Bethesda, Md. 20014 (USA).

Recent Advances in Studies on Cardiac Structure and Metabolism
Volume 9
The Sarcolemma
Edited by Paul-Emile Roy and Naranjan S. Dhalla

EFFECTS OF OUABAIN AND NOREPINEPHRINE ON THE MECHANICAL PERFORMANCE OF THE NORMAL AND FAILING HEART OF THE MYOPATHIC HAMSTER

A. A. ALOUSI, and L. SANTA BARBARA

Department of Pharmacology,
Sterling-Winthrop Research Institute,
Rensselaer, New York, USA

SUMMARY

The effects of ouabain and norepinephrine on the mechanical efficiency of the normal and failing heart of the myopathic Syrian hamster were studied in a modified Langendorff preparation. Ouabain administration to the normal and failing heart caused a significant increase in heart work and energy utilization per beat with no significant change in the mechanical efficiency of either heart. Norepinephrine caused a significant increase in heart work, energy utilization, and mechanical efficiency of the normal and failing heart. It was concluded that the isolated perfused hamster heart from normal and myopathic hamsters could provide a new model for the demonstration of the effect of cardiotonic agents on the mechanical performance of the heart.

Numerous investigators have studied the effects of cardiac glycosides and catecholamines on heart work and myocardial oxygen consumption. The results have been in complete disagreement.

The early observations that the cardiac glycosides do not increase myocardial oxygen consumption (3, 4, 13, 17) would not appear to be compatible with more recent findings that the glycosides increase the velocity of myocardial fiber shortening (21, 23), an important determinant of myocardial oxygen consumption (11, 16, 20).

The ability of digitalis to increase the strength of myocardial contraction is characterized by its capacity to increase both maximal force development and the rate of development of that force. Digitalis increases the developed force somewhat more than the velocity function (21). This is in contrast to the response to catecholamines which have a greater influence on the velocity function (20).

The myopathic Syrian hamster has been used for many years as a model for the study of congestive heart failure (1). Although a large number of studies have been performed using this model, there are no reports on heart work or efficiency. The present study was designed to investigate the effects of ouabain and norepinephrine on the mechanical performance of the isolated perfused heart from normal hamsters and myopathic Syrian hamsters in a late stage of cardiac decompensation.

METHODS

Male myopathic Syrian hamsters (BIO 14.6), weighing 120–140 gm, obtained from Telco Laboratory, were used in this study. The hamsters were 8–9 months old and were in a late stage of cardiac decompensation. Male hamsters of the same age (BIO RP, random breed) were used as controls.

The modified Langendorff preparation was described in detail previously (1). The isometric contraction of the left ventricle was measured from left ventricular wall tension exerted on a latex balloon filling the ventricular cavity. The balloon was inserted into the left ventricle through a small incision made in the left auricle. The balloon was then filled with water and connected to a Statham P23Ac pressure transducer (Statham Instruments, Inc., Oxnard, Calif.) using a PE 100 tube. End diastolic pressure in the left ventricle was maintained at 2 mm Hg. The pressure transducer was connected to a Grass polygraph (Grass Instruments, Quincy, Mass.) and left ventricular pressure was recorded continuously.

Myocardial oxygen uptake was obtained by measuring oxygen concentration in the perfusate before and after it circulated through the heart. Oxygen concentration was calculated from oxygen partial pressure (pO_2) measured polarographically by means of a Beckman pO_2 macroelectrode. Samples of perfusate from the aortic cannula and from the pulmonary artery were fed into a Beckman modular cuvette connected to a Beckman physiological gas analyzer 160. Continuous recording of arterial and venous pO_2 was obtained using a Grass polygraph.

Coronary flow rate was measured by means of a EP-300, 1/16-inch electromagnetic flow probe and a two-channel electromagnetic flowmeter (model 320, Carolina Medical Electronics, King, N.C.). The flow probe was placed between the reservoir and the aortic cannula. Calculation of the efficiency of the perfused heart was as follows:

$$\text{Efficiency} = \frac{\textit{work performed (joules/sec)}}{\textit{energy utilized (joules/sec)}}$$

$$\text{Heart word} = \text{left ventricular pressure} \times \text{heart rate} \times \text{coronary flow}$$

$$\text{Energy utilized} = \frac{\textit{Arteriovenous oxygen difference} \times \textit{coronary flow}}{100}$$

The heart was allowed to equilibrate for 30 min before any drug was administered. Ouabain was added to the reservoir in a concentration of 5×10^{-5} M to the normal heart and 1×10^{-6} M to the failing heart. The solution was recirculated for 50 min. A solution of l-norepinephrine-HCl was infused through a side arm into the aortic cannula at a rate of 0.01 μg/0.2 ml/min for 5 min in the normal heart and 0.05 μg/0.2 ml/min in the failing heart. Control hearts from normal and myopathic hamsters were treated similarly using Chenoweth's solution instead of drugs.

RESULTS

Effects of ouabain

In the normal heart the administration of ouabain (5×10^{-5} M) resulted in a gradual rise in left ventricular pressure which reached a maximum within 30 min of drug administration then plateaued for the rest of a 50-min ouabain administration time (fig. 1).

In the myopathic heart, ouabain (5×10^{-5} M) caused severe arrhythmias and myocardial depression. Thus it was necessary to lower the ouabain concentration to 1×10^{-6} M, a concentration which caused no significant increase in left ventricular pressure of the normal heart. At this concentration ouabain caused a very slow but significant rise in left ventricular pressure of the failing heart. A maximum increase in myocardia contraction was reached 40 min after the initiation of ouabain treatment. In most cases the continuation of ouabain treatment for 50 min resulted in a decline in left ventricular pressure toward control values as shown in figure 1.

Treatment with ouabain caused a significant decrease in heart rate of the normal and failing heart (fig. 2). Ouabain caused no significant change in coronary flow of either heart, and no change in oxygen uptake of the normal heart. However, oxygen uptake of the failing heart was increased by 100 percent (fig 3). This increase led to a significant increase in total energy utilization of the failing heart only (fig. 4).

Calculation of energy utilization per beat resulted in a significant increase in energy utilization per beat in both hearts. The data are illustrated in figure 4, *B* and *D*.

The effect of ouabain on heart work is shown in figure 5. Total work of the normal heart and the failing heart decreased gradually during ouabain circulation; however, heart work per beat increased moderately at the peak effect of ouabain in both hearts (fig. 5, *B* and *D*). Ouabain caused no significant change in the mechanical efficiency of the failing heart during its therapeutic effect but a significant decrease during the last 10 min of its infusion. No significant change

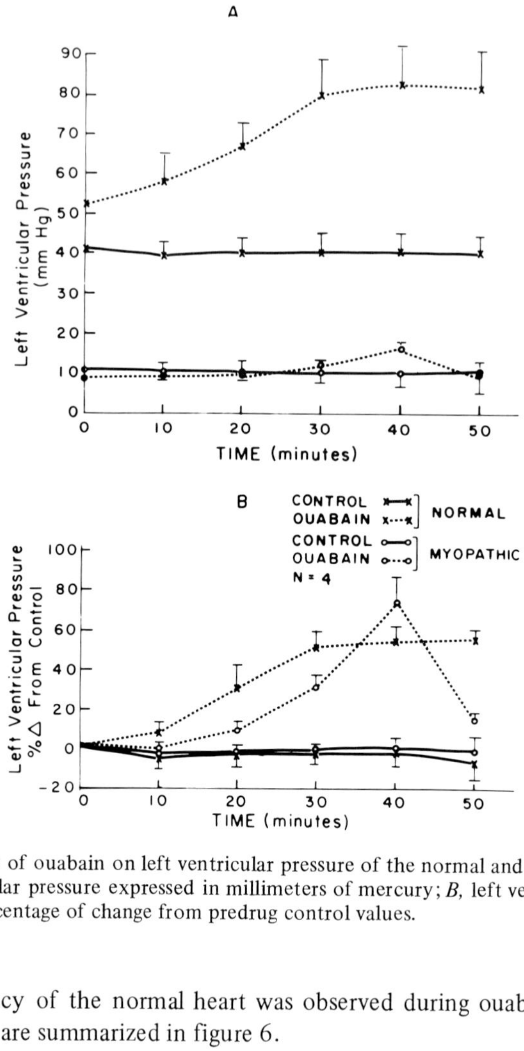

Figure 1 Effects of ouabain on left ventricular pressure of the normal and myopathic heart. *A*, Left ventricular pressure expressed in millimeters of mercury; *B*, left ventricular pressure expressed in percentage of change from predrug control values.

in the efficiency of the normal heart was observed during ouabain administration. The data are summarized in figure 6.

Effects of norepinephrine

A 5-min infusion of l-norepinephrine-HCl in the normal heart (0.01 μg/min) caused a marked increase in left ventricular pressure. This concentration of norepinephrine had no significant effect on left ventricular pressure of the failing heart. Norepinephrine infusion (0.05 μg/min for 5 min) in the failing heart caused a significant increase in left ventricular pressure as shown in figure 7. There was a significant increase in heart rate of the failing preparation during

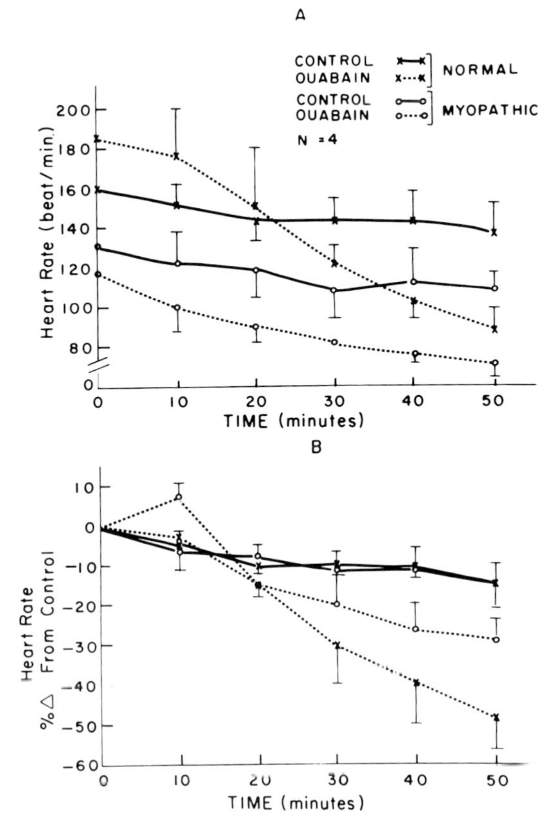

Figure 2 Effects of ouabain on heart rate of the normal and myopathic heart. *A*, Heart rate expressed in beats per min; *B*, heart rate expressed in percentage of change from predrug control values.

norepinephrine infusion but no significant changes in the rate of the normal heart as shown in figure 8.

Coronary flow rate did not change significantly in either heart during norepinephrine infusion. However, oxygen uptake was markedly increased in both hearts (fig. 9). This latter effect led to an increase in the calculated total energy utilization of both hearts. When energy utilization per beat was calculated for each heart, the failing heart showed no significant change in its energy utilization per beat, whereas in the case of the normal heart, this parameter increased markedly. The data are summarized in figure 10.

Norepinephrine infusion in the normal and failing heart caused a marked increase in the calculated heart work and heart work per beat in both hearts as

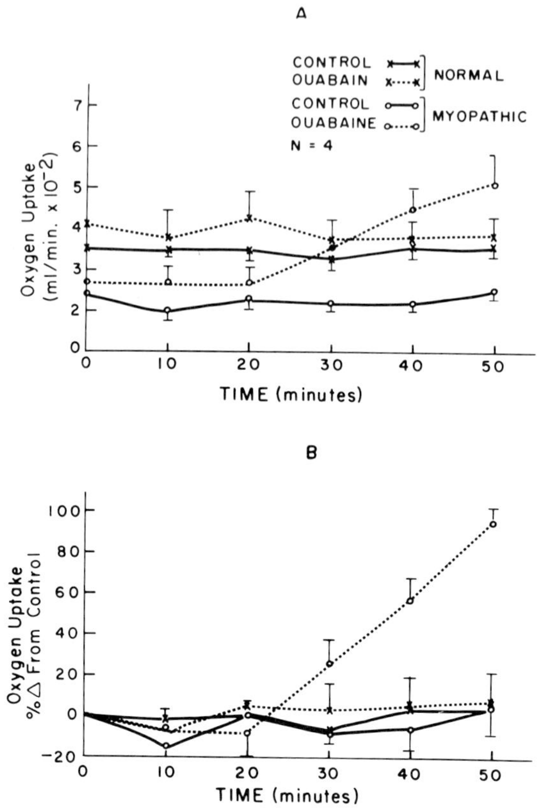

Figure 3 Effects of ouabain on oxygen uptake of the normal and myopathic heart. *A*, oxygen uptake expressed in milliliters per min × 10^{-2}; *B*, oxygen uptake expressed in percentage of change from predrug control values.

shown in figure 11. The mechanical efficiency of the normal heart treated with norepinephrine (0.01 μg/min for 5 min) was increased by more than 100 percent, whereas that of the failing heart treated with 5 times the concentration of norepinephrine used in the normal heart was increased by 280 percent (fig. 12).

DISCUSSION

For many years it has been taught that digitalis increased the efficiency of myocardial contraction; that there is an increase in cardiac contractile force without an increase in myocardial oxygen consumption. This is not in complete agreement with numerous reports.

In most experiments the oxygen uptake of fresh slices and of homogenates of heart muscle and heart mitochondria did not appear to be affected by digitalis (9, 24). In the isolated cat papillary muscle, ouabain first increased contractility without affecting oxygen uptake, but oxygen uptake then rose as the peak inotropic effect was reached and continued to rise as the muscle went into contracture (10). Coleman (7) reported that an augmentation of contractile state characterized by an increased V_{max} resulted in an increased oxygen consumption when the cat papillary muscle was contracting both isometrically and isotonically in the presence of acetyl strophanthidin.

Peters and Visscher (14) and Moe and Visscher (12) showed that cardiac glycosides tended to increase myocardial oxygen consumption in the canine heart-lung preparation in which end diastolic volume was held constant, al-

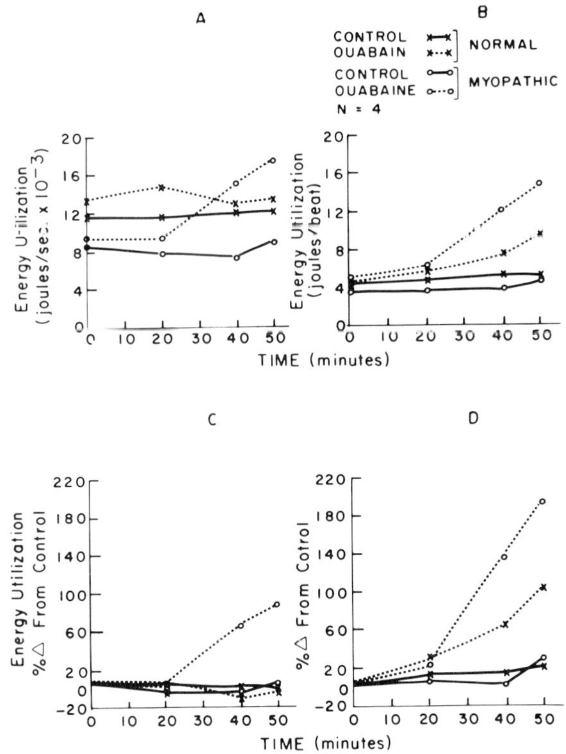

Figure 4 Effects of ouabain on energy utilization of the normal and myopathic heart. *A*, Energy utilized expressed in joules per sec × 10^{-3}; *B*, energy utilized expressed in joules per beat; *C*, energy utilized expressed in percentage of change in *A* from predrug control values; *D*, energy utilized expressed in percentage of change in *B* from predrug control values.

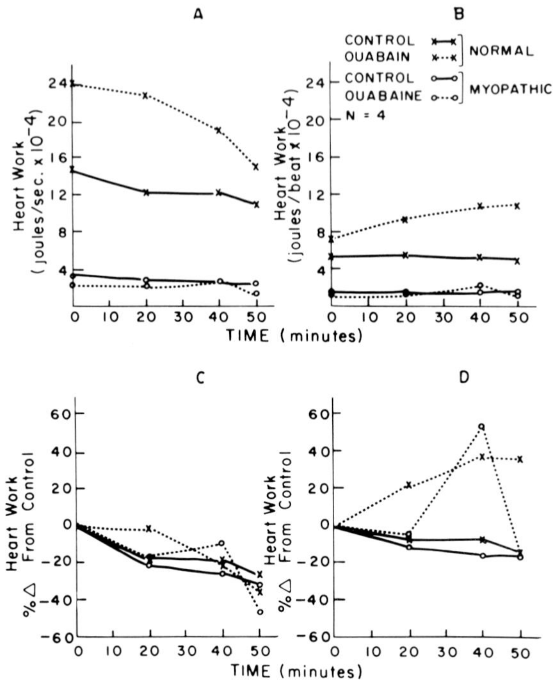

Figure 5 Effects of ouabain on heart work of the normal and myopathic heart. *A*, Heart work expressed in joules per sec $\times$ 10^{-4}; *B*, heart work expressed in joules per beat $\times$ 10^{-4}; *C*, heart work expressed in percentage of change in *A* from predrug control values; *D*, heart work expressed in percentage of change in *B* from predrug control values.

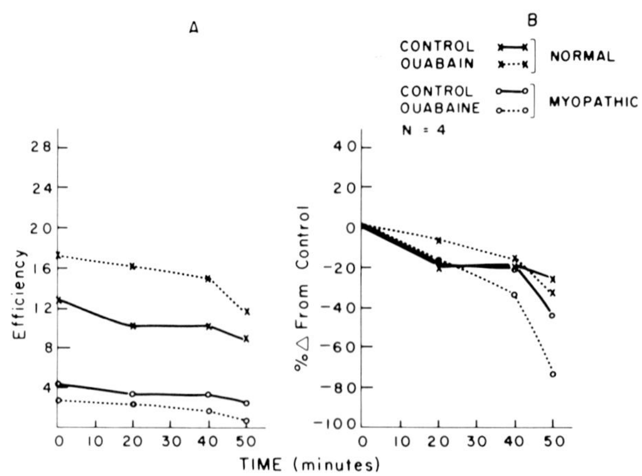

Figure 6 Effect of ouabain on the mechanical efficiency of the normal and myopathic heart. *A*, Efficiency expressed in absolute values; *B*, efficiency expressed in percentage of change from predrug control values.

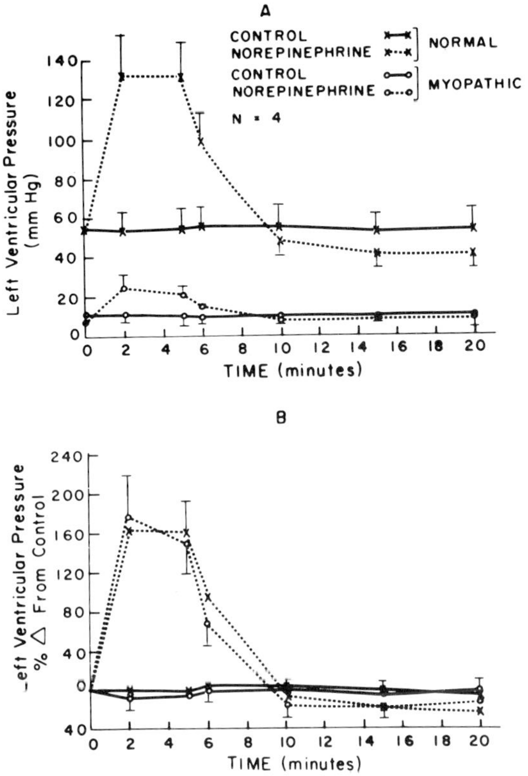

Figure 7 Effects of norepinephrine on left ventricular pressure of the normal and myopathic heart. *A*, Left ventricular pressure expressed in millimeters of mercury, *B*, left ventricular pressure expressed in percentage of change from predrug control values.

though in some experiments no change in oxygen uptake occurred. More recently, Covell *et al.* (8) reported that in the intact canine heart, strophanthidin produced increments in oxygen consumption that correlated directionally with changes in the extent and velocity of shortening.

On the other hand, several studies support the contention that digitalis does not affect myocardial oxygen consumption or even decreases it. Olson *et al.* (13) reported that acetyl strophanthidin did not alter myocardial oxygen consumption in the failing intact dog heart and Bing and his colleagues (3) observed no alteration in oxygen consumption when strophanthidin or lanatoside C was administered to normal subjects or to patients with heart failure. In the isolated supported canine heart, Sarnoff (17) observed a marked increase in myocardial contractility without a change in either myocardial oxygen consumption or external efficiency.

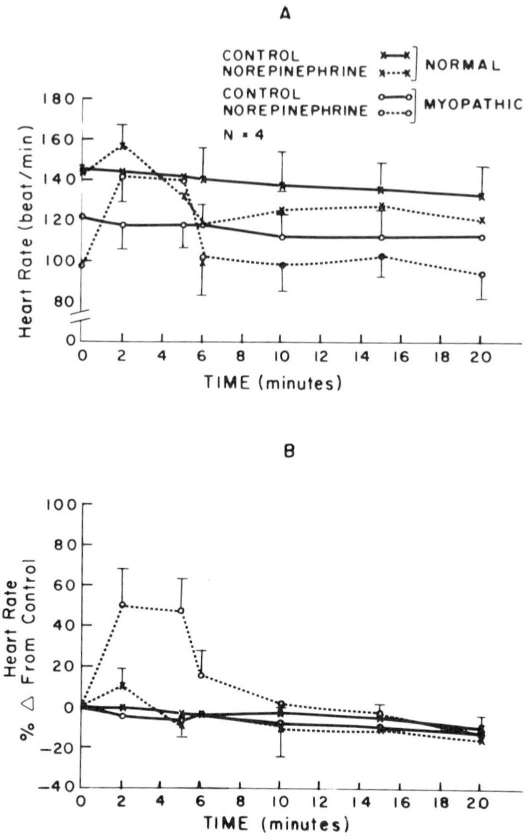

Figure 8 Effects of norepinephrine on heart rate of the normal and myopathic heart. *A*, Heart rate expressed in beats per min; *B*, heart rate expressed in percentage of change from predrug control values.

Thus, it appears that discrepancies in conclusions concerning the action of cardiac glycosides might be based on the specific technique employed for measuring the contractile properties of the heart or the parameters used for the calculation of cardiac efficiency.

Although external work can be estimated readily from flow times pressure, the actual myocardial work is not assessed accurately by this widely used method. This results from the relationship of La Place which take cognizance of the fact that the tension in the wall of a sphere is directly related to the pressure in and the radius of the cavity, and is inversely related to the thickness of the wall (22). Earlier observations which demonstrated an increased cardiac work with no

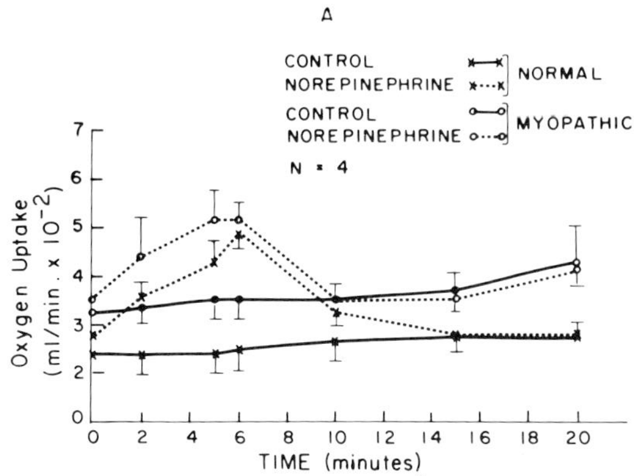

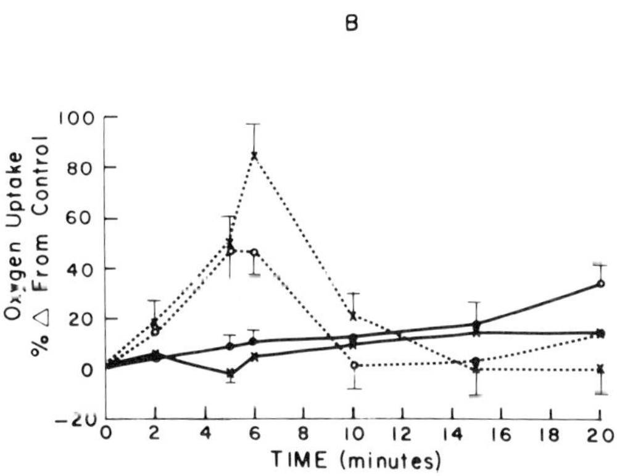

Figure 9 Effects of norepinephrine on oxygen uptake of the normal and myopathic heart. *A*, Oxygen uptake expressed in milliliters per min × 10^{-2}; *B*, oxygen uptake expressed in percentage of change from predrug control values.

change in myocardial oxygen consumption during the inotropic response to digitalis failed to recognize the importance of this consideration and overlooked the fact that the hearts under study manifested a decrease in left ventricular and diastolic pressure with a reduction of ventricular dilation as the inotropic response to digitalis developed. When hearts were studied at comparable ventricular volumes a clear increase in myocardial oxygen consumption was measured consistent with the increase in myocardial contractility (8).

The results of this investigation are in accord with the earlier findings that showed that digitalis glycosides produced an increase in myocardial oxygen consumption that correlates with the increase in heart work (6–8, 12, 23). The

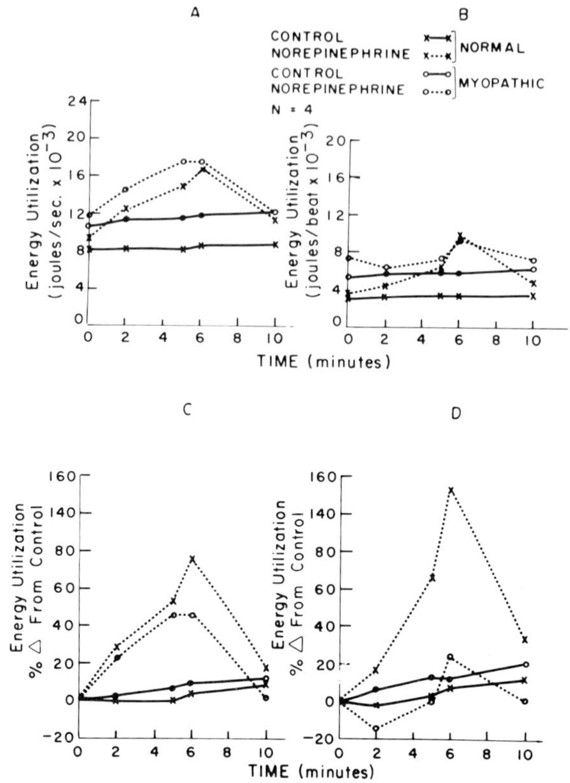

Figure 10 Effects of norepinephrine on energy utilization of the normal and myopathic heart. *A,* energy utilized expressed in joules per sec $\times 10^{-3}$; *B,* energy utilized expressed in joules per beat; *C,* energy utilized expressed in percentage of change in *A* from predrug control values; *D,* energy utilized expressed in percentage of change in *B* from predrug control values.

present technique has differed from previous studies in that heart work and efficiency of the isolated heart of normal and myopathic hamsters were investigated for the first time.

Analysis of the present data in terms of mechanical performance of the isolated perfused hamster heart reveals that the failing heart of the myopathic hamster appears to be markedly sensitive to ouabain. The highest concentration of ouabain in the perfusion fluid that could exert a significant positive inotropic effect on the failing heart without inducing cardiac arrhythmias was 1×10^{-6} M, as determined by dose-ranging experiments. In the normal heart, a dose ranging indicated that a concentration of 5×10^{-5} M ouabain in the perfusion fluid is

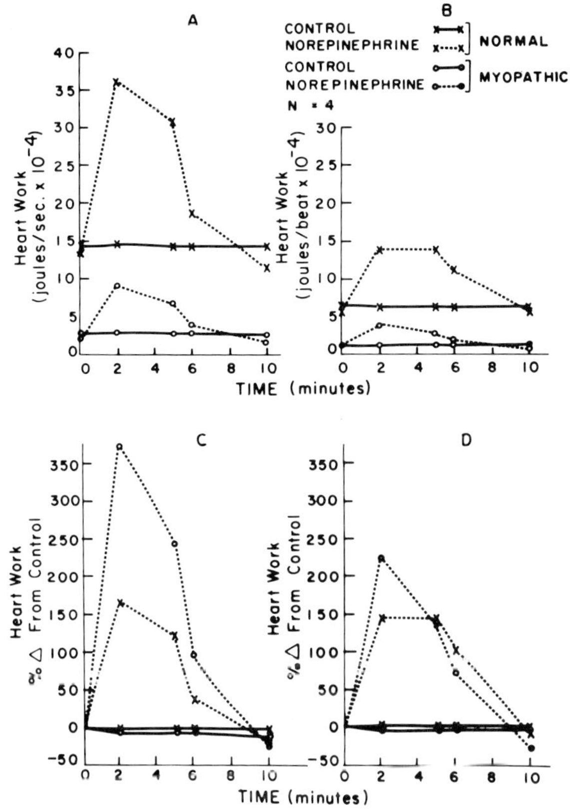

Figure 11 Effects of norepinephrine on heart work of the normal and myopathic heart. *A*, Heart work expressed in joules per sec × 10^{-4}; *B*, heart work expressed in joules per beat × 10^{-4}; *C*, heart work expressed in percentage of change in *A* from predrug control values; *D*, heart work expressed in percentage of change in *B* from predrug control values.

the lowest concentration that could cause a significant increase in left ventricular pressure. Therefore, neither equal doses nor doses yielding equal effects could be studied.

It is important to emphasize that the purpose of this investigation was to determine the effect of a therapeutic dose of ouabain and norepinephrine on the mechanical performance of the isolated perfused heart from normal and myopathic hamster in a late stage of cardiac decompensation, rather than to compare the effects of an equal dose of each of these agents on the performance of a normal versus a failing heart.

The data obtained in this study showed that ouabain caused a significant

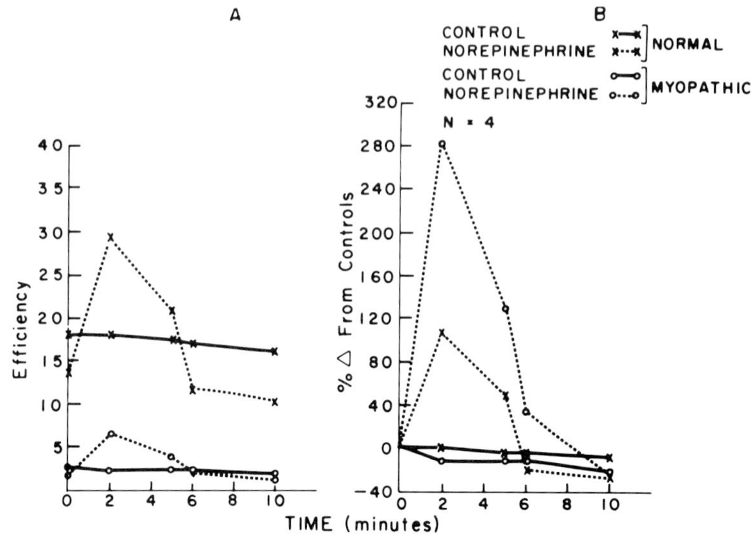

Figure 12 Effect of norepinephrine on the mechanical efficiency of the normal and myopathic heart. *A,* Efficiency expressed in absolute values; *B,* efficiency expressed in percentage of change from predrug control values.

increase in the amount of cardiac work per beat in the normal and failing heart. This increase in performance per beat was always accompanied by an increment in energy utilization per beat in both hearts.

The decrease in the actual total work performed per min was obviously attributable to the reduction in heart rate, an important contributor in the calculation of total heart work per min. The effect of ouabain on heart work and mechanical efficiency of the isolated perfused heart stimulated electrically at a constant rate during ouabain administration will be determined.

The decrease in heart work per min (fig. 5, *A* and *C*) has led to the lack of effect of ouabain on the mechanical efficiency of the isolated perfused heart in this study.

Although it is well established that norepinephrine augments myocardia contraction and causes substantial increases in myocardial oxygen consumption (19, 20), the mechanism by which norepinephrine augments myocardial oxygen consumption remains a matter of controversy.

Many investigators considered the increase in myocardial contractility induced by catecholamines to be accompanied by a substantial increase in myocardial oxygen consumption and to produce a decrease in efficiency or an "oxygen wasting" effect (15). It is, however, recognized that certain hemodynamic alterations which frequently accompany catecholamines administration could cause an increase in myocardial oxygen consumption and oxygen wasting effect (18).

The data presented in this study indicated that in the isolated perfused heart, norepinephrine caused very marked increase in myocardial contractility of the normal heart. This led to a significant increase in heart work despite the lack of a significant change in heart rate.

In the failing heart norepinephrine caused a significant increase in left ventricular pressure and heart rate which caused an even larger increase in heart work. The increase in the rate of the failing heart during norepinephrine infusion could be due to the large dose of norepinephrine used or to the hypersensitivity of the catecholamine-depleted failing heart (5).

The marked increase in heart work of the normal and failing heart and the less marked increase in myocardial energy utilization caused by norepinephrine led to a very significant increase in the mechanical efficiency of both hearts.

REFERENCES

1. ALOUSI, A. A. 1974. Cardiovascular effects of dopamine in the normal and failing heart of experimental animal. *In* Recent Advances in Studies on Cardiac Structures and Metabolism. University Park Press, Baltimore.
2. BAJUSZ, E., and LOSSNITZER, K. 1968. A new disease model of chronic congestive heart failure: Studies on its pathogenesis. Trans. N. Y. Acad. Sci. 30: 939–948.
3. BING, R. J., MARAIST, F. M., DAMMANN, J. F., JR., DRAPER, A., JR., HEINBECKER, R., DAILEY, R., GERALD, R., and CALAZEL, P. 1950. Effect of strophanthus on coronary blood flow and cardiac oxygen consumption of normal and failing human hearts. Circulation 2: 513–516.
4. BLAIN, J. M., EDDLEMAN, E. E., SIEGEL, A., and BING, R. J. 1956. Studies on myocardial metabolism. V. The effects of lanatoside C on the metabolism of the human heart. J. Clin. Invest. 35: 314–321.
5. CHIDSEY, C. A., BRAUNWALD, E., and MORROW, A. G. 1965. Catecholamine excretion and cardiac stores of norepinephrine in congestive heart failure. Amer. J. Med. 39: 442.
6. CLANCY, R. L., GRAHAM, T. P., JR., POWELL, W. J., JR. and GILMORE, J. P. 1967. Inotropic augmentation of myocardial oxygen consumption. Amer. J. Physiol. 212: 1055–1061.
7. COLEMAN, H. N. 1967. Role of acetyl stophanthidin in augmenting myocardial oxygen consumption: Relation of increased oxygen consumption to changes in velocity of contraction. Circ. Res. 21: 487–495.
8. COVELL, J. W., BRAUNWALD, E., ROSS, J., JR., and SONNENBLICK, E. H. 1966. Studies on digitalis. XVI. Effects on myocardial oxygen consumption. J. Clin. Invest. 45: 1535–1542.
9. LANGEMANN, H., BRODY, T. M., and BAIN, J. A. 1953. *In vitro* effects of ouabain on slices and mitochondrial preparations from heart and brain. J. Pharmacol. Exp. Therap. 108: 274–280.
10. LEE, K. S. 1953. A new technique for the simultaneous recording of oxygen consumption and contraction of muscle: The effect of ouabain on cat papillary muscle. J. Pharmacol. Exp. Therp. 109: 304–312.
11. McDONALD, R. H., JR. 1966. Developed tension: A major determinant of myocardial oxygen consumption. Amer. J. Physiol. 210: 351–356.

12. MOE, G. K., and VISSCHER, M. B. 1938. Studies on the native glycosides of digitalis lanata with particular reference to their effects upon cardiac efficiency and their toxicity. J. Pharmacol. Exp. Therap. 64: 65–85.
13. OLSON, R. E., ROUSH, G., and LIANG, M. L. 1955. Effect of acetyl strophanthidin upon the myocardial metabolism and cardiac work of normal dogs and dogs with congestive heart failure. Circulation 12: 755.
14. PETERS, H. C., and VISSCHER, M. B. 1936. Energy metabolism of the heart and the influence of drugs upon it. Amer. Heart J. 11: 273–280.
15. RABB, W., VANLITH, P., LEPESCHIN, E., and HERRLICH, H. C. 1962. Catecholamine-induced myocardial hypoxia in the presence of impaired coronary dilatibility independent of external cardiac work. Amer. J. Cardiol. 9: 455–470.
16. ROSS, J., JR., SONNENBLICK, E. H., KAISER, G. A., FRAMMER, P. L., and BRAUNWALD, E. 1965. Electroaugmentation of ventricular performance and oxygen consumption by repetitive application of paired electrical stimuli. Circ. Res. 16: 332–342.
17. SARNOFF, S. J., GILMOR, J. P. WALLACE, A. G., SKINNER, N. S., JR., MITCHELL, J. H., and DAGGETT, W. M. 1964. Effect of acetyl strophanthidin therapy on cardiac dynamics, oxygen consumption and efficiency in the isolated heart with and without hypoxia. Amer. J. Med. 37: 3–13.
18. SARNOFF, S. J., GILMOR, J. P., WEISFELDT, M. L., DAGGETT, W. M. and MANSFIELD, P. 1965. Influence of norepinephrine on myocardial oxygen consumption under controlled hemodynamic conditions. Amer. J. Cardiol. 16: 217–226.
19. SKELETON, C. L., LEVEY, G. S., and EPSTEIN, S. E. 1970. Positive inotropic effects of dibutyl cyclic adenosine 3′-5′-monophosphate. Circ. Res. 26: 35–44.
20. SONNENBLICK, E. H., ROSS, J., JR., COVELL, J. W., KAISER, G. A., and BRAUNWALD, E. 1965. Velocity of contraction as a determinant of myocardial oxygen consumption. Amer. J. Physiol. 209: 919–927.
21. SONNENBLICK, E. H., WILLIAMS, J. F., JR., GLICK, G., MASON, D. T., and BRAUNWALD, E. 1966. Studies on digitalis. XV. Effects of cardiac glycosides on myocardial force-velocity relations in the non failing human heart. Circulation 34: 532–539.
22. SONNENBLICK, E. H. 1974. Myocardial ultrastructure in the normal and failing heart, pp. 3–13. *In* E. Braunwald (ed.), The Myocardium, Failure and Function. H. P. Publishing Co., Inc., New York.
23. SPANN, J. F., JR., SONNENBLICK, E. H., COOPER, T., CHIDSEY, C. A. WILLMAN, V. L., and BRAUNWALD, E. 1966. Studies on digitalis. XIV. Influence of cardiac norepinephrine stores on the response of heart muscle to digitalis. Circ. Res. 19: 326–331.
24. WOLLENBERGER, A. 1947. Metabolic action of cardiac glycosides. I. Influence on respiration of heart muscle and brain cortex. J. Pharmacol. Exp. Therap. 91: 39–51.

Mailing address: Dr. A. A. Alousi,
Department of Pharmacology,
Sterling-Winthrop Research Institute,
Rensselaer, N. Y. 12144 (USA).

Recent Advances in Studies on Cardiac Structure and Metabolism
Volume 9
The Sarcolemma
Edited by Paul-Emile Roy and Naranjan S. Dhalla

GLUCOSE AND PALMITATE UPTAKE IN THE MYOCARDIUM OF ISOLATED HEARTS FROM ADRENALECTOMIZED CATS[1]

A. C. BEARDSLEY, H. C. KUTCHAI, and A. M. LEFER

Department of Physiology,
University of Virginia School of Medicine,
Charlottesville, Virginia, USA

SUMMARY

Hearts from chronically adrenalectomized (ADX) cats deprived of any steroid support for 9–12 days were isolated and perfused in a Langendorff apparatus at a constant pressure of 95 mm Hg. The perfusion medium was Krebs-Henseleit buffer with either 10 mM glucose or 0.4 mM palmitate complexed to 3 percent albumin. Labeled ^{14}C substrate was used and the transient rate of glucose and palmitate uptake was measured. Oxygen consumption and [^{14}C]palmitate incorporation into CO_2 and heart lipids were also measured. ADX hearts showed an enhanced glucose uptake rate compared with controls, 65 ± 11.3 μmoles/gm to 16.2 ± 6. However, the qO_2 was not significantly different from control hearts. Palmitate uptake, O_2 consumption, and $^{14}CO_2$ were significantly lower in ADX hearts perfused with fatty acid as the energy substrate. Fatty acid uptake decreased from 9.7 ± 1.0 to 3.6 ± 1.1 and lipid fractions in the heart showed significant decreases in [^{14}C]palmitate incorporated into triglycerides ($p < 0.001$) and monoglycerides ($p < 0.01$). The ADX heart does not appear to have any impairment to glucose uptake but does show an impairment to fatty acid uptake. Because the heart uses lipid as the primary energy source, the impairment probably is not the primary factor responsible for cardiac failure in adrenal insufficiency because of its capability of using other available substrates for energy.

Hypotension and a severe reduction in cardiac function are recognized consequences of adrenalectomized animals which have been deprived of maintenance doses of corticosteroids (12, 19, 24). The principal cause of the hypotension and decreased cardiac output appears to be impairment of myocardial performance (13, 21, 25). Chronically adrenalectomized animals show no changes in either the myocardial intracellular water and electrolyte concentrations (16) or in cardiac catecholamine levels (10). The decreased cardiac glycogen concentration (1, 5, 6) and the ultrastructural disorganization of myofibrils (8, 22) resulting

[1] This research was supported by National Institutes of Health Grant H1-09924 from the National Heart and Lung Institute.

from adrenalectomy do not appear to be the major contributing causes of the circulatory collapse. The decrease in myosin ATPase activity reported previously by Rovetto *et al.* (20) appears to be one factor responsible for the decline in cardiac function but does not adequately explain the total impairment. The present study was designed to determine the contribution of myocardial metabolism to the impairment in myocardial performance observed in adrenal insufficiency. Glucose and palmitic acid as exogenous energy substrates were used to determine whether any major metabolic alterations could be observed.

METHODS

Animals

Healthy adult cats of either sex weighing 2.2–3.5 kg were used. The animals were anesthetized with sodium pentobarbital (30 mg/kg). Animals that were chronically adrenalectomized required only 15–20 mg/kg of anesthesia to reach a comparable surgical plane as those of control animals. Adrenalectomies were performed under aseptic conditions via a midventral laparotomy. Immediately after surgery the animals were given 50 mg of erythromycin (Gallimycin and 1 mg/kg of dexamethasone (Decadron) intramuscularly. The following day an additional dose of dexamethasone (1 mg/kg) was administered, and thereafter all steroid support was withdrawn. ADX cats were given free access to Purina cat chow and water containing 0.9 percent sodium chloride. The ADX animals were killed 9–12 days after withdrawal of steroid support. Before excision of the heart, the femoral artery was cannulated in all animals, and mean arterial blood pressures were recorded using a Statham P-23Db pressure transducer (Statham Instruments, Inc., Oxnard, Calif.). Each animal was then given 500 U/kg of heparin intravenously.

Preparation and perfusion of heart

Hearts from control and ADX cats were excised and placed in a beaker of Krebs-Henseleit solution containing 10 mM glucose at 4°C. Within 2 min, the heart was attached to the perfusion cannula of a Langendorff apparatus.

The apparatus used to perfuse the heart was similar to that described by Neely *et al.* (14), but modified for a cat heart. After the aorta was cannulated, retrograde perfusion was started at a pressure of 70 mm Hg for 5 min to allow washout of blood from the coronary bed. During washout, the pulmonary artery was attached to a cannula, thus allowing a portion of the effluent from the coronary bed to pass through a Radiometer oxygen electrode (Radiometer, Cleveland, Ohio) for continuous readings of PO_2 on the venous side of the heart. Each heart was electrically paced with a Grass S-4 stimulator (Grass Instruments, Quincy Mass.), at a frequency of 120 Hz, 10 msec duration, and at voltage 10 percent above threshold. After washout, the heart was then switched to a closed circuit system containing 100 ml of perfusion medium and perfused at 95 mm Hg pressure. At the end of the perfusion period, the heart was placed on the open circuit system for approximately 1 min to wash out any exogenous radioactive substrates.

Perfusion medium

The perfusion medium used in all experiments was Krebs-Henseleit (K-H) bicarbonate buffer (11) equilibrated with 95 percent O_2 and 5 percent CO_2 at 37°C. In experiments on glucose uptake, 10 mM [1-^{14}C] glucose (0.25 μCi/ml) was used as substrate. For the lipid experiments, 0.4 mM [1-^{14}C] palmitate (0.25 μCi/ml) was used as the substrate. The palmitic acid was solubilized as the potassium salt and bound to albumin (3 percent in K-H) by injecting the salt mixture into the albumin medium. The albumin-fatty acid complex was then dialyzed overnight against K-H. No glucose was added to the medium in the fatty acid experiments. Before each experiment, the buffer was filtered through an 8-μm Millipore filter.

O_2 and CO_2 determinations

The oxygen consumption was estimated by taking the arterial-venous difference in PO_2 and the rate of coronary flow at regular intervals during the course of the experiment. The venous PO_2 was monitored continuously, whereas arterial PO_2 was measured by collecting samples of the oxygenated perfusate and processing them through an O_2 electrode. CO_2 determinations were made by collecting the coronary effluent and oxygenated perfusate under heptane. The samples were then acidified and the $^{14}CO_2$ released was trapped in hyamine and the radioactivity counted in a scintillation counter.

Glucose and fatty acid uptake determinations

The transient rate of glucose and fatty acid uptake was determined from the decrease in radioactivity of the substrate in the perfusion medium during the perfusion time. Samples of the perfusion medium were collected at various timed intervals. In the glucose experiments, the glucose was isolated by paper chromatography using butanol-pyridine-water (6:4:3), and the glucose spot was counted for radioactivity. For the fatty acid experiments, lipids were extracted by the method of Folch (7) and thin layer chromatography (TLC) was used with an ether-petroleum ether-acetic acid (20:80:1) solvent to separate the lipids. The free fatty acid spot was then counted for radioactivity.

Experimental protocol

Two groups of animals were used, control and chronically adrenalectomized cats. The hearts from both groups were perfused with K-H containing either 10 mM glucose or 0.4 mM palmitate complexed with 3 percent albumin. In the hearts perfused with glucose, samples were collected over a 60-min perfusion period and those perfused with palmitate for 25 min. After the perfusion, each heart was rinsed with fresh K-H, blotted dry, and weighed. All values except glucose and fatty acid uptake rates were tested for significance by the Student *t*-test. Glucose and palmitate uptake rates were tested for significance by confidence limits of the linear regression lines by the least squares fit method.

RESULTS

Effects of adrenalectomy on coronary vascular bed of isolated perfused cat hearts

Adrenalectomized cat hearts do not appear to have any large alterations to the responsiveness of the coronary bed to perfusion at 95 mm Hg. Table I summarizes these results. There was no significant differences in the average coronary flow between control and ADX hearts perfused with either glucose or palmitic acid.

Effects of adrenalectomy on glucose uptake and O_2 consumption of perfused hearts

Control hearts perfused with 10 mM glucose, exhibited mean glucose uptakes of 16.2 ± 6.0 μmoles/gm of weight/hr (fig. 1), whereas ADX hearts took up 65.0 ±

Table I Coronary flow and perfusion pressure in isolated hearts of control and 9–12-day postadrenalectomized cats perfused with glucose (10 mM) or palmitate (0.4 mM, 3 percent albumin) as energy substrate in Krebs-Henseleit buffer*

Group	Coronary flow (ml/g·min)	Perfusion pressure (mm Hg)
Glucose		
Control (6)	2.98 ± 0.69†	95
ADX (5)	3.62 ± 0.86†	95
Palmitate		
Control (8)	3.75 ± 0.49†	95
ADX (6)	2.59 ± 0.55†	95

*All values are means ± S.E.M. Numbers in parentheses indicate number of hearts.

†Not significant.

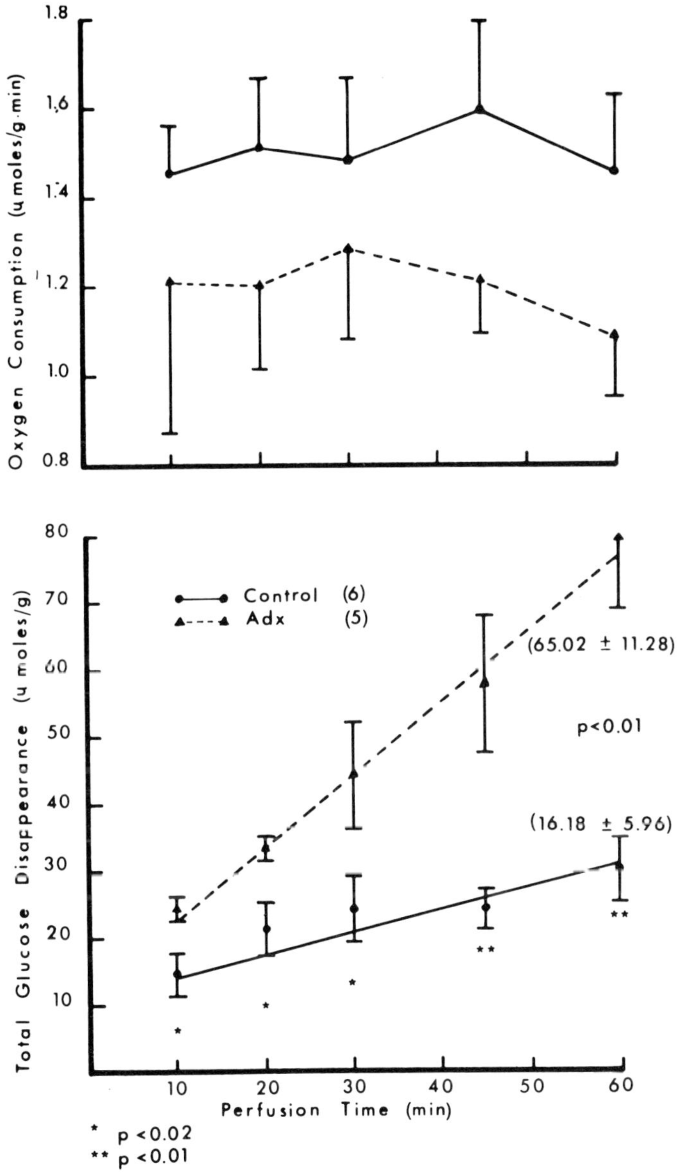

Figure 1 Time course of total glucose disappearance from the perfusion media and oxygen consumption rates during 1 hr of perfusion, at 95 mm Hg pressure, in isolated perfused cat hearts of control and 9–12-day postadrenalectomized cats. Each point is mean ± S.E.M. for six controls and five ADX animals. Glucose uptake rates were determined by linear regression of all values in each group. *Numbers in parentheses near the regression lines* represent mean uptake ± S.D. p values between regression lines represent significance between the two uptake rates.

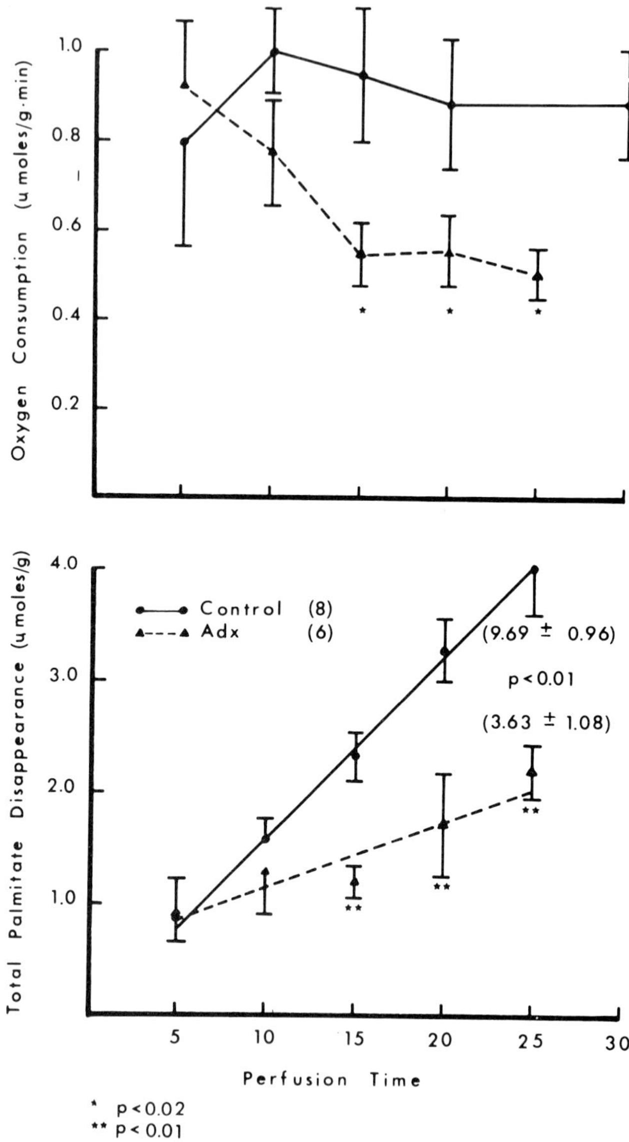

Figure 2 Time course of total palmitate disappearance from the perfusion medium and oxygen consumption rates during 25 min of perfusion at 95 mm Hg pressure in isolated perfused cat hearts of control and 9–12-day postadrenalectomized cats. Each point is mean ± S.E.M. for eight control and six ADX animals. Palmitate uptake rates were determined by linear regression of all values in each group. *Numbers in parentheses near the regression lines* represent mean uptake ± S.D. *p* values between regression lines represent significance between the two uptake rates.

11.3 μmoles/gm/hr, an increase of 400 percent. At every sampling point during the perfusion period, the amount of glucose extracted by ADX hearts was significantly higher than that of controls ($p < 0.02$). Oxygen consumption rates, however, were not significantly different from control values over the entire perfusion period, although mean values for ADX hearts were somewhat less than those of controls.

Effects of adrenalectomy on free fatty acid utilization, O_2 consumption, and CO_2 production in perfused hearts

Mean palmitate uptake in control hearts was 9.7 ± 1 compared with 3.6 ± 1.1 for ADX hearts (fig. 2). The total disappearance of palmitate from the medium was not significantly different between control and ADX hearts in the first 10 min of perfusion, but thereafter a significant ($p < 0.01$) difference was present. Oxygen consumption was significantly decreased in ADX hearts ($p < 0.02$). qO_2 values at the termination of the perfusion period were 0.88 ± 0.12 for control, and 0.51 ± 0.05 for ADX hearts, a decrease of 43 percent. The oxidation of palmitate into $^{14}CO_2$ (fig. 3) also shows that in the ADX heart there is a significant decrease of $^{14}CO_2$ to 15 percent of control values.

To further evaluate the deficit in FFA uptake in ADX hearts, the heart lipids were chromatographed on TLC plates to determine the rate of [1-^{14}C]palmitate incorporation into heart lipids (fig. 4). The major fate of palmitic acid in heart

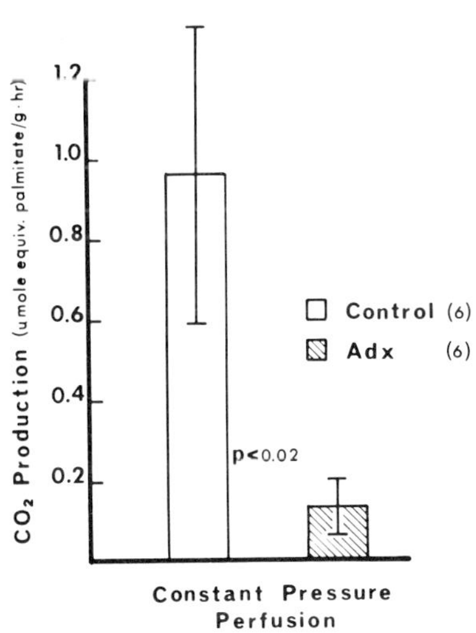

Figure 3 Average rate of $^{14}CO_2$ production from exogenous [1-^{14}C]palmitate in isolated perfused hearts from control and 9–12-day postadrenalectomized cats during 25 min of perfusion at 95 mm Hg pressure. *Numbers in parentheses* represent the number of animals in each group. *Bar* represents mean ± S.E.M.

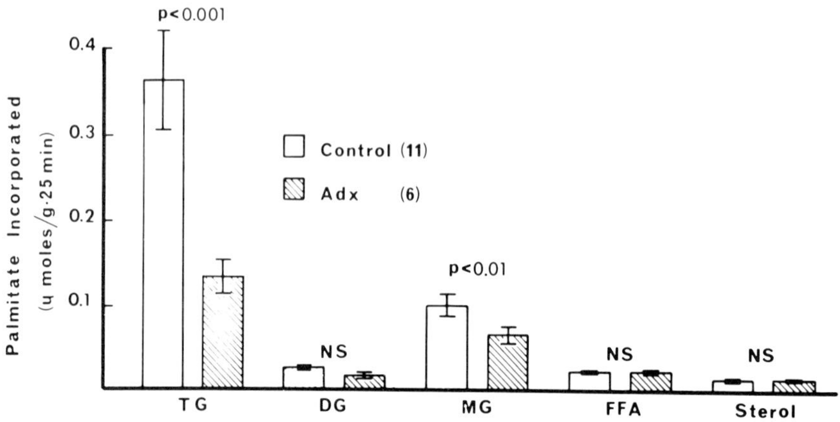

Figure 4 Incorporation of exogneous [1-^{14}C] palmitate into triglyceride (*TG*), diglyceride (*DG*), monoglyceride (*MG*), free fatty acid (*FFA*), and sterol ester (Sterol) fractions of cardiac lipid in isolated perfused hearts from control and 9–12 day postadrenalectomized cats. Perfusion was for 25 min at 95 mm Hg pressure: *Numbers in parentheses* represent numbers of animals in each group. *Bars* represent mean ± S.E.M. *NS,* not significant.

lipids was triglycerides and ADX hearts showed significant decreases in both the triglyceride ($p < 0.00$) and the monoglyceride pools ($p < 0.01$)

DISCUSSION AND CONCLUSIONS

These experiments represent the first detailed study on myocardial metabolism in animals suffering from chronic adrenal insufficiency. Very few previous studies have been reported which give information relative to myocardial metabolism in adrenalectomized animals. Ullrick *et al.* (23) used myocardial strips from the left ventricle of rats to measure respiratory rates and found an increased oxygen consumption in the adrenalectomized heart. However, decreased oxygen consumption has been reported in rat hearts (4, 9), and in dogs (26). However, no previous data are available regarding substrate utilization in the perfused or *in situ* heart in chronic adrenal insufficiency. Our data suggest that there is no defect in the ability of the adrenalectomized heart to take up glucose when an adequate amount is supplied to it. The reason for the increased glucose uptake by adrenalectomized hearts cannot be explained on the basis of the present data. The possibility that the increased glucose uptake may be due to an increased glycogen synthesis rate is unlikely because in this model there are no significant changes in cardiac glycogen between adrenalectomized and control hearts (20). The decrease in glycogen in adrenalectomized cat hearts occurs only after severe impairment to the myocardium. Additionally, Berdanier *et al.* (3) has shown that serum insulin and glucose levels in adrenalectomized rats do not change significantly from controls; therefore, preconditioning of the adrenalectomized heart *in vivo* before isolated perfusion does not appear to be a factor. Moreover, no such enhanced glucose uptake was observed in heart slices obtained from adrenalectomized cats (2).

In contrast, myocardial lipid metabolism does appear to be impaired in the adrenalectomized cat. The data show a marked decrease in fatty acid uptake, oxygen consumption and CO_2 production, when lipid is the only exogenous substrate available to the heart. Additionally, the triglyceride and monoglyceride fractions of cardiac lipid are decreased in hearts obtained from adrenalectomized cats. Whether the impairment in fatty acid utilization is caused by a defect in the metabolism or to the transport of the exogenous fatty acid cannot be determined from the available data. The specific activity of the labeled fatty acid synthesized into myocardial triglycerides may be diluted and thus reflect an apparent decrease in the total pool. However, preliminary data (2) indicate that the total pool is indeed smaller in the adrenalectomized hearts. A decreased $^{14}CO_2$ production may reflect an impairment on β-oxidation of labeled fatty acid or an increased preference towards synthesis of labeled fatty acids to triglycerides. However, since the activity of the triglyceride pool is decreased, the decreased $^{14}CO_2$ production reflects a decreased β-oxidation of fatty acids.

Since the heart uses lipids as the primary energy substrate, the findings reported here on lipid metabolism may represent an important alteration in the heart. However, although the impairment of fatty acid utilization may contribute to the hypofunction of the heart, it probably cannot account for the cardiac failure in adrenal insufficiency since the heart is capable of utilizing a variety of substrates for energy metabolism (15, 17, 18) and may not be able to shift to the other substrates. Further studies are necessary to determine whether changes reported here in lipid metabolism are critical to the functional capacity of the hypodynamic heart in adrenal insufficiency.

REFERENCES

1. BARTA, E., and PAVLOVICÔVA, H. 1965. Role of adrenals in maintaining the level of carbohydrate metabolism in the failing heart. Cor Vasa 7: 60–70.
2. BEARDSLEY, A. C. 1974. Myocardial metabolic alterations in chronic adrenal insufficiency in the cat. Ph.D. thesis, University of Virginia. University microfilms, Ann Arbor, Michigan.
3. BERDANIER, D. D., and MOSER, P. B. 1973. Metabolic responses of adrenalectomized rats to exercise. Proc. Soc. Exp. Biol. Med. 141: 490–493.
4. BOTTOMS, G., and GOETSCH, D. D. 1968. Effects of corticosterone on oxidative metabolism in different tissues of the rat. Gen. Comp. Endrocrinol. 10: 310–314.
5. BRITTON, S. W., SILVETTE, H., and KLINE, R. F. 1938. Adrenal insufficiency in American monkeys. Amer. J. Physiol. 123:705–711, 1938.
6. DAW, J. C., LEFER, A. M., and BERNE, R. M. 1968. Influences of corticosteroids on cardiac glycogen concentration in the rat. Circ. Res. 22: 639–647.
7. FOLCH, J., LEES, M., and STANLEY, G. H. S. 1957. A simple method for the

isolation and purification of total lipides from animal tissue. J. Biol. Chem. 226: 497–509.
8. GLENN-BOTT, M., IMMS, F. J., JONES, M. T., and PAPADAKI, L. 1970. The influence of adrenocortical insufficiency on the ultrastructure of cardiac muscle. J. Anat. 106: 187–188.
9. GOULDING, J., and BELLAMY, D. 1971. Effects of adrenalectomy on daily rhythms in oxygen consumption of tissues in the growing rat. J. Endocrinol. 51: 203–208.
10. KAYE, M. P., JELLINEK, M., COOPER, T., and HANLON, C. R. 1964. Effect of adrenalectomy on myocardial catecholamines, J. Surg. Res. 4: 257–259.
11. KREBS, H. A., and HENSELEIT, K. 1932. Untersuchungen uber harnstoffbildung im tierkorper. Hoppe-Seyler's Z. Physiol. Chem. 210: 33–66.
12. LEFER, A. M., and VERRIER, R. L. 1970. Role of corticosteroids in the treatment of circulatory collapse states. Clin. Pharmacol. Therap. 11: 630–655.
13. LEFER, A. M., VERRIER, R. L., and CARSON, W. W. Cardiac performance in experimental adrenal insufficiency in cats. Circ. Res. 22: 817–827.
14. NEELY, J. R., LIEBERMEISTER, H., BATTERSBY, E. J., and MORGAN, H. E. 1967. Effect of pressure development on oxygen consumption by isolated rat heart. Amer. J. Physiol. 212: 804–814.
15. NEELY, J. R., ROVETTO, M. J., and ORAM, J. F. 1972. Myocardial utilization of carbohydrate and lipids. Prog. Cardiovasc. Dis. 15: 289–329.
16. NOBLE, R. L. 1950. Physiology of the adrenal cortex, pp. 67–180. *In* G. Pincus and K. V. Thiemann (eds.), The Hormones, Vol. II. Academic Press, New York.
17. OPIE, L. H. 1968. Metabolism of the heart in health and disease. Part I. Amer. Heart J. 76: 685–698.
18. OPIE, L. H. 1969. Metabolism of the heart in health and disease. Part II. Amer. Heart J. 77: 100–122.
19. REIDENBERG, M. M., OHLER, E. A., SEVY, R. W., and HARAKAL, C. 1963. Hemodynamic changes in adrenalectomized dogs. Endocrinology 72: 918–923.
20. ROVETTO, M. J., LEFER, A. M., and MURPHY, R. A. 1971. Alterations in myocardial cell function in adrenal insufficiency. Pfluger's Arch. ges. Physiol. 329: 59–71.
21. SOLOMON, N., TRAVIS, R., and SAYERS, G. 1959. Corticosteroids and the functional capacity of the rat heart-lung preparation. Endocrinology 64: 535–541.
22. SUZUKI, T. 1967. Electron microscopic study on the effects of adrenalectomy on the heart muscle. Tohoku J. Exp. Med. 91: 239–248.
23. ULLRICK, W. C., BRENNAN, B. B., and WHITEHORN, W. V. 1961. Characteristics of cardiac tissue from adrenalectomized rats. Amer. J. Physiol. 200: 117–121.
24. VERRIER, R. L., ROVETTO, M. J., and LEFER, A. M. 1969. Blood volume and myocardial function in adrenal insufficiency. Amer. J. Physiol. 217: 159.
25. WEBB, W. R., DEGERLI, I. U., HARDY, J. D., and UNAL, M. 1965. Cardiovascular responses in adrenal insufficiency. Surgery 58: 273–282.
26. WEBB, W. W., and DEGERLI, I. U. 1968. Myocardial metabolism: Effects of adrenalectomy and acute cortisol replacement. J. Surg. Res. 8: 73–77.

Mailing address: Dr. A. M. Lefer,
Department of Physiology,
Jefferson Medical College,
Philadelphia, Pa. 19107 (USA).

Recent Advances in Studies on Cardiac Structure and Metabolism
Volume 9
The Sarcolemma
Edited by Paul-Emile Roy and Naranjan S. Dhalla
Copyright 1976 University Park Press Baltimore

RADIATION-INDUCED ENZYME EFFLUX FROM RAT HEART: SEDENTARY ANIMALS[1]

L. D. MacWILLIAM and N. M. G. BHAKTHAN

Department of Kinesiology,
Simon Fraser University,
Burnaby, British Columbia, Canada

SUMMARY

Serum levels of lactate dehydrogenase, creatine kinase, and glutamate oxaloacetate transaminase show initial elevations within 12 hr of exposure to 2,000 rads of γ-radiation to the thoracic region of rats. Significant decreases in heart muscle homogenate levels of these enzymes parallel initial elevations in the serum and may suggest that enhanced leakage of enzymes is a consequence of radiation injury to heart muscle. Insignificant alterations in mitochondrial glutamate oxaloacetate transaminase levels after exposure indicate that *in vivo* injury to the mitochondria from therapeutic levels of γ-radiation is questionable. The results support the contention that ionizing radiation instigates alterations in the dynamic permeability of membranes, allowing leakage of biologically active material out of the injured cell.

One of the biochemical indicators of acute radiation injury is an increase in serum levels of lactate dehydrogenase (LDH). This has been documented by Hawrylewicz and Blair (14), Hori *et al.* (18, 19) and Kärcher (21), who have shown independently transient elevations of total LDH and the various isoenzymes after irradiation. In addition, research involving serum glutamate oxaloacetate transaminase (GOT) as a radiobiological marker has shown that in some mammalian species sublethal exposures led to significant, transient increases in serum activities (2–4, 26, 30). The data from Brent *et al.* (4) suggests that the response is varied between species. Irradiated rabbits exhibited significant elevations of serum GOT within 24 hr of treatment, whereas rats subjected to either whole-body or local irradiation of the liver did not show any substantial GOT elevation after 1 day. Although little is known about the potential of creatine kinase (CK) as a radiobiological marker, observable serum elevations occurred in patients

[1] This work was supported by a grant from the National Research Council of Canada and the B. C. Heart Foundation, Vancouver, Canada.

(post-treatment) receiving incidental cardiac irradiation (27). Since CK is located in skeletal and cardiac muscle in high concentration, its appearance in the blood may indicate radiation injury to these tissues.

Much of the work to date has involved the enzymological effects of whole-body irradiation, where cell death and disintegration of radiosensitive tissues may contribute greatly to the observed serum elevation. Little work, however, has been published outlining the effects of localized irradiation of radioresistant tissues, where therapeutic doses do not involve massive cell disruption. The present study was undertaken to investigate possible changes in serum levels of LDH, GOT, and CK as possible indicators of radiation injury to the heart, and to attempt to correlate these changes with alterations in tissue levels of the enzyme.

MATERIALS AND METHODS

Irradiation procedure

Sprague-Dawley, male rats of 150–200 gm were placed under light ether anesthesia 24 hr before irradiation and marked laterally across the top of the scapula to define the target area. At the time of irradiation, the animals were placed into ventilated Plexiglas tubes to restrict movement. The field exposure area was adjusted to cover 3 cm posterior to the scapula markings, as this was found to be optimal in minimizing radiation damage to peripheral tissues and organs, while allowing irradiation of the entire heart. Rats were irradiated in groups of four, using an Eldorado 8 (Atomic Energy of Canada) therapeutic cobalt unit. The dose rate was 281 rads/min at the surface and exposure time was corrected for an absorbance loss of 15.7 percent at an average depth of 3 cm to given an acute dosage of 2,000 rads. Sham-irradiated controls were treated in a manner identical to the experimental animals.[2]

Sampling procedure

In order to minimize serum turbidity the rats were fasted for 6 hr before death. At predetermined intervals rats were placed under mild ether anesthesia and an incision was made across the abdomen. Blood was withdrawn from the inferior vena cava using a nonheparinized syringe and immediately transferred to sealed centrifuge tubes. Samples were allowed to clot for 30 min at 25°C and were then placed on ice for approximately 30 min until centrifuging at 2,000 × *g* for 15 min (41).

Serum was collected and divided into individual vials for the enzyme assays. All samples for LDH determinations were stored at 4–6°C and assays were run

[2] All animals were treated in accordance with guidelines established in *Care of Experimental Animals–A Guide for Canada,* published by the Canadian Council on Animal Care, Ottawa.

within 40–48 hr. LDH samples were not frozen because of known inactivation of the enzyme (36). Samples for CK and GOT determinations were frozen in liquid nitrogen and stored on Dry Ice until assayed. It has been shown that CK and GOT activity levels remain stable for several weeks when stored at −20°C (Lederle Diagnostics, American Cyanamid Co., Bulletin 27819).

Heart muscle preparation

Rat hearts were excised, trimmed of fatty and connective tissue and plunged into ice-cold 0.15 M KCl. After rinsing, the hearts were quickly blotted dry and weighed. The tissue was finely minced with scissors and washed five times in ice-cold 0.15 M KCl and twice in a multiple salts buffer modified from Chappell and Perry (6). This medium contains 0.1 M KCl, 0.005 M $MgSo_4$, and 0.001 M ethylenediaminetetraacetate (EDTA) in 0.05 M Tris-HCl buffer, pH 7.4. The minced muscle was then homogenized in 2 volumes of the above medium using a ground-glass, Ten-Broeck-type homogenizer, and appropriately diluted to make a 10 percent homogenate. Homogenate samples were centrifuged at 650 X *g* for 5 min after which the supernatant was collected and the operation repeated to remove cellular debris (10). The resultant supernatant was centrifuged at 6,000 X *g* for 15 min to obtain a crude mitochondrial pellet, which was resuspended in fresh buffer and centrifuged as before. Although higher centrifuge speeds increase the product yield, it was found that 6,000 X *g* offered a considerably purer fraction. This has been shown independently by Holton *et al.* (16) and Hedman (15). The supernatant resulting from the first mitochondrial fraction was collected, divided into individual vials, and stored in the same manner as were serum samples. The washed mitochondrial pellet was rinsed three times in 0.15 M KCl and suspended in 1.0 ml of this solution. The final mitochondrial suspension was frozen in liquid nitrogen and stored on Dry Ice until assayed.

In the above procedure all manual operations were carried out on ice and centrifugations were performed at 0°C in a Beckman model J21 centrifuge using JA-20, fixed angle rotor.

Enzyme assays

Creatine kinase assays employed the forward reaction or Rosalki technique (32) using the following reagents (given as final concentrations in the assay medium): 5×10^{-2} M PIPES buffer, 11.5×10^{-3} M magnesium salt, 1.45×10^{-2} M glucose, 9.1×10^{-3} M AMP, 1.8×10^{-2} M creatine phosphate, 1.1×10^{-3} M ADP, 8.67×10^{-3} M reduced glutathione, 5.0×10^{-4} M NADP, 3 IU/assay of hexokinase, 1 IU/assay of glucose 6-phosphate dehydrogenase (all reagents were pre-packaged by Calbiochem, San Diego, Calif.). Serum assays were run using freshly thawed, undiluted samples. Homogenate samples were diluted to 0.2 percent after thawing, using modified Chappell-Perry medium. Measurement of background utilized NADP, using creatine phosphate-free control solutions

(Calbiochem), was predetermined for both serum and homogenate preparations, and was found to be negligible. Since turbidity errors for both assays were also found negligibly small, assays were calibrated against a distilled water blank at 340 nm and run for 12 min at 30°C by addition of 10–20 μl of enzyme solution to 3.0 ml of assay mixture.

Lactate dehydrogenase assays were performed using the reverse reaction developed by Wroblewski and LaDue (42) in which the substrate, pyruvate, is reduced to lactate. The assay system contained 0.08 M potassium phosphate buffer, pH 7.5, 7.5×10^{-4} M sodium pyruvate, and 1.1×10^{-4} M NADH. Prepared reagents were purchased from Calbiochem. Serum was assayed in undiluted form; homogenate samples were diluted to 1.0 percent with modified Chappell-Perry buffer. Background error was predetermined, using a substrate-free control solution of 1.1×10^{-4} NADH (Calbiochem) in 0.08 M potassium phosphate buffer, pH 7.5. The analysis showed no detectable utilization of NADH, indicating no need for parallel control runs. Turbidity errors in serum and homogenate were found to be negligibly small, allowing assays to be calibrated at 340 nm against a distilled water blank. Assays were run for 2–3 min at 30°C, after adding 10–20 μl of enzyme solution to 3.0 ml of assay mixture.

Glutamate oxaloacetate transaminase was determined by the ultraviolet method of Karmen (22) on serum, homogenate and heart mitochondrial fractions. The assay system contained 0.1 M phosphate buffer, pH 7.4, 4.0×10^{-2} M L-aspartate, 1.2×10^{-2} M NADH, 0.25 mg of malate dehydrogenase/ml, 0.25 mg of lactate dehydrogenase/ml, and 0.25 α-oxoglutarate. Prepared reagents were purchased from Boehringer/Mannheim Corporation. To 3.2 ml of assay mixture was added 0.5 ml of enzyme solution, and the degradation of NADH was followed at 366 nm for 4–6 min at 25°C. Serum was assayed in undiluted form; homogenate samples, however, were diluted with buffer to 0.1 percent. Because of the impermeability of bilogical membranes to many substrates and cofactors (31), mitochondrial samples were ruptured by addition of 1 percent Triton X-100 in 0.1 M Tris-acetate buffer, pH 6.8 (1, 13, 39). Although this assay employs an ammonia-free system, possible contamination by glutamate dehydrogenase (GLDH) in homogenate and mitochondrial preparations was checked. Enzyme analysis by the method of Schmidt (35) revealed only negligible levels of the enzyme in rat heart. This has been further shown by Kochakian (24) *et al.* who found attempts to determine GLDH levels in mice and rat hearts yielded questionable results. Background degradation of NADH, using substrate-free controls for the GOT system, indicated an average error of less than 3 percent in serum. No detectable background activity was found in homogenate and mitochondrial preparations. Because of considerable serum turbidity, samples were calibrated against a serum blank containing 0.50 ml undiluted serum in 3.2 ml of 0.1 M phosphate buffer, pH 7.4. Because turbidity errors for homogenate and mitochondrial samples were negligible, transmittance was calibrated against a blank of 0.1 M phosphate buffer.

Protein determination

In order to relate the calculated mitochondrial enzyme activities to the concentration of mitochondria, determinations of total mitochondrial protein were obtained using the biuret method (25). To 0.4–0.8 ml of mitochondrial suspension, 0.6–0.2 ml of 5 percent deoxycholate (20) was added to remove turbidity. After 10 min 4 ml of biuret reagent (12) was added and the mixture was allowed to stand for 30 min at room temperature. Absorbance values were recorded at 550 nm against a blank containing 4 ml of biuret reagent with 1.0 ml of distilled water. Protein concentrations were obtained by reference to a calibration curve established by serial dilution of bovine serum albumin.

Data analysis

Reaction order kinetics for each enzyme system were determined previously and reaction rates were followed on a Beckman DB-GT spectrophotometer fitted with a Phillips model PM 8100 flat bed recorder. Reaction rate data was then fitted to the equation for the corresponding order, and rates were computed by a least squares fit using a Digital Equipment Company PDP-8 computer. Mean values and individual variances were determined for each sample group after subjecting the samples to a Z-test analysis. Application of the F-test for an analysis of variance between groups showed no significant differences in the population variances, and warranted the determination of a pooled estimate of variance (7). *t*-tests between control and experimental values were run for one-tailed, directional hypothesis, using the pooled estimate of variance (8).

RESULTS

Serum assays

In all cases serum enzyme levels reached maximal values within 24 hr after thoracic irradiation. Peak elevations occurred 6–12 hr postirradiation, after which was noted a general trend to return to normal levels. These results are in agreement with other studies involving whole-body irradiation (2, 3, 18, 19, 38).

Lactate dehydrogenase serum levels rose 139 percent above control levels 6 hr after exposure (table I). Statistical analysis showed this increase to be significant ($p < 0.05$). Twelve and 24 hr after exposure, serum LDH activity decreased, respectively, to 48 percent and 21 percent above control level. Although the 12-hr sample shows significant elevation, the 24-hr level is not statistically different from control level. On day 3 a second increase in serum LDH activity to 25 percent above control showed marginal significance at the 5 percent level, and returned to normal on days 6 and 9. This general pattern has also been observed by Hori and Nishio (17). A third transient increase to 79 percent above control activity occurred 12 days after irradiation. The mean LDH activity on day 15 does not show significant deviation from the control (table I).

Table I Serum levels of lactate dehydrogenase at various intervals after exposure to 2,000 rads of γ-radiation*

Group	No. samples	Mean activity $\times 10^3$ (mU/ml)	Significant difference from control ($p < 0.05$)	% Elevation from control
Control	5	1.31		
6 hr	6	3.14	+	139
12 hr	6	1.94	+	48
24 hr	5	1.59	−	21
3 day	4	2.00	+	52
6 day	4	1.27	−	(−)3.3
9 day	3	1.36	−	3.0
12 day	3	2.35	+	79
15 day	2	1.89	−	44

*Significant differences between control and test means were determined by t-test, using a pooled estimate of variance. Pooled variance: 356,561; pooled S.D.: 597.

Creatine-kinase serum activity rose 131 percent above the control value 6 hr after irradiation ($p < 0.05$). At 12 hr a 68 percent elevation was observed ($p < 0.05$), which decreased to within control level by 24 hr (table II). A secondary elevation to 110 percent above control activity occurred on day 3 ($p < 0.05$), but returned to normal by day 6. A third significant elevation occurred on day

Table II Serum levels of creatine kinase at various intervals after exposure to 2,000 rads of γ-radiation*

Group	No. samples	Mean activity (mU/ml)	Significant difference from control ($p < 0.05$)	% Elevation from control
Control	5	208		
6 hr	5	481	+	131
12 hr	6	350	+	68
24 hr	6	299	−	44
3 day	4	437	+	110
6 day	4	244	−	17
9 day	3	272	−	31
12 day	3	439	+	111
15 day	2	365	−	76

*Significant differences between control and test means were determined by t-test, using a pooled estimate of variance. Pooled variance: 13,812; pooled S.D.: 118.

Table III Serum levels of glutamate oxaloacetate transaminase at various intervals after exposure to 2,000 rads of γ-radiation*

Group	No. samples	Mean activity (mU/ml)	Significant difference from control	% Elevation from control
Control	6	49.2		
6 hr	6	75.5	+	54
12 hr	5	81.4	+	66
24 hr	6	53.5	–	9
3 day	4	47.9	–	(–)2.6
6 day	4	29.2	+†	(–)41
9 day	3	36.3	–	(–)26
12 day	3	48.9	–	(–)0.6
15 day	3	60.1	–	+22

*Significant differences between control and test means were determined by *t*-test, using a pooled estimate of variance. Pooled variance: 105; pooled standard deviation: 10.3.

†Where activity levels differed significantly from control levels and did not agree with the directional hypothesis, a nondirectional, two-tailed *t*-test was used to determine the level of significance.

12 which, as in the case of the serum LDH pattern, remained elevated at day 15. However, because of the small sample size of day 15, the elevation is not significantly different from the control.

Assays of serum GOT activities revealed an increase of 54 percent at 6 hr postirradiation ($p < 0.05$). Maximal elevation occurred after 12 hr, showing an activity increase of 66 percent ($p < 0.05$). After 24 hr serum GOT levels returned to normal (table III). An unexpected drop in activity to 59 percent of the control level on day 6 was not consistent with the directional hypothesis. However, since the probability value for a nondirectional test showed significance at the 5 percent level, a possible rationale for this observed decrease will be presented elsewhere in this chapter. Serum GOT levels for days 9 and 12 after exposure do not show significant deviations from control values. However, a general trend of elevation does occur from day 6 to day 15, showing definitive increases in enzyme levels. Such secondary increases, as noted in all enzyme systems studied, may be indicative of the manifestation of major biochemical lesions induced by the ongoing metabolism of the injured tissue cells.

HOMOGENATE ASSAYS

Lactate dehydrogenase activity in heart muscle homogenates (table IV) decreased 36 percent by 12 hr postirradiation ($p < 0.05$). This decrease remained statistically significant after 24 hr, showing a 33 percent mean loss, returning to

Table IV Heart homogenate levels of lactate dehydrogenase at various intervals after exposure to 2,000 rads of γ-radiation*

Group	No. samples	Mean activity (mU/mg muscle)	Significant difference from control ($p < 0.05$)	% Decrease from control
Control	3	457		
6 hr	3	365	+	20
12 hr	3	291	+	36
24 hr	3	305	+	33
3 day	3	462	–	(–)1.0
6 day	3	426	–	6.9
9 day	3	362	+	21
12 day	3	470	–	3.0
15 day	3	470	–	3.0

*Significant differences between control and test means were determined by t-test, using a pooled estimate of variance. Pooled variance: 2,414; pooled S.D.: 49.1.

normal after 3 days. Activity levels from 3 to 6 days after exposure displayed a general decreasing trend, however, were not statistically different with respect to the control activity. Day 9 showed a secondary loss of homogenate activity significantly different from the control at the 5 percent level. However, activities returned to normal by day 12.

Creatine kinase levels in heart homogenates (table V) reflected similar changes with an observed loss of 21 percent from control level at 12 hr postirradiation. Activity levels after 24 hr reflected a 25 percent mean loss which was not statistically different from the 12-hr measurement. Although enzyme levels returned to normal during day 3, a secondary drop in homogenate activity, occurring during days 6 and 9, showed statistically significant losses of 17 and 25 percent, respectively.

Glutamate oxaloacetate transaminase loss from homogenate preparation shows a pattern similar (table VI) to the previous system studied. The mean postirradiation activity loss after 12 hr is 30 percent ($p < 0.05$), but returns to within control limits by 24 hr ($p > 0.05$). Homogenate GOT levels from day 3 to day 15 fluctuate within control limits, and do not display statistically significant alterations with respect to control activities.

Mitochondrial assays

Loss of mitochondrial GOT activity, induced by an acute radiological dosage of 2,000 rads, is not significant at the 5 percent level of probability (table VII).

Table V Heart homogenate levels of creatine kinase at various intervals after exposure to 2,000 rads of γ-radiation*

Group	No. samples	Mean activity (mU/mg muscle)	Significant difference from control ($p < 0.05$)	% Decrease from control
Control	3	690		
6 hr	2	594	−	14
12 hr	3	545	+	21
24 hr	3	519	+	25
3 day	3	616	−	11
6 day	3	573	+	17
9 day	3	518	+	25
12 day	3	629	−	8.8
15 day	2	635	−	7.9

*Significant differences between control and test means were determined by *t*-test, using a pooled estimate of variance. Pooled variance: 2,712; pooled standard difference: 52.1.

Table VI Heart homogenate levels of glutamate oxaloacetate transaminase at various intervals after exposure to 2,000 rads of γ-radiation*

Group	No. samples	Mean activity (mU/mg muscle)	Significant difference from control ($p < 0.05$)	% Decrease from control
Control	3	132		
6 hr	3	106	−	20
12 hr	3	92.3	+	30
24 hr	3	114	−	13
3 day	3	142	−	(−)8.0
6 day	3	156	−	(−)18
9 day	2	107	−	19
12 day	3	134	−	(−)2.0
15 day	2	149	−	(−)13

*Significant differences between control and test means were determined by *t*-test, using a pooled estimate of variance. Pooled variance: 414; pooled standard deviation: 20.3.

Table VII Mitochondrial levels of glutamate oxaloacetate transaminase at various intervals after exposure to 2,000 rads of γ-radiation*

Group	No. samples	Mean activity $\times 10^{-3}$ (mU/mg buiret protein)	Significant difference from control ($p < 0.05$)	% Decrease from control
Control	3	1.31		
6 hr	3	0.87	–	36
12 hr	3	1.09	–	17
24 hr	3	1.14	–	13
3 day	3	1.64	–	(–)25
6 day	3	1.34	–	(–)2.0
9 day	3	1.49	–	(–)13
12 day	3	1.11	–	16
15 day	3	1.30	–	8.0

*Significant differences between control and test means were determined by *t*-test, using a pooled estimate of variance. Pooled variance: 133,812; pooled standard deviation: 366.

Although a mean activity loss of 36 percent occurred after 6 hr postirradiation, the large sample variance limits the precision of measurement and may obscure the significance of such small alterations.

DISCUSSION

Although the heart has often been regarded as one of the most radioresistant organs of the body (5), the onset of radiation-induced heart disease has been recognized as an important manifestation of thoracic exposure (37). The present investigation documents biochemical evidence of initial cardiac injury in response to the acute administration of ionizing radiation to the thoracic area of rats.

One of the biochemical indicators in acute radiation injury is an elevation of serum lactate dehydrogenase activity after whole-body exposure (18). Lactate dehydrogenase, located in most organs, has an important function in carbohydrate metabolism, and under normal conditions remains at a fairly constant level. Glutamate oxaloacetate transaminase and creatine kinase are both found in relatively high concentrations in cardiac muscle. GOT, involved in the mediation of protein metabolism, and CK in the restoration of available energy for muscular contraction, play important roles as biochemical markers for the clinical detection of myocardial injury (9, 35), and were therefore selected as potential indicators for the study of radiation injury to cardiac muscle.

Serum levels of all three enzymes revealed transient elevations within 12 hr of thoracic irradiation, returning to within control levels by 24 hr. Similar response patterns have been documented in whole-body exposure (1, 3, 19, 38), indicating that such patterns appear to be representative of a general radiological response and are not related to the radiosensitivity of the system studied. Lactate dehydrogenase and creatine kinase serum levels displayed greater than 2-fold initial elevations while GOT showed an increase of only 50 percent above normal. Although these elevations are significant, they are not remarkable. Because of the smallness of the changes and the sample variance within a population, clinical application of these systems as biochemical markers for radiation injury to the cardiac muscle does not appear likely unless one had preirradiation serum levels on the subject. In addition, studies by Brent and coworkers (4) reveal that no apparent correlation exists between induced serum elevations and the likelihood of death.

All serum enzyme patterns indicate a trend toward normality after reaching initial maximal levels. Such a pattern suggests a possible reinstatement of membrane integrity after a transient alteration of membrane permeability. However, the effect might equally well be mediated by a lowering of the efflux rate of enzymatic material from the cell because of a decrease of the concentration gradient between cellular and extracellular levels caused by the depletion of the cellular constituents. Such factors, possibly operating in concert, combined with the rapid removal of the extracellular enzymes through active uptake in the extracellular fluid (35) may account for a "pseudo" transient nature of the initial elevations. Note that the restoration of tissue enzyme levels to control values on day 3 (possibly attributable to enhanced synthesis of the enzymes) tends to restore the concentration gradient. This could lead subsequently to enhanced leakage and a secondary elevation of serum LDH and CK if the normal permeability of the cell membrane had not yet been restored. Although tissue homogenate GOT levels returned to normal on day 3, the serum level of GOT, unlike LDH and CK, did not increase significantly above normal at that time. This may indicate a significant restoration of membrane permeability with respect to GOT. Some overcompensatory repair mechanism for GOT permeability, whether at the cellular or hormonal level, or an enhanced inactivation of the serum enzyme could conceivably account for the noted transient reduction in serum GOT levels as shown in table III.

Trends in the serum enzyme levels occurring after initial elevations suggest the possible manifestation of secondary radiation effects resulting from the formation of metabolic or biochemical lesions in the target organ. However, a strict interpretation of the serum elevations with respect to change to the heart must be approached with extreme caution since radiation damage to peripheral tissues cannot be eliminated. In addition, physiological alterations mediated by hormonal action may further distort the relationship.

The significant decreases in the homogenate levels of the enzymes studied parallel the initial serum elevations, indicating that the observed serum elevations are not a result of enhanced tissue levels of the enzymes, but rather are a consequence of radiation-induced leakage of biologically active molecules in post-irradiated cells and organelles have been recorded (23, 28, 33). Secondary serum elevations, however, do not appear to be related to any significant secondary depletions of homogenate activities. It is unlikely that direct inactivation of the enzymes could account for the noted initial decreases in the homogenate activities since it has been demonstrated that most enzymes are relatively radioresistant (29, 34, 40), requiring substantially larger doses to initiate any significant inactivation.

It appears, at the biochemical level, that *in vivo* injury to the mitochondria with 2,000 rads of γ-radiation is not significant at the 5 percent confidence level. The apparent resistance of the mitochondrial membrane to damage, in comparison with the cell membrane, may result from the double membrane character of the mitochondria or physiochemical differences in the structure of the membrane systems. It is, moreover, probable that the GOT concentration gradient between mitochondria and cytoplasm is not nearly as great as the cytoplasma gradient. Thus, the motive force and resultant leakage of the enzyme from the damaged mitochondria would not reflect as accurately the injury sustained from radiation damage to the organelle.

Acknowledgments We are grateful to Dr. R. Cornelson of the British Columbia Cancer Institute for providing radiation facilities and to Mr. T. Legault for his help in the statistical analysis.

REFERENCES

1. ABOOD, L. G., and ALEXANDER, L. 1957. Oxidative phosphorylation by an enzyme complex from disrupted brain mitochondria. J. Biol. Chem. 227: 717–725.
2. ALMONTE, T. J., BARNES, J. H., BAUTISTIA, S. C., DE LACRUZ, B., LANSANGAN, L., and PHINEDA, P. A. 1965. Serum transaminase in mice after whole-body x-irradiation and the effect of a chemical protector. Int. J. Radiat. Biol. 9: 37–42.
3. BECKER, R., WILLIAMS, R. B., JR., VOOGD, J., DOWLING, J., and ISTOCK, J. 1962. Studies of the effect of total body x-irradiation upon the level of serum glutamic oxalacetic-transaminase. Radiat. Res. 14: 450.
4. BRENT, R. L., McLAUGHLIN, M. M., and STABILE, J. N. 1958. The effects of irradiation on the serum glumatic-oxalacetic transaminase level. Radiat. Res. 9: 24–32.
5. BURCH, G. E., SOHAL, R. S., SUN, S., MILLER, G. C., and COLCOLOUGH, H. L. 1968. Effects of radiation on the human heart: An electron microscopic study. Arch. Intern. Med. 121 (3): 230–234.
6. CHAPPELL, J. B., and PERRY, S. V. 1954. Biochemical and osmotic properties of skeletal muscle mitochondria. Nature 173: 1094–1095.
7. DIXON, W. J., and MASSEY, F. J., JR. 1957. Introduction to Statistical Analysis, Ed. 2, pp. 109–110. McGraw-Hill, New York.

8. DOWNIE, N. M., and HEATH, R. W. 1970. Basic Statistical Methods. Ed. 3, pp. 183–185. Harper & Row, New York.
9. DREYFUS, J. C., SCHAPIRA, G., RESNAIS, J., and SCEBAT, L. 1960. La creatin kinase serique dans le diagnostic de l'infarctus myocardique. Rev. Franc. Etud. Clin. Biol. 5: 386.
10. ERNSTER, L., and NORDENBRAND, K. 1967. Skeletal muscle mitochondria, pp. 86–94. *In* R. W. Estabrook and M. E. Pullman (eds.), Methods of Enzymology, Vol. 10. Academic Press, New York.
11. FAJARDO, L. F., STEWART, J. R., and COHN, K. E. 1968. Morphology of radiation-induced heart disease. Arch. Pathol. 86(5): 512–520.
12. GORNALL, A. G., BARDAWILL, C. J., and DAVID, M. M. 1949. Determination of serum proteins by means of the biuret reaction. J. Biol. Chem. 177: 751–766.
13. GREVILLE, G. D., and CHAPPELL, J. B. 1959. The latent rhodanese of isolated rat liver mitochondria. Biochim. Biophys. Acta. 33: 267–269.
14. HAWRYLEWICZ, E. J., and BLAIR, W. H. 1966. Effect of gamma and proton irradiation on lactic dehydrogenase isoenzymes. Radiat. Res. 28: 538–547.
15. HEDMAN, R. 1965. Properties of isolated skeletal-muscle mitochondria. Exp. Cell Res. 38: 1–12.
16. HOLTON, F. A., HULSMANN, W. C., MYERS, D. K., and SLATER, E. C. 1957. A comparison of the properties of mitochondria isolated from liver and heart. Biochem. J. 67: 579–594.
17. HORI, Y., and NISHIO, K. 1962. Effects of x-irradiation on plasma lactic dehydrogenase activity in rabbit and mouse. Annual Report, Radiation Centre, Osaka Prefecture 3: 130–132.
18. HORI, Y., TAKAMORI, Y., and NISHIO, K. 1968. The effect of x-irratiation on the lactate dehydrogenase level in plasma and in various organs of mice. Radiat. Res. 34: 411–420.
19. HORI, Y., TAKAMORI, Y., and NISHIO, K. 1970. The effects of x-irratiation on lactate dehydrogenase isoenzymes in plasma and in various organs of mice. Radiat. Res. 43: 143–151.
20. JACOBS, E. E., JACOB, M., SANADI, D. R., and BRADLEY, L. B. 1956. Uncoupling of oxidative phosphorylation by cadmium ion. J. Biol. Chem. 223: 147–170.
21. KARCHER, K. N. 1971. Enzymological examinations: An indication of radiation effects in experiments and clinical practice, pp. 277–283. *In* D. H. Hill (ed.), Biochemical Indicators of Radiation Injury in Man, Proceedings of the International Atomic Energy Agency. Unipublishers, New York.
22. KARMAN, A. 1955. A note on the spectrophotometric assay of glutamic-oxalacetic transaminase in human serum. J. Clin. Invest. 34: 131–133.
23. KAWASAKI, S., and SAKURAI, K. 1969. Effects of x-irradiation on mouse liver mitochondria (Part 1). Nippon ACTA Radiol. (Nippon Soc. Radiol) 28(*12*): 1672–1676.
24. KOCHAKIAN, C. D., ENDAHL, B. R., and ENDAHL, G. L. 1959. Influence of androgens on the transaminases of glutamic dehydrogenase of tissues. Amer. J. Physiol. 197: 129–134.
25. LAYNE, E. 1957. Spectrophotometric and turbidimetric methods for measuring proteins, pp. 447–454. *In* S. P. Colowick and N. O. Kaplin (eds.), Enzymology, Vol. 8. Academic Press, New York.
26. MILCH, L. J., and ALBAUM, H. G. 1956. Serum transaminase activity in x-irradiated rabbits. Proc. Soc. Exp. Biol. 93(3): 595–596.
27. MUGGIA, F. M. 1970. Creatine phosphokinase and other serum enzymes during radiotherapy. J.A.M.A. 211: 1345–1350.
28. OKADA, S., and PEACHEY, L. S. 1957. Effect of gamma irradiation on the desoxyribonuclease II Activity of isolated mitochondria. J. Biophysic. and Biochem. Cytol. 3(2): 239–251.
29. ORD, M. G., and STOCKEN, L. A. 1953. Biochemical aspects of the radiation syndrome. Physiol. Rev. 33: 356–386.

conversion from aerobic to anaerobic metabolism. Despite the acceleration of glycogenolysis and glycolysis which occurs when the heart is exposed to hypoxic conditions, these metabolic pathways are less efficient than oxidative phosphorylation in providing the energy requirements of the myocardium. To improve cardiac function during hypoxia and/or after reoxygenation, it is necessary to increase substrate supply and to enhance the utilization of available substrates. Scheurer and Stezoski (22), using an isolated rat heart preparation, demonstrated that hearts of animals pretreated with reserpine contained higher levels of cardiac glycogen, were associated with greater glycolytic production of ATP, and exhibited improved cardiac function during hypoxia. On the other hand, a recent report from our laboratory demonstrated that enhancement of energy utilization by the hypoxic rat heart can be seen when the polypeptide, glucagon, is added to the perfusion medium during the hypoxic insult (3). The addition of glucagon results in improved recovery from hypoxia which was suggested to be related to enhanced glycogenolysis via the activation of phosphorylase. This activation of posphorylase is secondary to an increased intracellular concentration of cyclic AMP, which has itself been implicated as a mediator of the positive inotropic response (16).

Recent evidence (8, 9) has also implicated another cyclic nucleotide, cyclic GMP, as a fundamental chemical mediator in the heart. The action of cyclic GMP may be related to induced negative inotropism as seen with cholinergic agents or possibly other cardiac depressants.

Several laboratories (4, 8, 9, 11, 25) have reported that both physiological and pathological changes in cardiac function may be mediated by the relative intracellular concentrations of cyclic AMP and cyclic GMP. In view of these findings, this present investigation was undertaken to determine whether the protective effect of glucagon on the isolated perfused hypoxic rat heart was associated with any changes in the myocardial levels of these two cyclic nucleotides.

METHODS

Male Sprague-Dawley rats (250–300 gm) were decapitated and their hearts rapidly removed and attached to an Anderson perfusion apparatus by tying the outflow of the perfusion apparatus into the aorta. Care was taken to insure that there was no obstruction of the coronary ostia. Coronary perfusion was conducted with modified Tyrode's solution equilibrated with 95 percent O_2 and 5 percent CO_2 (pO_2 450–500 mm Hg) and maintained at 37°C and pH 7.4. An hypoxic perfusion medium (pO_2 10–20 mm Hg, pH 7.4) was produced by equilibrating a similar Tyrode's solution with 95 percent N_2 and 5 percent CO_2 for 30 min. Both oxygenated and hypoxic perfusion media were set in parallel systems so that with the turn of a stopcock, either medium could be used selectively for immediate coronary perfusion. The ionic composition of the

Tyrode's solution in millimoles per liter was NaCl, 120; KCl, 5; $CaCl_2$, 4; $MgSO_4$, 30; and KH_2PO_4, 0.6. All hearts were perfused with an open end system so that none of the perfusate was recirculated. Aortic filling pressure was maintained at 70 cm of H_2O with an average coronary flow of 7 ml/min. Force of contraction was measured with a calibrated Grass Ft 0.03 force displacement transducer, (Grass Instruments, Quincy, Mass.), connected to the heart by a silk suture attached to a hook placed into the apex. The transducer was mounted on a movable assembly in such a way that tension could be placed on the hearts by movement of the transducer. In this manner, sufficient tension was placed on the heart to place the contraction at the peak of its length-tension curve. Force and rate of contraction were recorded on a Grass model 7 polygraph. Heart rate was maintained constant at 20 percent above the intrinsic rate by square wave stimuli (4–8 V, 4 msec duration) delivered from an E & M stimulator through hook electrodes placed on the apex and right atrium.

Rat hearts were divided into two major groups: 1) those perfused with glucose-free Tyrode's solution and 2) those perfused with Tyrode's solution containing 5 mM glucose. Rat hearts in each group were equilibrated for a period of 10 min with oxygenated Tyrode's solution, then perfused for 3 min with hypoxic medium, and finally reperfused for 5 or 10 min with oxygenated Tyrode's. This protocol was followed in both glucose-free and glucose-containing perfusions. Experimental hearts were treated in exactly the same manner as described above except that during the period of hypoxia, they were perfused with a subinotropic concentration of glucagon (8.7×10^{-9} M). Glucagon was only present in the hypoxic perfusion medium. Preliminary studies in this laboratory indicate that concentrations of glucagon at 8.7×10^{-9} M or lower do not exert an inotropic effect in the normal rat heart.

In an additional series of experiments, hearts equilibrated with glucose-free Tyrode's solution were perfused with an hypoxic medium containing glucagon in concentrations ranging from 4.3×10^{-8} M to 8.7×10^{-7} M. Recovery of force of contraction during the period of reoxygenation was determined.

In a parallel series of experiments, instead of allowing the hearts to recover during reoxygenation, they were quickly frozen at 15 sec, 60 sec, and 3 min of hypoxic perfusion with Wollenberger clamps that had been precooled in liquid nitrogen. The hearts were then stored at –80°C until assayed for their content of cyclic AMP and cyclic GMP. The analyses for cyclic AMP and cyclic GMP were according to the methods employed previously (9). All values are expressed as nanomoles of cyclic nucleotide per kg of heart tissue.

RESULTS

The results of the perfusion of rat hearts with glucose-free hypoxic Tyrode's solution are presented in figure 1. Within 3 min of hypoxic perfusion, there is a rapid dimunition in force of contraction to approximately 10 percent of control.

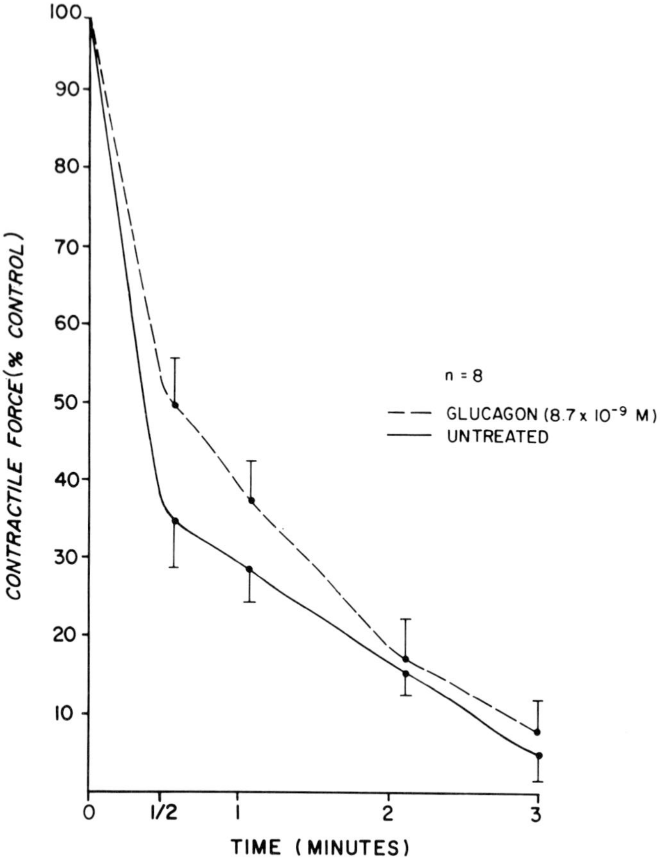

Figure 1 Rate of decline in contractile force during exposure to hypoxia in untreated and glucagon-perfused isolated rat hearts. The *abscissa* represents the time during which the hearts were perfused with glucose-free hypoxic medium as described under Methods. A subinotropic dose of glucagon (8.7×10^{-9} M) was included in the hypoxic perfusion medium where indicated.

The most rapid decline occurs within 30 sec with a less marked but yet progressive decrement from 30 sec to 3 min. During the hypoxic period both untreated hearts and hearts perfused with the subinotropic dose of glucagon (8.7×10^{-9} M) exhibited a similar rate and degree in decline of contractile force.

Figure 2 depicts the recovery of force of contraction of rat hearts perfused with glucose-free Tyrode's solution during a 5-min period of reoxygenation that followed the 3 min of hypoxic insult. Hearts perfused with glucagon (8.7×10^{-9} M) during the initial hypoxic period exhibited a significantly enhanced recovery during the reoxygenation period at all time intervals measured ($p < 0.05$). The rate of recovery in glucagon-perfused hearts was significantly greater than untreated hearts. These glucagon-perfused hearts regained 38 percent of original contractile force at 30 sec, whereas hearts not perfused with glucagon attained only 24 percent of their original contractile force in this same time interval. Furthermore, glucagon-perfused hearts generated greater contractile force at all time intervals measured. Maximum force of contraction was seen between 2 and 3 min in both untreated and glucagon-perfused hearts.

The loss of force of contraction when both untreated hearts and glucagon-perfused (8.7×10^{-9} M) hearts were exposed to glucose-containing hypoxic

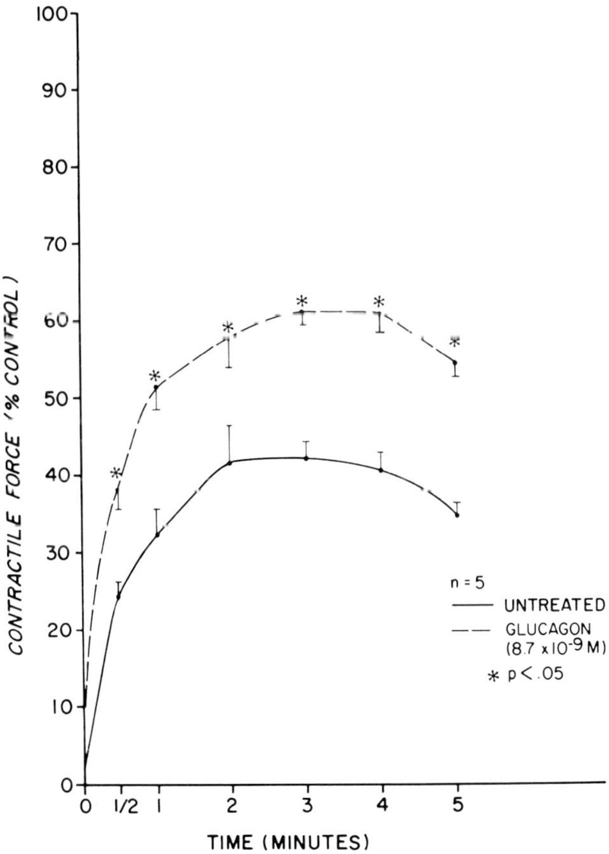

Figure 2 Recovery of contractile force after exposure to hypoxia. The abscissa represents the time of reoxygenation of rat hearts after 3 min of perfusion with glucose-free hypoxic medium. When included in the hypoxic medium, glucagon, at a subinotropic concentration of 8.7×10^{-9} M afforded a significant increase ($p < 0.05$) in recovery of contractile force at each interval.

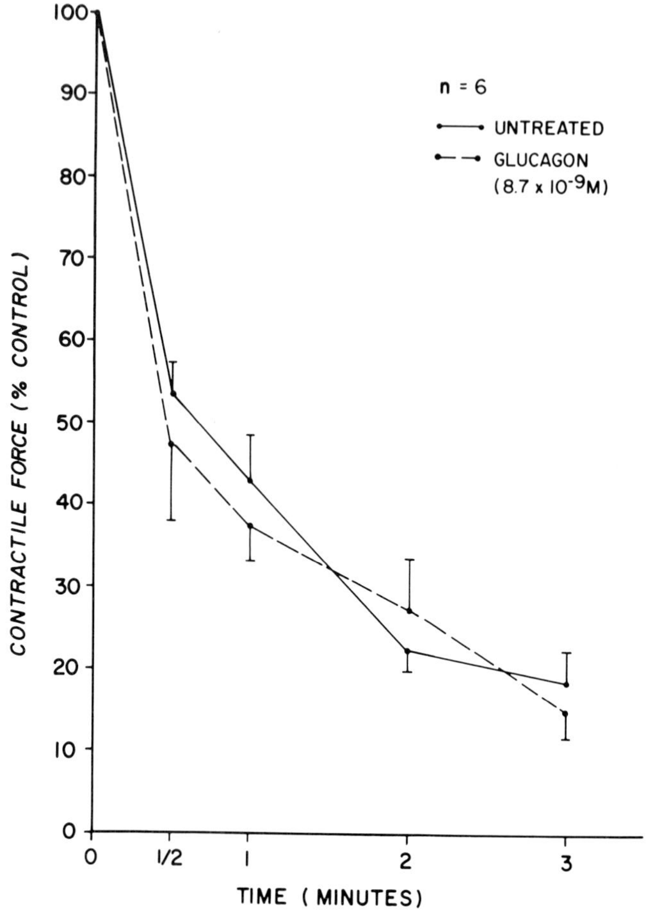

Figure 3 Rate of decline of contractile force of glucose-perfused hearts during exposure to hypoxia. The abscissa represents the time during which the hearts were perfused with hypoxic Tyrode's solution containing 5 mM glucose. A subinotropic dose of glucagon (8.7×10^{-9} M) was included in the hypoxic perfusion where indicated.

Tyrode's solution is shown in figure 3. As in the experiments performed with glucose-free medium, glucagon was administered only during the hypoxic period. Although the depression in contractile force was still characterized by a rapid and slow phase similar to the glucose-free perfusions (fig. 1), the extent of the depression in contractility was not as great. At 3 min of hypoxic perfusion with glucose-containing Tyrode's solution, hearts attained approximately twice the force of contraction seen in glucose-free perfusions. There was no difference between the performance of untreated and glucagon-perfused hearts during the 3 min of hypoxia.

Figure 4 shows the recovery of contractile force of glucose-perfused hearts. Hearts perfused with the subinotropic concentration of glucagon (8.7×10^{-9} M) during the initial hypoxic period performed better throughout the period of reoxygenation than untreated hearts. Maximum force of contraction occurred at 5 min and was 68 percent and 105 percent of control for untreated and glucagon-perfused hearts, respectively.

The effects of glucagon on cardiac cyclic nucleotide levels and contractile force after hypoxia in rat hearts perfused without glucose are shown in table I. Control hearts were equilibrated for 13 min and then frozen rapidly. All other hearts were frozen after a 3-min period of hypoxia which followed 10 min of equilibration. Experimental hearts were perfused during the hypoxic period with glucagon in concentrations ranging from a subinotropic dose (8.7×10^{-9} M) to 8.7×10^{-7} M. Hypoxic hearts regained only 35 percent of contractile force at 5 min after reoxygenation. In contrast, hypoxic hearts perfused with glucagon in

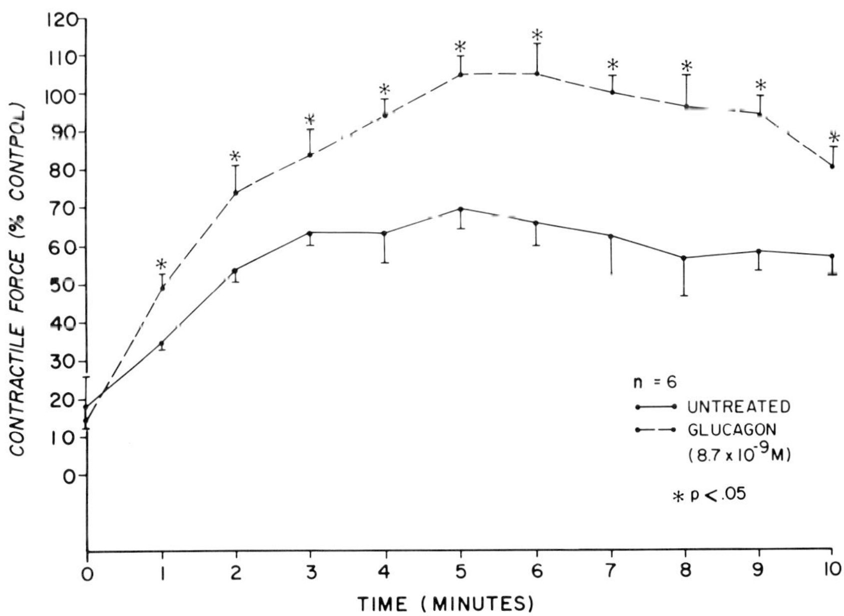

Figure 4 Recovery of contractile force after exposure to hypoxia. The *abscissa* represents the time of reoxygenation of rat hearts after 3 min of perfusion with glucose-containing hypoxic medium. When included in the hypoxic medium, glucagon at a subinotropic concentration of 8.7×10^{-9} M afforded a significant increase ($p < 0.05$) in recovery of contractile force at each interval.

Table I Effects of glucagon on cardiac cyclic nucleotide levels and contractile force after hypoxia

Experimental conditions	*n*	Contractile force (% control at 5 min reoxygenation	Cyclic GMP (nanomoles/kg tissue before reoxygenation, with 3 min hypoxia)	Cyclic AMP (nanomoles/kg tissue before reoxygenation)
Control	8	100 ± 5	235 ± 18	497 ± 31
Hypoxia alone	8	35 ± 1*	372± 20*	503 ± 29
Hypoxia + glucagon (8.7×10^{-9} M)	5	60 ± 3* †	70 ± 7* †	470 ± 13
Hypoxia + glucagon (4.3×10^{-8} M)	5	61 ± 7* †	90 ± 10* †	503 ± 46
Hypoxia + glucagon (8.7×10^{-8} M)	5	53 ± 5* †	130 ± 16* †	500 ± 39
Hypoxia + glucagon (4.3×10^{-7} M)	5	53 ± 8* †	125 ± 2* †	606 ± 6* †
Hypoxia + glucagon (8.7×10^{-7} M)	5	26 ± 6*	298 ± 34	737± 91

*Significantly different from control ($p < 0.05$).
†Significantly different from hypoxia alone ($p < 0.05$). Each value represents the mean ± S.E. of the mean.

concentrations from 8.7×10^{-9} M to 4.3×10^{-7} M attain a contractile force of 53–61 percent of control at this same time period of reoxygenation. This enhanced recovery seen in glucagon-perfused hearts was associated with a 40–75 percent decrease in cyclic GMP levels at the time immediately before reoxygenation (3 min of hypoxia) when compared with hypoxic hearts not perfused with glucagon. Glucagon at a concentration of 8.7×10^{-7} M exerted no significant protective effect on cardiac recovery, and cyclic GMP levels were not significantly different from hypoxic hearts perfused without glucagon. There were no significant differences in the cyclic AMP concentrations of untreated hypoxic hearts and glucagon-perfused hypoxic hearts except at the concentration of 4.3×10^{-7} M. Ratios between the levels of cyclic AMP and cyclic GMP were determined. At 3 min of hypoxia this ratio was found to be 1.4 in hearts not perfused with glucagon and 5.2 in hearts that were perfused with glucagon. The change in the ratio was apparently due to the large change in cyclic GMP concentrations.

Cyclic nucleotide levels in hearts perfused with Tyrode's solution containing both glucagon and glucose during hypoxia are shown in table II. In untreated hearts exposed to 15 sec, 60 sec, or 3 min of hypoxia, cyclic GMP levels were found to be 125, 198, and 123 nmoles/kg of tissue, respectively. However, in hearts perfused with the subinotropic concentration of glucagon during hypoxia,

Table II Effect of glucagon on cyclic nucleotide levels in glucose-perfused rat hearts during hypoxia*

Experimental conditions	n	Cyclic AMP (nanomoles/kg tissue)	Cyclic GMP (nanomoles/kg tissue)
Control	8	500 ± 120	84 ± 5
Hypoxia 15 sec	8	884 ± 106†	125 ± 5†
Hypoxia 60 sec	7	899 ± 92 †	198 ± 27†
Hypoxia 3 min	8	481 ± 73	123 ± 10†
Hypoxia 15 sec with glucagon (8.7×10^{-9} M)	6	630 ± 111	88 ± 6‡
Hypoxia 60 sec with glucagon (8.7×10^{-9} M)	6	538 ± 46‡	82 ± 13‡
Hypoxia 3 min with glucagon (8.7×10^{-9} M)	5	530 ± 36	79 ± 7‡

*Control hearts equilibrated with oxygenated Tyrode's solution. Each value represents the mean ± S.E. of the mean.

†Significantly different from control ($p < 0.05$).

‡Significantly different from untreated hypoxic hearts at respective time interval ($p < 0.05$).

the levels of cyclic GMP were 88, 82, and 79 nmoles/kg at 15 sec, 30 sec, and 3 min, respectively. These levels were not significantly different from the control value of 84 nmoles/kg. Levels of cyclic AMP in untreated hearts were 884, 899, and 481 nmoles/kg for the same periods of hypoxia. Although glucagon-perfused hearts exhibit increases in the levels of cyclic AMP, they were not as large. The ratio of cardiac cyclic AMP to cyclic GMP levels during glucagon-perfusion was approximately 1–1.5 times greater than that obtained from hearts not perfused with glucagon (table I). This increase in the cyclic AMP to cyclic GMP ratio was also primarily attributable to a decrease in the level of cyclic GMP.

DISCUSSION

The results of these studies demonstrate that perfusion of the isolated rat heart with glucagon during hypoxia enhances the recovery of contractile force during reoxygenation. This improved recovery during reoxygenation was seen in hearts which were perfused both with and without glucose. However, the glucose-free hearts, as a whole, did not perform as well as the hearts perfused with Tyrode's solution containing glucose. Perhaps an explantion of this is that the only source of energy production available to glucose-free hearts was endogenous glycogen and lipids. It has been clearly shown that supplying isolated hearts with exogenous carbohydrate maintains myocardial levels of high energy phosphates (12). Thus, in the present studies the availability of glucose in the hypoxic medium could result in improved cardiac performance.

The precise mechanism by which glucagon enhances recovery during reoxygenation is unclear. Although glucagon has been shown to produce positive inotropic and chronotropic effects in isolated papillary muscle (10), closed chested dog preparations (19), and cats with chronic heart failure (18), our present studies suggest that the protective effect of this hormone is unrelated to its direct inotropic action. The enhanced recovery after hypoxia which was observed with several concentrations of glucagon was more sustained at a subinotropic concentration. It should be pointed out that when inotropic concentrations of glucagon between 4.3×10^{-8} M and 4.3×10^{-7} M were used, a higher maximal force of contraction was obtained after reoxygenation. However, this increase was followed by a more rapid decline in performance than that seen in the hearts treated with a subinotropic concentration of glucagon. This rapid decline in force of contraction seen with the higher doses of glucagon may be the result of a rapid depletion of energy-producing substrates in response to an increased metabolic demand of the heart during positive inotropism.

A possible mechanism for the protective effect of glucagon is that this hormone stimulates glycolysis and glycogenolysis, providing additional energy to the hypoxic myocardium at a critical period. Indeed, it has been reported previously (6) that glycogen stores are not completely depleted when the rate of glycogenolysis is significantly reduced. Glucagon stimulation of glycolysis in the perfused rat heart has been reported (15). However, the maintenance of accelerated glycolysis *in vivo* may not occur since a decrease in intracellular pH, secondary to lactate accumulation, would be inhibitory to phosphofructokinase activity (5).

Based on the present cyclic nucleotide data, an alternate mechanism whereby glucagon protects the myocardium might be suggested. This mechanism involves a critical modulation in the cardiac levels of cyclic AMP and cyclic GMP by glucagon. A role for cyclic AMP in mediating the positive inotropic effect of catecholamines (23) and glucagon (17) has been suggested. On the other hand, increased levels of cyclic GMP have been associated with actions of negative inotropic agents (8, 9). A number of studies have indicated that reciprocal changes in the levels of both of these nucleotides may mediate the physiological effects of adrenergic and cholinergic agents on the heart (8, 9, 11). Additionally, it has been reported that cardiac cyclic AMP levels are elevated during the systolic phase of the cardiac cycle (1), whereas cyclic GMP levels are lowered during this period (25).

Preliminary studies have shown that in isolated perfused hearts, hypoxia produces an increased release of lysosomal enzymes that is related to a lowering in the cyclic AMP to cyclic GMP concentration ratio (2). This finding is relevant since it has been demonstrated that a selective discharge of lysosomal enzymes from neutrophils can be directly influenced by cyclic GMP and cyclic AMP (13). Thus it is possible that the relative concentrations of cyclic AMP and cyclic GMP influence the response of the myocardium to various pathological stimuli such as ischemia and hypoxia. Our data strongly support this hypothesis in that the improved recovery during reoxygenation seen in glucagon-perfused hearts was associated consistently with a higher ratio in the relative concentrations of cyclic AMP to cyclic GMP during the hypoxic insult. Both untreated and glucagon-perfused hearts had a cyclic nucleotide ratio of about 7:1 at the beginning of the hypoxic insult. After 3 min of hypoxic perfusion, the ratio of cyclic AMP to cyclic GMP in untreated hearts decreased to 4:1. This finding is in contrast to glucagon-perfused hearts which maintained a ratio of 7:1. Furthermore, although the concentrations of both cyclic nucleotides were changed in response to hypoxia, it is clearly apparent that glucagon-perfused hearts did not show an increase in cyclic GMP levels in response to hypoxia as did the untreated hearts. It appears to be primarily this reduction in cyclic GMP accumulation which maintained the cyclic AMP to cyclic GMP ratio of 7:1 in glucagon-perfused hearts.

Although it has been well documented (17) that glucagon is capable of activating adenyl cyclase to form cyclic AMP, there have been few reports in the literature regarding the role of this polypeptide in the regulation of cyclic GMP accumulation. However, this latter possibility must be considered. It has been proposed by Illiano *et al.* (14) that there may exist an insulin receptor interaction at the cell membrane which might transform the substrate specificity of adenyl cyclase into a form which expresses specificity for GTP. Thus, the rate of synthesis of both cyclic AMP and cyclic GMP would be determined by the relative occupancy of receptors that favor the ATP form of the enzyme over the GTP form, respectively. Such a regulation in syntheses of cyclic nucleotides would support reciprocal relationships existing between the concentrations and actions of cyclic AMP and cyclic GMP (11).

The results of these studies are in excellent agreement with the "yin-yang" theory of Goldberg (11) in which it has been suggested that a "dualism" between the two cyclic nucleotides occurs after cellular events induced by drugs or conditions, which in a broad sense, modulate the tissue accumulation of cyclic GMP or cyclic AMP. The glucagon administration or hypoxic perfusion may indeed represent two interacting stimuli that exert their effects via such a modulation in cardiac cyclic nucleotide levels. Further studies will be necessary to clarify the precise biochemical mechanism by which this modulation of the intracellular levels of cyclic AMP and cyclic GMP occurs.

REFERENCES

1. BROOKER, G. 1973. Change in myocardial adenosine 3′,5′-cyclic monophosphate during the contraction cycle, pp. 207–212. *In* N. S. Dhalla (ed.), Recent Advances in Studies on Cardiac Structure and Metabolism, Vol. 3. University Park Press, Baltimore.
2. BUSUTTIL, R. W., IGNARRO, L., and GEORGE, W. J. 1975. The protective effect of methylprednisolone sodium succinate on the isolated perfused hypoxic rat heart.
3. BUSUTTIL, R. W., PADDOCK, R. J., and GEORGE, W. J. 1974. Protective effects of glucagon on the isolated perfused rat heart. Clin. Res. 22: 54.
4. BUSUTTIL, R. W., PADDOCK, R. J., and GEORGE, W. J. 1974. Effects of glucagon on cyclic nucleotide levels and cardiac performance in isolated perfused rat hearts following hypoxia. Circ. Res. 33: 344.
5. COOPER, E. G., BOOKER, W. M., and JOHNSON, J. B. 1973. Pharmacological interventions in experimental myocardial infarction, pp. 831–847. *In* N. S. Dhalla (ed.), Recent Advances in Studies on Cardiac Structure and Metabolism, Vol. 3. University Park Press, Baltimore.
6. CORNBLATH, M., RANDLE, P. J., PARMEGGINI, A., and MORGAN, H. E. 1963. Effects of glucagon and anoxia on lactate production, glycogen content and phosphorylase activity in the perfused isolated rat heart. J. Biol. Chem. 238: 1592–1597.
7. DHALLA, N. S., YATES, H. C., WALZ, D. A., McDONALD, V. A., and OLSEN, R. E. 1972. Correlation between changes in the endogenous energy stores and myocardial function due to hypoxia in the isolated perfused rat heart. Canad. J. Physiol. Pharmacol. 50: 333–345.
8. GEORGE, W. J., POLSON, J. B., O'TOOLE, A. G., and GOLDBERG, N. D. 1970. Evaluation of guanosine 3′,5′ cyclic phosphate in rat hearts after perfusion with acetylcholine. Proc. Nat. Acad. Sci. U. S. A. 66: 398–403.
9. GEORGE' W. J., WILKERSON, R. D., and KADOWITZ, P. J. 1973. Influence of acetylcholine on contractile force and cyclic nucleotide levels in the isolated perfused rat heart. J. Pharmacol. Exp. Therap. 184: 228–235.

10. GLICK, G., PARMLEY, W. W., WECHSLER, A. S., and SONNENBLICK, E. H. 1968. Glucagon–its enhancement of cardiac performance in the cat and dog and persistence of its inotropic action despite Beta-receptor blockade with propranolol. Circ. Res. 22: 789–799.
11. GOLDBERG, N. D., O'DEA, R. F., and HADDOX, M. K. 1973. Cyclic GMP. Advan. Cyclic Nucl. Res. 3: 155–223.
12. HEARSE, D. J., and CHAIN, E. B. 1973. Effect of glucose on enzyme release from, and recovery of, the anoxic myocardium, pp. 763–772. *In* N. S. Dhalla (ed.), Recent Advances in Studies on Cardiac Structure and Metabolism, Vol. 3. University Park Press, Baltimore.
13. IGNARRO, L. J., and GEORGE, W. J. 1974. Hormonal control of lysosomal enzyme release from human neutrophils: Elevation of cyclic nucleotide levels by autonomic neurohormones. Proc. Nat. Acad. Sci. U. S. A.
14. ILLIANO, G., TELL, G. P. E., SIEGEL, M. I., CUATRECASAS, P. 1973. Guanosine 3',5'-cyclic monophosphate and the action of insulin and acetylcholine. Proc. Nat. Acad. Sci. U. S. A. 70: 2443–2447.
15. KRIESBERG, R. A., and WILLIAMSON, J. R. 1964. Metabolic effect of glucagon on the perfused rat heart. Amer. J. Physiol. 207: 347–354.
16. LEVEY, G. S., and EPSTEIN, S. E. 1969. Activation of adenyl cyclase by glucagon in cat and human heart. Circ. Res. 24: 151–156.
17. MAYER, S. E., NAMM, D. H., and RICE, L. 1970. Effect of glucagon on cyclic 3',5'-AMP, phosphorylase activity, and contractility of heart muscle of the rat. Circ. Res. 26: 225–233.
18. NOBEL-ALLEN, N., KERSCH, M., and LUCCHESE, B. R. 1974. Glucagon: Its enhancement of cardiac performance in the cat with chronic heart failure. J. Pharmacol. Exp. Therap.
19. REGAN, T. J., LEHAN, P. H., HENNEMAN, D. H., BEHAR, A., and HELLEMS, H. K. 1964. Myocardial metabolic and contractile response to glucagon and epinephrine. J. Lab. Clin. Med. 63: 638–646.
20. ROVETTO, M. J., WHITMER, J. T., and NEELY, J. R. 1973. Comparison of the effects of anoxia and whole heart ischemia on carbohydrate utilization in isolated working rat hearts. Circ. Res. 32: 699–711.
21. SCHEUER, J., and STEZOSKI, W. S. 1968. Effect of high energy phosphate depletion and repletion on the dynamics and electrocardiogram of isolated rat heart. Circ. Res. 23: 519–530.
22. SCHEURER, J., and STEZOSKI, W. S. 1970. Protective role of increased myocardial glycogen stores in cardiac anoxia in the rat. Circ. Res. 27: 835–849
23. SUTHERLAND, E. W., ROBISON, G. A., and BUTCHER, R. W. 1868. Some aspects of the biological role of adenosine 3',5'-monophosphate (cyclic AMP). Circulation 37: 279–306.
24. WILLIAMSON, J. C. 1966. Glycolytic control mechanisms. II. Kinetics of intermediate changes during aerobic-anoxic transition in perfused rat heart. J. Biol. Chem. 241: 5026–5036.
25. WOLLENBERGER, A., BABSKII, E. B., KRAUSS, E. G., GENZ, S., BLOHM, D., and BOGDANOVA, E. V. 1973. Cyclic changes in levels of cyclic AMP and cyclic GMP in frog myocardium during the cardiac cycle. Biochem. Biophys. Res. Commun. 55: 446–452.

Mailing address: Dr. W. J. George,
Department of Pharmacology,
Tulane University School of Medicine,
New Orleans, La. 70112 (USA).

Recent Advances in Studies on Cardiac Structure and Metabolism
Volume 9
The Sarcolemma
Edited by Paul-Emile Roy and Naranjan S. Dhalla

EFFECTS OF ISCHEMIA AND REPERFUSION ON MYOCARDIAL UPTAKE OF TRITIATED DIGOXIN[1]

G. A. BELLER,[2] W. B. HOOD, JR., and T. W. SMITH[3]

Cardiac Unit, Massachusetts General Hospital and
Cardiology Division, Boston City Hospital; and
Departments of Medicine, Harvard and Boston University Medical Schools,
Boston, Massachusetts, USA

SUMMARY

The effect of coronary reperfusion on the uptake of cardiac glycosides by ischemic myocardium was studied in 17 open chested dogs undergoing anterior wall infarction produced by snaring confluent branches of the left coronary system. Epicardial electrograms delineated ischemic zones of S-T elevation, border, and nonischemic zones. Animals were reperfused by snare release 1, 2, and 6 hr after occlusion. After 15 min of reperfusion, 1.0 mg of [^{3}H]digoxin was given intravenously, and 2 hr later the hearts were excised and endocardial (endo) and epicardial (epi) samples from each zone were analyzed for [^{3}H]digoxin concentration. In five dogs occluded for 1 hr and reperfused, [^{3}H]digoxin uptake was comparable in endo and epi layers of all three zones. In six dogs reperfused after 2 hr of occlusion, mean (±S.E.) [^{3}H]digoxin concentrations (nanograms per gm) were significantly reduced by 54 percent in endo (111 ± 18) and 35 percent in epi (151 ± 23) layers of the ischemic zone as compared with the mean nonischemic concentration (endo 249 ± 34; epi 239 ± 34). Border zone endo and epi [^{3}H]digoxin uptake was reduced by 21 and 17 percent, respectively. In six dogs reperfused after 6 hr of occlusion, [^{3}H]digoxin uptake in the ischemic zone was markedly reduced by 85 percent in endo (34 ± 4) and 60 percent in epi (86 ± 12) layers as compared with the nonischemic concentration (endo 232 ± 19; epi 217 ± 15). Border zone uptake was decreased by 54 percent in endo and 38 percent in epi regions. We conclude that coronary reperfusion between 2 and 6 hr of coronary occlusion is associated with markedly reduced myocardial digoxin uptake, more pronounced in subendocardial regions of ischemic tissue. This alteration in digoxin binding by reperfused ischemic myocardium is consistent with ischemia-induced structural or functional alterations in the putative digitalis receptor, ($Na^+ + K^+$)-ATPase.

One of the most essential functions of the myocardial cell membrane is the maintenance of physiological cellular Na^+ and K^+ concentrations by active

[1] This research was supported in part by United States Public Health Service contracts NHLI 71-2498 and NO 1 HV-71443 and by Grants NHLI 14325, 14646, and HE 07299.
[2] Special Research Fellow of the National Heart and Lung Institute.
[3] Established Investigator of the American Heart Association.

transport processes. Normal gradients of these monovalent cations are required for a transmembrane potential to exist so that excitation can occur. Water movement and the regulation of cell volume are linked at least in part to the movement of solutes across cells, and cell swelling may take place when active monovalent cation transport is inhibited.

There is convincing evidence that one of the molecular mediators of active transport of Na^+ and K^+ is a membrane-bound, Na^+ and K^+-activated ATP phosphohydrolase enzyme complex which is thought to function as a "sodium pump" (17). It has been shown that ($Na^+ + K^+$)-ATPase activity and monovalent cation transport are specifically inhibited by digitalis glycosides (6, 9, 13) and several investigators have suggested that the enzyme is the pharmacological receptor for digitalis glycosides (1, 6, 7, 9, 13–15, 18).

There is little information concerning possible alterations in the myocardial ($Na^+ + K^+$)-ATPase system after myocardial ischemia and the role, if any, of these changes in the pathogenesis of irreversible myocardial cell injury. Clearly, lack of ATP as a result of ischemia-induced inhibition of oxidative phosphorylation will result in loss of normal cation transport activity, and myocardial intracellular potassium loss (5) and sodium gain (12) are well known concomitants of the ischemic state. Prolonged ischemia might also result in structural alterations in the cell membrane and inactivation of (Na^+-K^+)-ATPase. The functional consequences of these changes would also include a net increase in intracellular sodium concentration with consequent increase in intracellular water content, resulting in myocardial cell swelling, occlusion of capillary lumina, further limitation of blood flow, and accelerated tissue injury (8).

The aim of the experiments described below was to evaluate indirectly the integrity of myocardial membrane ($Na^+ + K^+$)-ATPase *in vivo* after periods of coronary occlusion and reperfusion by assessing the ability of myocardial tissue to bind tritiated digoxin.

METHODS

Myocardial uptake of tritiated digoxin ([^{3}H]digoxin) was determined in four groups of dogs after varying periods of coronary occlusion and reperfusion, in addition to a sham-operated group of control animals.

Experimental Model

An open chested canine model was employed in these studies. Animals were anesthetized with intravenous pentobarbital (30 mg/kg) after which left thoracotomy was performed. Respiration was maintained with a positive pressure Bennett respirator (Bennett Respirator Products, Inc., Santa Monica, Calif.). All animals breathed 40 percent oxygen throughout the experiment. The heart was suspended in a pericardial cradle after which snares were placed upon multiple confluent branches of left anterior descending and circumflex coronary arteries supplying an area of the apex and anterior wall of the left ventricle (fig. 1,*A* and *B*).

For each dog, 10–12 mapping areas were designated on the anterior and lateral surface of the left ventricle (fig. 1*A*). Electrocardiographic epicardial surface mapping at each site was performed utilizing a stainless steel cylinder 8 cm long and 2.5 mm in diameter, the tip of which was smoothly ground and polished. The cylinder was attached to the exploring electrode of the conventional electrocardiographic lead system. This electrode was held lightly at a right angle to the epicardium during the surface mapping procedure. Epicardial electrograms were standardized at 1 mm/mV stylus deflection.

A femoral artery was isolated and a catheter was passed into the central aorta. Arterial pressure was monitored with a Statham P23Db pressure transducer. Both arterial pressure and electrocardiographic recordings were made on a two-channel Sanborn recorder at a paper speed of 50 mm/sec.

After control recording, serial coronary arterial branch ligations were performed at 10–15-min intervals until approximately one-third of the anterior

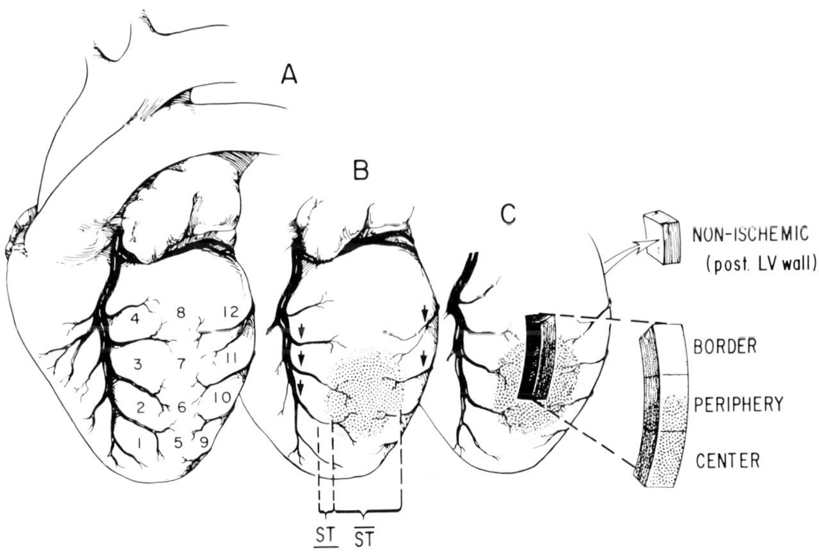

Figure 1 A. Diagrammatic representation of the canine heart. Twelve epicardial mapping sites are designated on the anterior and lateral surfaces of the left ventricle. *B*, Acutely infarcted left ventricle (*LV*) showing central and peripheral zones of ST segment elevation ($\overline{ST}$) and border zone ST segment depression ($\underline{ST}$). *Arrows* indicate sites of ligations or snaring of branches of the left coronary system in this example. *C*, Myocardial sampling technic. Specimens taken from the center and periphery of the ischemic zone, the contigous border zone, and from nonischemic posterior wall were analyzed for [^{3}H]digoxin concentration. (Reproduced from Beller, Smith, and Hood (2) by permission of the American Heart Association.)

surface of the left ventricle exhibited S-T segment elevation in the epicardial electrocardiograms. This required from three to six ligatures. After ligations were completed, surface mapping was again performed at the same sites mapped previously in the control period. Postocclusion electrograms usually revealed three distinct zones (fig. 1*B*): 1) central and peripheral zones of ST segment elevation ($\overline{ST}$) greater than 1 mV; 2) a border zone adjacent to the zone of $\overline{ST}$ and, 3) a nonischemic zone within which ST segments were unchanged from control. In most animals ST segment depression was noted in the border zone area.

Occlusion studies

This group comprised seven dogs which underwent coronary occlusion in the manner described above. One hour after the last ligation, 1 mg of [^{3}H] digoxin (specific activity 123 μCi/mg) diluted in 3.0 ml of 95 percent ethanol was administered via a foreleg vein. Two hours later (3 hr after occlusion) repeat epicardial electrograms were recorded. The hearts were then rapidly excised, and multiple full thickness myocardial samples weighing between 0.5 and 1.0 gm from ischemic, border, and nonischemic zones were obtained (fig. 1*C*). These samples were then divided into endocardial and epicardial halves for myocardial [^{3}H] digoxin determinations.

Reperfusion studies

In these studies the uptake of [^{3}H] digoxin after acute ischemia and reperfusion was examined in 17 open chested dogs underoing anterior wall infarction in the manner described above. Animals were reperfused by snare release 1 (n = 5), 2 (n = 6), or 6 (n = 6) hr after occlusion. After 15 min of reperfusion, 1.0 mg of [^{3}H] digoxin (specific activity 143 μCi/mg) was given intravenously and 2 hr later the hearts were excised and endocardial and epicardial samples from each zone were analyzed for [^{3}H] digoxin concentrations.

Sham-operated control dogs

This group included four dogs that received 1 mg of [^{3}H] digoxin intravenously in a foreleg vein after a left thoracotomy performed under pentobarbital anesthesia. As performed in the other groups of experimental animals, 120 min after [^{3}H] digoxin was given, the animals were killed and the hearts removed. Endocardial and epicardial specimens were obtained from the apex, base, and posterior wall of the left ventricle for myocardial [^{3}H] digoxin concentration determinations.

Myocardial [^{3}H] digoxin measurements

The method utilized for measuring left ventricular [^{3}H] digoxin uptake was as described previously (2). Briefly, left ventricular endocardial and epicardial samples (fig. 1*C*) were blotted dry, washed, weighed, and immediately refrigerated at −20°C. Within 1–2 weeks the specimens were thawed, diced, and placed into vials containing 15 ml of the liquid scintillation medium described by Bray (4). After vigorous shaking for 48 hr at 37°C to elute [^{3}H] digoxin from the samples, the vials were counted in a liquid scintillation spectrometer. Completeness of elution of [^{3}H] digoxin was assured by comparison with samples fully solubilized with NCS. (Amersham/Searle, Des Plaines, Ill.). Quenching correction was achieved by addition of internal standards of [^{3}H] digoxin.

Statistical analysis

All data are presented as mean values ± S.E.M., significance of differences was assessed by the paired *t* test with each dog serving as its own control.

RESULTS

Occlusion studies

Endocardial and epicardial digoxin concentrations (nanograms per gm of wet weight ± S.E.M.) in the center and periphery of the ST zone of ischemia, in the border zone, and in the nonischemic posterior wall region are shown in table I. There was a significant decrease in digoxin uptake in the endocardial and epicardial halves of the center and periphery of the ischemic as compared with the border and nonischemic zones ($p < 0.05$). A significant transmural gradient of digoxin was demonstrable between endocardial and epicardial layers of ischemic zones, with less uptake in endocardium ($p < 0.05$).

There was no statistically significant difference in digoxin uptake of either endocardial or epicardial layers of the border zone as compared with the nonischemic region; however, the endocardium of the border zone showed a lower mean digoxin concentration as compared with the epicardium of this zone ($0.05 < p < 0.10$). In nonischemic myocardium, endocardial [^{3}H] digoxin

Table I Myocardial [^{3}H] digoxin concentrations (nanograms per gm ± S.E.) in dogs undergoing coronary occlusion for 1 hr

Site	Nonischemic	Border	Ischemic periphery	Ischemic center
Endocardial	256 ± 22*	217 ± 21	51 ± 13*†	27 ± 11*†
Epicardial	220 ± 20	242 ± 15	112 ± 13†	58 ± 10†
Endocardial to epicardial ratio	1.16	0.89	0.46	0.46

*$p < 0.05$ compared with epicardium.
†$p < 0.05$ compared with nonischemic tissue.

concentration was slightly but significantly greater than epicardial concentration with an endocardial to epicardial ratio of 1.16.

Thus, in the acutely infarcted canine left ventricle there was a decreasing concentration of digoxin from nonischemic to border to ischemic zones, which was more pronounced in endocardial layers. The transmural gradients of the glycoside observed in ischemic and border zones were the reverse of those found in nonischemic portions of the myocardium.

Reperfusion studies

[^{3}H] Digoxin concentrations in endocardial and epicardial samples of ischemic, border, and nonischemic myocardial zones after coronary reperfusion in dogs occluded for 1, 2, and 6 hr are shown in table II.

In five dogs occluded for 1 hr and reperfused, [^{3}H] digoxin was slightly, but not significantly, reduced in the endocardial layers of the center of the ischemic zone. Transmural [^{3}H] digoxin distribution as assessed by endocardial to epicardial ratios was comparable in all four zones. In six dogs reperfused after 2 hr of occlusion, mean (± S.E.) digoxin concentrations (nanograms per gm) were significantly reduced ($p < 0.001$) by 54 percent in endocardial (111 ± 18) and 35 percent in epicardial (151 ± 23) layers of the central ischemic zone as compared with the mean nonischemic concentrations (endocardial, 249 ± 34; epicardial, 239 ± 34). The endocardial to epicardial ratio in this zone was reduced to 0.73 as compared with 1.05 in the nonischemic region. In the peripheral zone of ischemia endocardial [^{3}H] digoxin uptake was significantly reduced by 38 percent (153 ± 36) and epicardial uptake was reduced by 27 percent (169 ± 19) as compared with nonischemic [^{3}H] digoxin concentrations ($p < 0.05$). Border zone endocardial and epicardial [^{3}H] digoxin uptake was reduced by 21 percent and 17 percent, respectively ($p < 0.05$).

In six dogs reperfused after 6 hr of occlusion, [^{3}H] digoxin uptake in the ischemic zone was markedly reduced by 85 percent in endocardial (34 ± 4) and 60 percent in epicardial (86 ± 12) layers ($p < 0.001$) as compared with the nonischemic concentration. This reduction was significantly greater than that seen in the 2-hr occluded and reperfused group. The endocardial to epicardial ratio of digoxin uptake in the center of the ischemic zone was reduced to 0.40 as compared with 1.07 in nonischemic myocardium. In the periphery of the ischemic zone, [^{3}H] digoxin uptake was also significantly reduced: by 72 percent in endocardial (69 ± 24) and 50 percent (109 ± 10) in epicardial layers ($p < 0.001$) with a transmural digoxin concentration gradient of 0.59. Border zone uptake after reperfusion was decreased by 54 percent in endocardial and 38 percent in epicardial regions ($p < 0.001$).

Digoxin concentrations in the apex, base, and posterior wall of the four sham-operated control animals are shown in table III. There was no significant difference in digoxin uptake in these three regions, indicating a homogenous

Table II Myocardial [^{3}H] digoxin concentrations (nanograms per gm ± S.E.) after reperfusion in dogs occluded for 1, 2, and 6 hr

Duration of occlusion	Site	Nonischemic	Border	Ischemic periphery	Ischemic center
1 hr ($n = 5$)	Endocardial	278 ± 9	252 ± 21	238 ± 28	230 ± 28
	Epicardial	241 ± 17	227 ± 24	237 ± 24	234 ± 55
	Endo/epi ratio	1.17	1.12	1.10	1.00
2 hr ($n = 6$)	Endocardial	249 ± 34	187 ± 22*	153 ± 36*	111 ± 18*
	Epicardial	239 ± 34	189 ± 18*	169 ± 19*	151 ± 23*
	Endo/epi ratio	1.05	0.99	0.87	0.73
6 hr ($n = 6$)	Endocardial	232 ± 19	101 ± 21*	69 ± 24*	34 ± 4*
	Epicardial	217 ± 15	133 ± 11*	109 ± 10*	86 ± 12*
	Endo/epi ratio	1.07	0.78	0.59	0.40

*$p < 0.05$, as compared with nonischemic [^{3}H] digoxin concentrations.

Table III Left ventricular myocardial [^{3}H] digoxin concentrations (nanograms per gm ± S.E.M.) in sham-operated dogs

Site	Apex (n = 4)	Base (n = 4)	Posterior wall (n = 4)
Endocardial*	246 ± 41	249 ± 31	257 ± 30
Epicardial	227 ± 21	231 ± 27	236 ± 28
Endocardial to epicardial ratio	1.08	1.08	1.09

*p < 0.05 for mean endocardial concentration from apex, base, and posterior wall (n = 12) compared with epicardial mean for all three regions.

distribution of the glycoside in the intact canine left ventricle. There was, however, significantly greater uptake of digoxin in endocardial layers of these regions with an average endocardial to epicardial ratio of 1.08 ($p < 0.05$).

DISCUSSION

This study shows that coronary occlusion in the canine model described here results in marked reduction of binding of intravenously infused [^{3}H] digoxin. Furthermore, coronary reperfusion instituted after 2 and 6 hr of coronary occlusion is also associated with markedly reduced myocardial digoxin uptake. Impaired binding of the glycoside *in vivo* after reperfusion was more severe the longer the ischemic period before flow restoration. When reperfusion was carried out after 6 hr of coronary occlusion, [^{3}H] digoxin uptake in ischemic and border myocardial zones was markedly reduced and similar to that observed in the 1-hr occluded animals in which reperfusion was never instituted. In addition, subendocardial concentrations of [^{3}H] digoxin were reduced significantly more than subepicardial concentrations, as evidenced by the reversal of the normal endocardial to epicardial transmural gradients.

The mechanism for impairment of left ventricular myocardial digoxin uptake in animals undergoing coronary occlusion probably reflects, at least in part, altered regional flow to ischemic and nonischemic tissue. It may be, however, that the marked decrease in digoxin binding in ischemic tissue is also related to a primary alteration of the ability of membrane ($Na^+ + K^+$)-ATPase to bind the glycoside. This might result from substrate (ATP) limitation, since ATP is known to be required for optimal binding of cardiac glycosides to isolated and partially purified ($Na^+ + K^+$)-ATPase (14), and intact metabolic machinery is required for optimal digitalis binding to myocardial tissue in the models studied by Marks and his colleagues (10). Alternatively, denaturation or loss of the enzyme might occur under the conditions existing in ischemic myocardium.

The reperfusion experiments were undertaken in order to determine the potential reversibility of the alterations in myocardial digoxin binding observed during ischemia. In a group of animals occluded for 1 hr, reperfusion resulted in almost complete restoration of myocardial digoxin binding in all ischemic and border zones. There was, however, some possible impairment of digoxin uptake in endocardial layers of the center and periphery of the ischemic zone. This reduction (15 percent) was not statistically siginificant. Reperfusion carried out after 2 and 6 hr of coronary occlusion resulted in a persistent reduction of digoxin binding by ischemic myocardium. The magnitude of this reduction was greater in 6 hr than in 2-hr occluded and reperfused animals. In addition, impairment of digoxin uptake was greatest in the center of the ischemic zone (which showed the highest ST segment elevation during the occlusion period) in both groups of animals. Interestingly, in the group of dogs occluded for 6 hr and then reperfused, [^{3}H] digoxin uptake patterns were comparable with those observed in the 1-hr occluded group in which reperfusion was not carried out. In the seven dogs occluded for 1 hr and not reperfused, mean [^{3}H] digoxin concentration was reduced by 90 percent compared with the nonischemic control concentration. In the seven dogs occluded for 6 hr but reperfused, mean [^{3}H] digoxin concentration in endocardial layers of the center of the ischemic zone was reduced to 85 percent of the control value. Thus, restoration of coronary flow after 6 hr of coronary occlusion appeared to have had little or no effect in reversing the ischemia-induced alteration in digoxin binding.

Results of these experiments therefore suggest that when coronary blood flow is restored to areas of myocardium rendered acutely ischemic for more than 1 hr, myocardial tissue in the ischemic zone shows diminished binding of digoxin, which might possibly be due to structural or functional alterations in cell membrane digitalis receptors. The drug may have reduced access to membrane receptor sites because of partial or complete capillary occlusion resulting from ischemic cell swelling.

It has recently been suggested that ischemic swelling of myocardial cells may play a major role in the pathogenesis of irreversible cell death after acute myocardial infarction (8). Cell swelling might accelerate damage to ischemic tissue by obstruction of small blood vessels with further diminution in myocardial blood flow to affected areas. The mechanism whereby ischemia produces cell swelling is postulated by Leaf (8) to be as follows: 1) after coronary occlusion oxygen and substrate delivery to the cell is reduced; 2) the supply of metabolic energy becomes insufficient to maintain the usual rate of extrusion of sodium from the cell; 3) the accumulation of sodium exceeds the loss of potassium and a net gain of intracellular solute occurs which osmotically draws water into the cell which then swells. This sequence of events, then, might be contributed to by ischemia-induced derangement of the cell membrane (Na^+ + K^+)-ATPase medium, with diminished rates of active transport of sodium and potassium.

The extent to which the integrity of the sodium pump or ($Na^+ + K^+$)-ATPase is affected by myocardial ischemia has not been fully defined. Schwartz *et al.* (16) were not able to demonstrate a reduction in myocardial ($Na^+ + K^+$)-ATPase activity in the first 24 hr of ischemia in dogs undergoing circumflex coronary artery ligation. In the present study in which large areas of the anterior wall of the canine heart were made ischemic, coronary occlusion for 1 hr was associated with altered *in vivo* binding of [^{3}H] digoxin consistent with ischemia-induced structural or functional alterations in digitalis-binding components of the myocardium. After 6 hr of coronary occlusion these changes appear to be irreversible since restoration of coronary flow did not reverse the impaired digoxin binding. We cannot, however, exclude the possibility that longer periods of reperfusion would be more effective in reversing the ischemia-induced alterations in glycoside binding.

REFERENCES

1. AKERA, T., LARSEN, F. S., and BRODY, T. M. 1970. Correlation of cardiac sodium- and potassium-activated adenosine triphosphatase activity with ouabain-induced inotropic stimulation. J. Pharmacol. Exp. Therap. 173: 145–151.
2. BELLER, G. A., SMITH, T. W., and HOOD, W. B., JR. 1972. Altered distribution of tritiated digoxin in the infarcted canine left ventricle. Circulation 46: 572–579.
3. BELLER, G. A., SMITH, T. W., and HOOD, W. B., JR. 1974. Altered regional myocardial blood flow following coronary reperfusion in acute myocardial ischemia. Clin. Res. 22: 262A.
4. BRAY, G. A. 1960. A simple efficient liquid scintillator for counting aqueous solutions in a liquid scintillation counter. Anal. Biochem. 1: 279–285.
5. CASE, R. B., NASSER, N. G., and CRAMPTON, R. G. 1969. Biochemical aspects of early myocardial ischemia. Amer. J. Cardiol. 24: 766–775.
6. GLYNN, I. M. 1964. The action of cardiac glycosides on ion movement. Pharmacol. Rev. 16: 381–407.
7. LANGER, G. A. 1972. Effects of digitalis on myocardial ionic exchange. Circulation 46: 180–187.
8. LEAF, A. 1973. Cell swelling: A factor in ischemic tissue injury. Circulation 48: 455–458.
9. LEE, K. S., and KLAUS, W. 1971. The subcellular basis for the mechanism of inotropic action of cardiac glycosides. Pharmacol. Rev. 23: 193–261.
10. MARKS, B. H. 1972. Factors that affect the accumulation of digitalis glycosides by the heart, pp. 69–93. *In* B. H. Marks and A. M. Weissler (eds.), Basic and Clinical Pharmacology of Digitalis. Charles C Thomas, Springfield, Ill.
11. MATSUI, H., and SCHWARTZ, A. 1968. Mechanism of cardiac glycoside inhibition of the (Na^+ + K^+)-dependent ATPase from cardiac tissue. Biochim. Biophys. Acta 151: 655–663.
12. NAYLER, W. G., STONE, J., CARSON, V. 1971. Effect of ischemia on cardiac contractility and calcium exchangeability. J. Molec. Cell. Cardiol. 2: 125–143.
13. REPKE, K. 1963. Metabolism of cardiac glycosides, pp. 47–73. *In* W. Wilbrandt (ed.),

New Aspects of Cardiac Glycosides. First International Pharmacology Meeting, Stockholm, 1961, Vol. 3. Pergamon Press, Oxford.
14. SCHWARTZ, A., MATSUI, H., and LAUGHTER, A. H. 1968. Tritiated digoxin binding to (Na^+ + K^+)-activated adenosine triphosphatase: Possible allosteric site. Science 160: 323–325.
15. SCHWARTZ, A., ALLEN, J. C., and HARIGAYA, S. 1969. Possible involvement of cardiac Na^+,K^+-adenosine triphosphatase in the mechanism of action of cardiac glycosides. J. Pharmacol. Exp. Therap. 168: 31–41.
16. SCHWARTZ, A., WOOD, J. M., ALLEN, J. C., BARNET, E. P., ENTMAN, M. L., GOLDSTEIN, M. A., SORDAHL, L. A., and SUZUKI, M. 1973. Biochemical and morphologic correlates of cardiac ischemia. I. Membrane systems. Amer. J. Cardiol. 32: 46–61.
17. SKOU, J. C. 1965. Enzymatic basis for active transport of sodium and potassium across cell membrane. Physiol. Rev. 45: 596–617.
18. SMITH, T. W., and HABER, E. 1973. Medical progress: Digitalis. New Eng. J. Med. 289: 945–952; 1010–1015; 1063–1072; 1125–1129.

Mailing address: Dr. G. A. Beller,
Cardiac Unit, Massachusetts General Hospital,
Fruit Street, Boston, Mass. 02114 (USA).

Author Index

Subject Index